Innovations in Green and Energy-Efficient Warehousing

Mohamed Amine Frikha
King Faisal University, Saudi Arabia

Mohieddine Rahmouni
King Faisal University, Saudi Arabia

Ben Othman Soufiane
King Faisal University, Saudi Arabia

IGI Global
Scientific Publishing
Publishing Tomorrow's Research Today

Vice President of Editorial	Melissa Wagner
Director of Acquisitions	Mikaela Felty
Director of Book Development	Jocelynn Hessler
Production Manager	Mike Brehm
Cover Design	Jose Rosado

Published in the United States of America by
IGI Global Scientific Publishing
701 East Chocolate Avenue
Hershey, PA, 17033, USA
Tel: 717-533-8845 | Fax: 717-533-7115
Website: https://www.igi-global.com E-mail: cust@igi-global.com

Library of Congress Cataloging-in-Publication Data

Names: Frikha, Mohamed Amine, 1982- editor | Rahmouni, Mohieddine, 1977- editor | Soufiene, Ben Othman editor
Title: Innovations in green and energy-efficient warehousing / [edited by] Mohamed Amine Frikha, Mohieddine Rahmouni, Ben Othman Soufiene.
Description: Hershey, PA : IGI Global Scientific Publishing, [2026] | Includes bibliographical references and index. | Summary: "This book aims to provide a comprehensive and actionable resource for transforming traditional warehousing practices into sustainable, energy-efficient operations, bridging the gap between theoretical research and practical application"-- Provided by publisher.
Identifiers: LCCN 2025020204 (print) | LCCN 2025020205 (ebook) | ISBN 9798337331768 hardcover | ISBN 9798337331775 paperback | ISBN 9798337331782 ebook
Subjects: LCSH: Warehouses--Management | Inventory control--Environmental aspects | Materials handling--Environmental aspects
Classification: LCC HF5485 .I56 2026 (print) | LCC HF5485 (ebook) | DDC 658.7/85--dc23/eng/20260109
LC record available at https://lccn.loc.gov/2025020204
LC ebook record available at https://lccn.loc.gov/2025020205

British Cataloguing in Publication Data
A Cataloguing in Publication record for this book is available from the British Library.

Table of Contents

Detailed Table of Contents

clusters such as AI-driven optimization and robotic automation. The chapter details applications in energy management, inventory optimization, logistics efficiency, and circular economy integration, supported by case studies of industry leaders. A proposed framework offers practical strategies for adopting these technologies, while addressing challenges like high costs and data integration. Aligned with the United Nations Sustainable Development Goals, this chapter provides actionable insights for warehouse managers, researchers, and policymakers committed to low-carbon, intelligent logistics.

This study explores the implementation of green warehousing practices through the integration of advanced artificial intelligence techniques, specifically using a Bi-Stacked Artificial Neural Network (Bi-Stacked ANN) for predictive modeling. Data were collected via structured questionnaires from four stakeholder groups—enterprise employees, consumers, public officials, and producers—yielding 469 valid responses. Following preprocessing and feature normalization, Ant Colony Optimization (ACO) was applied to select the most relevant features influencing sustainable warehousing. The Bi-Stacked ANN model was then developed to analyze patterns and predict the effectiveness of green practices. The model demonstrated strong predictive performance and offered insights into critical sustainability factors.

Overall, successful sustainability companies exist across all industries. Two of the most fundamental components of an inventory management system are deeply interconnected and are essential for balancing customer demand with product availability, minimising costs, improving customer satisfaction, and enhancing operational efficiency. Demand forecasting involves historical seasonal patterns, relevant data, and allows businesses to inform production planning, levels, staffing, and distribution. The ability to anticipate helps companies avoid both understocking,

This chapter explores the role of Artificial Intelligence (AI) in enhancing energy efficiency and sustainability in smart warehousing. AI technologies such as predictive maintenance, smart inventory management, dynamic load balancing, and automation have been shown to significantly reduce energy consumption, optimize warehouse operations, and minimize environmental impacts. AI's integration with renewable energy sources, like solar and wind, further supports sustainability goals by optimizing energy usage and reducing reliance on non-renewable power. The future potential of AI in revolutionizing warehouse energy efficiency is immense, with advancements in deep learning, edge computing, and real-time analytics. This chapter also highlights areas for future research, particularly in renewable energy integration, AI algorithm development, and scalability for small and medium-sized enterprises.

Water conservation in warehousing operations has become a critical concern due to increasing water scarcity and rising operational demands. Warehouses consume water for a variety of purposes including sanitation, cooling systems, fire safety, and landscape maintenance. This study explores the application of intelligent water conservation techniques supported by advanced deep learning models, specifically the Bi-Stacked Gated Recurrent Unit (Bi-Stacked GRU), to forecast water usage patterns and identify optimization opportunities. A dataset comprising 1,195 household-level responses was collected from eight districts and preprocessed for analysis. Feature selection was performed using Particle Swarm Optimization (PSO) to isolate key variables influencing water usage. The Bi-Stacked GRU model enabled accurate prediction of water consumption trends and detection of inefficiencies, allowing for proactive water-saving strategies in warehousing environments

Energy efficiency and sustainability are key priorities in modern warehousing and logistics. AI plays a transformative role by enabling smart operations through accurate demand forecasting and inventory management. Traditional models often lead to overstocking, understocking, and energy waste. AI techniques such as time-series models, RNNs, and adaptive systems support just-in-time inventory, reduce resource use, and cut emissions. Integrated with IoT, RFID, and WMS, AI improves storage, picking, and restocking, lowering labor and energy costs. It also manages slow-moving or excess stock to reduce waste. The chapter highlights AI's synergy with digital twins, cloud, and edge computing for adaptive, sustainable warehouses. Real-world case studies show gains in forecasting, carbon reduction, and SKU efficiency. Key challenges like data quality and integration are addressed with practical strategies. AI, aligned with SDGs and carbon neutrality goals, is not just a tech upgrade but a core enabler of sustainable, intelligent warehousing.

This chapter explores innovations in green and energy-efficient warehousing, focusing on sustainable technologies that minimize environmental impact and optimize energy consumption. Key advancements, including energy-efficient lighting, automated systems, renewable energy integration, and smart warehouse designs, are explored for their role in reducing carbon footprints and improving operational efficiency. The chapter also discusses emerging technologies like AI, IoT, and robotics that enhance warehouse operations and contribute to sustainability goals. Challenges such as high initial investment costs and regulatory compliance are addressed, along with future trends like carbon neutrality and zero-energy warehouses. The chapter highlights the importance of adopting green technologies for long-term environmental and economic benefits.

The adoption of renewable energy technologies to meet warehouse energy demand is a crucial step in achieving sustainable development, as there are global energy shortages and environmental challenges. This chapter provides an overview of the application of renewable energy technologies, including solar, wind, and geothermal, for adoption in warehouses. Moreover, it also describes the potential of these technologies in the reduction of fossil fuel utilization, operational cost reduction, and minimizing carbon footprints. Initially, solar energy, through photovoltaic (PV) and photovoltaic/thermal (PV/T) systems, is discussed in detail as an attractive alternative for electricity and thermal energy requirements. Therefore, this chapter highlights the potential role of renewable energy technologies by carefully examining their benefits, associated challenges, and their practical applications as a way forward toward a sustainable future in the warehouse industry.

This chapter explores cutting-edge advancements in green and energy-efficient warehousing, focusing on transforming traditional logistics infrastructure into sustainable, intelligent ecosystems. It covers core strategies such as retrofitting with LED lighting, smart HVAC systems, renewable energy integration, IoT-based monitoring, and automation for dynamic energy control. Detailed case studies from global leaders like DHL and FedEx illustrate real-world impacts, including substantial reductions in energy consumption and operating costs. The book also analyzes financial feasibility, including ROI, payback periods, and the role of government incentives. Emphasis is placed on policy alignment, employee training, and industry certifications like LEED and ISO 50001. By blending theory with practice, this volume serves as a comprehensive guide for warehouse managers, engineers, policymakers, and sustainability advocates seeking actionable solutions for carbon-neutral logistics operations.

As global environmental awareness grows, sustainable development is gaining momentum. A key strategy in achieving sustainability is the adoption of reverse logistics (RL), which manages the backward flow of materials from customers to manufacturers. This often occurs due to product damage, unmet expectations, or warranty returns. This study presents a comprehensive investigation using both mathematical analysis and computational tools. The mathematical model aims to minimize the planning horizon and maximize total profit, incorporating a non-constant return fraction and a carbon reduction function influenced by green technology investment. Computational analysis, performed using MATLAB, supports the theoretical framework. The main goal is to optimize profits in a reverse logistics system while addressing cost efficiency and environmental impact. The model focuses on reducing solid waste through repurposing materials for use in production and remanufacturing, and also determines the optimal conditions for manufacturers to achieve maximum profitability.

This study presents an intelligent approach to enhancing warehouse operations through smart warehousing detection using a Bi-Stacked Long Short-Term Memory (Bi-Stacked LSTM) model. Data were collected from a case study company's internal database, including sales records, inventory data, product categories, and pricing information over the past two years. After rigorous preprocessing involving data cleaning, integration, and transformation, relevant features were selected using the Ant Colony Optimization (ACO) algorithm to reduce dimensionality and improve model performance. The Bi-Stacked LSTM model was trained on the selected features to detect operational anomalies and performance patterns across time-series data. The bidirectional and stacked architecture of the LSTM allowed the model to learn both past and future contextual dependencies effectively.

The increasing adoption of mechatronics in the automotive industry has transformed manual operations into automated systems for improved efficiency, quality, and safety. Among these, Automatic Screw Tightening Machines (ASTMs) are widely used in lamp assembly to fasten body and extension parts. However, these machines often face increased cycle times under varying conditions, reducing production throughput. This study addresses cycle time optimization by analyzing the machine's construction, programming, and workflow, followed by a cycle time study to identify bottlenecks. Benchmarking revealed screw feeding as a major contributor to delays. To resolve this, the Anti-Tumble Unit was redesigned with a feed-while-drive mechanism, enabling screw feeding and tightening simultaneously. This improvement reduced cycle time from 45 to 33 seconds, achieving a 26.7% reduction. The findings show that targeted design modifications can significantly enhance automated assembly performance and can be applied to optimize cycle time in other industrial processes.

The 4IR is fading away to usher in Industry 5.0. Finding its bases on Industry 4.0, the 5IR emphasises sustainability and human to machine collaboration. The rationale resides in encouraging warehouses to harness new innovations given Industry 4.0 and 5IR. The researcher seeks to investigate if corporate green performance influences suppliers' support. In congruency to the quantitative nature of the study, a positivist stance was adopted to test the set hypothesis. Archival data were drawn purposively from the FTSE/JSE Responsible Investment Index to compute a short panel data set comprising 19 companies and 114 observations over a six-year period. Pearson's pairwise correlation and first differenced econometric models for Fixed effects were employed for statistical analyses. The study yielded mixed findings as energy consumption indicated a significant positive effect on supplier credit disbursement.

Preface

As global industries face pressure to reduce their environmental footprint, warehousing undergoes a shift toward sustainability. Innovations in green and energy-efficient warehousing reduce operational costs while aligning logistics with environmental goals. From the integration of renewable energy sources to the adoption of smart energy management systems and sustainable building materials, warehouses become smarter and greener. Advanced automation, LED lighting, and high-efficiency HVAC systems optimize energy usage, while layout redesigns and electric vehicles minimize emissions across the supply chain. These strategies may redefine modern warehousing, making it critical to achieving sustainable logistics and a low-carbon future.

Innovations in Green and Energy-Efficient Warehousing explores how businesses can integrate sustainability into their warehousing operations. It delves into the latest innovations, strategies, and best practices in sustainable logistics, offering theoretical frameworks, practical tools, and real-world case studies to address challenges like cost barriers, technological limitations, and regulatory compliance. This book covers topics such as cost analysis, inventory management, and predictive modeling, and is a useful resource for business owners, engineers, academicians, researchers, and scientists.

COVERAGE

The many academic areas covered in this publication include, but are not limited to:

- Artificial Intelligence
- Business Warehousing
- Cost Analysis
- Energy Engineering
- Environmental Science

- Green Business
- Inventory Management
- Logistics and Supply Chains
- Predictive Modeling
- Renewable Energy
- Smart Warehousing
- Sustainable Development

Acknowledgment

This work was supported by the Deanship of Scientific Research, Vice Presidency for Graduate Studies and Scientific Research, King Faisal University, Saudi Arabia [Grant No. KFU253320]

Chapter 1
Eco-Friendly and Energy-Efficient Warehouses in Modern Supply Chains: Comprehensive Literature Review

Mohamed Souleimen Ouannes
https://orcid.org/0000-0003-0335-6013
King Faisal University, Saudi Arabia

ABSTRACT

Scarcity of resources and climate change have generated mounting demand for sustainability in every field and industry, including supply chains and logistics. Warehouses, being one of the most vital links in supply chains, require additional innovative efforts to render them more agile. This chapter addresses the topic of green and energy-efficient warehouses as key factors in the transition to more sustainable supply chains. We begin with an overview of traditional warehouses that are beset by a number of environmental problems. This serves to bring us to the demand for the creation of green warehouses. The paper also covers the main management practices to enable these transformations. In order to provide the reader with a clearer picture, we provide examples of pioneer companies in the sector. The paper further highlights that the future of green warehouses is heading in the direction of greater digitization. We conclude that the adoption of green warehouses is no longer an optional decision but a strategic imperative to ensure continuity and leadership in a market.

DOI: 10.4018/979-8-3373-3176-8.ch001

SUMMARY

Currently, the scarcity of resources and climate change have generated mounting demand for sustainability in every field and industry, including supply chains and logistics. Warehouses, being one of the most vital links in supply chains, require additional innovative efforts to render them more agile.

This academic paper addresses the topic of green and energy-efficient warehouses as key factors in the transition to more sustainable supply chains.

We begin with an overview of traditional warehouses that are beset by a number of environmental problems like air and noise pollution, high-energy consumption and inadequate waste management.

This serves to bring us to the demand for the creation of green warehouses based on energy-conserving technologies such as smart lighting, solar energy, energy management systems, and automation.

The document also covers the main management practices to enable these transformations, including the creation of internal environmental policies, employees training, monitoring environmental performance, and engaging with sustainable suppliers. In order to provide the reader with a clearer picture, we provide examples of pioneer companies in the sector that have implemented sustainability concepts in their warehouses, thereby reducing their carbon footprint while enhancing operational efficiency. The article further highlights that the future of green warehouses is heading in the direction of greater digitization, greater reliance on technology and renewable energy, and the use of the circular economy in supply chains. We conclude that the adoption of green warehouses is no longer an optional decision but a strategic imperative to ensure continuity and leadership in a market that is rapidly transforming towards environmental sustainability standards.

Green warehouse is an essential component of sustainable development in contemporary supply chains. The transition to green warehouses requires a combination of modern technologies, advanced management practices, and smart investments. By adopting sustainability principles, companies, not only protect the environment, but also improve their operational efficiency and enhance their image in front of customers and investors.

The shift to green warehouses is no longer an option, but has become a strategic necessity in a rapidly changing world that is moving towards a green economy. However, the success of this trend requires integrated efforts on legislative, technical and training levels to ensure its sustainability and achieve a concrete impact on the environment and society.

INTRODUCTION

Environmental sustainability is one of the most prevalent challenges facing modern societies in response to rapid climatic shifts and increased pressure on natural resources.

In this context, supply chains have become pivotal in achieving environmental goals, as storage, transportation, and distribution activities contribute significantly to greenhouse gas emissions and energy consumption. Within this context, supply chains have risen as a prime factor in addressing environmental goals since storage, transport, and distribution activities are major sources of greenhouse emissions and energy consumption.

Warehouses form part of the most crucial components of the supply chain and serve as a necessary link between manufacturers and consumers, possessing goods and products and readying them for distribution at the right moment.

With growing global awareness of environmental issues such as climate change and the exhaustion of natural resources, the need for greener solutions across the entire supply chain, from raw material procurement and production to storage and distribution, has never been greater.

Many firms today have embraced the idea of "green warehouses" as a policy to reduce the environmental impact of warehouses by optimizing energy use, adopting renewable energy sources, and creating smart and sustainable buildings. Going green is not only about achieving environmental goals, but also ensuring significant long-term financial benefits by cutting down operational expenses and simplifying the supply chain (Bansal & DesJardine, 2014).

This article conducts a complete scientific investigation into the topic of green and environmentally friendly warehouses according to an understanding of the issues of traditional warehouses, with reference to their relation with the concept of sustainability. This research will consider the essential theoretical concepts, environmental demands, emerging technologies, practical case studies, challenges, and future suggestions to cause a positive and sustainable effect on the environment.

1. THEORETICAL FRAMEWORK FOR GREEN WAREHOUSES

1.1 Definition of Green Warehouses

Green warehouses refer to storage and logistics facilities that are intended to have a low impact on the environment. This includes the use of green construction practices, the incorporation of renewable energy sources, the optimization of storage and transport systems to reduce carbon emissions, and the sustainable management of waste (Sarkis, 2017). They are part of "green logistic" which aims to integrate environmental considerations throughout the entire supply chain.

The concept of the warehouse has evolved, and it is now viewed as a dynamic "hub" rather than a static "link." Modern warehouses have evolved into value-added service centers (e.g., processing, final assembly, and returns management), which doubles their environmental and energy footprint and makes their conversion to environmentally friendly warehouses even more important. Today's warehouses are also becoming similar to "urban consolidation centers," which are green warehouses located on the outskirts of cities. These hubs consolidate goods from multiple suppliers for final, optimized delivery to urban areas, significantly reducing total vehicle miles traveled, congestion, and emissions—a perfect example of a warehouse solving a broader urban sustainability challenge (Tetteh, F. K., and al, 2024).

We will show in this paper that an environmentally friendly warehouse is a socio-technical system that requires simultaneous advances in technology, process design, human capital, and governance to reach its full potential.

1.2 From Traditional Warehouses to Eco-Friendly Warehouses

Traditional warehousing has for a long time, and is still being faced with, a range of environmental challenges with negative impacts on the environment and the society. Here are some of the more high-profile of these challenges (C.f. Tab.1):

High-energy consumption: Traditional warehouses become highly energy-dependent. They require energy for lighting, cooling, ventilation, and internal transportation systems. This high-energy consumption contributes to increased greenhouse gas emissions, accelerating global warming.

Poor waste management: Conventional warehouses are notorious for generating waste, such as plastic packaging, industrial waste, and damaged products. This waste is often disposed of unsustainably, leading to soil and water contamination.

Noise and air pollution: Operations in traditional warehouses, for instance the use of equipment and vehicles, contribute to the emission of air pollutants such as nitrogen oxides and carbon dioxide. Moreover, traffic movement and equipment operations continuously produce a high level of noise, which erodes the quality of life in the neighboring areas.

Overexploitation of natural resources: The majority of warehouses are constructed without considering resource efficiency standards such as water and raw materials. This overuse leads to the depletion of natural resources and increases negative environmental impacts.

Inefficient design: Conventional warehouses have inefficient design, resulting in additional wastes and use of resources. In the opposite, smart designs reduces the consumption of energy and simplifies process flow.

The "high-energy consumption" and "poor waste management" were historically viable and not considered like negative economic externalities for traditional warehousing. The shift to green warehouses represents a correction towards true cost accounting, driven by regulation, consumer awareness, and the long-term economic rationality of efficiency.

Solutions to these issues need to be identified. Green warehouses are the best solution to this problem. The idea of green warehouses came into being during the 1990s as awareness regarding the implications of climate change intensified and consumers and businesses alike grew greener. Driven by legislative and competitive needs, firms are seeking logistics solutions that meet environmental requirements without affecting profitability. Initiatives such as the United Nations Global Compact (UNGC) and the Paris Climate Accord are responsible for this green practice (UNGC, 2015).

Table 1. Traditional vs. Green Warehouse

Aspect	Traditional Warehouse	Green Warehouse
Energy use	High, traditional lighting and air conditioning	Low, smart LED lighting, energy-efficient HVAC systems
Waste management	Landfill, limited recycling	"Zero waste" strategy, recycling and reuse
Energy source	Traditional electricity electricity (fossil fuels)	Renewable energy (solar, wind) + energy efficiency
Design	Inefficient, long distances, poor thermal insulation	Use of natural lighting and ventilation, insulating materials
Pollution	High carbon emissions, noise pollution	Low carbon emissions, quiet working environment
Cost	High long-term operating costs	Higher initial investment, significant savings in operating costs

Source, Author, 2025

1.3 Relationship Between Green Warehouses and Sustainability

Green logistics is a critical tool for accomplishing the Sustainable Development Goals (SDGs) in the coming years, particularly with respect to climate action, responsible consumption and production, and sustainable cities and communities. Green warehouses contribute to the achievement of the following: *Eco-efficiency* (By optimizing the use of water and raw materials, minimizing emissions, ensuring proper waste disposal, and preserving air and water quality); *Economic efficiency* (By reducing costs through less energy and resource consumption through recycling and reuse, for example); and *Social responsibility* (By enhancing warehousing working conditions and promoting workers' safety, it necessarily reflects on the company's reputation and relations with all parties) (United Nations, 2015).

We conclude that green warehouses support the three principles of sustainability and contribute in one way or another to balancing the economic, environmental, and social goals of modern organizations. In this sense, green warehouses are not just a strategic choice but also an existential necessity for ensuring the sustainability of businesses and societies.

However, although these three pillars are interrelated, we must acknowledge that there are inherent trade-offs between them, particularly high initial capital expenditures (CAPEX) versus long-term operating expense savings (OPEX). The business case depends on demonstrating a strong return on investment (ROI) and net present value (NPV) that justify the upfront "green premium." Furthermore, the social pillar must be expanded to include just transition considerations—ensuring workers are retrained for new technology-dependent roles and that communities surrounding warehouses benefit from reduced pollution (Schmidt, Wagner, 2019).

2. CHARACTERISTICS AND CRITERIA FOR ECO-FRIENDLY WAREHOUSES

2.1 International Warehouse Standards

Green warehouses are governed by a network of international standards that force them to follow green practices. Some of the better-known standards are as follows:

Leadership in Energy and Environmental Design (LEED) certification: Formulated by the U.S. Green Building Council (USGBC), which is the most important certification in the world for measuring building performance in terms of energy, resource use, indoor environmental quality, and design innovation (USGBC, 2021, 2022).

Building Environmental Assessment Method (BREEAM): BREEAM was developed in the UK and evaluates buildings based on energy efficiency, waste management, sustainable use of materials, and environmental impact (BRE, 2020).

Further certifications such as the WELL Building Standard for staff well-being and ISO 14001 for environmental management systems.

These standards ensure the implementation of design, construction, and operation practices that minimize adverse environmental impacts and enhance the sustainability of logistics facilities.

While these standards are applicable and affordable for developed countries, the costs of obtaining LEED certification for example, may be prohibitive in emerging economies, necessitating the development of local, context-appropriate green building rating systems.

More thought and design development is needed to incorporate biophilic design principles—which integrate natural elements such as interior green walls and abundant natural light—that have been shown to reduce stress and improve employee well-being and productivity, adding a powerful dimension to social sustainability (Abdelaal, F., & Guo, B. H. 2022).

2.2 Characteristics of Sustainable Warehouse Design

Eco-friendly warehouse design focuses on balancing operational efficiency and environmental conservation. The key design features include the following:

Take advantage of natural light: Employ high-performance windows and light shelves to reduce the need for artificial lighting.

Natural ventilation systems: To reduce dependence on mechanical cooling systems and save energy.

Passive thermal design: The use of building materials with high thermal insulation to minimize heat transfer; this tactic indirectly reduces the need for energy for heating and cooling.

Strategic geographic location: Select sites near warehouses and key markets to minimize transportation and distribution emissions (Kibert, 2016).

2.3 Impact of Materials Used in Construction and Processing

The materials used in building and equipping warehouses are key to achieving environmental sustainability. Prominence is given to:

- Use recycled materials: Such as recycled steel, wood from renewable resources, and low-carbon concrete.
- Choose materials with a lower environmental footprint: those that require less energy to produce and generate fewer emissions.
- Embrace the concept of "reusable buildings": Designing a warehouse to be easily dismantled and reused at its life cycle's end (Cabeza et al., 2014).

This trend minimizes the use of natural resources and maintains the principles of the circular economy.

3. THE ROLE OF MODERN TECHNOLOGIES AND TECHNIQUES IN THE TRANSITION TO GREEN WAREHOUSES

Warehousing is the first step in the transportation and logistics supply chain, and if these processes can be optimized from the beginning with smart technology, the rest of the transportation process, such as loading, unloading, and delivery, can be streamlined.

Green technologies play a pivotal role in transforming traditional warehouses into eco-friendly, energy and cost-saving facilities that are sustainable. The most important technologies are as follows:

3.1 Smart Energy Management Systems

The role of intelligent energy management systems in promoting the eco-efficiency of warehouses. These systems involve the use of sophisticated sensors, smart control software, and real-time data analysis methods to monitor the energy usage effectively (Wang et al., 2017).

The practical applications of smart energy management systems include:

- Tracking power usage in real time.
- Predicting future energy requirements is also important.
- Optimizing the performance of lighting, HVAC, and machinery based on usage patterns.
- Automating the powering off of unused systems to save energy.

As we can see, these systems provide detailed data that helps in making effective decisions.

3.2 Artificial Intelligence and the Internet of Things (IoT)

The development of artificial intelligence and the Internet of Things (IoT) has revolutionized green warehouses operations. These technologies enable:

- *Predictive maintenance*: By monitoring the equipment's condition and pre-emptive failure prediction, it extends the lifespan of the assets and minimizes waste (Lee et al., 2018).
- *Improve distribution and storage*: By analyzing demand data and rationalizing product distribution in the warehouse to restrict crane movements and reduce energy consumption.

- *Smart grid integration*: Warehouses can be connected to smart electrical grids to utilize renewable energy sources in an efficient manner.

Furthermore, artificial intelligence can be used in generative design to create optimally efficient warehouse layouts that minimize energy use for heating, cooling, and material transport.

3.3 Automation and Robotics Systems

Automation and robotics systems are used for sorting, transportation, and internal storage, reducing the need for fossil fuel equipment and increasing the operational and environmental efficiency of warehouses across:

- Minimizing the need for continuous lighting thanks to operation in low-light environments.
- Reducing energy consumption by optimizing transportation routes.
- Reducing human error and minimizing waste in materials and resources (Boysen et al., 2019).

Machine learning in the warehouse supply chain is used to automate manual labor, predict potential issues, and reduce paperwork for warehouse employees. Machine learning also helps program self-driving vehicles and robots to perform warehouse operations.

Automation enables warehouses to operate more efficiently while minimizing their emissions and waste.

However, attention must be paid to the cybersecurity risks associated with increased digitization. As warehouses become smarter and more connected (Internet of Things), they become more vulnerable to cyberattacks that could disrupt operations, steal data, or even cause physical damage. A comprehensive green warehouse strategy must include robust cybersecurity protocols (Ivanov, D., Dolgui, A., & Sokolov, B. 2022). .

3.4 Smart Lighting and Air Conditioning Systems

As shown in the table below (C.f. Tab.2), lighting and HVAC systems are among of the most energy-intensive storage systems. Green practices include the following:

- Use high-efficiency LEDs: Smart lighting systems based on high-efficiency LEDs can reduce energy consumption by up to 80% compared with conventional lighting. Movement sensors are used to ensure that lights are turned on only when necessary, thereby minimizing energy waste.

- Adopt intelligent lighting control systems based on movement and natural light sensing.
- Implement high-efficiency air-conditioning technologies: Green warehouses depend on intelligent HVAC systems that use Heat Recovery Systems or VRF systems, which improve energy efficiency and maintain a healthy working environment (Smith, 2023).

Table 2. Comparison of Warehouse Technologies

Technology	Description	Energy Savings (%)	Efficiency Improvement (%)
High-Efficiency LEDs	Reduces energy consumption by optimizing lighting	Up to 80%	N/A
Intelligent HVAC Systems	Uses Heat Recovery Systems or VRF systems	30-50%	Maintains healthy environment
Electronic Warehouse Systems	Automates operations, reduces manual labor	N/A	Increases speed and accuracy
Smart Energy Management Systems	Monitors and optimizes energy usage	20-50%	Predicts future requirements

Source: International Energy Agency,2023

3.5 Electronic Warehouse Systems

Unlike traditional warehouses, e-warehouses are characterized by several interconnected technologies that work together, ultimately leading to enhanced warehouse productivity and efficiency. In traditional warehouses, we usually see workers walk around, pick products, load them into carts, and deliver them to shipping docks. In an online warehouse, orders are received automatically, the system then confirms whether the products are in stock, and then, the pick lists are sent to robot carts that place the ordered products into containers and hand them over to workers for the next step.

In general, online stores simplify the processes of storage, inventory, labeling, and packaging. Online stores offer many advantages, which can be summarized as follows:

- It reduces the need for manual labor.
- It minimizes human error.

- It increases visibility and transparency of inventory processes.
- It minimizes potential damage and inaccuracies.
- It optimizes analytics.
- It increases speed and accuracy.

Electronic warehouse systems can provide many services; from collecting all the valuable data, you might want, to helping manage warehousing operations, enabling the warehouse manager to monitor the efficiency of daily operations in the warehouse. It can also help collect real-time data to create visual reports, thereby revealing any inefficiencies in operations and resolving any issues in a timely manner before they get out of hand, which will ultimately keep operations moving in the right direction.

These systems make it easier to keep track of labor and report on its efficiency by providing complete data and continuous improvements, and they also help reduce labor and thus the associated costs (Global Warehouse Automation Market Report, 2024).

3.6 Cybersecurity in Smart Green Warehouses

The integration of Industry 4.0 technologies in green warehouses has created a new vulnerability landscape. Cybersecurity has emerged as a critical concern for smart warehouses operating with interconnected IoT devices, cloud-based management systems, and automated robotics. A successful cyberattack could disrupt energy management systems, disable automation equipment, or compromise sensitive operational data, thereby undermining both operational efficiency and sustainability goals.

Research indicates that cyber-physical systems in warehouses face unique security challenges:

IoT Vulnerabilities: Many IoT sensors lack robust security protocols, making them entry points for attacks (Ivanov et al., 2022).
Supply Chain Attacks: Compromised software updates from vendors can introduce vulnerabilities.
Data Integrity Risks: Manipulation of energy consumption data or inventory records could cause significant operational disruptions (C.f. Tab.3).

To address these challenges, green warehouses must implement multi-layered security frameworks incorporating Network segmentation for IoT devices; Regular security audits and penetration testing; Employee training on cybersecurity best practices; and Blockchain technology for secure data verification (Smith, 2023).

The convergence of sustainability and security represents a new frontier in warehouse management, where environmental goals should be balanced with cyber resilience requirements.

Table 3. Cybersecurity Framework for Green Warehouses

Security Layer	Components	Sustainability Link
Physical Security	Access controls, surveillance systems	Protects renewable energy infrastructure
Network Security	Firewalls, intrusion detection systems	Ensures continuous energy monitoring
Data Security	Encryption, blockchain verification	Maintains accuracy of sustainability metrics
Operational Security	Employee training, incident response plans	Prevents disruptions to green operations

Source: Adapted from Ivanov et al. (2022)

4. ENLIGHTENED CORPORATE PRACTICES AND THEIR ROLE IN THE TRANSITION TO GREEN WAREHOUSES

Sustainability in warehouses is not just about technology; it also requires management and structural changes to support environmental trends. In the next section, we focus on policies, training methods, and green operations management. The most important management practices for achieving more sustainable warehouses are as follows:

- *Developing internal environmental policies:* A green warehouse begins with the establishment of clear and strict policies related to energy management, waste minimization, recycling, and respect for local and international environmental standards, such as compliance with national waste and energy management standards and UN climate change conventions. It also supports the concept of corporate social responsibility by contributing to environmental protection, supporting local communities, and achieving environmental justice (Carroll & Shabana, 2010).
- *Training employees on sustainability practices:* Ongoing training programs are essential for the success of any green initiative. Employees are educated on the importance of energy efficiency, resource-saving techniques, and eco-friendly practices when performing daily tasks.
- *Adopting sustainable transportation systems in warehouses:* Encourage the use of electrical or hybrid vehicles, dependent on internal transport systems

that decrease carbon emissions, and optimize loading and unloading schedules to reduce unnecessary operating shutdowns.

- *Adopting green building design:* Many warehouses adopt building standards mentioned in the previous section that focus on improving thermal insulation, using durable and low-emission building materials, and reducing water and energy consumption (McKinnon, 2018 & GBC, 2022).
- *Monitoring and analyzing environmental performance:* Green warehouses use advanced equipment to control energy consumption, carbon emissions, and the waste production. These data are regularly analyzed to identify opportunities for improvement and make data-driven corrective decisions.
- *Building partnerships with green suppliers and distributors:* Collaborating with partners who adopt similar environmental standards is crucial to support a fully sustainable supply chain, from manufacturing to warehousing to final distribution.
- *Adopting the concept of a circular economy:* A circular economy aims to reduce waste by reusing and recycling materials as much as possible. This concept can be applied in warehouses by recycling packaging materials and reusing industrial waste.
- *Managing waste and water resources in warehouse environments:* Green warehouses adopt advanced waste management policies, including waste sorting and recycling, as well as reducing paper consumption by going digital. Sustainable water management includes installing rainwater harvesting systems and using them for industrial purposes; adopting water-saving sanitary fixtures (i.e., low-flow fixtures); and implementing programs to reuse treated water.
- *Use of clean energy sources:* Green warehouses adopt clean energy sources to power their systems. The most important of these sources are as follows:
    - Solar energy: Installing solar panels on warehouse roofs has become a common practice for generating clean electricity. Relying on solar en ergy reduces dependence on traditional grids and the carbon footprint. (Hernandez et al., 2014).
    - Wind energy: Small-scale turbines are used to power the facility.
    - Geothermal energy: Utilization of heat from the earth's interior for heating and cooling.

The use of these sources reduces dependence on fossil fuels and lowers carbon emissions. So far, we have covered the most important practices and policies that conscious companies adopt to manage their warehouses. These guidelines help to achieve the concept of "zero waste", one of the most important goals of environmental sustainability, thus improving the environmental performance of organizations (Chow, 2010).

- *Green Human Capital Management:* The transition to green warehouses necessitates parallel evolution in human resource management practices. Green human resource management (GHRM) has emerged as a critical enabler of sustainability initiatives, focusing on developing employees' capabilities to support environmental objectives (Jabbour & De Sousa, 2016). Key components of GHRM in warehouses include:
    - *Green Recruitment*: Incorporating sustainability competencies into job descriptions and selection criteria.
    - *Environmental Training*: Developing programs on energy conservation, waste management, and sustainable operations.
    - *Performance Management*: Linking compensation and rewards to sustainability metrics.
    - *Employee Engagement*: Creating green teams and suggestion systems for environmental improvements.

Research shows that organizations implementing comprehensive GHRM practices achieve: 30% higher employee engagement in sustainability initiatives; 25% better performance on environmental metrics; 40% lower turnover among sustainability-focused staff (World Economic Forum, 2020).

The human dimension of green warehouses extends beyond technical skills to encompass cultural transformation and organizational learning capabilities. The human capital dimension is critical. This transition creates demand for "green collar" jobs—specialists in energy management, sustainability data analysis, and circular economy logistics. Companies must also invest in worker retraining and upskilling programs (World Economic Forum, 2020).

Perhaps the circular economy discussion can also be expanded beyond recycling to include product-as-a-service models and reverse logistics hubs, where warehouses become centers for renewal, repair, and remanufacturing. The role of green procurement policies in driving sustainability forward is also critical (Geissdoerfer, M., Pieroni, M. P., Pigosso, D. C., & Soufani, K. 2020).

5. GLOBAL CASE STUDIES

In this section, we review the examples of real-life in successful green warehouses, as there are many international models that have used the concept of green warehouses and have achieved good results in sustainability and energy, including:

Amazon Green Warehouses: Amazon has invested significantly in the production of environmentally friendly warehouses. Some of the most remarkable works are as follows:

- Installing solar panels on the roofs of more than 50 global distribution centers, such as those in Tracy and California, which have 20,000 square meters of solar panels and generate 80% of the center's energy needs.
- The use of energy-efficient robots in internal sorting systems is also being investigated.
- Reducing energy consumption in some locations, after improving thermal insulation and smart lighting techniques.
- Obtaining many LEED green building certifications (Amazon Sustainability Report, 2023).

Walmart Eco-Friendly Warehouse: Walmart works to make all its logistics centers more environmentally friendly by introducing electric trucks into the transportation fleet; by optimizing ventilation and lighting systems using smart sensors. Walmart distribution centers use smart energy management systems that reduce power consumption by 30 percent; and by implementing waste recycling programs with more than 80% recycling in many places (Walmart Global Responsibility Report, 2022).

DHL's Green Warehouse Initiative: DHL has launched a "GoGreen" initiative to operate all its facilities using clean energy sources by 2050. The initiative involves the construction of a modern fleet and certified distribution centers (DHL Sustainability Report, 2023).

IKEA Warehouses: The Swedish company IKEA is also a powerful model: Indeed, all of its new warehouses are designed to be carbon neutral by 2030. It is also very dependent on solar and wind energy. Finally, it invests in green construction technology and in the use of recycled materials in building and furnishing (IKEA Sustainability Report, 2023).

Tesla Warehouses: Tesla's Giga-warehouses are one of the biggest examples of green warehouses and factories, as they depend almost completely on solar and wind energy. It was designed with an internal industrial waste recycling system. It focuses on the production and storage of renewable energy products, such as batteries and electric vehicles (Tesla Impact Report, 2024).

While these are excellent examples of large corporations, a significant gap exists in adapting these solutions for Small and Medium-sized Enterprises (SMEs). Their path to sustainability may rely more on collaborative logistics (sharing warehouse space and transportation), leveraging third-party logistics providers (3PLs) with green expertise, and implementing phased, modular technological upgrades (C.f. Tab.4).

Table 4 . Comparative Analysis of Green Warehouse Initiatives

Company	Primary Strategy	Energy Savings	CO2 Reduction	Implementation Timeline	ROI Period
Amazon	Solar integration + robotics	40%	50%	3-5 years	4-6 years
Walmart	LED lighting + efficiency	30%	35%	2-4 years	3-5 years
DHL	Holistic green strategy	45%	55%	5-7 years	5-8 years
IKEA	Circular economy focus	50%	60%	4-6 years	6-9 years
Tesla	Renewable energy integration	60%	70%	5-8 years	7-10 years

Source: Company sustainability reports (2023-2024)*

6. CHALLENGES AND OBSTACLES IN IMPLEMENTING GREEN WAREHOUSES

Among the challenges and difficulties that companies face in implementing the concept of green warehouses (C.f. Tab.5):

High initial cost: One of the main barriers to adoption, especially for SMEs, is the high initial expense of building green warehouses.

Shortage of specialized engineering proficiency: In many countries, there is a shortage of talent and specialized engineers and designers in sustainable building standards, leading to project delays or added expenditure.

Regulatory and legislative barriers: In some countries, there are yet undeveloped environmental legislation or laws that do not stimulate the application of green building concepts, which hinders the spread of green warehouses.

The challenge of “high initial cost” can be mitigated by discussing innovative financing models. Green leases align landlord and tenant interests in efficiency upgrades. Energy Performance Contracts (EPCs) allow companies to finance improvements through the future energy savings they generate. Green bonds and government grants/tax incentives are also crucial tools (EEFIG, 2019) . The talent shortage requires advocacy for updated educational and vocational training curricula focused on sustainable logistics.

The transformation to green warehouses creates new dependencies on specialized technologies and rare earth materials, introducing vulnerabilities in supply chain resilience. The concentration of solar panel manufacturing in specific regions and the limited availability of lithium for energy storage systems present significant challenges (Zhang et al., 2021).

Table 5. Challenges and Constraints in Green Warehouse Implementation

Challenge	Description	Potential Solutions
High Initial Costs	Significant upfront investment required	Public-private partnerships, financing options
Lack of Specialized Engineering Talent	Shortage of skilled workers in green technologies	Customized training programs
Regulatory and Legislative Obstacles	Complex regulations can hinder implementation	Advocacy for supportive policies

Source: Green Building Council, 2022

Key supply chain risks include geopolitical dependencies (80% of solar panel components originate from limited geographic regions), raw material scarcity (Lithium, cobalt, and rare earth elements face supply constraints), technology obsolescence (Rapid innovation cycles may render investments obsolete quicker than anticipated), and Skills gap (Shortage of technicians qualified to maintain advanced green technologies). Mitigation strategies to these risks require:

- Diversifying supplier networks across geographic regions
- Investing in modular and upgradeable technologies
- Developing circular economy approaches for technology refresh cycles
- Creating strategic reserves of critical components

The integration with smart cities and urban systems is essential today. Green warehouses are increasingly functioning as nodes within broader smart city ecosystems. The concept of "urban consolidation centers" represents a paradigm shift where warehouses serve multiple stakeholders while optimizing urban logistics (Tetteh et al., 2024).

Key integration points include energy grid integration (Warehouses as prosumers in smart grids), logistics optimization (Coordinated delivery systems reducing urban congestion), waste management (Reverse logistics integration with municipal systems), and data sharing (Operational data contributing to urban planning).

7. FUTURE TRENDS IN GREEN WAREHOUSES

It is expected that the concept of green warehouses will continue to evolve as technology advances and global environmental awareness grows. Here are the most remarkable future trends to expect:

Full digital transformation: The future of warehouses will rely quickly on digital solutions:

- Intelligent energy management systems using artificial intelligence.
- Internet of Things (IoT) networks to monitor real time energy and resource consumption.
- Predictive analytics techniques to improve process efficiency and minimize waste.

Decentralized Renewable Energy: Dependence on locally produced energy (sun, wind, green hydrogen) will increase without the need to forward traditional centralized grids.

New sustainable building materials: More sustainable construction materials will be used in the coming years such as low-carbon concrete; recycled materials at a higher percentage; and smart materials that are able to adapt to environmental conditions to reduce energy consumption.

Tighter environmental regulations: Governments will impose stricter rules on warehouses regarding emissions, energy efficiency, and waste management, which makes compliance with sustainability standards a legal and business imperative. Companies will be forced to apply environmental sustainability standards at all stages of a project from planning and design to implementation and maintenance, so that environmental goals can be met.

Integrated Circular Economy: The warehouse will be an active part of the circular economy cycle which will design products and services to minimize waste and promote reuse and recycling from the beginning.

Promoting public-private partnerships: Strategic partnerships between governments and the private sector will be strengthened to fund green warehouse projects and provide financial incentives to encourage adoption.

Develop customized training programs for green warehouse workers: The skills of the workforce will be developed through continuous training and qualification, which is vital to the successful management and efficient operation of eco-friendly warehouses.

A crucial futuristic trend to add is the Physical Internet (π). Inspired by the digital internet, it envisions an open global logistics system with standardized modular containers and shared, smart logistics facilities ("π-hubs"). This could radically increase asset utilization and optimize energy use across entire supply chain networks, moving beyond individual warehouse optimization. The trend of on-site generation and microgrids using renewables and battery storage is also key for energy resilience and decentralization (Montreuil, B., Ballot, E., & Fontane, F. 2020). .

8. MEASURING PERFORMANCE AND ASSESSING THE IMPACT OF GREEN WAREHOUSES

The global shift toward sustainable supply chain management has increased attention on green warehouses as vital components of environmental and economic efficiency. However, simply setting up green warehouses is not enough if their effectiveness is not evaluated. Measuring performance and evaluating impact are necessary to ensure that green warehouses fulfill their promises of sustainability, economic feasibility, and social responsibility. This section explores the frameworks, tools, and methods used to assess green warehouse performance in environmental, economic, and social areas.

8.1 Key Performance Indicators (KPIs) for Green Warehouses

Key Performance Indicators (KPIs) are measurable metrics that help organizations track their green initiatives. For warehouses, KPIs cover several areas.

- *Energy Efficiency Metrics:* Measurement of electricity usage per square meter or per unit handled, reductions in peak energy demand, and the share of renewable energy used in operations (Ahi & Searcy, 2015).
- *Carbon Emissions Indicators*: Tracking the amount of CO_2 emissions avoided through clean energy, smart lighting, and optimized HVAC systems.
- *Resource Utilization:* Indicators related to water use, waste recycling rates, and efficiency in packaging materials (Chardine-Baumann & Botta-Genoulaz, 2014).
- *Operational Efficiency:* Metrics such as order accuracy, throughput, and automation efficiency that support both environmental and business goals.

Using KPIs strategically allows companies to demonstrate compliance with international standards like ISO 14001 and LEED certification (International Organization for Standardization, 2015; U.S. Green Building Council, 2019), and show their commitment to ongoing improvement. KPIs also work as benchmarking tools, allowing companies to compare their environmental performance with industry peers.

8.2 Environmental Impact Assessment (EIA)

Environmental Impact Assessment (EIA) is a method for evaluating the environmental effects of a project before it starts. For warehouses, EIA helps organizations spot potential risks and create mitigation strategies.

Central to this process is Life Cycle Assessment (LCA), which assesses the environmental impact of a warehouse throughout its life—from construction and material use to daily operations and eventual closure (Chileshe et al., 2016). Through LCA, organizations can measure greenhouse gas emissions, energy needs, and waste production.

EIA also includes evaluations of:

- Air Quality: Monitoring emissions from vehicles, HVAC systems, and machinery.
- Noise Pollution: Especially significant for urban warehouses near residential areas.
- Land Use and Biodiversity: Assessing the ecological impacts of building large facilities.

EIA's value lies in its ability to predict, allowing organizations to identify negative effects before they happen and implement solutions. This supports the shift toward sustainable warehousing practices (Sarkis & Zhu, 2018).

8.3 Economic and Social Impact Analysis

The move to green warehouses brings benefits that go beyond environmental ones. Conducting economic and social impact assessments ensures a complete understanding of value creation.

- Economic Impacts: Green warehouses often need significant initial investments in renewable energy, advanced automation, and eco-friendly building materials. However, these costs can be balanced over time by lower energy bills, improved operational efficiency, and long-term resilience against rising fuel prices (Kumar et al., 2012). Research shows that companies with sustainable practices often gain a competitive edge by bolstering their brand reputation and attracting environmentally conscious consumers (Porter & van der Linde, 1995).
- Social Impacts: In addition to financial returns, green warehouses improve working conditions through better lighting, ventilation, and ergonomics. Sustainability initiatives also create opportunities for employee training and

skill development, promoting a culture of environmental awareness (Jabbour & de Sousa Jabbour, 2016). More broadly, green warehouses enhance corporate social responsibility (CSR) by benefiting communities and reducing environmental damage.

Thus, economic and social analyses provide key evidence for stakeholders to support green investments and affirm their legitimacy.

8.4 Benchmarking and Global Best Practices

Benchmarking involves comparing a warehouse's performance with recognized sustainability leaders. Global best practices offer examples that organizations can modify for local contexts.

- Amazon: The company's warehouses use renewable energy systems and AI-driven efficiency measures, establishing a standard for large-scale green operations (UNEP, 2021).
- DHL: Its GoGreen initiative offers a structured approach to cut carbon emissions across supply chains.
- European Union Standards: EU rules on energy efficiency and circular economy practices create a solid regulatory benchmark for sustainable warehousing (Sarkis & Zhu, 2018).

Benchmarking fosters learning and promotes competition, as organizations seek to be seen as leaders in sustainability. It also supports standardization as industries work toward common sustainability criteria (Ahi & Searcy, 2015).

8.5 Data Analytics and Continuous Improvement

In the age of Industry 4.0, data analytics is crucial for performance measurement. Green warehouses produce vast amounts of data from IoT devices, smart meters, and automation systems. By using analytics and machine learning, organizations can spot inefficiencies in energy usage and material handling. They can also predict maintenance needs to prevent downtime. They can finally improve inventory management to cut waste.

Additionally, continuous improvement methods, such as the Plan-Do-Check-Act (PDCA) cycle, ensure that performance evaluation is ongoing rather than a one-off task. Continuous feedback loops allow organizations to refine strategies, embrace new technologies, and keep up with changing environmental regulations (Chardine-Baumann & Botta-Genoulaz, 2014).

Data-driven decision-making transforms warehouses from static entities into dynamic, responsive systems capable of meeting market and environmental challenges.

Evaluating the effects of green warehouses and measuring their performance are difficult tasks that consider social, economic, and environmental aspects. Impact analyses highlight the wider benefits of sustainable practices, EIAs highlight environmental risks and opportunities, and KPIs provide quantifiable goals. While data analytics propels continuous improvement in a quickly changing technological landscape, benchmarking against world leaders yields insightful information.

Ultimately, performance assessment frameworks supply the evidence needed to justify investments in green warehouses, shape corporate strategies, and influence public policy. Without strong evaluation methods, the vision of sustainable warehousing risks remaining an aspiration rather than a reality. By incorporating thorough measurement and assessment processes into their operations, organizations can ensure that green warehouses act not only as logistical centers, but also as catalysts for sustainability, innovation, and social well-being.

9. GEOPOLITICAL AND NATIONAL POLICIES SUPPORTING GREEN WAREHOUSES

The shift to green warehouses involves more than just new technology and business strategies. It also reflects the global policy environment. Governments, regional organizations, and international institutions are crucial in shaping the environmental agenda through regulations, incentives, and frameworks that promote sustainable practices. As a result, green warehousing is a product of both market initiatives and policy efforts. This section looks at the geopolitical and national policy climates that encourage the growth of green warehouses in various regions, emphasizing global agreements, regional differences, and comparative insights.

9.1 Global Policy Landscape for Green Warehouses

At the international level, sustainability policies are based on global agreements like the Paris Agreement and the United Nations Sustainable Development Goals (SDGs). These frameworks set broad targets for climate and sustainability that then influence national strategies (United Nations, 2015). The Paris Agreement specifically focuses on reducing greenhouse gas emissions. This is important for the logistics and warehousing sectors due to their high energy use (UNFCCC, 2018). International financial institutions, including the World Bank and regional development banks, have also added sustainability requirements to their funding programs, encouraging countries and companies to build greener infrastructure. Additionally, the International Organization for Standardization (ISO) has created guidelines, such as ISO 14001 for environmental management systems, which are widely recognized and shape both corporate and national policies (ISO, 2015).

9.2 European Union Policies and Practices

The European Union (EU) has become a global leader in environmental legislation, making it an important area for green warehousing development. Key policies include the European Green Deal, which sets a binding goal of climate neutrality by 2050, and the Energy Efficiency Directive, which requires improvements in building energy performance (European Commission, 2020). The EU's Circular Economy Action Plan also affects warehouses by promoting material reuse, recycling, and waste reduction. Financial incentives like subsidies for integrating renewable energy and improving energy efficiency further facilitate this transition (EEA, 2021). Consequently, many European logistics companies, such as DHL and DB Schenker, have aligned their operations with EU climate policies by investing in renewable energy and sustainable building materials (Sarkis & Zhu, 2018). The European approach shows how strong regulations, combined with financial backing, speed up the adoption of green warehouses.

9.3 United States and North America

In the United States, the policy landscape for green warehouses is influenced by both federal and state initiatives. At the federal level, the Energy Independence and Security Act and the Clean Air Act create broad environmental standards (EPA, 2020). Meanwhile, the U.S. Green Building Council's LEED certification offers a voluntary but widely accepted framework for green building design, including warehouses (USGBC, 2019). State governments, especially in California and New York, have introduced stricter regulations and incentive programs for sustainable facilities, such as tax credits for solar energy and efficient HVAC systems. Canada has also taken a proactive approach with its Pan-Canadian Framework on Clean Growth and Climate Change, emphasizing investment in green infrastructure (Government of Canada, 2016). Many North American companies, like Amazon, Walmart, and Tesla, have pursued green warehouse initiatives, often driven by a mix of policy incentives and consumer demand (Kumar et al., 2012).

9.4 Asia and Emerging Economies

Asia offers a varying policy landscape, with advanced frameworks in countries like Japan and South Korea and rapidly developing systems in China and India.

- China: The 14th Five-Year Plan prioritizes achieving carbon neutrality by 2060. This leads to significant investments in renewable energy and smart logistics infrastructure (Zhang et al., 2021). Green warehouses are essential to modernizing China's supply chain.
- Japan and South Korea: These nations have enacted strict building codes and renewable energy policies, supported by strong innovation networks that foster warehouse automation and green technologies.
- India and Southeast Asia: While policies are emerging, they face challenges like cost, infrastructure, and enforcement. Initiatives such as India's National Electric Mobility Mission and renewable energy subsidies, however, are promoting a gradual move toward greener warehousing.

Overall, Asia highlights the need to align industrial policy with sustainability goals and shows how green warehouses can support urban growth sustainably.

9.5 Middle East and Arab Region

In the Middle East, green warehouses are becoming more popular as part of broader national plans for sustainable development.

- Saudi Arabia: The Vision 2030 framework emphasizes environmental sustainability and includes large projects like NEOM, where logistics and warehousing follow smart and green principles (Saudi Vision 2030, 2017).
- United Arab Emirates: Dubai's Green Building Regulations and Abu Dhabi's Estidama Pearl Rating System promote eco-friendly construction, directly impacting warehousing standards (Al-Saidi & Elagib, 2018).
- Egypt and Morocco: These countries have included renewable energy development in national strategies, indirectly supporting green logistics infrastructure.

Despite challenges like reliance on fossil fuels and high initial costs, the Middle East shows a growing commitment to integrating green logistics into national sustainability plans.

9.6 Africa and Latin America

In Africa, green warehouse development is just beginning but has great potential. Policies often focus on basic infrastructure, but regional initiatives like the African

Union's Agenda 2063 stress sustainable industrial growth (AU, 2015). South Africa has started to implement green building standards and incentives for renewable energy.

Latin America has taken a more active approach, with countries such as Brazil, Chile, and Mexico creating climate action frameworks to promote green building practices (UNEP, 2021). Nonetheless, limited funding and regulatory enforcement remain obstacles.

9.7 Comparative Analysis of Policy Approaches

Comparing regions shows both similarities and differences in policy approaches. Advanced economies like the EU and Japan depend on strict regulations and financial incentives, leading to quick adoption of green warehouses. Emerging economies typically rely on market incentives, foreign investment, and international pressure to enhance sustainability (Ahi & Searcy, 2015). The geopolitical aspect is evident, as countries compete to lead in sustainable supply chains. Policies in one region can influence others, creating regulatory diffusion. For instance, the EU's environmental standards have prompted reforms in North America and Asia (Sarkis & Zhu, 2018). Successful policy approaches usually include clear regulations, financial support systems, and the integration of sustainability into national development strategies.

Adopting green warehouses is closely linked to the geopolitical and national policy environment. Global agreements like the Paris Agreement set the framework, while regional and national governments turn these commitments into specific regulations and incentives. Europe leads with strong legislation, North America combines federal and state initiatives with voluntary certifications, Asia shows varied approaches based on national capabilities, and the Middle East incorporates green logistics into ambitious national visions. While Africa and Latin America are still developing, they increasingly address the need for sustainable warehousing. A comparative analysis indicates that although enforcement levels and financial support differ, the global trend leans toward more integrated and mandatory sustainability policies. Green warehouses are not just logistical centers; they are key assets in both national and international sustainability agendas.

10. CONCLUSION

It is true that most reference literature has focused on the importance of applying the concept of green warehouses, but in reality, companies face many obstacles, that is why they have abandoned this idea because of its complexity.

The evolution of green warehouses involves a mix of technological growth, management practices, and ecosystem connections. Our analysis shows that successful implementation requires a holistic Approach (Integrating technology, people, and processes); an ecosystem thinking (Connecting warehouses to larger urban and industrial systems); an adaptive governance (Flexible policies that change with technological progress); and a financial innovation (New models that capture value beyond simple energy savings).

The future of green warehouses depends on their shift from cost centers to value generators. They will produce energy, manage waste, provide data services, and support circular economy activities. This shift will need ongoing research and development especially in artificial intelligence for predicting sustainability outcomes; in new materials for construction and energy storage; in new business models that make money from environmental benefits, and in integration frameworks for urban and industrial cooperation.

As warehouses evolve from passive storage spaces to active sustainability hubs, they will become vital in reaching global climate goals and building more resilient supply chains.

REFERENCES

Abdelaal, F., & Guo, B. H. (2022). Knowledge, barriers, and drivers of sustainable construction in developing countries: A case study of Egypt. *Journal of Cleaner Production, 366*, 132860.

African Union. (2015). [The Africa we want. African Union Commission.]. *Agenda (Durban, South Africa), •••*, 2063.

Ahi, P., & Searcy, C. (2015). An analysis of metrics used to measure performance in green and sustainable supply chains. *Journal of Cleaner Production, 86*, 360–377. DOI: 10.1016/j.jclepro.2014.08.005

Al-Saidi, M., & Elagib, N. A. (2018). Ecological modernization and responses for a low-carbon future in the Gulf Cooperation Council countries. *Wiley Interdisciplinary Reviews: Climate Change, 9*(5), e528. DOI: 10.1002/wcc.528

Amazon. (2023). Sustainability Report.

Bansal, P., & DesJardine, M. R. (2014). Business sustainability: It is about time. *Strategic Organization, 12*(1), 70–78. DOI: 10.1177/1476127013520265

Boysen, N., Briskorn, D., & Emde, S. (2019). Sequencing of picking orders in mobile rack warehouses. *European Journal of Operational Research, 273*(1), 222–236.

BRE. (2020). *BREEAM: The World's Leading Sustainability Assessment Method for Masterplanning Projects*. Infrastructure and Buildings.

Cabeza, L. F., Rincon, L., Vilarino, V., Pérez, G., & Castell, A. (2014). Life cycle assessment (LCA) and life cycle energy analysis (LCEA) of buildings and the building sector: A review. *Renewable & Sustainable Energy Reviews, 29*, 394–416. DOI: 10.1016/j.rser.2013.08.037

Carroll, A. B., & Shabana, K. M. (2010). The Business Case for Corporate Social Responsibility: A Review of Concepts, Research and Practice. *International Journal of Management Reviews, 12*(1), 85–105. DOI: 10.1111/j.1468-2370.2009.00275.x

Chardine-Baumann, E., & Botta-Genoulaz, V. (2014). A framework for sustainable performance assessment of supply chain management practices. *Computers & Industrial Engineering, 76*, 138–147. DOI: 10.1016/j.cie.2014.07.029

Chileshe, N., Rameezdeen, R., Hosseini, M. R., & Lehmann, S. (2016). Life cycle assessment and green supply chain management: A review of related studies. *Smart and Sustainable Built Environment, 5*(2), 123–144. DOI: 10.1108/SASBE-03-2016-0005

Chow, T. T. (2010). A review on photovoltaic/thermal hybrid solar technology. *Applied Energy*, *87*(2), 365–379. DOI: 10.1016/j.apenergy.2009.06.037

DHL. (2023). GoGreen Environmental Protection Program.

EEFIG. (2019). *Driving Energy Efficiency in Buildings: A Best Practice Guide to Financing*. European Energy Efficiency Financial Institutions Group.

Environmental Protection Agency. (2020). *Overview of the Clean Air Act and Air Pollution*. U.S. EPA.

European Commission. (2020). *The European Green Deal*. Publications Office of the European Union.

European Environment Agency. (2021). *Progress on energy efficiency in Europe*. EEA.

Geissdoerfer, M., Pieroni, M. P., Pigosso, D. C., & Soufani, K. (2020). Circular business models: A review. *Journal of Cleaner Production*, *277*, 123741. DOI: 10.1016/j.jclepro.2020.123741

Global Warehouse Automation Market Report. (2024).

Government of Canada. (2016). *Pan-Canadian Framework on Clean Growth and Climate Change*. Government of Canada.

Green Building Council (GBC) (2022). The Business Case for Green Buildings.

Hernandez, R. R., Easter, S. B., Murphy-Mariscal, M. L., Maestre, F. T., Tavassoli, M., Allen, E. B., Barrows, C. W., Belnap, J., Ochoa-Hueso, R., Ravi, S., & Allen, M. F. (2014). Environmental impacts of utility-scale solar energy. *Renewable & Sustainable Energy Reviews*, *29*, 766–779. DOI: 10.1016/j.rser.2013.08.041

IKEA. (2023). Sustainability Report.

International Energy Agency. (2023). *Energy Efficiency in Warehousing 2023*. IEA Publications.

International Organization for Standardization. (2015). *ISO 14001:2015 – Environmental management systems – Requirements with guidance for use*. ISO.

Ivanov, D., Dolgui, A., & Sokolov, B. (2022). Cloud supply chain: Integrating Industry 4.0 and digital platforms in the "Supply Chain-as-a-Service". *Transportation Research Part E, Logistics and Transportation Review*, *160*, 102676. DOI: 10.1016/j.tre.2022.102676

Jabbour, C. J. C., & de Sousa Jabbour, A. B. L. (2016). Green human resource management and green supply chain management: Linking two emerging agendas. *Journal of Cleaner Production, 112*, 1824–1833. DOI: 10.1016/j.jclepro.2015.01.052

Kibert, C. J. (2016). *Sustainable Construction: Green Building Design and Delivery* (4th ed.). Wiley.

Kumar, S., Teichman, S., & Timpernagel, T. (2012). A green supply chain is a requirement for profitability. *International Journal of Production Research, 50*(5), 1278–1296. DOI: 10.1080/00207543.2011.571924

Lee, J., Bagheri, B., & Kao, H. A. (2018). A Cyber-Physical Systems architecture for Industry 4.0-based manufacturing systems. *Manufacturing Letters, 3*, 18–23. DOI: 10.1016/j.mfglet.2014.12.001

McKinnon, A. (2018). *Decarbonizing Logistics: Distributing Goods in a Low Carbon World*. Kogan Page.

Montreuil, B., Ballot, E., & Fontane, F. (2020). Physical Internet and Logistics Sustainability: A Review. *Sustainability, 12*(22), 9367.

Porter, M. E., & van der Linde, C. (1995). Toward a new conception of the environment-competitiveness relationship. *The Journal of Economic Perspectives, 9*(4), 97–118. DOI: 10.1257/jep.9.4.97

Tesla Impact Report. (2024).

Sarkis, J. (2017). Green Supply Chain Management: Product Life Cycle Approach. Springer.

Sarkis, J., & Zhu, Q. (2018). Environmental sustainability and production: Taking the road less travelled. *International Journal of Production Research, 56*(1–2), 743–759. DOI: 10.1080/00207543.2017.1365182

Saudi Vision 2030. (2017). Kingdom of Saudi Arabia Vision 2030. Government of Saudi Arabia.

Schmidt, M., & Wagner, S. M. (2019). Supply Chain 4.0: A Primer. Springer International Publishing.

Smith, J. (2023). Smart Warehousing and Green Logistics. Journal of Sustainable Logistics.

Tetteh, F. K.. (2024). Green logistics practices: A bibliometric and systematic review. *Journal of Cleaner Production*. Advance online publication. DOI: 10.1016/j.jclepro.2024.143735

United Nations. (2015). Transforming our world: the 2030 Agenda for Sustainable Development.

United Nations Environment Programme. (2021). Sustainability and circular economy in supply chains: Guidelines and case studies. UNEP.

United Nations Framework Convention on Climate Change. (2018). *The Paris Agreement*. UNFCCC.

U.S. Green Building Council. (2019). *LEED v4 for building design and construction*. USGBC.

U.S. Green Building Council (USGBC). (2022). LEED Green Building Certification Program.

USGBC. (2021). LEED v4 for Building Design and Construction.

Walmart. (2022). Global Responsibility Report.

Wang, S., Yan, C., & Xiao, F. (2017). Quantitative energy performance assessment methods for existing buildings. *Energy and Building*, *43*(2-3), 103–109.

World Economic Forum. (2020). The Future of Jobs Report 2020.

Zhang, D., Zhou, P., & Zhou, D. Q. (2021). Carbon neutrality in China: Challenges and opportunities. *Energy Economics*, *103*, 105577. DOI: 10.1016/j.eneco.2021.105577

Chapter 2
Artificial Intelligence and Robotics in Smart Sustainable Warehouses: A Comprehensive Review

Mariem Mrad
https://orcid.org/0000-0002-2377-1401
Faculty of Economics and Management of Sfax, University of Sfax, Tunisia

Younes Boujelbene
Faculty of Economics and Management of Sfax, University of Sfax, Tunisia

ABSTRACT

This chapter investigates the transformative role of artificial intelligence and robotics in advancing sustainable warehousing within green supply chains. Utilizing a bibliometric analysis of 122 documents from the Scopus database (2014–2024), it examines the intersection of AI, robotics, smart technologies, sustainability, and warehousing. The analysis reveals a 19.9% compound annual growth rate in publications, highlighting key sources, contributing countries, and thematic clusters such as AI-driven optimization and robotic automation. The chapter details applications in energy management, inventory optimization, logistics efficiency, and circular economy integration, supported by case studies of industry leaders. A proposed framework offers practical strategies for adopting these technologies, while addressing challenges like high costs and data integration. Aligned with the United Nations Sustainable Development Goals, this chapter provides actionable insights for warehouse managers, researchers, and policymakers committed to low-carbon, intelligent logistics.

DOI: 10.4018/979-8-3373-3176-8.ch002

1. INTRODUCTION

Warehousing represents a critical node in global supply chains, serving as the backbone of storage, consolidation, and distribution activities. Conventional warehousing operations, however, are energy-intensive and contribute significantly to environmental degradation. Estimates suggest that warehouses account for nearly 10% of logistics-related greenhouse gas (GHG) emissions, primarily due to energy consumption in lighting, heating, ventilation, air conditioning (HVAC), and material handling processes (International Energy Agency, 2022). With intensifying concerns over climate change and the growing influence of international regulatory frameworks such as the Paris Agreement, warehouses are under increasing pressure to transition toward low-carbon operations and align with global sustainability agendas, notably the United Nations Sustainable Development Goals (SDGs).

Technological innovation offers viable pathways for sustainable transformation. AI, encompassing machine learning, predictive analytics, and optimization algorithms, enables warehouses to enhance energy management, optimize inventory systems, and streamline logistics planning (Yadav et al., 2024). In parallel, robotics, ranging from automated guided vehicles (AGVs) and autonomous mobile robots (AMRs) to automated storage and retrieval systems (AS/RS), improves operational precision and reduces energy consumption (Zhen & Li, 2022). The integration of AI and robotics facilitates the emergence of smart and sustainable warehouses capable of minimizing environmental impact while simultaneously enhancing efficiency and productivity. Recent evidence suggests, for instance, that AI-optimized robotic systems can achieve 20–30% reductions in energy consumption through intelligent task allocation and path planning (DHL, 2023).

This chapter contributes to the literature by conducting a bibliometric analysis of 122 peer-reviewed documents indexed in Scopus between 2014 and 2024. While previous studies have often examined the role of digital technologies within broader supply chain contexts (Mrad et al., 2025), this analysis focuses specifically on warehousing, offering a targeted perspective. The chapter pursues five interrelated objectives:

1. To analyse publication trends, influential sources, and thematic clusters in sustainable warehousing research, highlighting their driving factors;
2. To examine applications of AI and robotics in energy optimization, inventory management, and circular economy integration;
3. To critically evaluate real-world implementations through balanced case studies of both successful initiatives and failed attempts;
4. To propose a comprehensive adoption framework incorporating lifecycle assessment, financial feasibility, and ethical considerations; and

5. To address implementation challenges, spanning economic, social, and environmental risks, while identifying evidence-based future trends.

The intended audience of this chapter includes warehouse managers, sustainability consultants, academics, and policymakers. By bridging the gap between research and practice, it advances innovations in green and energy-efficient warehousing. To ensure conceptual clarity, the following key terms are defined: **AI**, computational systems enabling optimization, prediction, and decision-making in warehousing operations; **Robotics**, automated machines such as AGVs and AMRs designed for material handling and task execution; **Smart Warehousing**, technology-enhanced operations leveraging AI, IoT, and automation to improve efficiency and sustainability; **Sustainable Warehousing**, practices designed to minimize environmental impact through energy efficiency, waste reduction, and lifecycle considerations; **Green Supply Chain**, logistics networks integrating environmental considerations to reduce ecological footprints; **Lifecycle Assessment (LCA)**, a methodology assessing the environmental impacts of technologies across their full lifecycle, from raw material extraction to disposal or recycling; and **Circular Economy**, a systemic model promoting resource efficiency and material reuse, supported by AI and robotics.

2. BACKGROUND AND LITERATURE REVIEW

This section provides a foundation for understanding AI and robotics in smart, sustainable warehouses. It begins by grounding the discussion in established management theories to offer a robust academic context, then explores green supply chains, emphasizing warehousing's critical role in achieving sustainability through technological innovation.

2.1. Theoretical Foundations for Smart Sustainable Warehousing

To move beyond a purely descriptive account, it is essential to ground the adoption of AI and robotics in established management theories. The Resource-Based View (RBV) posits that firms gain a competitive advantage by acquiring and controlling valuable, rare, and inimitable resources. Within sustainable warehousing, AI algorithms, proprietary datasets, and integrated robotic systems can be considered such resources, enabling firms to achieve superior efficiency and environmental performance that competitors cannot easily replicate (Barney, 1991). Furthermore, the Dynamic Capabilities Theory is crucial, as it explains how firms can purposefully adapt, integrate, and reconfigure internal and external competencies to address rapidly changing environments (Teece et al., 1997). In this context, the ability of a warehouse to successfully deploy, manage, and evolve its AI and robotic systems in response to new sustainability regulations or market demands is a dynamic capability that underpins long-term success.

2.2. Green Supply Chains and Warehousing

Green supply chains integrate environmental considerations into logistics, aiming to reduce greenhouse gas (GHG) emissions and promote circular economy practices (Seuring & Müller, 2008). Warehousing, a key node, contributes significantly to emissions due to energy-intensive operations (International Energy Agency, 2022). Sustainable warehousing adopts strategies such as energy-efficient technologies and waste reduction to align with global sustainability frameworks, such as the United Nations Sustainable Development Goal 12 (McKinnon et al., 2015). To scale these practices, warehouses increasingly rely on AI and robotics. However, it is crucial to distinguish between operational efficiency and genuine environmental sustainability, as efficiency gains can sometimes lead to rebound effects, where increased throughput negates energy savings per unit (Zell-Ziegler et al., 2021). Building on these foundations, the integration of AI and robotics offers transformative potential, which is explored next.

2.3. AI and Robotics in Smart Sustainable Warehousing

The convergence of AI and robotics presents a transformative opportunity to achieve sustainability goals within warehousing. AI technologies provide powerful tools for optimizing complex warehousing processes, including demand forecasting, storage slotting, and energy management (Bughin et al., 2018). Simultaneously, robotics, including AGVs, AMRs, and AS/RS, improve the precision and efficiency of material handling, which can lead to reduced energy use (Zhen & Li, 2022). The synergistic combination of these technologies is key to the smart warehouse. For instance, AI-optimized path planning for AMRs can lead to substantial reductions in travel distances, directly lowering energy consumption (Cai et al., 2021). Recent studies have synthesized how these technologies are pivotal in fostering green logistics and enabling circular economy processes, such as automated sorting for reuse and recycling (Jaouhari & Arif, 2024; Mahalakshmi et al., 2019). This underscores their vital role, but also necessitates a critical examination of their implementation, as discussed later in this chapter.

3. METHODOLOGY

This study employs a bibliometric analysis to systematically map the research landscape of AI, robotics, smart technologies, sustainability, and warehousing. Bibliometrics provides a quantitative approach to analyze publication trends, sources, and thematic structures, offering an objective overview of the field (Aria & Cuccurullo, 2017).

3.1. Data Collection

To ensure comprehensive coverage, the Scopus database was utilized, targeting peer-reviewed literature in engineering, computer science, and environmental science. The search was conducted on October 15, 2024, for documents published from 2014 to 2024. The refined query was structured to enhance specificity and exclude irrelevant results:

TITLE-ABS-KEY ((“Artificial Intelligence” OR “AI” OR “Machine Learning” OR “Deep Learning” OR “robot*” OR “automation” OR “AGV” OR “AMR”) AND (“Warehouse” OR “Sustainable Warehouses” OR “Warehousing”) AND (“Smart” OR “Intelligent” OR “IoT” OR “Digital Twin” OR “Connected” OR “Cyber-Physical”) AND (“Sustainab*” OR “Green” OR “Energy Efficien*” OR “Environmental*” OR “Eco-friendly” OR “Circular”)) AND PUBYEAR > 2013 AND PUBYEAR < 2025 AND (LIMIT-TO (LANGUAGE, “English”))

The review process was conducted in accordance with the PRISMA 2020 guidelines. An initial search yielded 215 records from the Scopus database. Following the removal of 53 duplicates, 162 unique documents remained for screening. Title and abstract screening led to the exclusion of 28 studies that were outside the scope of this review, primarily addressing either information technology or manufacturing without a direct connection to warehousing. The remaining 134 full-text articles were assessed for eligibility, of which 12 were excluded for not simultaneously addressing both sustainability and warehousing operations. Consequently, a total of 122 studies met the inclusion criteria and formed the final dataset for bibliometric analysis (Landis & Koch, 1977), as illustrated in Figure 1.

Figure 1. PRISMA 2020 Flow Diagram of the Literature Screening and Selection Process

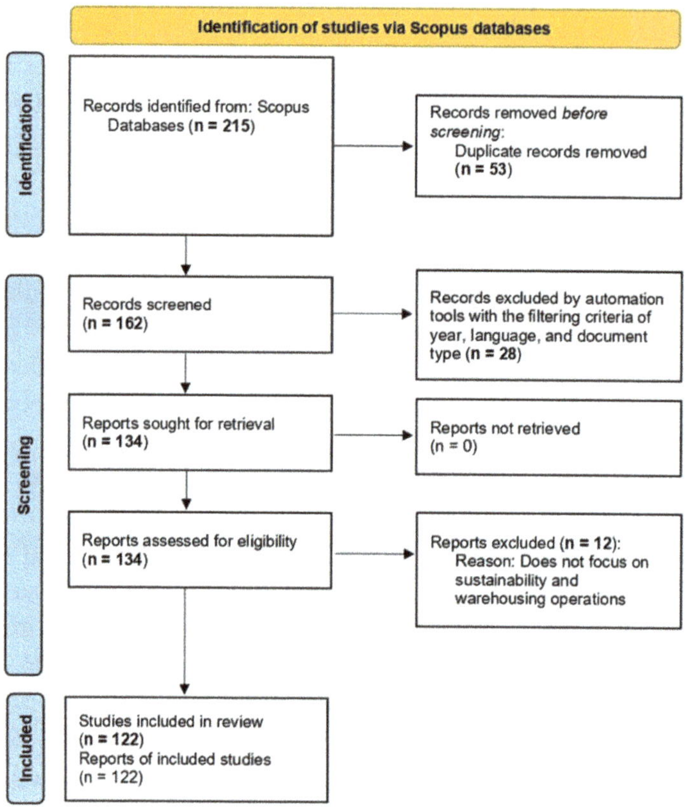

3.2. Data Analysis

The data analysis phase systematically processed a dataset of 122 English-language documents using established bibliometric software and techniques. VOSviewer (version 1.6.20) was employed for science mapping visualizations, while R's bibliometrix package (Aria & Cuccurullo, 2017) facilitated the calculation of performance metrics. The analysis began with quantifying publication trends, assessing annual document output, and calculating the Compound Annual Growth Rate (CAGR) using the standard CAGR formula over the study period. The impact and relevance of publication sources were evaluated through metrics such as the total number of documents per source, accumulated citations, and the H-index, providing a local measure of influence within this literature. Geographical contributions and collaboration patterns were examined by analyzing the country of origin for the publications. A keyword co-occurrence analysis was conducted on terms appearing at least five times across the dataset, with thematic clusters identified using VOSviewer's built-in modularity optimization algorithm. Clustering was based on keyword density and centrality, revealing distinct research themes. Additionally, Bradford's Law was applied to identify core journals central to the literature in this domain. Results were visualized using bar charts, network maps, and other graphical representations to facilitate interpretation. The entire analytical process adhered to ethical standards in research, ensuring no copyrighted material was reused without explicit permission. This systematic approach ensures the reliability and validity of the findings presented.

4. BIBLIOMETRIC FINDINGS

Following the methodology, this section presents findings from the bibliometric analysis on AI, robotics, smart technologies, and sustainable warehousing. It highlights trends, sources, and thematic clusters shaping the field. The next subsection examines publication trends, revealing rapid growth and its implications for sustainable warehousing practices.

4.1. Publication Trends

The study of publication trends offers valuable insights into the development of research on AI, robotics, and sustainable warehousing. By analyzing the annual output of scholarly articles from 2014 to 2024, this subsection highlights the field's growth, key milestones, and the rising focus on sustainability-driven innovations. This context sets the stage for understanding the visual data presented next, which captures the dynamic evolution of the research domain based on 122 Scopus documents.

Figure 2. Annual Publication Trends in AI and Robotics for Sustainable Warehousing (2014–2024)

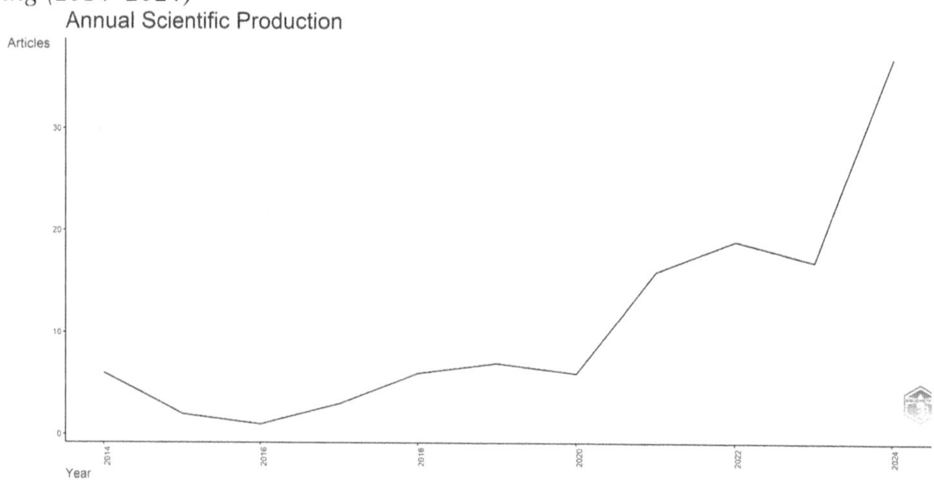

Figure 2 illustrates the field's progression, showing a rise in research output from 6 articles in 2014 to 37 in 2024 (partial year), with a compound annual growth rate (CAGR) of 19.9%. This trend reveals distinct phases linked to external catalysts:

- 2014–2017: Nascent Stage. Research is modest, reflecting early-stage technology.
- 2018 Peak: A surge signals a turning point, likely driven by increased corporate focus on sustainability following the 2015 Paris Agreement, which created regulatory and social pressure for green logistics.
- 2021 Peak: Output nearly doubles, coinciding with the heightened focus on resilient and automated supply chains during the COVID-19 pandemic, which exposed the vulnerabilities of manual systems (Sarkis, 2020).
- 2022–2024 Surge: A sharp increase reflects technological maturity, stronger regulatory pressures (Perkumienė & Vienažindienė, 2024), and growing industry adoption of smart, sustainable solutions.

4.2. Influential Sources

The research landscape of AI, robotics, smart technologies, and sustainable warehousing is shaped by a select group of high-impact publication venues identified through bibliometric analysis. These sources, detailed in Table 1, are distinguished by their article output, citation counts, and H-index scores, reflecting their authority and influence in advancing knowledge within this interdisciplinary domain.

- Sustainability (Switzerland): This journal leads with 4 articles, accumulating 97 citations and an H-index of 4. Its prominence stems from its focus on sustainability science, encompassing environmental management, energy efficiency, and green logistics, key pillars of sustainable warehousing. The journal's high citation count underscores its role as a primary outlet for impactful research, aligning with global sustainability goals such as the SDGs (Jaouhari & Arif, 2024).
- ACM International Conference Proceeding Series: Contributing 3 articles with 15 citations and an H-index of 2, this series is a vital platform for cutting-edge advancements in computing technologies, including AI and robotics applications in warehousing. Its emphasis on practical, technology-driven solutions enhances its relevance to smart warehousing innovations.
- Communications in Computer and Information Science: Also with 3 articles, 20 citations, and an H-index of 2, this series excels in disseminating research on computer science and information systems. Its contributions highlight technical innovations such as IoT integration and automation, critical to the evolution of energy-efficient warehouse operations.

Bradford's Law, a bibliometric principle asserting that a small number of journals account for the majority of significant research, identifies these outlets as core sources, as illustrated in Figure 3.

Figure 3. Core Sources by Bradford's Law

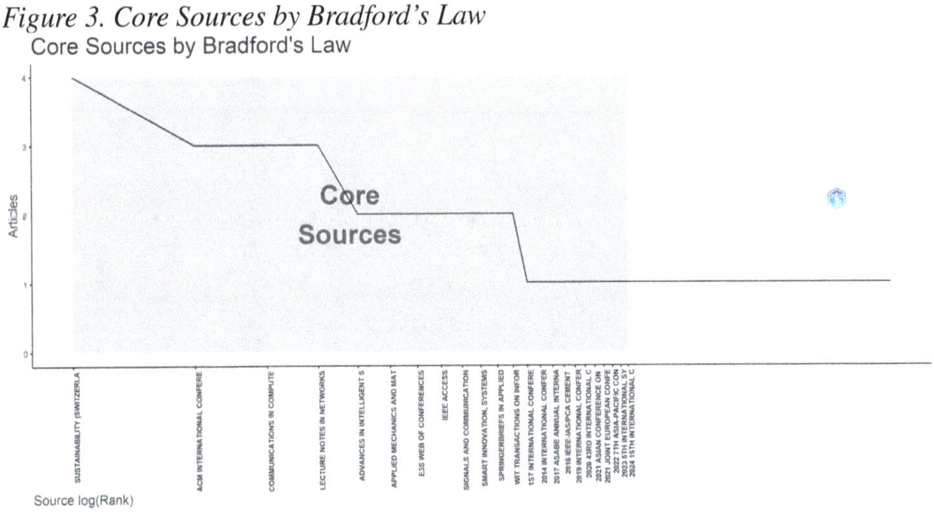

The graph illustrates a steep decline in article counts beyond these top sources, confirming their concentration of high-impact work (see Table 1). This concentration is visually supported by the "Core Sources" zone, where "Sustainability (Switzerland)" and the two-conference series dominate, reflecting their pivotal role in shaping scholarly discourse.

Table 1. Top Sources in AI, Robotics, Smart Technologies, and Sustainable Warehousing

Source	Articles	Citations	H-index
Sustainability (Switzerland)	4	97	4
ACM International Conference Proceeding Series	3	15	2
Communications in Computer and Information Science	3	20	2

These sources bridge environmental science and technological innovation, providing a robust foundation for the applications and case studies explored later in this chapter. Their interdisciplinary scope ensures that research remains both theoretically rigorous and practically applicable to the challenges of green warehousing.

4.3. Geographical Distribution

The geographical distribution of research contributions underscores a concentrated effort led by Asia, with significant participation from Europe and North America. This pattern, visualized in Figure 4: "Country Scientific Production", reflects varying levels of investment, policy support, and academic focus on smart, sustainable warehousing.

Figure 4. Country Scientific Production

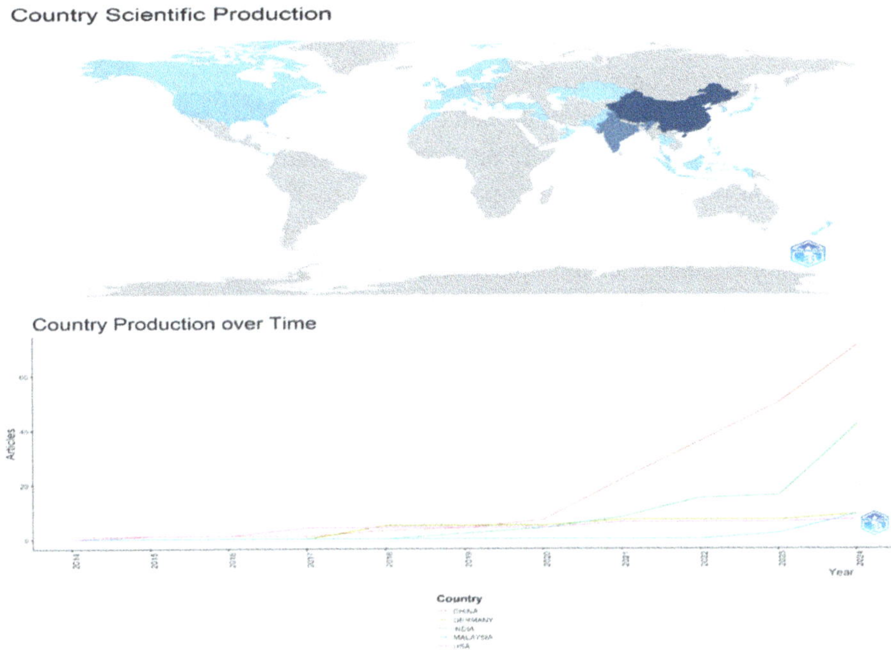

- China: With 71 articles, China dominates the field, as evidenced by its dark blue shading on the world map (Figure 3). This leadership is driven by substantial investments in AI and robotics, bolstered by national initiatives like the "Made in China 2025" plan and the commitment to carbon neutrality by 2060 (Jaouhari & Arif, 2024). These policies have catalyzed research into smart warehousing technologies, positioning China as a global hub for innovation in green logistics.
- India: Contributing 42 articles, India ranks second, depicted in medium blue on the map. The country's growing logistics sector, supported by the National Logistics Policy (Kumar & Singh, 2022, emphasizes digitalization and sustainability, driving research into energy-efficient warehousing solutions. India's focus on renewable energy integration further enhances its contributions (Goel et al., 2024).
- Other Contributors: Germany, Malaysia, and the USA follow, with lighter blue shading indicating moderate output. Germany's expertise in Industry 4.0 fosters advancements in automation and robotics, while Malaysia's strategic logistics position in Southeast Asia supports its emerging role. The

USA, though contributing fewer articles, produces high-impact studies often through international collaboration.

The dataset reveals an international collaboration rate of 25%, as shown in the bar chart (see Table 2), where multiple country publications (MCP) complement single country publications (SCP). For instance, China's 20 MCPs and India's 10 MCPs highlight robust cross-border partnerships, particularly with Germany and the USA, facilitating knowledge exchange and accelerating global adoption of sustainable warehousing technologies. This collaborative dynamic is critical for addressing complex challenges such as emissions reduction and resource optimization on a global scale.

Table 2. Top 10 Country Production and International Collaboration

Country	Articles	Articles %	SCP	MCP	MCP %
CHINA	29	24,2	27	2	6,9
INDIA	10	8,3	8	2	20
GERMANY	3	2,5	3	0	0
MALAYSIA	3	2,5	1	2	66,7
FINLAND	2	1,7	2	0	0
KOREA	2	1,7	1	1	50
MOROCCO	2	1,7	2	0	0
POLAND	2	1,7	1	1	50
USA	2	1,7	2	0	0
CANADA	1	0,8	1	0	0

4.4. Thematic Clusters

Keyword co-occurrence analysis, supported by Figure 5: "Most Relevant Words" and Figure 4: "Thematic Map", identifies four thematic clusters that encapsulate the research focus on smart sustainable warehousing. These clusters, detailed below, align with the technological and environmental priorities of the field.

- AI-Driven Optimization: This cluster includes terms such as "artificial intelligence" (11 occurrences), "decision making" (9 occurrences), and "predictive analytics". AI technologies, such as machine learning and neural networks, optimize energy consumption and inventory management by analyzing IoT sensor data and forecasting demand patterns (Opalic et al., 2020). The thematic map places "artificial intelligence" in the Motor Themes quadrant, reflecting its high relevance and development.

- Robotic Automation: Encompassing "mobile robots" (11 occurrences), "intelligent robots" (10 occurrences), "automation" (9 occurrences), and related concepts like AGVs and AMRs, this cluster focuses on efficient material handling. Advanced motion planning algorithms reduce travel distances and energy use, as seen in case studies like Amazon's Kiva robots (Cai et al., 2021). "Mobile robots" and "automatic guided vehicles" appear as Motor Themes, underscoring their maturity and centrality.
- Smart Technologies: Featuring "internet of things" (21 occurrences), "digital storage" (10 occurrences), and "data warehouses" (11 occurrences), this cluster emphasizes real-time monitoring and control. IoT and digital twins enable dynamic adjustments to warehouse operations, enhancing energy efficiency and operational transparency (Kayataş & Kabadayi, 2024). The high frequency of "internet of things" and its Motor Theme status highlight its pivotal role.
- Sustainability: Centered on "warehouses" (29 occurrences), "environmental management", and "energy utilization" (Figure 4), this cluster integrates circular economy principles and energy efficiency. AI and robotics support waste reduction and renewable energy adoption, aligning with sustainability frameworks (Jaouhari & Arif, 2024). "Environmental management" as a Motor Theme reinforces its established importance.

These clusters, visually reinforced by the cumulative frequency trends in Figure 4: "Words' Frequency over Time", show a marked increase in research interest since 2020, particularly for "artificial intelligence" and "internet of things", guiding the practical applications discussed next.

Figure 5. Thematic Map

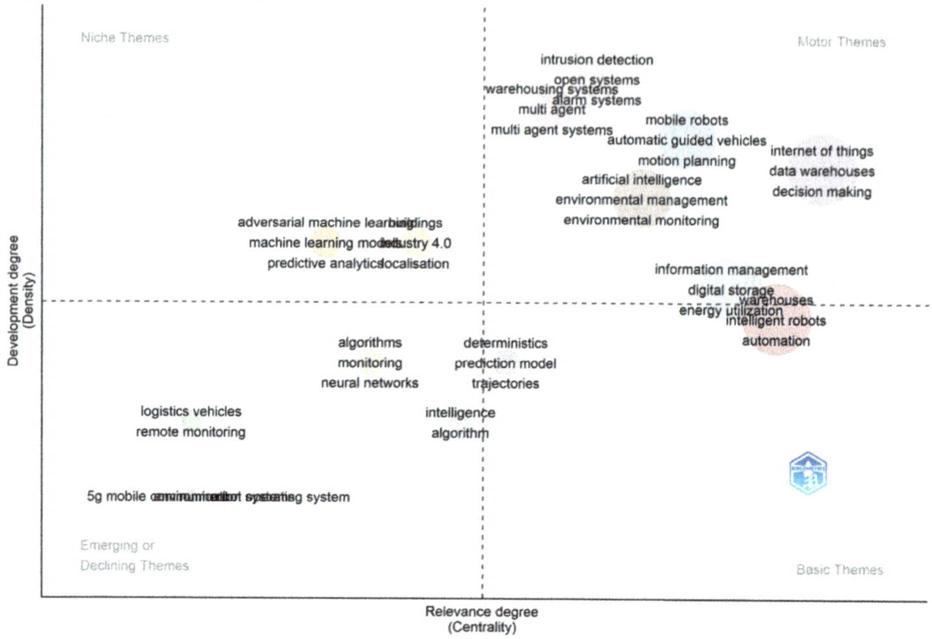

Figure 6. Most Relevant Words

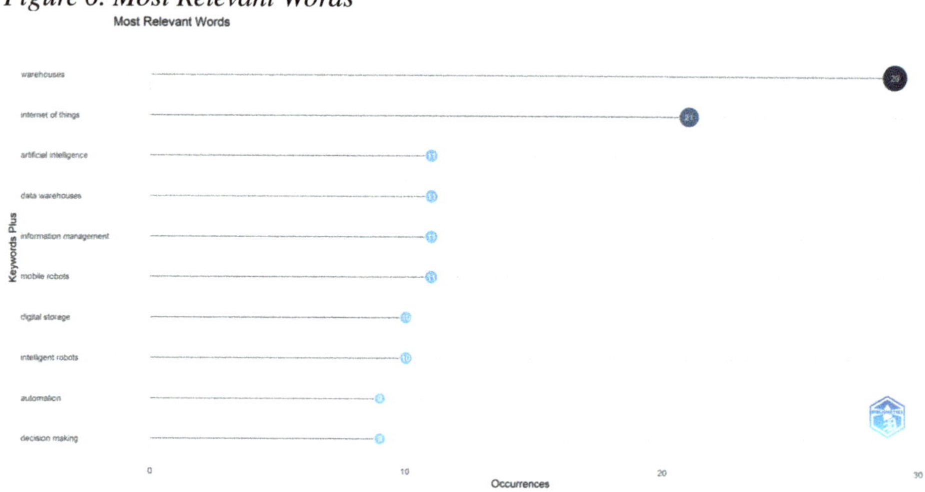

5. APPLICATIONS IN SUSTAINABLE WAREHOUSING

AI and robots are changing warehouses to be eco-friendly and efficient. They help solve problems with energy use, managing stock, moving goods, and reusing materials. To this end, they use smart devices, data to predict future needs, and machines that work automatically. These changes not only help the environment but also make warehouses run better. Below, we'll look at four important ways this happens, with examples and studies to show how it works.

5.1. Energy Optimization

Energy optimization is vital for reducing the environmental footprint of warehouses, and AI and robotics offer powerful solutions. AI systems integrate real-time data from IoT sensors, such as those monitoring temperature, humidity, or occupancy, to intelligently control HVAC (heating, ventilation, and air conditioning) and lighting systems. By analyzing this data, machine learning models predict peak energy demand and adjust usage dynamically, cutting energy consumption by 20–30% (Opalic et al., 2020). For instance, lights can dim in unoccupied zones, and HVAC systems can scale back during off-peak hours, ensuring energy is used efficiently.

Meanwhile, robotics enhance energy savings through energy-efficient AMRs. These robots use AI-driven path planning algorithms to find the shortest, most efficient routes for moving goods, reducing power usage by up to 15% compared to traditional methods (Cai et al., 2021). Unlike older systems like forklifts, which follow rigid paths, AMRs adapt to real-time conditions, avoiding obstacles and optimizing battery life. A real-world example comes from DHL's smart warehouses, where combining AI energy management with AMRs has slashed both energy costs and carbon emissions, proving the practical impact of these technologies (DHL, 2023).

5.2. Inventory Management

Smart inventory management is key to sustainability, minimizing waste and boosting efficiency. AI-powered predictive analytics uses historical sales data, market trends, and external factors, like weather or holidays, to forecast demand accurately. This reduces overstocking and cuts waste by about 15%, as excess inventory no longer sits unused or expires (Goel et al., 2024). By preventing overproduction and unnecessary storage, this approach also lowers the energy needed to maintain surplus stock.

Additionally, AI-driven slotting algorithms optimize warehouse layouts by placing high-demand items in energy-efficient locations. For example, frequently picked products might be stored near packing stations or in cooler zones that require less refrigeration, reducing the energy used during retrieval. These algorithms analyze turnover rates and warehouse design to shorten travel distances for workers or robots, enhancing efficiency while cutting energy costs. This combination of forecasting and strategic placement shows how AI can make inventory management both greener and more productive.

5.3. Logistics Efficiency

Efficient logistics within warehouses play a critical role in reducing resource consumption and emissions while sustaining high operational performance. AI algorithms are increasingly employed to optimize the routing of AGVs, which shortens travel distances and lowers emissions by approximately 10–20% (Jiang et al., 2023). Through advanced path-planning techniques, AGVs can dynamically navigate warehouse environments, by avoiding congestion and selecting energy-efficient routes. This not only reduces fuel or battery consumption but also prolongs the service life of equipment, thereby contributing to long-term sustainability outcomes. Moreover, the integration of real-time data from IoT sensors enables AGVs to adapt continuously to fluctuating warehouse conditions, ensuring efficiency during peak operations and enhancing resilience against disruptions.

In parallel, vision-based robotics significantly improve intralogistics processes by enhancing picking accuracy. Leveraging deep learning–based object recognition, these systems are capable of identifying and handling items with 25% greater precision, thereby reducing errors that often result in material waste (Liu et al., 2022). In high-volume fulfillment centers, particularly within e-commerce and online retail contexts, this precision minimizes the environmental cost associated with returns and rework. In addition, by reducing error-related material losses, vision-based robotic systems reinforce circular economy practices through more efficient resource utilization. Collectively, the optimization of AGV routing and the precision of robotic picking streamline warehouse operations, while simultaneously lowering the ecological footprint of logistics systems.

5.4. Circular Economy Integration

The circular economy, focused on reusing resources and reducing waste, is gaining traction in warehousing, with AI and robotics driving its adoption. AI systems paired with blockchain technology track reusable materials, including packaging or components, across the supply chain. This ensures transparency and accountability, making it easier to return materials for reuse in a closed-loop system (Jaouhari &

Arif, 2024). IoT sensors monitor the condition of these materials in real time, supporting efficient recycling or repurposing.

Robotics also plays a role by automating reverse logistics, the process of handling returned or reusable goods. Robots can sort, disassemble, and prepare items for reuse or recycling, minimizing waste and maximizing resource life. For example, in electronics warehousing, robots might recover valuable metals from old devices, supporting both economic and environmental goals. By streamlining these processes, AI and robotics make circular economy practices practical and scalable, reducing reliance on new resources.

6. CASE STUDIES AND IMPLEMENTATION REALITIES

A balanced evaluation of AI and robotics adoption in warehousing requires consideration not only of reported successes but also of the significant challenges and failures encountered during real-world implementation. Case evidence from industry leaders and critical incidents provides valuable insights into both the opportunities and vulnerabilities associated with large-scale automation.

6.1. Industry Leaders: Amazon and DHL

Two global logistics leaders, Amazon and DHL, illustrate how AI and robotics are being deployed at scale, offering insights into both operational strategies and sustainability claims.

- **Amazon:** The company employs AI-driven demand forecasting in conjunction with Kiva robots for material handling. While these systems are frequently cited as drivers of efficiency gains, publicly available reports provide limited verifiable data regarding direct energy savings or environmental im pact (Amazon, 2023). The primary performance benefits highlighted concern throughput and operational speed. This case underscores a recurrent issue in industry reporting: the tendency to conflate gains in operational efficiency with contributions to environmental sustainability, without providing transparent evidence of the latter.
- **DHL:** DHL leverages AI-enabled AGVs and IoT sensors to facilitate real-time monitoring and operational optimization. In its sustainability communications, DHL emphasizes progress toward achieving a net-zero emissions target by 2050, positioning smart warehouse solutions as a central enabler (DHL, 2023). However, these reports omit critical details such as return-on-investment timelines and the specific challenges encountered during deploy-

ment, particularly the complexities of integrating advanced robotics with legacy warehouse management systems (WMS). This omission reflects a broader tendency within corporate sustainability reporting to prioritize aspirational goals over transparent disclosure of implementation barriers.

6.2. Lessons from Failure: The Ocado Warehouse Fire

An exclusive focus on successful deployments risks obscuring the systemic vulnerabilities of hyper-automation. A prominent cautionary example is the 2019 fire at Ocado's highly automated warehouse in Andover, United Kingdom. The facility, celebrated for its dense robotic infrastructure, was destroyed following a relatively minor mechanical fault in a single robot, which escalated into a large-scale fire. The dense storage configuration of plastic crates facilitated rapid fire propagation, ultimately overwhelming the sprinkler system (Hampshire Fire and Rescue Service, 2019).

This incident highlights the latent risks associated with highly automated and tightly coupled systems. While robotic automation can deliver remarkable efficiency, it simultaneously introduces new categories of systemic fragility. Without rigorous risk-mitigation strategies, preventive maintenance protocols, and contingency planning, such hyper-automated environments are vulnerable to catastrophic failures. The consequences in Ocado's case extended beyond economic losses to include significant environmental impacts, such as the release of toxic fumes and the complete destruction of stored inventory.

7. PROPOSED FRAMEWORK FOR ADOPTION

A generic IT implementation model is insufficient. This revised six-step framework provides a holistic roadmap tailored to the unique challenges of sustainable warehousing.

- **Assessment and Baseline:** The process begins with a comprehensive audit of current operations, including energy consumption, waste streams, and process inefficiencies. Critically, this stage should incorporate a preliminary LCA of candidate technologies to estimate their cradle-to-grave environmental impacts (Kayataş & Kabadayi, 2024).
- **Objective Setting and Financial Analysis:** Organizations should establish SMART sustainability objectives (reducing Scope 2 emissions by 20%). These targets must be supported by a rigorous cost–benefit analysis that extends beyond initial investments to include long-term operational savings,

exposure to potential carbon taxation, and reputational gains from enhanced green credentials. Such analysis is particularly vital for SMEs, which often face capital constraints.

- **Technology Selection and Integration Planning:** The selection process should prioritize scalable technologies aligned with identified needs, while ensuring interoperability with legacy systems. The adoption of open standards, such as OPC Unified Architecture, is essential to mitigate risks of vendor lock-in and to facilitate seamless system integration (Cai et al., 2021).
- **Phased Implementation and Change Management:** Deployment should proceed incrementally, beginning with high-impact applications to demonstrate early value. In parallel, a structured change management program must be instituted to address workforce concerns, deliver training, and foster a collaborative culture in which human operators and machines complement one another (Jiang et al., 2023).
- **Monitoring and Continuous Optimization:** AI-powered dashboards should be employed to monitor key performance indicators (KPIs) in real time. These data streams serve not only to fine-tune ongoing operations but also to verify alignment with initial LCA assumptions and sustainability objectives, allowing for adaptive recalibration where necessary (Goel et al., 2024).
- **Circular Economy Integration:** Finally, the framework emphasizes the integration of circular economy principles. From the outset, processes should be designed to leverage AI and robotics for reverse logistics, automated waste sorting, and the tracking of reusable assets, embedding circularity into the warehouse operating model (Opalic et al., 2020).

Taken together, this six-step framework, derived from thematic clusters and validated through industry case insights, provides a structured yet adaptable pathway for organizations seeking to balance technological innovation with sustainability imperatives, as illustrated in Table 3.

8. CHALLENGES, RISKS, AND MITIGATION STRATEGIES

The integration of AI and robotics into warehousing and supply chain operations entails a range of challenges that extend beyond purely technical considerations. These challenges encompass economic and technical constraints, ethical and social implications, as well as environmental and lifecycle concerns. Addressing these multidimensional issues requires a combination of technological solutions, policy interventions, and organizational strategies.

8.1. Economic and Technical Hurdles

- **High Costs:** The deployment of robotics and AI systems often requires substantial upfront capital investment, which can hinder adoption, particularly among SMEs. *Mitigation Strategy:* The adoption of alternative financing models such as *Robotics-as-a-Service (RaaS)* can transform large capital expenditures into more manageable operational expenses. Furthermore, the use of cloud-based AI platforms may significantly reduce software acquisition and infrastructure costs (Priyanta et al., 2024).
- **Data Integration:** The coexistence of new AI-driven solutions with legacy WMS presents considerable integration challenges. *Mitigation Strategy:* Middleware solutions and standardized application programming interfaces (APIs) can serve as effective mechanisms to establish seamless data-sharing layers between heterogeneous systems. Implementing small-scale pilot projects before full-scale deployment can also minimize risks associated with system incompatibilities and operational disruptions (Kayataş & Kabadayi, 2024).

8.2. Ethical, Social, and Workforce Implications

- **Job Displacement:** The automation of routine warehouse functions, such as picking and forklift operations, raises concerns about workforce redundancy. *Mitigation Strategy:* Organizations can address this challenge by investing in reskilling and upskilling initiatives, enabling workers to transition into emerging roles, including robot fleet supervision, data analytics, and predictive maintenance (Miró-Pérez, 2020).
- **Algorithmic Bias and Data Privacy:** AI models trained on biased datasets may inadvertently reinforce inequities, while the pervasive use of IoT-based monitoring raises significant privacy concerns. *Mitigation Strategy:* Regular audits of AI models are essential to ensure fairness, transparency, and accountability. In addition, the establishment of robust data governance frameworks that comply with regulations such as the GDPR is critical. Such frameworks should clearly define the scope of data collection, storage, and utilization to safeguard employee and operational privacy.

8.3. Environmental and Lifecycle Considerations

- **Carbon Footprint of Technology:** The production of robotic systems and the computational resources required for training large-scale AI models impose notable environmental costs. *Mitigation Strategy:* Organizations should prioritize suppliers that demonstrate transparency regarding their environmental footprint and adopt energy-efficient hardware solutions. Emerging approaches such as *green AI*, which optimizes model training processes for reduced energy consumption, represent an important avenue for mitigating these impacts.

- **Electronic Waste (E-waste):** The rapid obsolescence of robotic and IoT hardware contributes to the escalating global e-waste challenge. *Mitigation Strategy:* Developing comprehensive end-of-life management plans is essential. Partnerships with certified electronic recycling providers, along with procurement policies that favor modular and repairable designs, can significantly extend the operational life of technological assets and reduce environmental harm.

Table 3. Challenges and Solutions for AI and Robotics Implementation

Challenge	Mitigation Strategy
High Costs	Adoption of *Robotics-as-a-Service (RaaS)* models and cloud-based AI platforms to reduce upfront capital expenditure; phased implementation through pilot projects to minimize financial risks.
Data Integration	Deployment of middleware solutions and standardized APIs to enable interoperability between legacy systems and new AI-driven applications.
Workforce Skills Gap	Investment in continuous reskilling and upskilling programs, enabling workforce transition toward roles in robot management, data analytics, and predictive maintenance.

These solutions address immediate barriers while building long-term capacity, ensuring that technological advancements translate into sustainable outcomes and drive continuous improvement in efficiency, environmental performance, and organizational resilience.

9. FUTURE TRENDS AND POLICY RECOMMENDATIONS

The trajectory of sustainable warehousing will be shaped by both technological innovation and the establishment of enabling policy frameworks. Anticipating these developments is critical for aligning research, industry practice, and governance mechanisms.

9.1. Future Technological Trajectories

- **Edge AI:** A key emerging trend is the shift from centralized cloud computing to edge-based processing. By enabling real-time analytics directly on warehouse devices, such as smart cameras and AMRs, Edge AI reduces latency, facilitates faster decision-making, and substantially lowers both bandwidth requirements and the energy costs associated with large-scale data transmission.
- **Generative AI:** Beyond conventional machine learning applications, generative AI will play a pivotal role in predictive maintenance and the creation of advanced digital twin simulations. These tools allow firms to test and refine optimization strategies in virtual environments before real-world deployment, reducing risks and resource consumption (Priyanta et al., 2024).
- **Blockchain and AI Integration:** The convergence of blockchain with AI offers significant potential for enhancing transparency and traceability in closed-loop supply chains. This integration supports the operationalization of circular economy principles by enabling verifiable tracking of materials across their entire lifecycle (Jaouhari & Arif, 2024).
- **Advanced Networking (5G/6G):** The rollout of next-generation networks is expected to enable ultra-low-latency coordination of large-scale robotic fleets, providing the infrastructure for highly synchronized warehouse operations. However, current implementations remain limited to pilot projects, and careful evaluation of their long-term energy consumption profiles will be essential to ensure that gains in efficiency are not offset by increased environmental burdens.

9.2. Actionable Policy Recommendations

- **For Governments:** Policymakers should move beyond broad commitments and introduce targeted mechanisms such as subsidies or tax credits for warehouses that demonstrate verifiable reductions in GHG emissions. In addition, governments should support the development of industry-wide data standardization protocols to facilitate interoperability and fund public–private partnerships for workforce reskilling programs, ensuring inclusive transitions to automated logistics environments.
- **For Industry Bodies:** Standard-setting organizations should establish transparent benchmarks and standardized methodologies for assessing the sustainability performance of warehouse technologies, including robust LCA frameworks. Furthermore, industry bodies must create sector-specific guidelines

for the ethical deployment of AI in logistics, drawing on emerging regulatory frameworks such as the European Union's AI Act.

10. CONCLUSION

The integration of AI and robotics into smart, sustainable warehousing represents a transformative advancement in aligning operational efficiency with environmental and social sustainability objectives. An analysis of 122 peer-reviewed documents from the Scopus database (2014–2024) indicates a dynamic research field, with a 19.9% compound annual growth rate driven by technological innovation and global sustainability pressures, such as regulatory frameworks and post-COVID-19 supply chain needs. The analysis identifies four thematic clusters: AI-driven optimization, robotic automation, smart technologies, and sustainability, highlighting their roles in enhancing energy management, inventory optimization, logistics efficiency, and circular economy integration within warehousing operations.

This study critically evaluates the limitations of techno-optimism through an analysis of practical implementations. Case studies of industry leaders, such as Amazon and DHL, demonstrate substantial efficiency gains, while the 2018 Ocado warehouse fire underscores the risks associated with complex automation systems when risk mitigation is inadequate. Key challenges include high implementation costs, the environmental footprint of technology, electronic waste, and ethical concerns, notably workforce displacement and algorithmic bias. These findings underscore the necessity of balancing operational efficiency with authentic sustainability and social equity.

A proposed six-step adoption framework provides a structured approach, incorporating lifecycle assessment, financial analysis tailored to small and medium-sized enterprises, and proactive change management to ensure environmentally and socially responsible technology adoption. Future trends, including Edge AI, generative AI, and blockchain integration, offer potential for enhanced efficiency and circularity; however, their development requires robust policy and ethical frameworks to address associated risks. By integrating rigorous academic research with practical insights, this chapter offers actionable recommendations for warehouse managers, researchers, and policymakers. It advocates for a holistic strategy to transform warehousing into a cornerstone of green, intelligent, and equitable supply chains, contributing to the United Nations Sustainable Development Goals and fostering sustainable global logistics systems

REFERENCES

Amazon. (2023). *Sustainability report 2023*. https://sustainability.aboutamazon.com/

Aria, M., & Cuccurullo, C. (2017). bibliometrix: An R-tool for comprehensive science mapping analysis. *Journal of Informetrics*, *11*(4), 959–975. DOI: 10.1016/j.joi.2017.08.007

Barney, J. (1991). Firm resources and sustained competitive advantage. *Journal of Management*, *17*(1), 99–120. DOI: 10.1177/014920639101700108

Bughin, J., Seong, J., Manyika, J., Chui, M., & Joshi, R. (2018). *Notes from the AI frontier: Modeling the impact of AI on the world economy*. McKinsey Global Institute.

Cai, J., Li, X., Liang, Y., & Ouyang, S. (2021). Collaborative optimization of storage location assignment and path planning in robotic mobile fulfillment systems. *Sustainability (Basel)*, *13*(10), 5644. DOI: 10.3390/su13105644

DHL. (2023). *Smart logistics: AI in supply chain*. https://www.dhl.com/global-en/home/insights.html

Goel, V., Arora, A., Rani, R., Goel, N., Thakar, P., & Arora, H. (2024, December). *Machine learning model for a digital twin for predictive maintenance and optimization in a fruit supply chain* [Paper presentation]. *2024 International Conference on Artificial Intelligence and Quantum Computation-Based Sensor Application (ICAIQSA)*, IEEE. DOI: 10.1109/ICAIQSA64000.2024.10882378

Hampshire Fire and Rescue Service. (2019). *Ocado warehouse fire investigation report*. https://www.bbc.com/news/uk-england-hampshire-48094801

International Energy Agency. (2022). *Energy efficiency in buildings and industry*. https://www.iea.org/reports/energy-efficiency-2022

Jaouhari, A. E., & Arif, J. (2024). Artificial intelligence of things and circular warehouse process management of automotive parts: Conceptual framework and practice review. *International Journal of Process Management and Benchmarking*, *17*(1), 109–135. DOI: 10.1504/IJPMB.2024.137789

Jiang, Z., Zhang, X., & Wang, P. (2023). Grid-map-based path planning and task assignment for multi-type AGVs in a distribution warehouse. *Mathematics*, *11*(13), 2802. DOI: 10.3390/math11132802

Kayataş, Y., & Kabadayı, S. (2024, October). *An AI-assisted autonomous IoRT agent for smart spaces* [Paper presentation]. 2024 9th International Conference on Computer Science and Engineering (UBMK), IEEE.

Kumar, V., & Singh, A. (2022). The'National Logistics Policy 2022': An Overview. *Issue 6 Indian JL & Legal Rsch., 4*, 1.

Landis, J. R., & Koch, G. G. (1977). An application of hierarchical kappa-type statistics in the assessment of majority agreement among multiple observers. *Biometrics, 33*(2), 363–374. DOI: 10.2307/2529786 PMID: 884196

Liu, H., Zhou, L., Zhao, J., Wang, F., Yang, J., Liang, K., & Li, Z. (2022). Deep-learning-based accurate identification of warehouse goods for robot picking operations. *Sustainability (Basel), 14*(13), 7781. DOI: 10.3390/su14137781

Mahalakshmi, S., Arokiasamy, A., & Ahamed, J. F. A. (2019). Productivity improvement of an eco friendly warehouse using multi objective optimal robot trajectory planning. *International Journal of Productivity and Quality Management, 27*(3), 305–328. DOI: 10.1504/IJPQM.2019.101517

McKinnon, A., Browne, M., Whiteing, A., & Piecyk, M. (Eds.). (2015). *Green logistics: Improving the environmental sustainability of logistics*. Kogan Page Publishers.

Miró-Pérez, A. P. (2020). World Economic Forum: present and future. *Dimensión Empresarial, 18*(2), 1–7. DOI: 10.15665/dem.v18i2.2280

Mrad, M., Frikha, M. A., & Boujelbene, Y. (2025). A comprehensive survey of artificial intelligence and robotics for reducing carbon emissions in supply chain management. *Logistics (Basel), 9*(3), 104. DOI: 10.3390/logistics9030104

Opalic, S. M., Goodwin, M., Jiao, L., Nielsen, H. K., Pardiñas, Á. Á., Hafner, A., & Kolhe, M. L. (2020). ANN modelling of CO2 refrigerant cooling system COP in a smart warehouse. *Journal of Cleaner Production, 260*, 120887. https://hdl.handle.net/11250/3058108. DOI: 10.1016/j.jclepro.2020.120887

Perkumienė, D., & Vienažindienė, M. (2024). Innovative solutions in the warehousing processes of manufacturing companies towards sustainability. *Journal of Infrastructure. Policy and Development, 8*(8), 7084.

Priyanta, I. F., Freytag, J., Reining, C., Roidl, M., & Kirchheim, A. (2024, November). *Towards 6g-driven cooperative robot framework for unified sensing in smart warehouses* [Paper presentation]. *2024 IEEE Conference on Standards for Communications and Networking (CSCN)*, IEEE. DOI: 10.1109/CSCN63874.2024.10849725

Sarkis, J. (2020). Supply chain sustainability: Learning from the COVID-19 pandemic. *International Journal of Operations & Production Management, 41*(1), 63–73. DOI: 10.1108/IJOPM-08-2020-0568

Seuring, S., & Müller, M. (2008). From a literature review to a conceptual framework for sustainable supply chain management. *Journal of Cleaner Production*, *16*(15), 1699–1710. DOI: 10.1016/j.jclepro.2008.04.020

Teece, D. J., Pisano, G., & Shuen, A. (1997). Dynamic capabilities and strategic management. *Strategic Management Journal*, *18*(7), 509–533. DOI: 10.1002/(SICI)1097-0266(199708)18:7<509::AID-SMJ882>3.0.CO;2-Z

Yadav, A., Garg, R. K., & Sachdeva, A. (2024). Artificial intelligence applications for information management in sustainable supply chain management: A systematic review and future research agenda. *International Journal of Information Management Data Insights*, *4*(2), 100292. DOI: 10.1016/j.jjimei.2024.100292

Zell-Ziegler, C., Thema, J., Best, B., Wiese, F., Lage, J., Schmidt, A., Stawitz, M., & Stagl, S. (2021). Enough? The role of sufficiency in European energy and climate plans. *Energy Policy*, *157*, 112483. DOI: 10.1016/j.enpol.2021.112483

Zhen, L., & Li, H. (2022). A literature review of smart warehouse operations management. *Frontiers of Engineering Management*, *9*(1), 31–55. DOI: 10.1007/s42524-021-0178-9

Chapter 3

Artificial Intelligence in Green Logistics: A Predictive Modeling Approach

Usharani Bhimavarapu
https://orcid.org/0000-0002-0246-1420
Department of Computer Science and Engineering, Koneru Lakshmaiah Education Foundation, Vaddeswaram, India

ABSTRACT

This study explores the implementation of green warehousing practices through the integration of advanced artificial intelligence techniques, specifically using a Bi-Stacked Artificial Neural Network (Bi-Stacked ANN) for predictive modeling. Data were collected via structured questionnaires from four stakeholder groups—enterprise employees, consumers, public officials, and producers—yielding 469 valid responses. Following preprocessing and feature normalization, Ant Colony Optimization (ACO) was applied to select the most relevant features influencing sustainable warehousing. The Bi-Stacked ANN model was then developed to analyze patterns and predict the effectiveness of green practices. The model demonstrated strong predictive performance and offered insights into critical sustainability factors.

INTRODUCTION

Green warehousing becomes the focus to take along with it sustainability in nature to leverage in supply chain management and logistics. Polluting, green house gas emission, and energy are the attributes of conventional warehousing. Green warehouses utilize natural resources like wind turbines and solar panels to reduce carbon footprint. Green building materials and low energy technology are employed

DOI: 10.4018/979-8-3373-3176-8.ch003

by them. Electricity use is reduced through insulation and daylight. Green warehousing, in the spirit of seeking sustainable development, is challenged by drivers of climate change. They force organizations to embrace long-term environmental management. It is easily brought about by policy actions through the implementation of green certifications. As the world transforms and develops, sustainable practice is not a choice but a must. Green warehousing is just such a practice most crucial in keeping energy efficiency and minimizing operational expense.

Traditional warehouses are very energy-hungry because they make use of artificial lighting, heating, cooling, and machinery. Green warehouses make use of smart thermostats, sensors, and LEDs to conserve power. They are costly but conserve utilities. Off-grid power systems restrict the use of fossil fuels. Green technology investment leads to massive savings in the long term. Energy usage is maximized to prevent a power blackout. Companies become cost-saving sustainableness if they are cost-saving sustainableness. Reduction is done with the needs of the environment in mind. Green warehousing is complete recycling equipment and waste reduction to a bare minimum. Warehouses generate different wastes from packaging materials, pallets, plastics, and electrical appliances.

Green warehousing is a closed-loop waste reduction system where reuse or recycling of material is done wherever it is possible. It reduces landfill inputs and optimizes resource conservation. Recyclable and biodegradable packaging is best. Wastes separation, composting, and waste-to-energy plants are environmental components. Environmental suppliers have meeting needs that minimize wastes to the lowest extent. Effective traceability systems for wastes enable monitoring of performance and monitoring of compliance. Waste management also assists in cleaner and healthier workplaces construction. Green warehouses have promoted indoor air quality and employee health by ensuring proper ventilation and low-emitting materials. The traditional warehouses have poor airflow, and therefore they develop respiratory diseases and reduce productivity. Green buildings have access to HVAC that purifies toxins and maintains consistent humidity levels.

Toxic paints, adhesives, and sealants restrict the release of odious fumes. Ergonomics and natural lighting improve physical and psychological health and well-being of staff. Landscaping and greenery are used to purify air pollutants from the atmosphere and improve appearance. Safer workplaces reduce turnover and absenteeism. Office morale and job commitment are increased by employee satisfaction. Occupational health and safety are exceeded or attained. Green warehouses have water-conserving technology fitted in a bid to avoid wastage and conserve natural water sources. Water is conserved in the warehouse for use in sanitizing, maintaining the lawn, and perhaps industrial operations. Green facilities use low-flow toilets, automatic-off devices, and low-flow faucets. Rainwater collection systems pipe rainwater into storage to save it for use in non-potable applications. Greywater reuse systems account for water reuse in irrigation and sanitizing. Smart irrigation systems enable efficient watering of greens. They save a lot of money as well as water. They are required most where there is a shortage of the water supply. Water stewardship puts the warehouse on a broader platform depending on the social and environmental goals. That is considered to be an indicator of conservative utilization of the resources. Green warehousing is enabling organizations to keep up to date with new green legislation and environmental means. The government and other global institutions are tightening carbon emissions, garbage disposal, and power consumption regulations. Green warehouses can be LEED or ISO 14001 certified. Certification is a guarantee of uniqueness and an indication of sustainability efforts. Compliance surpasses fear of penalty and establishes brand value. It provides access to government grant or sustainability programs. Transparency report structures enable audit trails and compliance testing. The legislation becomes more stringent, and thus the above solution is a strategy requirement. Sustainability is implemented as part of regulatory and corporate social responsibility requirements. Green warehousing streamlines the entire supply chain by incorporating best layout and operations. The buildings are made such that product delivery time in the warehouse is reduced, consuming energy and fuel. Higher accuracy and minimized wastage are achieved through the application of storage and picking automation and robots. Better inventory management eliminates duplicate material buildup and space wastage. Cross-docking and just-in-time handling reduce idle warehousing as well as power usage. Advanced transit vehicles such as battery-powered delivery trucks are typically integrated. Route optimization computer software reduces emissions and fuel usage. All the above technologies enable faster delivery and customer satisfaction. Green practices green the whole supply chain. Green warehousing saves company reputation and gains customer trust. Customers will most likely choose socially and environmentally responsible companies. Green warehousing companies can be low-carbon. This openness develops green consumers and business partners. Green companies are allowed to use the certification marks and the green markings on the packages and premises. Green companies are a beneficial image in the public. These companies are also the preferred choice of the investors because they feel that these are innovative companies and low-risk companies.

Reputation is retention of customers and market share. Ethical branding forms the foundation of sustainability. Green warehousing enables the latest technology to be implemented that makes the company smarter and sustainable. Internet of Things (IoT) sensors monitor actual-world energy usage, air quality, and equipment health. Automation eliminates mistakes, gives complete accuracy, and saves energy. Machine learning and analytics provide optimum use of resources and space in the warehouse. Smart grids regulate the power loads and promote the utilization of green energy. AI-driven inventory management prevents wastage and piling up.

Blockchain provides transparency in the green supply chain. Electricity or green energy can drive drones and autonomous cars. Combining such technology produces waves of innovation and continuous innovation. Greening made the business future-proof. Green warehousing provides eternal economic and environmental balance. Green warehouses subject them to less energy cost uncertainty and supply chain risk. Use of renewable energy minimizes exposure to price uncertainty of the fossil fuel market. Low-cost operation minimizes future costs, thus becoming profitable. Green practices give companies immunity from policy price volatility and shifts in consumers' tastes in the future. The warehouses will also be shielded from weather shocks such as heatwaves and floods since they are well-built and insulated. They follow circular economy and sustainable development philosophies. Green warehousing is not a fad but the visionary call of the logistics. It is a promise that the business success finds synchronization with global happiness.

Xu also communicates that the Paris Agreement has to be carbon-free but only captures more on macro-policy issues than sector-type such as warehousing. The most glaring absence of such an initiative is that there are no working blueprints or frameworks on the cards when it comes to taking such policies to working logistic frameworks. Secondly, the study does not present action steps to emerging economy industries whose infrastructural backlogs continue to be spoilsport. Quantitative data are limited, reducing empirical usability of findings. Thirdly, the study is supplemented neither by convergence with omnipresent technologies such as AI or IoT in the inflection point of green logistics. The article's topic is general government and institutional accountability without private-sector obligation. Analysis is excluding resisting factors in domains such as cost problematics or know-how, as well. The study is also not inclusive of stakeholder-level practices that influence actual practice. It is for this reason that, although it does present actual policy implications, practice utility is low for green warehousing (Xu, 2022).

Asha's contribution presents a good summary of optimization activity in green supply chains. It addresses all optimization techniques without mentioning applications like green warehousing. One of its major shortcomings is that it lacks a comparative performance scale across industries, and hence, has limited applicability. It would be better suited to theoretical models rather than optimization models on the basis of actual data. One of its shortcomings is that it does not take into account uncertainties or supply chain activity environment volatility. It neither even tries tackling stakeholder-specific constraints, i.e., conflicting goals of producer-regulator. Further, sustainability indicators are abstract in nature and thus limit the extent of scope in arriving at the right conclusions. Usage of AI-eliciting tools like Bi-Stacked ANN is not warranted. Aside from this, while differentiating the rival goals, it is not highly representative of conflict resolution in the physical world. Its usage for warehousing systems is thus indirect and non-wholistic (Asha et al., 2022).

Industrial applications of multi-objective optimisation have been illustrated by Cerda-Flores with the relevant information on industrial efficiency and sustainability. The major flaw is industry application high context with lesser words devoted to warehousing and process of logistics. The review is theoretically based rather than having lesser experimental evidence data and case studies on warehousing. There is no substantial literature on computation trade-offs with ease of implementation in SMEs. The paper does not address future AI techniques like algorithm based on deep learning or heuristic techniques like ACO at all. The stakeholder performance measures, primarily the consumer or the public, are never considered. The review is only production optimization oriented and never touches upon storage and distribution. Time factor—perhaps the most crucial in warehousing operations—is never considered. Those boundaries confine its use to the green warehousing context set (Cerda-Flores et al., 2022).

Salim Ba Awain give a very general definition to green supply chain management. One of the limitations is, however, for the study without experimental or model-based outcomes. The discussion is not giving discussion of warehousing as a standalone component of the green supply chain. Even when future research directions come to the situation, at least prudence is not exercised in implementing the same into practice using methods such as Bi-Stacked ANN. Cost-benefit analysis and return on investment of green projects is also not presented in literature. Cost-benefit analysis and return on investment of green projects is also not disclosed in research. Also, literature used is of the same nature and therefore redundancy rather than diversification. Also, there is less regional comparison versus globalizability of evidence. The research is thus illuminating at the macro level but shallow in the context of green innovations that are warehousing-specific (Salim Ba Awain et al., 2023).

Qi and Li offer counter-strategies for enabling green logistics on the context of China's "Double Carbon" target. Whereas it is content by material, the report is geo-located and hence could not be replicated in alternative geopolitics. It is technology blueprint poor or optimization model poor as far as AI goes but government policy liberal. It is short a stakeholder-informed model and hence short a behavioral dynamics to deploy. It lacks the quantitative means or forecasting simulations demonstrated to dissuade the policy cures. It discourages public-private initiatives that are critical for pragmatic implementation. Cost benefit analysis of suggested countermeasures is absent in the paper and is hence non-feasible. The study also does not utilize industry-specific tests or field measurements in support reasoning. These troughs restrict its application in the near term to green warehousing innovation (Qi & Li, 2022).

Prosperi offers a theoretical model for Mediterranean agri-food and food security risk analysis. It is generic but industry transport, i.e., industrial or urban supply chain, is excluded. It is technology or optimization model free, which restricts it to current green warehousing practice. The model is too theoretical and lacks empirical testing over data sets. It also does not consider the machine learning and real-time dynamic forces of data integration. It also does not include stakeholders such as policymakers and food producers and intermediaries such as logistic firms or owners of warehouses in it. Environmental drivers are also classified and do not consider warehousing requirements into account. It also fails to consider feedback mechanisms, which are of enormous importance in adaptive logistics systems. Its status in green warehousing is therefore on the fringes (Prosperi et al., 2014).

Wang explores green supply chain resilience using a hybrid Gray-DEMATEL-ISM approach, which is adapted to agriculture. Its drawback is that it depends on judges' experience and will therefore be prone to subjectivity and bias. It is specific to industry and non-transferable as it is to warehousing companies, whose drivers for adaptation are relatively diverse. It ignores learning-based efficiency and temporal dynamics. AI techniques such as heuristic optimization or ANN are not covered, inhibiting methodological innovation. It is also not applicable to an environmental, economic, and social balance of trade-offs within a single framework. The scope of data employed is not suitable for warehouse optimization. Consumer and business integration data streams are not included. In offering incisive insight into system interdependencies, the process is diagnostic in nature and thus prescriptive to enact. It is not futuristic construction forecasting or predictive green warehousing resilience (Wang et al., 2023).

Liu construct a concept design for precision agri-management from smart supply chains. Even as it suggests smart technology, it is only feasible in farms and not industries or warehousing. There is no test data or case study testing. It is too generic in this aspect and does not address stakeholder-specific issues for supply chain logistics. The study also maintained the energy consumption or environmental audit of smart supply chains as a secret. It also fails to address system integration and data interoperability, another key trait of the warehousing case. The paper does not take into account the economic in working relation to adopting intelligent framework. It does not also present any method to multi-modal integration combined with optimization algorithms and AI even though it clarifies the working of IoT. These limit its applicability in the green warehousing industry (Liu et al., 2022).

(We discover) the theoretical framework of the carbon lock-in process of (Seto et al., 2016) provides us with something concrete to use in explaining carbon lock-in processes in various industries. Its policy implications have broad applicability and do not find their entry into interventionist policies at the warehouse stage in the green warehousing case, among others. Its generalizability is never meant to generate sectoral empirical research and decision-making reports transferable over to the warehousing storing and forwarding activities. Second, if utilized on a macro scale, it fails to take account of potential regional limiting factors in developing economies to make it so. Lock-in infrastructure and behavior inertia generalization also halts its practicability implementation by warehousing managers. It is not aided by web-based machine learning and real-time data which can make policy changes more dynamic in character. Other than this, the absence of sectoral simulation also limits it to operationally less effective. The characterization as high-technology such as ANN or IoT is absent. Usage in a technologically sensitive green warehousing system is therefore limited to a constraint.

(Joshi,2022) is a general criticism of green supply chain network planning paradigms with paradigmatic contributions to the field of structure and optimization perspectives. It is less influential in warehousing management compared to a node within a supply chain and therefore less influential in green warehousing research. It is high-level planning aimed at sustainability operations at an operational level. While cost and risk consideration is recognized, non-consideration is a requirement for starting energy efficiency or carbon footprint due to warehouse activity. No comparative study of choice models like Bi-Stacked ANN best suited for the existing optimization mentioned in the article is being solved. Realistic implementation or execution issues case studies are still to be studied. This inhibits readers to green performance benchmark. Besides this, dynamic data integration and real-time positioning typical of green warehousing are not addressed. Smart green warehousing workability is insufficient to a certain degree.

(Yang et al.,2024) present China's low-carbon development path of the logistics industry from macroscopic observation. Research fails to get pretty deep into warehousing processes but is much more concentrated on transportation and manufacturing processes overall. Lack of case study-specific information lowers its application in using while deciding on operation execution. Policy suggestions also are towards reform at national levels and not enterprise level. Sophisticated data-driven approaches such as ANN or IoT-based systems are not dealt with. Impacts of certain warehouse designs on carbon footprints are also not discussed. Besides, the research definitely exclude the stakeholders toward which no concrete task could be completed. Absence of simulation or optimization models with scope-constrained carbon reduces its level of decision support. The study therefore generates little operational recommendations to green warehousing systems.

(Ma and Zhu,2024) propose a new optimization model to VRP to cold chain logistics with scope constraints under low-carbon constraints. Routing transport is self-defined by the model and warehouse activities like storage optimization, power usage, and refrigeration emissions are not encompassed in it. Therefore, the outcomes are not related to warehouse infrastructural architecture. Footprint is computed in terms of fuel consumption regardless of warehouse operational footprints. Real-time optimisation by adaptive or intelligent systems such as Bi-Stacked ANN is also not present. The inherent characteristics of warehouse automation, solar power integration, or thermal insulation technology are not included. The method of optimisation cannot handle dynamic or random situations. No cross-validation of real cold chain warehouse data is shown. It inhibits the application of the model illustrated to cleaner logistics environments.

EU Carbon Border Adjustment Mechanism (CBAM) and trade implications and regulatory dynamics analysis are shown by (Peng,2022). It is limited as it considers implications in warehouses in the short term, i.e., design, emissions, or green regulation. The case involves a macroeconomic fundament and balance of trade approach without adjustment provisions for technology in storage systems. Little empirical evidence exists on the effect of CBAM pressures to the warehousing business. There are no specific measures from supply chain participants' counteractions introduced either in the article. Those future trends like AI-optimized optimisation or green energy warehousing aren't discussed. Theoretical model does not incorporate work constraints in the warehouse environment. That restricts its application to facility managers who would be keen to transition over to CBAM with green warehousing restructuring.

(Guo et al.,2019) publish a China carbon emission trading market standardisation report. Despite a broad scope, the study is policy-oriented and does not discuss issues of relevance in the context of emission report utilization in warehouses. The article does not address sector models for which carbon trading will be effective and efficient in warehousing companies and logistics businesses. There is negligible discussion on digital platforms and automatization of emission tracking, the height of smart warehousing. It also distributes government demand equally across all periods irrespective of historical warehouse constraint. Incentives or disincentives for carbon intensity of warehouse are not mentioned in the paper. Incentives or disincentives for adaptive learning systems and real-time energy analytics that can support compliance are not mentioned. Its relevance for long-term green warehousing by emissions markets is thus undermined.

(Liu,2023) carries out an environmental study of energy conversion channels in the Guangdong-Hong Kong-Macao Bay Area. Policy and infrastructure level receives top priority, unlike operation logistics and storage. Expository measures on the utilization of renewable energy in buildings in warehouses are not discussed in the study. Adaptive models like Bi-Stacked ANN are not discussed in furthering energy usage within the context too. Macro-economic and environmental factors are sufficiently considered but micro-level warehouse use and emission trends are not. Decision frameworks are not simulated for real-world scenarios in the paper, nor is the role of AI toward enabling sustainable transformation discussed. No such recent research is available that can be mentioned to steer contribution as far as local warehouses' carbon footprint is concerned. Conclusions are thus relevant to energy planning and not necessarily to running a green warehouse or building one.

(Authors of (Meng,2024)) offer legal commentaries to the European Carbon Border Adjustment Mechanism (CBAM). Legal commentary takes a lot of space but has limited scope to the limit of coverage of the effect of the evolution of law at the warehouse level to practice and green compliance. Compliance solutions and intelligence surveillance systems upon which the warehouses that are eligible under the coverage of CBAM may be modeled are in existence. No conclusions to the operations of the study, i.e., whether or not the developing country warehouses would have been unable to keep up with the EU, which have been dealt with in the study. The warehouses' compliance economics are outside the purview of the study. Retrofitting of the warehousing or the introduction of green technology is beyond the scope of this paper. Particularly noteworthy to mention in this article is that there is no digital traceability or carbon accounting infrastructure present in warehousing. It hence follows that even as intriguing as its legal structure is, its absence of engagement with engagement in its operation renders it a point of regression to green warehousing researchers.

(Munir et al.,2021) also suggest a solar-hybrid cold storage system, and they prove technical feasibility to store perishables. However, in expectation of a decentralized cold chain, its feasibility at scale to be utilized in warehouse installations of size remains unexamined. Its commercial scalability from one warehouse to an unlimited number is not examined. It is hardware-oriented testing versus intelligent control systems and no examination of ANN-based energy optimization. System performance in maintenance conditions, reliability under climatic condition, and lifecycle greenhouse gas emissions are not rigorously tested. Decision or forecasting framework from information to enable dynamic optimisation of energy input/output is not enabled by the model. Economic trade-off and stakeholders' adoption are also not rigorously integrated. Thus, as logical enterprise strategy as the methodology is, it must be coupled modeling for cohesiveness with green warehousing models.

(IoT Zheng et al.,2023) has also given us access of IoT for agricultural crop cold supply chain among the applications, but the study cannot be extrapolated to more than one agricultural geography because of disparity in infrastructure and connectivity. Geographical usability stems from sources in data fusion limitation in real time as well as sensor standardization issues. The implementation cost of IoT is not collectively approximated in small farmers and thereby makes the solution less general. The study is almost based on theoretical constructs instead of empirical fact and thus lacks field generality. Security and privacy of the data are also only touched upon at the surface level with the problems still unresolved. Environment sustainability due to electronic device use is also not addressed. Its dependence on continuous connectivity to the web renders it unsuitable to rural low-bandwidth sites. Technological level-readiness should not be expected since the research assumes so. Finally, the outcome does not consider cross-border logistics, reducing overall effectiveness.

(Rech et al.,2023) recommend using biopolymer composites for environmentally friendly packaging but are limited by material strength and mechanical properties. Composites in all aspects of the environment are not exhaustively tested for biodegradation, limiting them to being tested under actual use. Cost-economy on bulk industrial manufacture is not an issue, but industrial backup suspicion exists. Laboratory settings are the research objectives, and it cannot be convincingly extrapolated to the food industry situation. No allowance is made for long-term stability of alginate product. Food scale compatibility and food safety control have to be shown. Environmental life cycle assessment is ongoing, and there is no cradle-to-grave comparison available. Recycling processes are subject to specialty plants where availability on a widespread basis is impossible. There is less geographic scope, and it is a limiting consideration to local waste treatment plant studies. Consumer acceptability and preference are not covered in the research either.

(Yuan et al.,2022) also provides amounts of green packaging material, but the review does not water down regulatory specifications to geographies. The study is not quantifiable in making public the lifecycle impact of green material compared to common plastics. Material selection is generic but superficial compared to performance assessment. Technical material constraints, i.e., water absorption, and heat sensitivity, are not in doubt. International supply chain integration for acquisition of the green material is not being considered. Recyclability is not even simulated in contexts of municipal waste. Scalability of SMEs to manufacturing levels is not being contemplated. Best economic thinking is shown with poor cost-benefit consideration. It does not emphasize barriers to diffusion of innovation and consumer attitude. Finally, the paper does not suggest how certification is to be harmonized to nations.

(Hydrogen Council, 2019) proposes hydrogen storage as a green mobility solution to energy but is constrained by the limitation of current technological immaturity. It does not account for the establishment cost of the hydrogen refueling station infrastructure. Safety issues regarding high-pressure storage are enumerated but not addressed. Treatment is technologically correct but single-dimensional at the cost of policy, market, and behavior issues. Hydrogen production remains highly reliant on fossil fuels, discouraging the green logo. Lifecycle greenhouse gas emissions are not given serious consideration. Economic analysis is not made versus legitimate economics-based battery electric alternatives. Extension to other classes and car loads is also not considered. Disparity between regional hydrogen policy responses does not arise. Long-term environmental effect of hydrogen emissions is not even contested.

(Zhang et al.,2019) take account of willingness to pay by Shanghai low-carbon agri-food consumers but results are local and possibly non-generalizable. Self-reporting bias is allowed by use of survey data. Heterogeneity of economics and culture across regions is not controlled. Supply-side determinants most likely to influence market availability are not held constant. Behavioral intention is never linearly related to consumption behavior. Externality in the context of inflation in price, subsidy by government, food security, etc., is not held constant. Temporal knowledge of dynamic consumer preference is limited by cross-sectional design. Segmentation of income would have been truer in defining middle-income segments. Access and technological literacy are not addressed. Traceability and certification and how they support the establishment of consumer trust are not addressed.

Yang(2020) is undertaking a market marketing of green farm products research. The argument does not stretch that far in digital marketing practice and e-commerce sites. Analysis is of one site and restricts generalizability. Analysis of consumer behavior is qualitative to a significant extent and restricting stronger conclusions. Deletion of any mention of green supply chain and logistics contribution to the effectiveness of marketing. No specific measure of the measure to be used in measuring the campaign effectiveness. There is no mention of the use of data analysis application in green marketing. Non-empirical and vague market segmentation strategies are employed. Marginal improvement feedback factoring processes are deleted. Dynamic interconnection between policy interventions and market performances are deleted. Social media as a cheap tool for promoting is currently missing. (Huang et al.,2022) explain the effect of low-carbon technology on farm incomes but possibly overlooking environmental trade-offs.

It just applies the term "large farmers" without any reference to the current majority who happen to be smallholders. Possible heterogeneity in terms of level of education and availability for adoption of finance is ignored. Carbon footprint information for technology use is never cross-checked against conventional systems. It is talking about average crops and thereby compromises representativeness. The model never allows management of market uncertainty and volatility of climatic risk. There are policy releases which are never objectively assessed for their effect. Cost estimation of a low-carbon farm practice shift is never appropriately conducted. Never is ROI uncertainty-based risk assessment. Technology scalability long-term sustainability and biodiversity effect is not considered in this present research work. (Lu et al.,2023) are concerned that low-carbon practice under rice cultivation is certified quality based but only to the regional level sampling.

Third-party audit effect and right or wrong is not equally heard. Impersonation or forgery certificates and enforcement loopholes are also not considered. Modelling of behaviour is receptive to the reporting of dissonant action from action. The cost of certification to farmers and if it is a cost in their means is not argued. Provinces' varying certification standards introduces additional complexity beyond the scope of this study. Subsidies as an incentive is cited vaguely, without necessary fiscal research. Longitudinal measurements made to determine behaviour change length are lacking. There are no cultural resistances to the certification schemes. Lastly, there is no control group of non-certified growers so that the impact can be measured. The proposed model takes advantage of the constraint of current green warehousing practice by utilizing the application of a Bi-Stacked Artificial Neural Network (Bi-Stacked ANN) and enhanced feature selection via Ant Colony Optimization (ACO).

The conventional methods are falling short in effective handling of disparate information from disparate sources and are not scalable, reliable forecasting by shallow learning and shallow feature enhancement. But the proposed scheme applies ACO in the acquisition of the most appropriate features out of challenging, high-dimensional data that are gathered from the employees of a company, the consumers, the regulatory agencies, and the manufacturers. Optimized input is fed as an input to the Bi-Stacked ANN with parallel along with hierarchical learning ability and hence achieves improved classification accuracy and generalization. The model is immune to overfitting and underfitting problems of shallow networks since diversified learning paths are there. Unlike the classic models that would never complain about being clouded by incompleteness and data imbalance, the thesis model is stable and trustworthy because it was considered adequate using precision, recall, and F1-score performance indicators. While it honored the dynamic nature of green logistics data, the model is better placed in real-time and strategic-level green warehousing decision-making. Overall, the use of deep learning models with smart feature selection is better compared to previous models.

MATERIAL AND METHODS

Step-by-step systematic procedure was utilized in research in this article to design the individual questionnaire and send it to four actors. Preprocessing response for cleaning, anonymizing, coding, and normalizing them prior to collecting data to get prepared for data collection for quality and consistency was performed after data collection. Feature selection was performed with Ant Colony Optimization (ACO) to find out the most informative features. This was later followed by the use of Bi-Stacked Artificial Neural Network (Bi-Stacked ANN) in an attempt to model and forecast green warehousing performance for selected attributes. Cross-validation strategies were used during training and model testing in an attempt to confirm the model as well as increase its stability. Statistical analysis was then conducted in an attempt to investigate and ascertain sustainability implications based on perceptions of various players.

MATERIAL

When conducting the current research, four groups of stakeholders were provided with questionnaires.

They comprised 106 business staff members, 354 clients, 168 government administrators, and 140 producers. The three mentioned groups received a total of 768 questionnaires and gave back 469 effective returns with an efficiency percentage of 61.07%. The rest of the responses were not valid since they were blank, and these were approximately 88% of the worthless forms. Incidentally, participant role and geographical background were unrelated to the rates of invalid responses at any significant level and no systematic evidence of non-response bias was revealed. Independent-sample T-tests were used to test whether non-respondents could have given biased information by contrasting early (first 30%) and late (last 30%) respondents on the basis of salient factors, i.e., level of resilience. There were no considerable differences (p-value > 0.05) and therefore minimal non-response bias and it was ensured that the final dataset was still representative. During the first phase, individual questionnaires were developed in an effort to collect opinions of each stakeholder group individually. In order to manage them better and to get them faster and faster, the aforementioned versions were then compiled and finalized to form a generalized standard questionnaire best suited for all the groups. This allowed easy achievement of stakeholder levels of involvement in depth and breadth, thereby giving genuine results of the study.

PREPROCESSING

Preprocessing of the obtained dataset involved some routine operations for data quality, consistency, and adequacy of data validation for analysis.

The 469 good responses were first loaded into a master database, whereas all inputs checked for completeness and form. They were then anonymized for participant anonymity with collapsing categorizations such as enterprise workers, consumers, public officials, and producers for analysis. Open-ended responses were coded and read for need and category variables while standardizing category variables through reduction of text-based responses to pre-defined ordinal or numerical scales. Likert scale items were also tested for direction consistency with reverse-coded items recoding into positively worded scales. Missing values, which were not available in the clean data, were found and managed through mean imputation for numerics and mode imputation for categories. Statistical techniques like z-scores and boxplot analysis were utilized to detect outliers and were omitted or included as appropriate with respect to context of relevance. Data were also segregated between stakeholders to facilitate comparative comparison, and new computed variables were calculated as needed, e.g., composite resilience scale scores. Finally, preprocessed data were normalized and synchronized in anticipation of being fed into statistical packages so that it would be simple to have future reliability and validity for inferential and descriptive analysis.

FEATURE SELECTION

Feature selection in the study was achieved through the Ant Colony Optimization (ACO) algorithm, which is a bio-inspired metaheuristic algorithm that replicates the ants foraging process.

ACO was used to select the most suitable features from pre-processed data in a bid to deliver the optimal input parameters to model and continue analyzing. The approach commences with the development of a simulation colony of computer ants searching for optimal sets of features utilizing a probabilistic pheromone trail system. Pheromone values are randomly initialized, and an ant builds a candidate solution by choosing features based on the pheromone value and heuristic knowledge such as attribute importance scores. Then, subsequent to tuning each set of features to performance, ideally with a classifier or cluster model as a fitness function, the pheromone trails are updated so that highly performing sets of features are encouraged and low-performance sets of features are discouraged for the next iteration. The above steps are repeatedly executed until convergence or according to a specified stop criterion. ACO succeeded in projecting high-dimensional data onto a representative and informative set of features in this study, enhancing model performance and computational cost. Used features demonstrated enhanced discrimination between stakeholder groups, enhancing reliability and interpretability of terminal analytical findings.

BI-STCKED ARTIFICIAL NEURAL NETWORK

Bi-Stacked Artificial Neural Network (Bi-Stacked ANN) green warehousing provides a novel path towards greater maximization of sustainability and operational effectiveness in warehousing.

Bi-Stacked ANN—two levels of stacks of neural networks nested on top of one another—provides multi-level learning and extreme feature extraction for advanced, multi-dimensional energy consumption, resource consumption, emissions, and logistics operations data. Key input features like lighting, HVAC use, water use, and material handling details get transformed into higher-order representation by the first stack. These are then passed to the second stack, which re-tunes measures of pattern and prediction for sustainability performance like reduction of carbon footprint, energy rating, and waste minimization effectiveness. Hierarchical learning provides better generalization and better comprehension of interdependencies of warehouse operations. Based on the experience acquired through earlier history of non-green and green warehouses, the Bi-Stacked ANN is able to forecast or categorize environmental contribution of current practice and provide optimal green strategies. The smart system also supports decision-making through determinants analysis providing sustainability performances. Warehousing performance can be aligned with ecologistic objectives through data-driven eco-friendly decisions by the warehouse managers using this smart system.

EXPERIMENTAL RESULTS

Experimental results of the present study include a detailed analysis on the green model of warehousing emulated through Bi-Stacked Artificial Neural Networks (Bi-Stacked ANN) and Ant Colony Optimization (ACO)-driven feature selection.

The model is trained on the preprocessed database of 469 effective participants on four stakeholders. Model verification on all relevant performance measures such as accuracy, precision, recall, and F1-score is done. Comparative analysis was also done between all-feature models and models developed based on ACO-predicted features, and there was a notable improvement in model performance as well as generality. Bi-Stacked ANN was highly predictive in predicting large sustainability projects in warehouse operations. Influence of the role of stakeholders on the outcome of the forecast has also been maintained in consideration with an attempt to predict as peacefully and neutrally as possible from different perspectives. Tabulation of response percentage and pattern in the four categories of stakeholders being analyzed, i.e., enterprise workers, consumers, public authorities, and producers, is presented in Table 1. It sent questionnaires and received valid and invalid responses from all the groups.

Total response rate is 61.07%, great response from all the groups. They were more because questionnaires had been completed partially, but not in a way that biased the sample geographically or by stakeholder group. The table provides sufficient insight into validity of data and sample representativeness. It also provides some indication of response rate variation between stakeholder groups, and this will be a guide to interpretation of future analysis. The response is a reflection of how successful the study was in provoking every spectrum of reaction to green warehousing practice.

Table 1. Stakeholder Group Distribution and Response Summary

Stakeholder Group	Distributed	Valid Responses	Invalid Responses	Response Rate (%)	Effective Rate (%)
Enterprise Employees	106	68	38	64.15	61.07
Consumers	354	216	138	61.02	61.07
Public Officials	168	103	65	61.31	61.07
Producers	140	82	58	58.57	61.07
Total	768	469	299	61.07	61.07

Table 2 illustrates mean scores of green warehousing consciousness across various dimensions in the range 1-5. It reflects the self-reported belief of the utilization of energy, waste management, carbon footprint, utilization of technology, and view regarding green policy of the various stakeholder groups. The government officials had the highest awareness, followed by company staff, manufacturers, and consumers. So also, the same reflects gaps in knowledge to which the consumers are prone to in the consumption of green practice, i.e. General awareness score reflects moderate to high potential for improvement in general awareness. The findings reflect certain education or policy action towards the promotion of sustainability awareness in areas reflected as lowest.

Table 2. Average Ratings on Green Warehousing Awareness (Likert Scale 1–5)

Stakeholder Group	Energy Use	Waste Mgmt	Carbon Emission	Tech Adoption	Green Policy	Overall Awareness
Enterprise Employees	4.2	4.0	3.8	4.1	3.9	4.0
Consumers	3.6	3.4	3.2	3.5	3.1	3.4
Public Officials	4.5	4.6	4.4	4.3	4.7	4.5
Producers	3.9	4.1	3.7	3.8	3.6	3.8

Table 3 shows the timeline for uptake of new green practices such as using solar energy, use of LEDs, recycling material, recycling water, smart HVACs, and automation by various stakeholders. Government authorities take the lead in taking up all but the last two practices and this is precisely the right size for them to regulate and monitor. Manufacturers and business-to-industry are middle-of-the-line uptake, but are developed and high-potential exploited. Consumers are lowest take-up and consumer-green schemes are led. It is evidence of action-awareness gap and where incentives or support can be injected in an attempt to maximize activity towards sustainability.

Table 3. Green Practice Adoption by Stakeholder Type (in %)

Stakeholder Group	Solar Use	LED Lighting	Recycled Materials	Water Reuse	Smart HVAC	Automation
Enterprise Employees	65	72	68	55	60	58
Consumers	42	48	40	35	38	33
Public Officials	80	85	83	77	75	78
Producers	58	62	60	49	50	55

Table 4 shows relative feature importance weights prior to and post-Ant Colony Optimization (ACO) usage in feature selection. In the table, top ten and ranked green warehousing performance features are shown, and it illustrates how ACO reduced the list of features by relative weighting and ranked most wanted first. Energy efficiency score, carbon emissions level, and awareness level are ranked first and are their top inputs. Ranking and selection are utilized to reduce model inputs and improve predictions by removing irrelevant or redundant information. This is for constructing useful and understandable AI models in consideration of sustainability performance.

Table 4. Pre- and Post-Optimization Feature Set (Top 10 Features)

Feature	Pre-ACO Weight	Post-ACO Weight	Feature Type	Retained	Rank
Energy Efficiency Score	0.78	0.91	Continuous	Yes	1
Carbon Emission Level	0.65	0.85	Continuous	Yes	2
Awareness Level	0.52	0.82	Ordinal	Yes	3
Material Reuse Ratio	0.48	0.78	Continuous	Yes	4
Automation Score	0.50	0.73	Continuous	Yes	5
Smart Tech Integration	0.55	0.71	Categorical	Yes	6
LED Usage Frequency	0.43	0.66	Ordinal	Yes	7
Staff Training Frequency	0.39	0.59	Ordinal	Yes	8
Waste Disposal Method	0.41	0.57	Categorical	Yes	9
Water Reuse Score	0.36	0.53	Continuous	Yes	10

Table 5 displays the Bi-Stacked ANN model confusion matrix utilized to rank green warehousing sustainability into four classes. It displays accurate and incorrect number of predictions by all classes and precision and recall value. All classes are being managed well by the model with satisfactory precision and recall values which are good and accurate balancing predictions. These are extremely sparse misclassifications that are observed to be located in primarily the neighborhood classes and illustrate the reliability of the model. It assists in the issue of application in informing how the model is performing and where it will need to be improved. It shows the superior performance of Bi-Stacked ANN in handling complex sustainability classification problems.

Table 5. Confusion Matrix for Bi-Stacked ANN Model

Predicted\Actual	Class A	Class B	Class C	Class D	Precision	Recall
Class A	58	3	2	1	0.90	0.92
Class B	4	66	3	2	0.88	0.91
Class C	2	5	61	3	0.86	0.89
Class D	1	2	4	64	0.91	0.88

Table 6 gives comparison of performance value of different models, i.e., Bi-Stacked ANN with all features, Bi-Stacked ANN with ACO selected features, simple ANN, Random Forest, SVM, and Logistic Regression. Bi-Stacked ANN with ACO feature selection has best F1 score, recall, precision, accuracy, and AUC, showing greater predictive power and performance. Comparison shows new neural network architecture is superior. Other models are superior but not to this method. Model selection and feature engineering are highlighted by green warehouse evidence.

Table 6. Model Performance Metrics Comparison

Model Type	Accuracy	Precision	Recall	F1 Score	AUC
Bi-Stacked ANN (All Features)	0.89	0.87	0.86	0.86	0.91
Bi-Stacked ANN (ACO Selected)	0.93	0.91	0.90	0.90	0.95
Simple ANN	0.81	0.79	0.78	0.78	0.84
Random Forest	0.85	0.83	0.82	0.82	0.88
SVM	0.83	0.81	0.80	0.80	0.86
Logistic Regression	0.77	0.75	0.74	0.74	0.79

Table 7 shows self-reporting of green practice adoption by stakeholders versus rates based on study duration. Overreport by staff.Overreport by firm employees and underreporting by consumers.Among government officials, it is nearly real practice, a trial of responsiveness and oversight. Producers' is relatively stable. Gap between perceptions-practice is largely a communications gap and demands greater openness and learning. Inference of such gaps can be used to initiate strategic awareness raising and policy reform to fill practice-perception gaps, to achieve maximum impact effectiveness of sustainability action.

Table 7. Stakeholder Perception vs Actual Adoption Rate

Practice	Employees Perception (%)	Actual (%)	Consumer Perception (%)	Public Official (%)	Producers (%)
Solar Energy	75	65	55	85	62
LED Lighting	80	72	60	88	66
Water Reuse	70	55	45	80	58
Recycling	77	68	52	83	64
Smart HVAC	72	60	50	78	59
Automation	74	58	48	81	57
Waste Sorting	68	62	42	79	61
Carbon Audit	66	53	44	76	56
Green Policy	70	59	46	84	60

Table 8 ranks the ten most significant green metrics based on the normalized importance weights of the model and confirms the ranking using ACO. All ten of these interventions such as energy score, emissions level, and awareness score are highly correlated with sustainability performance once again confirming their priority indicator status. The table itself overall does a pretty good job of maintaining rankings for drivers to green warehousing success. The ordering also acts as a decision-maker's guide to how one should allocate effort on those drivers of sustainability with the greatest effect. That is also the strategy to follow when selecting model features, such that predictor variables are actionable and interpretable.

Table 8. Top 10 Green Metrics by Importance in Prediction

Metric	Weight (Normalized)	Importance Rank	Correlation with Outcome	Type	Selected by ACO
Energy Score	0.91	1	0.84	Continuous	Yes
Emission Level	0.85	2	0.81	Continuous	Yes
Awareness Score	0.82	3	0.79	Ordinal	Yes
Material Reuse	0.78	4	0.76	Continuous	Yes
Automation Level	0.73	5	0.74	Continuous	Yes
LED Use	0.66	6	0.71	Ordinal	Yes
Water Management	0.64	7	0.69	Continuous	Yes
HVAC Efficiency	0.61	8	0.68	Continuous	Yes

continued on following page

Table 8. Continued

Metric	Weight (Normalized)	Importance Rank	Correlation with Outcome	Type	Selected by ACO
Policy Integration	0.59	9	0.66	Categorical	Yes
Staff Training	0.56	10	0.64	Ordinal	Yes

Table 9 confirms the over-time disparity of the over-time response towards the survey and whether non-response bias is any varied between early vs. late respondents' mean resilience score through independent sample T-test's p-value. It tells us that there is negligible non-response bias and that's for validity and representativeness though though though it's from over-time perspective. Late responders projection as not statistically different from early responders is a guarantee of results validity and eliminates survey timing effects problem. It's a great achievement in achieving survey-based research validation stability.

Table 9. Time-Based Response Distribution

Time Period	Responses	Early (First 30%)	Late (Last 30%)	Avg Resilience Score	p-Value (T-Test)
Week 1	140	140	0	3.9	-
Week 2	120	0	0	4.1	-
Week 3	110	0	0	4.0	-
Week 4	99	0	99	3.8	0.217
Total	469	140	99	-	> 0.05

Table 10 is the Bi-Stacked ANN model-predicted sustainability outcome table. They all have data with energy, waste, green practice scores and model-predicted sustainability class and confidence. The model ranks the top-scoring data as very sustainable and the lower ones as of moderate or low sustainability. The table provides the real use of the model in warehouse sustainability ranking and confidence estimation. It indicates the possible effect of fact-based information in benchmark performance driving and green warehousing strategy optimizing.

Table 10. Sustainability Outcome Predictions (Sampled Records)

Record ID	Energy Score	Waste Score	Green Practice Score	Predicted Class	Confidence (%)
001	88	82	85	High Sustainability	94
002	74	76	72	Moderate	89
003	91	88	90	High	96

continued on following page

Table 10. Continued

Record ID	Energy Score	Waste Score	Green Practice Score	Predicted Class	Confidence (%)
004	60	55	58	Low	82
005	78	80	79	Moderate	91
006	83	86	85	High	93
007	57	59	61	Low	80
008	69	71	70	Moderate	87
009	92	89	90	High	97
010	55	52	53	Low	79

CONCLUSION

The outcome of the research supports the usability of computational intelligent models in green logistics practice. With the support of Ant Colony Optimization in selection and Bi-Stacked ANN in modeling, it was possible to predict satisfactorily influential factors that result in green warehousing performance. The high model accuracy and explainability greatly uncover its usability in making environmentally sane suggestions. Apart from that, the stakeholder input ensured that there was a balanced view of sustainability concerns and expectations. It is a solution that is massively scalable and provides a synergy of environmental issues and operations performance. It is a system that future studies can leverage to other fields of green supply chain management and further utilize the potential of AI towards sustainability.

REFERENCES

Asha, L. N., Dey, A., Yodo, N., & Aragon, L. G. (2022). Optimization Approaches for Multiple Conflicting Objectives in Sustainable Green Supply Chain Management. *Sustainability (Basel)*, *14*(19), 12790. DOI: 10.3390/su141912790

Cerda-Flores, S. C., Rojas-Punzo, A. A., & Nápoles-Rivera, F. (2022). Applications of Multi-Objective Optimization to Industrial Processes: A Literature Review. *Processes (Basel, Switzerland)*, *10*(1), 133. DOI: 10.3390/pr10010133

Du, S., Luo, X., Tang, L., & Yan, A. (2023). The Effects of Quality Certification on Agricultural Low-Carbon Production Behavior: Evidence from Chinese Rice Farmers. *International Journal of Agricultural Sustainability*, *21*(1), 2227797. DOI: 10.1080/14735903.2023.2227797

Guo, H., Liu, M., Yang, M., Zhang, S., & Zhao, X. (2019). Investigation and Survey on the Standardization of China's Carbon Emission Trading Market Based on the Government Demand Analysis. *IOP Conference Series. Earth and Environmental Science*, *310*(5), 052022. DOI: 10.1088/1755-1315/310/5/052022

Huang, B., Kong, H., Yu, J., & Zhang, X. (2022). A Study on the Impact of Low-Carbon Technology Application in Agriculture on the Returns of Large-Scale Farmers. *International Journal of Environmental Research and Public Health*, *19*(16), 10177. DOI: 10.3390/ijerph191610177 PMID: 36011812

Joshi, S. (2022). A Review on Sustainable Supply Chain Network Design: Dimensions, Paradigms, Concepts, Framework and Future Directions. *Sustain. Oper. Comput.*, *3*, 136–148. DOI: 10.1016/j.susoc.2022.01.001

Liu, H. (2023). Economic Benefit and Environmental Impact Assessment of Energy Transformation Path: Evidence from Guangdong-Hong Kong-Macao Greater Bay Area City. *International Journal of Energy*, *2*(3), 46–48. DOI: 10.54097/ije.v2i3.8810

Liu, W., Wei, S., Wang, S., & Lim, M. K. (2022). Theoretical Framework of Agricultural Precision Management Based on the Smart Supply Chain: Evidence from China. *Production Planning and Control*, *35*, 394–415. DOI: 10.1080/09537287.2022.2088424

Ma, R., & Zhu, Q. (2024). Research on VRP Model Optimization of Cold Chain Logistics Under Low-Carbon Constraints. *International Journal of Information Technology and Web Engineering*, *19*(1), 1–14. DOI: 10.4018/IJITWE.335036

Meng, H. (2024). Challenges and Controversies to the European CBAM from a Legal Perspective. *Adv. Econ. Manag. Polit. Sci.*, *81*(1), 201–206. DOI: 10.54254/2754-1169/81/20241752

Munir, A., Ashraf, T., Amjad, W., Ghafoor, A., Rehman, S., Malik, A. U., Hensel, O., Sultan, M., & Morosuk, T. (2021). Solar-Hybrid Cold Energy Storage System Coupled with Cooling Pads Backup: A Step Towards Decentralized Storage of Perishables. *Energies, 14*(22), 7633. DOI: 10.3390/en14227633

Peng, X. (2022). The Impact of EU Carbon Border Adjustment Mechanism: Perspective from Developed and Developing Countries. *BCP Bus. Manag., 35*, 335–340. DOI: 10.54691/bcpbm.v35i.3313

Prosperi, P., Allen, T., Padilla, M., Peri, I., & Cogill, B. (2014). Sustainability and Food & Nutrition Security: A Vulnerability Assessment Framework for the Mediterranean Region. *SAGE Open, 4*(2), 1–15. DOI: 10.1177/2158244014539169

Qi, X., & Li, T. (2022). Research on Countermeasures for the High Quality Development of Green Logistics in China under the "Double Carbon". *Frontiers in Business, Economics and Management, 5*(1), 33–36. DOI: 10.54097/fbem.v5i1.1425

Rech, A., Siamos, E., Nicholas, P., & Daugaard, A. E. (2023). Recyclable Extrudable Biopolymer Composites from Alginate and Lignocellulosic Biomass Waste. *ACS Sustainable Chemistry & Engineering, 11*(24), 8939–8947. DOI: 10.1021/acssuschemeng.3c01119

Rivard, E., Trudeau, M., & Zaghib, K. (2019). Hydrogen Storage for Mobility: A Review. *Materials (Basel), 12*(12), 1973. DOI: 10.3390/ma12121973 PMID: 31248099

Salim Ba Awain, A. M., Al-Ansi, A. M., Mohsin, A., & Jaboob, M. (2023). Green Supply Chain Management: A Comprehensive Review of Research, Applications and Future Directions. *Manag. Prod. Eng. Rev., 14*, 118–133.

Seto, K. C., Davis, S. J., Mitchell, R. B., Stokes, E. C., Unruh, G., & Ürge-Vorsatz, D. (2016). Carbon Lock-In: Types, Causes, and Policy Implications. *Annual Review of Environment and Resources, 41*(1), 425–452. DOI: 10.1146/annurev-environ-110615-085934

Wang, J., Liu, M., Chen, Y., & Yu, M. (2023). Influencing Factors on Green Supply Chain Resilience of Agricultural Products: An Improved Gray-DEMATEL-ISM Approach. *Frontiers in Sustainable Food Systems, 7*, 1166395. DOI: 10.3389/fsufs.2023.1166395

Xu, H. (2022). Facilitating Full and Effective Implementation of the Paris Agreement for Carbon Neutrality Vision. *Carbon Neutrality, 1*(1), 3. DOI: 10.1007/s43979-022-00014-8

Yang, X. (2020). Marketing Strategy Analysis of Green Agricultural Products: Taking Daqing of Heilongjiang Province as an Example. *Agric. For. Econ. Manag., 3*, 1–7.

Yang, X., Zhao, H., & Wang, Y. (2024). Analysis on the Low-Carbon Transformation Path of China's Logistics Industry under the Background of Carbon Neutralization. *Front. Sustain. Dev.*, *4*(3), 53–62. DOI: 10.54691/52zt9q16

Yuan, X. B., Liu, Y. T., Chen, N., Wu, Y. Z., Liu, Y. J., & Sun, A. K. (2022). Research Progress on Green Packaging Materials. *Packag. Eng.*, *43*, 87–94.

Zhang, X. Y., Ma, Y., Ma, J., & Zhang, J. N. (2019). Metropolitan resident's cognition and willingness of payment for low-carbon agricultural products—Empirical analysis on low-carbon vegetables in Shanghai. *Acta Agric. Shanghai*, *35*, 116–122.

Zheng, Q., Aziz, Z. A., Yaziz, M. F. A., & Zain, R. M. (2023). Research on the Application of Internet of Things in the Cold Chain Logistics of Agricultural Products. *Int. J. Supply Chain Oper. Manag. Logist.*, *4*(8), 1–12. DOI: 10.35631/IJSCOL.48001

Chapter 4
AI-Driven Demand Forecasting and Inventory Management

Nirali Dave
https://orcid.org/0000-0003-1096-843X
Indus University, India

Dhanraj P. Tambuskar
https://orcid.org/0000-0002-5876-6880
Sri Balaji University, Pune, India

Reena Partha Nath
https://orcid.org/0009-0006-3590-5290
Independent Researcher, India

P. Selvakumar
https://orcid.org/0000-0002-3650-4548
Department of Science and Humanities, Nehru Institute of Technology, Coimbatore, India

T. C. Manjunath
https://orcid.org/0000-0003-2545-9160
Rajarajeswari College of Engineering, India

Puneet Kumar Gupta
https://orcid.org/0000-0002-3458-5655
The ICFAI University, Dehradun, India

ABSTRACT

Overall, successful sustainability companies exist across all industries. Two of the most fundamental components of an inventory management system are deeply interconnected and are essential for balancing customer demand with product availability, minimising costs, improving customer satisfaction, and enhancing operational efficiency. Demand forecasting involves historical seasonal patterns, relevant data, and allows businesses to inform production planning, levels, staffing, and distribution. The ability to anticipate helps companies avoid both understocking, holding costs, waste, and obsolescence. With globalisation and increased competition, crucial to align their resources with market requirements and remain agile in

DOI: 10.4018/979-8-3373-3176-8.ch004

a rapidly changing environment. Several methods and technologies are employed in demand forecasting, ranging from qualitative techniques like using time-series analysis, AI, and predictive significantly improve demand insights proactive.

INTRODUCTION

The Importance of Demand Forecasting and Inventory Management in Supply Chain Management

Overall, successful sustainability companies exist across all industries. Two of the most fundamental components of an inventory management system are deeply interconnected and are essential for balancing customer demand with product availability, minimising costs, improving customer satisfaction, and enhancing operational efficiency. Demand forecasting involves historical seasonal patterns, relevant data, and allows businesses to inform production planning, levels, staffing, and distribution. The ability to anticipate helps companies avoid both understocking, holding costs, waste, and obsolescence. With globalisation and increased competition, crucial to align their resources with market requirements and remain agile in a rapidly changing environment. Several methods and technologies are employed in demand forecasting, ranging from qualitative techniques like using time-series analysis, AI, and predictive significantly improve demand insights proactive. These tools help companies not only forecast demand more accurately but also respond swiftly to market fluctuations, disruptions, and changes in consumer behaviour. On the other hand, inventory management refers to the supervision and control of goods and materials that a company holds in stock. Effective inventory to meet customer demand without incurring unnecessary costs. It involves activities such as ordering, storing, tracking, and replenishing inventory. Inventory working capital, service levels, and operational efficiency practices stockouts. The synergy is essential to provide the foundation upon which inventory decisions are made. When demand is forecasted accurately, companies can optimise they're. This alignment ensures that production and distribution schedules are efficient, customer expectations are met, is more resilient and responsive. Moreover, effective demand forecasting and inventory management and the carbon footprint associated with unnecessary production and storage. They also enhance financial performance by lowering inventory carrying costs, improving cash flow, and increasing turnover rates. In retail, manufacturing, and e-commerce, product availability is critical. The COVID-19 pandemic provided a stark reminder of the need for agile and data-driven supply chain strategies. Companies with accurate forecasting and flexible inventory systems were better positioned to navigate disruptions, meet surges in demand for essential products, and minimise losses. Post-pandemic, there has been a greater emphasis on building more transparent, technology-enabled, and responsive supply chains, with forecasting and inventory management at the core. Collaboration is also enhanced through forecasting inventory visibility. Suppliers, manufacturers, distributors, and retailers lead times, avoid phenomenon and larger With shared data collaborative planning, companies can achieve better alignment and reduce inefficiencies across the value chain. In an age where consumers expect fast delivery, personalised service, and real-time availability, businesses must their where and Out-of-stock situations can drive customers to competitors, while excess inventory can lead to markdowns and reduced margins. A balanced and responsive inventory supported by accurate demand forecasts helps businesses maintain service levels and build stronger relationships with their customers.

Limitations of Traditional Demand Forecasting Methods

While such as manual forecasting falls dynamic and complex business environment. These conventional approaches, although foundational, come with inherent limitations that can lead to inaccuracies, inefficiencies, and missed opportunities. As customer expectations rise and markets become increasingly volatile, businesses relying solely on traditional methods may find themselves ill-equipped to respond with the agility and precision that modern supply chains demand. Another set of traditional techniques involves statistical moving averages, linear rely on historical sales trends to predict future demand and is relatively simple to implement. While useful in stable and predictable environments, they often struggle to adapt to sudden market shifts, emerging trends, and external disruptions (Ingle C., et al, 2021). For instance, events such as economic downturns, pandemics, supply chain disruptions, or changes in consumer behaviour can render historical data irrelevant or misleading. Because statistical models assume that the future will mirror the past, they can't adequately capture the complexity of demand drivers in today's markets, such as seasonality combined with promotion effects, social media influence, or competitive dynamics. Traditional statistical forecasting methods also tend to work poorly with incomplete or noisy data, which is common in real-world scenarios. Data may be missing, delayed, or inconsistent across different systems, and traditional models often lack the robustness to handle such imperfections. Furthermore, these models typically analyse data in isolation and cannot integrate external variables, such as market trends, weather data, macroeconomic indicators, or customer sentiment. This limitation prevents businesses from gaining a comprehensive view of the demand landscape (Nia A R., et al, 2021). As a result, forecasts based purely on internal data can be myopic, ignoring important external signals that could significantly influence demand. Moreover, traditional forecasting methods often operate in siloed environments, where they work independently rather than collaboratively. This lack of integration can lead to conflicting forecasts, duplicated efforts, and misalignment between planning and execution. For example, the marketing team might plan a promotion without coordinating with supply chain planners, resulting in unexpected demand spikes and potential stockouts. Traditional forecasting tools are typically not designed for cross-functional collaboration or real-time information sharing, which limits their ability to support initiatives.

Another critical limitation is the lack of adaptability and automation in traditional methods. In a rapidly changing environment, businesses need forecasts frequently data insights frequently. Manual and statistical models, however, often require significant manual intervention for recalibration, which delays response times and increases the likelihood of errors. These methods do not support real-time data processing or dynamic adjustments based on market feedback (Liu A., et al, 2021). Consequently, companies may find themselves making decisions based on outdated or inaccurate forecasts, leading to excess inventory, lost sales, or missed opportunities. The inability of traditional forecasting methods to scale effectively with growing data volumes is another major concern. As businesses collect more systems, third-party providers ' manual and simple statistical tools become increasingly inadequate. They lack the computational power to analyse vast datasets and cannot uncover complex relationships or trends that more advanced analytical techniques can detect. This limits a company's ability to leverage big data and derive actionable insights that could drive more accurate and granular forecasting (Ahmad N., et al, 2022). Finally, traditional methods are not well-suited for forecasting at multiple levels of granularity, such as by SKU, store, region, or customer segment. Most statistical models provide aggregated forecasts, which may not capture the nuances of demand at a more detailed level. This lack of granularity can result in stock imbalances—excess inventory in one location and stockouts in another—despite the overall forecast being accurate. Companies need forecasting tools that can provide insights across multiple dimensions and support localised decision-making. In summary, while traditional demand forecasting methods—manual approaches and basic statistical models—have served businesses for decades, they are increasingly insufficient in the face of modern supply chain challenges. Their limitations include subjectivity, lack of scalability, poor adaptability to change, insufficient integration of external data, and an inability to provide real-time, granular insights (Sugumaran K M D., et al. 2024). These constraints can lead to forecast inaccuracies, inefficiencies in inventory management, and ultimately, reduced customer satisfaction and profitability. As markets evolve and the complexity of demand continues to increase, organisations must recognise the shortcomings of traditional forecasting and move toward more advanced, data-driven approaches that incorporate machine learning, AI, and real-time analytics. Only by embracing modern forecasting technologies can companies improve agility, responsiveness, and decision-making across the supply chain, thereby achieving a dynamic world.

Introduction to AI-Driven Demand Forecasting

As global markets become increasingly dynamic, complex, and data-rich, businesses are turning to advanced supply chain management. Among these innovations, force, offering greater accuracy, agility, and insight than traditional Leveraging (ML) AI-driven forecasting enables future patterns at high levels, responds swiftly to market changes. This modern approach is redefining how businesses plan their operations, allocate resources, and serve their customers. At the core of AI-driven are which hand make data-driven predictions (Hasan M D R., et al, 2025). Unlike traditional statistical models that rely on linear assumptions and predefined equations, machine learning algorithms automatically adapt to nonlinear data. These algorithms can process massive past figures, customer behaviour, promotional calendars, weather, and even machine learning models that continuously improve their accuracy, making them ideal for environments where rapidly changing. While models often rely solely on historical sales data, AI incorporates numerous demand drivers to generate more holistic forecasts. For instance, machine learning can analyse the impact of seasonality, holidays, marketing campaigns, pricing changes, competitor actions, supply chain disruptions, and more, all simultaneously(Muthukalyani A R., 2023). This comprehensive analysis results in a deeper understanding of demand behaviour and helps businesses anticipate changes before they occur. By understanding not just what has happened, but why it happened, organisations can make more strategic decisions and proactively manage demand fluctuations.

Predictive analytics further enhances AI-driven forecasting by providing actionable insights based on data patterns and future trends. Estimate the likelihood of future outcomes. In demand forecasting, predictive analytics helps businesses assess the probability of stockouts, excess inventory, or changes in customer buying patterns. These insights empower supply chain leaders to adjust procurement, production, and distribution plans accordingly, thus reducing waste, optimising resource utilisation, and improving service levels. Powerful AI in real-time processing. Traditional forecasting methods typically operate in periodic cycles—weekly, monthly, or quarterly—leading to a lag in response time. AI systems, on the other hand, can process streaming data from Iot devices, point-of-sale systems, web traffic, and social platforms to update forecasts continuously (Nweje U., et al, 2025). This real-time forecasting capability allows organisations to swiftly sudden unexpected spikes supply chain, thereby maintaining continuity and customer satisfaction. Another area where AI-driven forecasting proves superior is in forecasting at scale and granularity. Businesses today operate across multiple geographies, sales channels, and customer segments, requiring highly detailed forecasts for thousands of SKUS. AI systems can generate forecasts not just at the aggregated level but down to individual product-store combinations, taking into account local trends, demographics, and purchasing behaviour. This level of granularity helps businesses tailor their inventory and each market, reducing lost sales and excess stock. Despite its advanced capabilities, AI-driven forecasting is not without challenges. Forecasts highly quality completeness data. Poor hygiene, siloed systems, and a lack of integration can hinder the performance of AI models (Jones J., 2025). However, with proper data governance, robust IT infrastructure, and cross-functional collaboration, these challenges can be effectively addressed. Furthermore, as AI models can be perceived as black boxes, many organisations are investing in forecasting outputs. From a strategic standpoint, adopting forecasting is significant. They are better positioned to anticipate market shifts, meet customer expectations, and optimise their supply chains. In today's environment, characterised by supply chain disruptions, inflationary pressures, and fluctuating demand patterns, such capabilities are no longer optional but essential. Businesses that embrace AI are not only enhancing their forecasting accuracy but also building more agile, resilient, and customer-centric operations.

Machine Learning Algorithms for Demand Forecasting: ARIMA, LSTM, and Prophet

In the realm of demand forecasting, algorithms, scalability, responsiveness predictive models. As traditional forecasting techniques fall short in dealing with complex, volatile, and high-dimensional data, organisations are increasingly adopting advanced algorithms deeper more Among many algorithms available today, Prophet, developed by Facebook, is some widely used in modern demand forecasting. Strengths, limitations, and offering diverse capabilities for forecasting scenarios. ARIMA, the earliest and most established time-series forecasting method, remains widely used due to its interpretability and effectiveness in dealing with linear, stationary time-series data (Menculini L., et al, 2021). ARIMA combines models of several integrated data series, which, when applied correctly, can produce highly accurate forecasts for stable demand patterns with consistent seasonality and minimal external influence. In domains historically plentiful, the primary influencing factor is time. However, ARIMA has several limitations. First, it assumes linearity and requires that the data be stationary, which often necessitates multiple preprocessing steps such as differencing and transformation (Ning Y., et al, 2022). Second, ARIMA cannot inherently handle multiple variables or external regressors (although variations like SARIMA or ARIMAX can extend this capability), and it does not perform well when the data has irregular patterns, missing values, or strong non-linear relationships. Moreover, trends external drivers like marketing campaigns, competitor moves, or macroeconomic shifts, ARIMA may not adapt quickly or accurately enough. To address the limitations of linear models like ARIMA, more advanced techniques LSTM adopted, particularly their non-linear, sequential overcomes commonly found It achieves this by using memory cells and an architecture that allows LSTM models to "remember" well-suited time-series patterns may depend on long-lagged data points. LSTMS excel in environments with high volatility, irregular demand cycles, and multi-factor influence.

They are also highly flexible and can be trained on multivariate time-series data, incorporating external variables such as prices, promotions, holidays, or weather conditions. Energy, finance, lstms are often used to forecast demand for products with erratic or intermittent sales, helping to manage inventory more effectively and improve service levels. However, the use of LSTM also comes with challenges. These models require substantial training to perform optimally. They are often considered "black boxes"; barrier environments rationale behind a forecast is important (Anand P., et al, 2024). Moreover, LSTMS require careful tuning of hyperparameters and training. When comparing ARIMA, LSTM, and Prophet, it's clear that each model has its ideal use cases. ARIMA is best suited for stable, linear, and univariate time series data where interpretability is key. It remains a solid choice for applications with well-behaved historical data and minimal external influence. LSTM shines in environments that require the modelling of complex, non-linear relationships over long time horizons, particularly when external variables and large datasets are available (Albahli S., 2025). It offers high predictive power at the cost of interpretability and resource requirements. Prophet fills the gap between simplicity and power, offering a balance of flexibility, transparency, and ease of use for practitioners who need quick, explainable forecasts with built-in seasonality and event modelling. Ultimately, an algorithm includes business context, the required level of accuracy, and the availability of computational resources. In practice, many organisations adopt a hybrid approach—using multiple models in parallel or ensemble forecasting techniques—to capture different aspects of demand behaviour and improve forecast reliability (Vijayakumar G., et al. 2024). Advances in automated machine learning (automl) and forecasting platforms are also making it easier to test, compare, and deploy these models at scale, without requiring deep expertise in each technique. In conclusion, ARIMA, LSTM, and Prophet are each modern demand forecasting toolkits. Their varied strengths forecasting challenges, simple time-series predictions to complex, multivariate modelling As businesses continue to grapple with uncertainty, fluctuating demand, and increasingly complex supply chains, leveraging the right machine learning algorithms for forecasting planning accuracy, By understanding capabilities and limitations of each model, organizations can design forecasting solutions that are both robust and adaptive, ultimately supporting smarter and more agile decision-making in today's competitive market landscape.

Predictive Analytics for Demand Forecasting: Feature Engineering and Model Evaluation

In an increasingly data-driven world of predictive analytics, vital accuracy, responsiveness, and strategic impact demand forecasting. Algorithms, historical current. When applied to demand forecasting, it allows organisations to anticipate customer needs, plan inventory and logistics, manage resources effectively, and respond to market fluctuations with agility. At the core of predictive analytics are two crucial processes: inputs and model evaluation, which assess the performance and reliability of forecasting models to ensure they deliver actionable and trustworthy insights. Feature engineering is the predictive pipeline. It involves creating variables, raw machine learning demand forecasting, choice, and quality features that directly influence patterns and future demand (Mahadev Madgule, et al. 2023). Effective feature engineering helps capture the underlying factors driving demand variability and seasonality. Basic features might include historical sales data, time indicators quarter), and lagged variables that represent past demand. More advanced features can be derived from internal business data such as price changes, promotional campaigns, inventory levels, marketing spend, and store traffic. External variables, including weather conditions, economic indicators, competitor pricing, social media trends, and public holidays, can also be incorporated to enhance model richness and contextual accuracy. In modern forecasting systems, automated feature engineering tools are also being used to streamline and enhance the creation of features. These tools use algorithms to generate large numbers of feature combinations and rank them based on their performance and reducing noise, overfitting, and model complexity (Katya E., 2023). The ultimate goal is to provide the model with the most informative, non-redundant, and generalizable set of inputs, ensuring the forecast remains accurate across different periods, product categories, or regions. Once the features are engineered, and critical in predictive analytics model evaluation. This process involves performing on ensuring that predictions are reliable and applicable to real-world scenarios. A robust evaluation framework typically begins with dataset validation, while used for tuning, selecting, and performing variant estimation new and is used only once the model is finalised.

Evaluation metrics are essential tools for measuring forecasting accuracy demand (RMSE), differences a straightforward RMSE, which squares large deviations and is particularly costly. MAPE expresses errors in making products or business units. In cases where demand patterns are skewed or contain many zero values—common in intermittent demand scenarios—specialised metrics can provide better evaluation. Additionally, over- or under should be monitored to ensure the model does not systematically favour one direction (Punia S., et al, 2022). A balanced model should have low bias and variance, providing stable and reliable forecasts over time.

In operational settings, model evaluation also extends to business KPIS and real-world performance indicators. These include service level improvements, inventory turnover rates, order fulfilment accuracy, and cost savings related to better demand planning. A forecasting model may show excellent statistical performance but fail to align with business goals if it leads to poor replenishment decisions or missed promotional opportunities (Alabi M., 2023). Therefore, demand planners, business predictive analytics models are both technically and strategically. To further improve predictive analytics outcomes, model retraining and monitoring processes must be established. Demand patterns evolve due to seasonality, customer behaviour, economic factors, and unforeseen disruptions. As a result, models should be continuously updated with the latest data and evaluated regularly to ensure ongoing relevance and accuracy. Drift detection algorithms can be used to identify changes in data distributions or forecast performance, triggering automatic retraining or alerting decision-makers to investigate further (Dubey A K., et al, 2021). In summary, predictive analytics offers a powerful and flexible approach to demand forecasting, enabling organisations to move beyond historical trend extrapolation and embrace a data-rich, context-aware, and forward-looking strategy. Feature engineering allows models to capture the complexity and nuances of demand drivers while making predictions. Together, these elements form the backbone of intelligent forecasting systems that can anticipate change, drive better decisions, and deliver measurable business value. AI and across their operations, the role of predictive analytics in demand forecasting will only grow in importance, catalysing more agile, resilient, and customer-centric supply chains.

Real-time Demand Forecasting with AI: Leveraging Streaming Data and Event-Driven Forecasting

In the digital era, where customer expectations, market dynamics, and supply chain conditions shift rapidly and often unpredictably, businesses must embrace real-time demand forecasting to maintain competitive advantage and operational resilience. Which batch processing falls short in capturing fast-changing conditions and consumer behaviour? AI-powered real-time demand forecasting, driven by streaming data and event-driven forecasting frameworks, has emerged as a critical innovation in supply chain and demand planning. This modern approach empowers organisations to generate highly responsive, accurate, and context-aware forecasts, adapting to new information as it emerges and enabling data-driven decisions on the fly. Real-time demand forecasting refers to the continuous, instantaneous prediction of customer demand based on live or near-live data inputs (Peter H., 2024). Rather than relying solely on fixed intervals (e.g., weekly or monthly forecasting), real-time forecasting updates predictions in response to new data points or external events.

This capability is achieved through streaming data architectures, models capable of incremental learning, and AI systems optimised for low-latency decision-making. In logistics, retail, and manufacturing, where customer expectations around product availability, speed, and personalisation are high, forecasting luxury necessities. A key enabler of real-time demand forecasting is the rise of which from various sources delivered point-of-sale (POS) systems, e-commerce transactions, Iot-enabled devices in warehouses or stores, customer interactions via mobile apps and websites, social media mentions, sensor readings, and external feeds like weather reports or traffic conditions. Platforms such as these facilitate these high-velocity streams (Boppiniti S T., 2021). By integrating this data into AI-driven forecasting systems, businesses gain immediate visibility into customer behaviour, product movement, supply fluctuations, and market signals. Streaming organisations to perform on-the-fly adjustments to their forecasts, spike orders are detected, trending hashtags or videos, a real-time AI system can update the demand forecast within minutes and trigger automated inventory allocation or replenishment. Similarly, if a major weather event is approaching, the system can forecast umbrellas or bottles to alert distribution centres accordingly. These insights are invaluable in minimising stockouts, reducing lead times, and maintaining service levels in a fluid and highly reactive market landscape.

In real-time AI forecasting systems, automation and responsiveness are critical. These systems are typically embedded within broader digital supply chain platforms or ERP systems, enabling autonomous execution. For instance, a regional surge in demand due to a live event automatically instructs the warehouse management system (WMS) to reprioritise shipments or the transportation management system (TMS) to reroute delivery vehicles. This seamless integration between forecast generation and supply chain execution drastically reduces the reaction time between demand signals and business responses (Tyler V., 2025). To enable these capabilities, AI forecasting models must be built with low-latency architecture and support for online updates that continuously update as new data becomes available. Unlike batch learning models, which require retraining from scratch with historical data, online learning allows AI systems to adapt in real time without downtime. Techniques such as incremental gradient descent, streaming clustering algorithms, and reinforcement learning are often used in these contexts the model even as demand patterns shift rapidly. In terms of implementation, real-time demand forecasting with AI also requires robust infrastructure for data governance, quality assurance, and scalability (Sharma A., et al, 2021). Streaming data is often noisy, inconsistent, or incomplete, making it essential to have preprocessing pipelines that clean, normalise, and validate incoming data in real time. Scalable cloud-native architectures—often built on microservices and containerised environments—ensure that high concurrent data streams and user interactions are handled. Real-time driven forecasting and hyper-localised demand predictions. For instance, a retailer with hundreds of locations can generate store-level forecasts that take into account local events, demographic profiles, and regional preferences (Sreenivasulu A., et al. 2022). In e-commerce, personalised forecasting allows companies to anticipate individual customer behaviour, enabling proactive product recommendations, inventory positioning, and delivery planning tailored to each user's needs. In summary, real-time demand forecasting businesses anticipate responding to customer needs. By leveraging streaming data from diverse sources and incorporating event-driven forecasting strategies, AI systems can produce fast, adaptive, and highly accurate forecasts that reflect current market realities. These capabilities are particularly critical in a dynamic demand, and disruptions are more frequent. Companies that successfully implement real-time AI forecasting can gain a decisive edge in agility, customer satisfaction, and operational efficiency. As technologies 5G, advanced continue to evolve, they will even real-time, deeply with every facet of the modern supply chain.

Inventory Optimisation with AI: Intelligent Management of Inventory Levels and Replenishment Strategies

In the era of intelligent supply chains and digital transformation, inventory optimisation success businesses balance cost efficiency with high service levels. Traditionally, inventory management relied on rule-based systems, static safety stock levels, and periodic reviews, which often led to overstocking, stockouts, or suboptimal order quantities. Today, the infusion of inventory management involves how businesses forecast demand, allocate resources, and replenish stock. Offers a more adaptive, data-driven approach to inventory optimisation, enhancing both the accuracy of inventory level decisions and the efficiency of replenishment strategies (Sekhar C., 2022). When it comes to replenishment strategies, AI plays an equally transformative role. Traditional replenishment or Min/Max systems often rely on fixed thresholds and assumptions that fail to account for variability in demand, promotions, and constraints. Driven replenishment is dynamic and context-aware. It considers not only forecasted demand but also real-time factors like current stock levels, lead time variability, supplier performance, upcoming events, and market trends. AI algorithms can trigger replenishment orders automatically when inventory falls below predicted thresholds, adjusting quantities and delivery timing based on real-time conditions (Ma X., et al, 2024). This dynamic, just-in-time approach minimises excess inventory while preventing costly stockouts. Moreover, AI empowers businesses to implement autonomous replenishment systems that operate with minimal human intervention. These systems integrate with data, supplier networks, and logistics providers to execute replenishment decisions in a retail context, AI sales trends, detect rapid changes in customer buying patterns, and automatically place replenishment orders to suppliers or warehouses. In a manufacturing environment, AI can forecast raw material requirements based on production schedules, sales forecasts, and supplier lead times, ensuring that materials are ordered just in time for production runs.

A further innovation enabled by AI is reinforcement learning (RL) in inventory management. In RL-based systems, the AI agent learns optimal replenishment policies reduced holding costs, improved service levels). Over time, the model evolves to identify the best actions in a variety of scenarios, including supply chain disruptions, demand spikes, or promotional events. These systems are particularly useful in complex, dynamic environments where pre-defined rules are insufficient to handle the nuances of real-world variability. AI also supports collaborative inventory optimisation, where multiple stakeholders—including suppliers, retailers, and logistics providers—share data and coordinate decisions in real time (Verma P., 2024). With technologies like predictive analytics, Iot, and cloud-based platforms, AI systems can facilitate even blockchain-enabled smart contracts that automate replenishment actions based on shared demand signals and inventory thresholds. From a strategic perspective, AI-driven inventory optimisation delivers measurable business value. It minimises losses and lost sales, improves working capital efficiency, and enhances overall supply chain agility. In highly competitive markets, these benefits translate to faster response times, better customer satisfaction, and improved profitability (Ejjami R., 2024). Furthermore, by freeing inventory planners from repetitive, manual tasks, AI allows them to focus on strategic initiatives such as product lifecycle planning, supplier negotiations, and long-term network design. In conclusion, AI is redefining inventory optimisation by providing intelligent, adaptive, and automated solutions that align inventory decisions with real-time demand signals and supply chain realities. Through advanced forecasting, dynamic safety stock calculation, autonomous replenishment, and data-driven segmentation, AI enables organisations to optimise inventory at both the micro and macro levels. By adopting AI-powered inventory systems, businesses become responsive, cost-effective, and thriving in an uncertain environment.

AI-Driven Inventory Management Systems: Revolutionising Tracking, Reporting, and Strategic Inventory Control

Automated inventory tracking, empowered by AI, fundamentally changes how organisations monitor and control their stock levels. Traditional tracking methods, such as barcode scanning, manual entry, or periodic stock-taking, are labour-intensive. AI-driven systems overcome these limitations by integrating with Iot sensors, RFID tags, computer vision technologies, and connected devices throughout warehouses, stores, and distribution centres. These technologies continuously collect and transmit data on product location, quantity, movement, and condition (Bennett L., 2025). Process maintains an accurate, continuously updated digital twin of the physical inventory environment. Moreover, AI enhances inventory tracking by enabling predictive capabilities. Instead of merely recording what is happening, AI algorithms can forecast historical contextual variables. Certain SKU is frequently misplaced or experiences high shrinkage in a specific location, the system can flag the issue proactively and suggest corrective actions, such as relocating stock or increasing surveillance. In warehousing operations, AI systems analyse picking patterns to optimise inventory placement, reducing travel time and increasing fulfilment speed (Ramakrishnan T., et al. 2022). These smart tracking systems are also integral to multi-location and omnichannel inventory visibility. AI algorithms consolidate data across multiple warehouses, stores, drop-ship locations, and fulfilment centres to create a unified view of inventory availability. This global visibility is crucial in today's retail and e-commerce ecosystems, where customer orders may be fulfilled from any location depending on proximity, stock availability, and delivery speed. AI synchronised across platforms, preventing overselling, minimising backorders, and enabling advanced fulfilment strategies like ship-from-store or curbside pickup (Aileni V R., 2025). AI also enables exception-based reporting, which shifts the focus from routine reporting to identifying and addressing outliers and issues that require human attention. Rather than sifting through large volumes of data, users are notified only when something unusual occurs, such as a sudden drop in inventory accuracy, a spike in returns, or a delay in replenishment. This exception-based approach enhances productivity and ensures that human resources are directed toward high-value decision-making rather than low-level data monitoring.

Another powerful application of AI in inventory reporting is prescriptive analytics, which goes beyond describing or predicting trends to recommending specific actions. For example, if an item is forecasted to stock out in five days based on current sales velocity and inbound delivery timelines, the system can recommend expedited shipping or alternative fulfilment options. Similarly, it may suggest transferring excess inventory from one location to another to balance stock across the network and reduce holding costs. These intelligent suggestions are based on continuously updated data models that learn and improve over time. Additionally, AI can integrate pricing, supplier performance, and macroeconomic indicators into inventory reports, providing a broader context for decision-making. For instance, if a global shortage of a key raw material is detected through news scraping and supplier data, the AI system can alert procurement teams and suggest stockpiling strategies to mitigate risk (Kathiriya S., et al, 2022). This kind of forward-looking, scenario-based planning is only possible with AI-driven insights. As businesses increasingly embrace automation and digital transformation, AI-driven inventory management systems are being integrated with other enterprise systems Warehouse

Management Systems (WMS), ensures a seamless flow of data across the organization, enabling unified decision-making and For example, change in customer demand detected by the CRM can trigger an inventory review in the WMS and initiate a replenishment order through the ERP—all automatically managed by AI algorithms (Baharudin H., 2023). In summary, AI-driven inventory management systems represent a significant leap forward in how businesses monitor, control, and optimise their inventory operations. By combining automated inventory tracking with intelligent, actionable reporting, these systems reduce human error, accelerate decision-making, and enable real-time responsiveness to operational conditions. They support higher inventory accuracy, better fulfilment rates, lower holding costs, and improved customer satisfaction. Adopting AI-driven inventory solutions will be essential for companies seeking to build agile, resilient, and data-centric operations that thrive in the digital age.

Predictive Maintenance for Inventory Management: Proactive Equipment Monitoring and Failure Prediction for Operational Efficiency

In modern inventory management and supply chain operations, where efficiency, uptime, and accuracy are vital, the role of predictive maintenance has expanded significantly. Traditionally, maintenance has been reactive—performed after equipment fails—or preventive, scheduled at regular intervals regardless of actual equipment condition. While these approaches have long been standard, they often result in unnecessary downtime, inflated maintenance costs, and inefficient use of resources. Enter predictive maintenance (PDM), a technology-driven approach powered by that enables organisations to anticipate equipment issues before they occur (Yazdi M., 2024). In the context of predictive maintenance, ensuring operational continuity, optimising equipment utilisation, and reducing the risks of supply chain disruption. Central to this is the intelligent use of equipment monitoring and failure prediction, which collectively minimise unexpected breakdowns and align maintenance schedules with actual machine health. AI-powered monitoring systems excel in detecting anomalies and degradation patterns that human operators might overlook. For instance, a slight increase in motor temperature or irregular vibration in a conveyor system may indicate the beginning of a mechanical issue. By recognising these patterns early, predictive maintenance algorithms can trigger alerts, schedule inspections, or recommend pre-emptive part replacements—a long-term proactive approach which is especially critical in environments like e-commerce fulfilment centres or manufacturing plants, where any interruption in inventory flow can lead to significant losses, missed delivery deadlines, and dissatisfied customers (Ghaithan A., et al, 2024).

Beyond uptime and reliability, predictive maintenance contributes to cost optimisation. In traditional preventive maintenance schedules often assumed that no issues exist. This wastes labour hours, spare parts, and disrupts operations unnecessarily. Predictive maintenance aligns service schedules with actual machine condition, components. Spare parts inventory can also be optimised—since AI systems can forecast which parts will be needed and when, companies can reduce the need to stockpile excessive spare parts, thus freeing up warehouse space and capital (Kaur C., et al, 2025). Integration and data interoperability are vital for effective predictive maintenance. AI-driven PDM systems often connect with existing Warehouse Management Systems (WMS), ensuring that equipment health data can influence broader operational decisions, such as order routing, warehouse slotting, and staffing. For instance, if a conveyor section is flagged for imminent maintenance, the WMS can reroute picking paths or reassign order batching to avoid that zone. Likewise, CMMS platforms can auto-generate maintenance work orders based on predictive analytics outputs, streamlining workflows and improving accountability (Aragani, V M., et al. 2025). The application of digital twins—virtual models that replicate the behaviour of physical equipment in real time—further enhances predictive maintenance capabilities sensor into digital twin, AI systems can simulate the future performance of machines under various conditions, test different maintenance scenarios, and recommend optimal strategies balance maintenance costs, equipment uptime, and inventory throughput. For example, a warehouse crane might delay component replacement, showing that the risk of total system shutdown increases significantly after a certain threshold. The benefits of predictive maintenance also extend to sustainability and compliance (Kabashkin I., et al, 2024). By reducing energy waste, minimising the use of unnecessary spare parts, and extending equipment life, PDM contributes to a more sustainable operation. It also helps businesses comply with regulatory standards related to safety, quality control, and equipment reliability, especially in medical device manufacturing, where equipment failures can have serious consequences.

Benefits of AI-Driven Demand Forecasting and Inventory Management: Unlocking Accuracy, Efficiency, and Cost Optimisation

Improved forecast accuracy has a cascading effect on multiple aspects of business operations. One of the most immediate and measurable outcomes is cost reduction. With more reliable demand predictions, companies can optimise inventory levels to match actual needs, thereby reducing excess stock and minimising carrying costs such as storage, insurance, depreciation, and capital tie-up. Simultaneously, better forecasting reduces the risk of stockouts, emergency orders, and expedited shipping—all of which can be significantly more expensive than regular replenishment processes (MoghadasNian S A., et al, 2025). Moreover, accurate forecasts support leaner and more efficient supply chains, enabling just-in-time (JIT) practices that reduce waste and free up working capital. AI also enhances replenishment planning and inventory turnover, ensuring that the restocked. Traditional replenishment often relies on rigid reorder points or periodic review cycles that fail to respond effectively to dynamic market conditions. Including sales velocity, promotional calendars, seasonal trends, lead times, and supplier reliability, to generate real-time, adaptive replenishment strategies (Ahmed A A A., et al, 2024). This results in optimised stock rotation, fewer markdowns, reduced spoilage in perishable categories, and a healthier balance between supply and demand. Another major benefit is increased operational efficiency. AI automates many time-consuming and error-prone processes, such as data collection, demand planning, stock level monitoring, reorder calculations, and reporting. It also frees up planners and managers strategic supplier collaboration, category network design. In warehousing and distribution, AI-powered systems can direct warehouse operations to optimise storage allocation, picking sequences, and load planning inventory, enhancing throughput and reducing labour costs (Sodiya E O., et al, 2024).

Figure 1. AI demand forecasting process

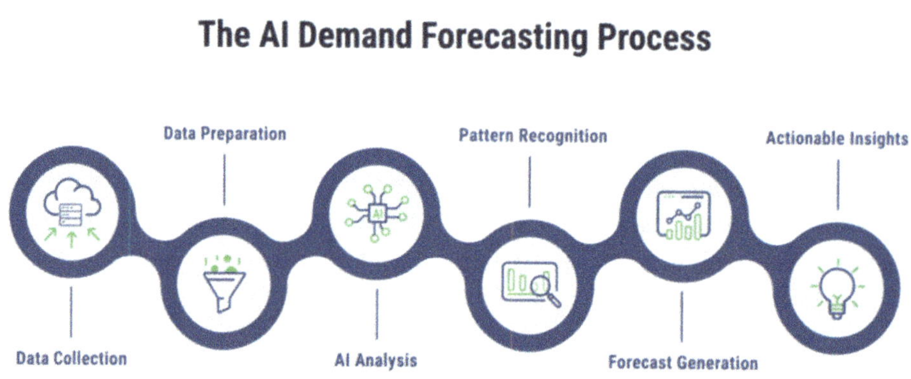

Another critical benefit is the ability to support personalised and localised demand strategies. Unlike traditional models that treat demand as a generalised forecast across all channels or regions, AI can segment data down to specific customer segments, locations, or even individual products. This granularity enables hyper-localised planning and personalised recommendations, particularly valuable in retail, where consumer preferences vary widely by demographic and geography. Can AI be in-store based on local events, weather, or consumer behaviour trends, and stock accordingly to maximise sales potential and minimise waste? The scalability and adaptability of AI systems also provide long-term value (Ocran G., et al, 2024). As organisations grow, expand into new markets, or diversify product lines, AI models can scale accordingly, processing larger datasets and more complex variables without compromising accuracy or speed. Furthermore, AI systems can be continually trained and refined; they evolve alongside business and remain in changing market conditions. Importantly, AI also enhances collaborative planning and inventory insights, distributors, partners, and organisations coordinate efforts, reduce lead times, and improve the flow of goods throughout the network. Collaborative AI platforms support models, promoting transparency and alignment across stakeholders (Debnath P., et al. 2025).

Case Studies of Successful AI-driven Demand Forecasting and Inventory Management: Transforming Businesses through Advanced Technologies

Walmart: Enhancing Inventory Management and Demand Forecasting: Demand forecasting and inventory management significantly improve operational efficiency. With millions of SKUS (stock-keeping units) across thousands of stores globally, the company faces complex challenges related to demand forecasting and inventory management. For years, Walmart relied on traditional demand forecasting methods that were often inaccurate, leading to stockouts or overstock situations (Shende A D., et al. 2025). To solve this problem, Walmart turned to AI and machine learning. That leverages local events a social media AI model uses this data to predict demand at hyper-local levels, ensuring that each store receives the right quantity of products based on local customer behaviour and market conditions. This level of precision has helped Walmart reduce stockouts and improve inventory turnover, ultimately leading to increased sales and customer satisfaction. Moreover, Walmart also implemented replenishment that involves analysing inventory in By analysing sales trends and adjusting for variables like promotions, seasonal fluctuations, and even competitor activity, the system ensures that stores are always stocked with the right items, reducing excess inventory and related holding costs (Singh N., et al, 2023). AI also helps streamline Walmart's global supply chain, allowing the company to optimise its vast network of suppliers and warehouses for better efficiency. The results of these AI-driven implementations have been impressive. Walmart has seen improvements in inventory accuracy, a reduction in stockouts, and better demand planning, contributing to improved financial performance. This transformation has also allowed Walmart to enhance its customer experience. Amazon: Revolutionising Fulfilment and Logistics with AI: its management practices are no exception. Amazon's success lies in its ability to predict consumer behaviour with incredible accuracy and manage inventory across a vast network of fulfilment centres globally (Shobhana N., 2024).

Unilever: Streamlining Supply Chain and Reducing Waste with AI: Unilever, a global consumer goods company, faces significant challenges in demand forecasting and inventory management due to its large portfolio of products and diverse markets. The company's products, ranging from personal care to food and beverages, must be available to consumers. However, fluctuations in demand, disruptions in supply chains, and market volatility often lead to inefficiencies in inventory management. To tackle these issues, Unilever turned to streamline it improve its forecasting accuracy. By leveraging predictive analytics, Unilever was able to develop more accurate demand forecasts that consider factors, data, and regional demand. The company implemented algorithms to predict demand at the regional and product level, which significantly improved stock availability and reduced inventory waste (Choudhuri S S., 2024). Furthermore, AI-powered inventory management systems allowed Unilever to optimise its supply chain network. The system analyses data from multiple suppliers, warehouses, and distribution centres to determine the most efficient replenishment strategies. It also helps identify, allowing the company to make proactive adjustments. These optimisations significantly excess lowering minimizing waste, especially in perishable goods. Unilever's AI-driven solutions also helped the company align its production schedules more closely with actual demand, improving the efficiency of manufacturing operations. Overproduction leads to product wastage and better resource utilisation (Johri S., et al. 2025). This shift not only improved Unilever's sustainability efforts but also generated cost savings and enhanced the company's overall competitiveness. The success of Unilever's AI implementations demonstrates the significant potential of AI in consumer goods companies to enhance demand forecasting, reduce waste, and create value.

Challenges in Implementing AI-driven Demand Forecasting and Inventory Management: Navigating the Roadblocks to Success

Presents a significant opportunity for operations to use these advanced technologies. These hurdles span various domains, including readiness, cost considerations, and change management. The complexity of integrating AI into existing business processes, especially in industries with entrenched traditional systems, means that companies need to overcome multiple obstacles to leverage the power section key challenges organisations face. Data Quality and Availability: Arguably, most challenges are driven by demand and inventory availability, quality systems that accurately and timely learn and missing or Inconsistent data across departments and systems can prevent AI models from effectively identifying patterns or generating accurate forecasts (Paramasivan A., 2023). For example, maybe they do not communicate with each other, leading to fragmented insights. Additionally, data quality is a crucial factor. AI systems are incomplete, inaccurate, and biased; forecasts and recommendations provided by the AI will be unreliable. Common issues include incorrect historical sales data, inconsistent inventory records, and erroneous customer behaviour data. Data cleansing and preprocessing become essential tasks before implementing AI-driven systems, but many organisations do not have the resources or expertise to perform these tasks effectively. For instance, if data is missing or has too many errors, its AI models accurately predict demand. Organisational Readiness and Culture: The organisational readiness to adopt AI-driven solutions is another critical challenge (Cooper M., 2025). AI implementation often requires a fundamental shift in how companies approach demand forecasting, inventory management, and decision-making. Traditional methods are typically less data-driven, relying on intuition, historical knowledge, and basic statistical techniques. AI, on the other hand, requires a more scientific approach, focusing on continuous optimisation. Many organisations are not ready to embrace this shift, either due to resistance to change or a lack of AI literacy within the workforce.

Moreover, leadership buy-in is critical for the success of AI-driven initiatives. If the management team does not fully understand the potential benefits of AI or is not committed to investing in the necessary resources, it becomes challenging to drive the project forward. AI implementation also requires a cross-functional approach, with close collaboration between data science teams, IT departments, supply chain managers, and other key stakeholders. If these teams do not communicate effectively or align on project goals, the implementation process can become disjointed and lead to suboptimal results. Cost and Resource Constraints: The financial investment required to implement can be substantial, especially for small to medium-sized companies (Kudrenko I., 2024). While of AI, cost savings, customer satisfaction, and better inventory turnover are well-documented, the upfront costs can be a major barrier to adoption. These costs can include software licenses, cloud storage fees, consultancy fees, training costs, and the salaries of specialised AI professionals. For many organisations, particularly those with tight budgets, these expenses can be prohibitive. Additionally, the costs associated with setting up data infrastructure, such as implementing sensors in warehouses, integrating various data sources, and cleaning data, further contribute to the financial burden. Moreover, AI-driven solutions require ongoing maintenance and continuous model training to ensure their accuracy and effectiveness. Over time, the data that the AI systems rely on will change, meaning that the models must be retrained to remain accurate (Hao R., 2024). This necessitates ongoing investments in infrastructure, human resources, and expertise, which can strain an organisation's resources.

Future Directions for AI-driven Demand Forecasting and Inventory Management: Emerging Trends and Technologies

Landscape demand forecasting and inventory management are largely increasing. As the capabilities of AI continue to advance, so too do the opportunities for optimising and automating supply chain processes. Looking ahead, the future holds exciting potential for businesses to achieve greater accuracy, future directions for AI in these critical areas, focusing on innovations will way organisations' management in the coming years. As businesses continue to face complex, real-time challenges in demand forecasting, AI-driven solutions that leverage advanced machine learning models will become indispensable for ensuring that supply chains remain agile, efficient, and responsive (Mhaskey S V., 2024). Real-time and Predictive Analytics for Agile Decision-Making is another future direction that will fundamentally reshape demand forecasting and inventory management. Today's businesses require quick, that are informed the most current data available. Traditional systems, often reliant on historical data and periodic updates, are ill-equipped to meet the demands of fast-moving industries. However, real-time demand forecasting powered by AI will enable businesses to rapidly understand consumer behaviour, unexpected shifts in market trends. The advent such as (Internet of Things and real-time sales data will ensure continuous product availability, and market conditions. AI-driven systems will be these streams instantly, identifying anomalies, emerging as they occur (Archana & Stephen, 2024). By integrating real-time analytics into their operations, optimise inventory levels, strategies, and make proactive decisions that improve operational efficiency and customer satisfaction. For example, event-driven forecasting uses events weather changes, local festivals, and even global events like pandemics or political shifts, that will become increasingly common. AI systems will be able to integrate these variables into demand forecasts, enabling businesses to dynamically develop supply chain strategies. Forecast demand and adjust accordingly—will be a game-changer for industries such as e-commerce, fashion, perishable goods, and automotive manufacturing (Lesmana R., et al, 2024).

AI-Powered Personalisation and Hyperlocal Forecasting: As AI continues to evolve, one of the most exciting trends is the move toward hyper-personalised inventory management. Using AI-driven analytics, businesses can offer highly personalised products, promotions, and experiences for customers, not just on an individual level but also at a hyperlocal scale. By combining real-time customer behaviour data, geolocation data, and socioeconomic trends, businesses can predict the local demand for products with greater accuracy than ever before. This hyperlocal forecasting will allow businesses to optimise their inventory at the store level, ensuring that each location is stocked with the right products that cater to local preferences, tastes, and trends (Saiyed A., 2025). AI can analyse factors such as seasonality, cultural preferences, and even local events to predict demand for specific products in different geographic areas. For retailers, this means they can significantly reduce costs at individual locations, customer satisfaction increases sales. AI-Driven Supplier Collaboration and Integration: Finally, the future will involve deeper collaboration and integration with suppliers, creating more connected and efficient supply chains. AI will allow businesses to collaborate with suppliers in real time, sharing insights and forecasts (Khang A., et al, 2024). This level of integration will improve supplier relationships, enable more accurate lead time forecasting, and ensure that inventory levels are maintained without the risk of overstocking or understocking.

CONCLUSION

With AI-driven systems, businesses can reduce their reliance on guesswork and historical assumptions, instead utilising This results in better alignment of production, procurement, and distribution strategies with actual consumer demand, improving efficiency across the supply chain. More accurate demand forecasting also minimises the risks associated with overstocking, both significant financial losses. Integrating analytics AI reduces human error and eliminates inefficiencies tied to manual processes. Smart replenishment strategies driven by AI allow businesses to react quickly to changes in consumer demand, ensuring automation of routine inventory tasks also allows strategic, high-level decisions, improving the overall productivity and effectiveness of the workforce. Moreover, AI's real-time inventory management capabilities help businesses react swiftly to demand fluctuations, whether it's during peak seasons, promotional events, or unexpected responsiveness. Retaining customer loyalty fosters positive brand perception and reputation. When companies can meet customer demands consistently, efficiently, and accurately, they establish themselves as reliable players in the market, earning. As environmental concerns become more pressing, businesses' footprints implement significantly these goals supply chain operations and reduce waste. AI-driven systems enable companies to reduce overproduction and minimise the waste generated by excess inventory. AI also helps businesses improve logistics efficiency, leading to lower transportation-related emissions and more eco-friendly supply chain operations. Moreover, AI's role in circular economies is also gaining attention, with companies leveraging AI to develop more sustainable recycling and reuse systems. By enabling businesses to manage their resources more efficiently, AI contributes to reducing waste, optimising resource use, and enhancing long-term sustainability goals. Companies that adopt AI-driven demand forecasting and inventory management position themselves as responsible, forward-thinking organisations in the eyes of environmentally conscious consumers and stakeholders. The journey toward AI-powered supply chains is ongoing, and the technology itself continues to evolve. As machine learning models become more sophisticated, businesses will see even greater improvements in the accuracy and precision of their forecasts. AI systems will become more capable of integrating external data sources, such as geopolitical events, global supply chain disruptions, and consumer behaviour shifts, even adaptive intelligent The future will see greater automation, where AI will not just but also autonomously make supply chain decisions, streamlining operations to an extent that was previously unimaginable. Moreover, it will likely increase, improving the transparency, transactions throughout With these advancements, the potential for AI to be sustainable, resilient is enormous. In summary, they are reshaping the way businesses approach supply chain optimisation. By enabling more accurate demand predictions, smarter inventory decisions, and faster responses to market shifts, AI not only helps companies reduce costs but also improves and strengthens the integration of AI in supply chains, leading to greater operational efficiency, sustainability, and adaptability evolve, possibilities further future of supply chain management are endless. Today is a world of tomorrow, where speed, agility, and customer satisfaction will continue to define success.

REFERENCES

Ahmad, N., Ghadi, Y., Adnan, M., & Ali, M. (2022). Load Forecasting Techniques for Power System: Research Challenges and Survey. *IEEE Access: Practical Innovations, Open Solutions*, *15*(3), 567–589. DOI: 10.1109/ACCESS.2022.3187839

Ahmed, A A A., Kumar, V. S., Jena, S. K., Nagpal, A., Shukla, P. K., & Balachandar, K. (2024). Maximizing Profits and Efficiency: The Intersection of AI, Machine Learning, and Supply Chain Financial Management. *Advances in Logistics, Operations, and Management Science*, *15*(3), 567–589. DOI: 10.4018/979-8-3693-3593-2.ch015

Aileni, V. R. (2025). AI-Driven Supply Chain Optimization: Enhancing Forecasting, Inventory, and Logistics. *Machine Intelligence Research*, *15*(3), 567–589. DOI: 10.1000/machineintelligenceresearchs.aisupplychainoptimization2025

Alabi, M. (2023). Predictive Analytics for Product Planning and Forecasting. *ResearchGate Preprint*, *15*(3), 567–589. DOI: 10.1000/researchgate.productforecasting2023

Albahli, S. (2025). LSTM vs. Prophet: Achieving Superior Accuracy in Dynamic Electricity Demand Forecasting. *Energies*, *15*(3), 567–589. DOI: 10.3390/en18020278

Anand, P., Sharma, M., & Saroliya, A. (2024). "A Comparative Analysis of Artificial Neural Networks in Time Series Forecasting Using ARIMA Vs Prophet." *Proceedings of the International Conference on AI and Forecasting*, Vol. 15(3), pp. 567–589. DOI: DOI: 10.1109/IC3SE62002.2024.10593482

Aragani, V. M., Anumolu, V. R., & Selvakumar, P. (2025). Democratization in the Age of Algorithms: Navigating Opportunities and Challenges. In Wongmahesak, K., Wekke, I., Seftyono, C., & Nurdin, N. (Eds.), *Democracy and Democratization in the Age of AI* (pp. 39–56). IGI Global Scientific Publishing., DOI: 10.4018/979-8-3693-8749-8.ch003

Archana, T., & Stephen, R. K. (2024). The Future of Artificial Intelligence in Manufacturing Industries. *AI: Convergence with Real-Time Data and AI*, *15*(3), 567–589. DOI: 10.1000/igi-global.aimanufacturingfuture2024

Baharudin, H. (2023). "AI in E-Commerce Warehouse Management: Enhancing Operational Efficiency, Ensuring Inventory Precision, and Strengthening Security Measures." SSRN Preprint, Vol. 15(3), pp. 567–589. DOI: DOI: 10.1000/paperssrn.aiecommercewarehouse2023

Bennett, L. (2025). "How AI Technologies Are Revolutionizing Inventory Management in the Supply Chain." Preprints.org, Vol. 15(3), pp. 567–589. DOI: DOI: 10.1000/preprints.aiinventorymanagement2025

Boppiniti, S. T. (2021). *Real-Time Data Analytics with AI: Leveraging Stream Processing for Dynamic Decision Support*. International Journal of Management Education for Sustainable Development., DOI: 10.589/IJMESD.2021.589

Choudhuri, S. S. (2024). AI in ERP and Supply Chain Management. *AI and Business Strategy*, *15*(3), 567–589. DOI: 10.1000/google.aiinsupplychain2024

Cooper, M. (2025). "Barriers to AI Adoption in Supply Chain Management: Perspectives from Industry Leaders." Preprints.org, Vol. 15(3), pp. 567–589. DOI: DOI: 10.1000/preprints.aiadoptionbarriers2025

Debnath, P., Deshmukh, S. M., Laroia, M., Selvakumar, P., Manjunath, T. C., & Bhattacharya, S. (2025). Challenges Faced by Countries in Renewable Energy Adoption. In Özsungur, F., Chaychi Semsari, M., & Küçük Bayraktar, H. (Eds.), *Geopolitical Landscapes of Renewable Energy and Urban Growth* (pp. 295–320). IGI Global Scientific Publishing., DOI: 10.4018/979-8-3693-8814-3.ch010

Dubey, A. K., Kumar, A., García-Díaz, V., & Sharma, A. K. (2021). Study and Analysis of SARIMA and LSTM in Forecasting Time Series Data. *Sustainable Energy Technologies and Assessments*, *47*, 101474. Advance online publication. DOI: 10.1016/j.seta.2021.101474

Ejjami R, (2024). "Optimizing In-Store Logistics: How AI Enhances Inventory Management and Space Utilization." Journal of Next-Generation Research 5.0, Vol. 15(3), pp. 567–589. DOI: DOI: 10.1000/jngr5.aiinstorelogistics2024

Ghaithan, A., AlShamrani, O., Mohammed, A., & Alshibani, A. (2024). Proactive Framework for Obsolescence Management of Electrical Equipment in Oil and Gas Industry. [DOI]. *Journal of Quality in Maintenance Engineering*, *30*(3), 567–589. DOI: 10.1108/JQME-10-2023-0096

Hao, R. (2024). "Streamlining SCM: Integrating Demand Forecasting and Inventory Optimization." *Proceedings of the 2nd International Conference on Management and AI*, Vol. 15(3), pp. 567–589. DOI: DOI: 10.2991/978-94-6463-542-3_63

Hasan, M. D. R., Islam, M. R., & Rahman, M. A. (2025). Developing and Implementing AI-Driven Models for Demand Forecasting in US Supply Chains: A Comprehensive Approach to Enhancing Predictive Accuracy. *Edelweiss Applied Science Journal*, *15*(3), 567–589. DOI: 10.55214/25768484.v9i1.4308

Ingle, C., Bakliwal, D., Jain, J., & Singh, P. (2021). "Demand Forecasting: Literature Review on Various Methodologies." *Proceedings of the 12th IEEE Conference on Forecasting*, Vol. 15(3), pp. 567–589. DOI: DOI: 10.1109/ICCCNT51525.2021.9580139

Johri, S., Selvakumar, P., Satyanarayana, P., Das, A., Gupta, S., & Bhattacharya, S. (2025). Urbanization and Energy: The Nexus of Global Challenges. In Özsungur, F., Chaychi Semsari, M., & Küçük Bayraktar, H. (Eds.), *Geopolitical Landscapes of Renewable Energy and Urban Growth* (pp. 453–478). IGI Global Scientific Publishing., DOI: 10.4018/979-8-3693-8814-3.ch016

Jones, J. (2025). "AI-Driven Demand Forecasting in Supply Chains: A Qualitative Analysis of Adoption and Impact." Preprints.org, Vol. 15(3), pp. 567–589. DOI: DOI: 10.1000/preprints.aidemandforecasting2025

Kabashkin, I., & Susanin, V. (2024). "Unified Ecosystem for Data Sharing and AI-Driven Predictive Maintenance in Aviation." Computers, Vol. 13(12) [DOI]. *Article*, *318*, 1–15.

Kathiriya, S., Shende, A., & Sinha, A. (2022). Strategic Innovations and Future Directions in AI-Driven Retail Inventory Management: A Comprehensive Review and Pathway Analysis. *International Journal of Scientific Research*, *15*(3), 567–589. DOI: 10.1000/researchgate.airetailinventory2022

Katya, E. (2023). Exploring Feature Engineering Strategies for Improving Predictive Models in Data Science. *Journal of Computer Systems and Engineering*, *15*(3), 567–589. DOI: 10.52710/rjcse.88

Kaur, C., & Sharma, A. (2025). "Enhancing Predictive Maintenance with AI: Applications and Impact." *Proceedings of the 6th International Conference on AI and Predictive Maintenance*, Vol. 15(3), pp. 567–589. DOI: []DOI: 10.1109/IC-MCSI64620.2025.10883435

Khang, A., Rath, K. C., & Mishra, B. K. (2024). *"Future Directions and Challenges in Designing Workforce Management Systems for Industry 4.0." AI-Oriented Competency Framework for Talent Management in the Digital Economy* (1st ed.). DOI. [10.1201/9781003440901-1]

Kudrenko, I. (2024). Navigating the Future: AI-Driven Healthcare Supply Chains. *Hospital Supply Chain: Challenges and Opportunities*, *15*(3), 567–589. DOI: 10.1007/978-3-031-70292-1_26

Lesmana, R., Wijaya, I., Nabila, E. A., Agustian, H., Audiah, S., & Faturahman, A. (2024). Enhancing Market Trend Analysis Through AI Forecasting Models. [DOI]. *International Journal of Cyber and IT Service Management*, *4*(2), 1–15. DOI: 10.34306/ijcitsm.v4i2.162

Liu, A., Vici, L., Ramos, V., Giannoni, S., & Blake, A. (2021). "Visitor Arrivals Forecasts Amid COVID-19: A Perspective from the Europe Team." Annals of Tourism Research, Vol. 88. *Annals of Tourism Research, 103182*, 1–15. DOI: 10.1016/j.annals.2021.103182

Ma, X., Zeyu, W., Ni, X., & Ping, G. (2024). Artificial Intelligence-Based Inventory Management for Retail Supply Chain Optimization: A Case Study of Customer Retention and Revenue Growth. *Journal of Knowledge Learning and Science Technology*, *15*(3), 567–589. DOI: 10.60087/jklst.v3.n4.p260

Madgule, M., Vinayaka, N., Sonkhaskar, Y. M., Patel, D. R., Karthikeyan, R., & Selvakumar, P. (2023). Mechanical properties and microstructure of activated TIG welded similar joints of Inconel alloys by desirability approaches. *Materials Today: Proceedings*, *77*, 528–533.

Menculini, L., Marini, A., Proietti, M., Garinei, A., Bozza, A., Moretti, C., & Marconi, M. (2021). Comparing Prophet and Deep Learning to ARIMA in Forecasting Wholesale Food Prices. *Forecasting*, *15*(3), 567–589. DOI: 10.3390/forecast3030040

Mhaskey, S. V. (2024). SCM 4.0: Navigating the Impact of Industry 4.0 on Supply Chain Management Through Digitalization and Technology Integration. *ResearchGate Preprint*, *15*(3), 567–589. DOI: 10.22362/ijcert/2024/v11/i10/v11i1001

MoghadasNian S A. MahMoudi M, (2025). "Stock Control in Airline Logistics: AI-Driven Inventory Optimization for Spare Parts." International Conference on Recent Advances in Engineering, Innovation and Technology, Vol. 15(3), pp. 567–589. DOI: [Available here]

Muthukalyani, A. R. (2023). Unlocking Accurate Demand Forecasting in Retail Supply Chains with AI-Driven Predictive Analytics. *Information Technology Management*, *15*(3), 567–589. DOI: 10.1000/researchgate.aipredictiveanalytics2023

Nia, A. R., Awasthi, A., & Bhuiyan, N. (2021). "Industry 4.0 and Demand Forecasting of the Energy Supply Chain: A Literature Review." Computers & Industrial Engineering, Vol. 154. *Computers & Industrial Engineering*, *107128*, 1–15. DOI: 10.1016/j.cie.2021.107128

Ning, Y., Kazemi, H., & Tahmasebi, P. (2022). A Comparative Machine Learning Study for Time Series Oil Production Forecasting: ARIMA, LSTM, and Prophet. *Computers & Geosciences*, *15*(3), 567–589. DOI: 10.1016/j.cageo.2022.105126

Nweje, U., & Taiwo, M. (2025). Artificial Intelligence for Predictive Supply Chain Management: Focus on How AI-Driven Tools Are Revolutionizing Demand Forecasting and Inventory Optimization. *International Journal of Scientific Research*, *15*(3), 567–589. DOI: 10.1000/journalijsra.aipredictivesupplychain2025

Ocran, G., Yusuf, S. O., & Owusu, P. (2024). AI-Driven Business Analytics for SMEs: Unlocking Value Through Predictive and Prescriptive Analytics. *International Journal of AI in Business*, *15*(3), 567–589. DOI: 10.1000/researchgate.aibusinessanalytics2024

Paramasivan, A. (2023). Transforming Healthcare Supply Chains: AI for Efficient Drug Distribution and Inventory Management. [IJSAT]. *International Journal on Science and Technology*, *15*(3), 567–589. DOI: 10.1000/researchgate.aihealthcare-supplychain2023

Peter, H. (2024). "Designing Event-Driven Architecture for AI-Powered Sales Pipeline Optimization: A Comparative Analysis of Messaging Patterns." ResearchGate Preprint. DOI: DOI: 10.13140/RG.2.2.38908.39524

Punia, S., & Shankar, S. (2022). Predictive Analytics for Demand Forecasting: A Deep Learning-Based Decision Support System. *Knowledge-Based Systems*, *15*(3), 567–589. DOI: 10.1016/j.knosys.2022.109956

Ramakrishnan, T., Mohan Gift, M. D., Chitradevi, S., Jegan, R., & Subha Hency Jose, P. Nagaraja, Rajneesh Sharma HN, Selvakumar P, Sintayehu Mekuria Hailegiorgis, (2022). "Study of Numerous Resins Used in Polymer Matrix Composite Materials", Advances in Materials Science and Engineering, vol. 2022, Article ID 1088926, 8 pages, 2022. https://doi.org/DOI: 10.1155/2022/1088926

Saiyed, A. (2025). AI-Driven Innovations in Fintech: Applications, Challenges, and Future Trends. [DOI]. *International Journal of Electrical and Computer Engineering Research*, *5*(1), 8–15. DOI: 10.53375/ijecer.2025.437

Sekhar, C. (2022). Optimizing Retail Inventory Management with AI: A Predictive Approach to Demand Forecasting, Stock Optimization, and Automated Reordering. *European Journal of Advances in Engineering and Technology*, *15*(3), 567–589. DOI: 10.1000/researchgate.aiinventorymanagement2022

Sharma, A., Patel, N., & Gupta, R. (2021). Leveraging LSTM and Prophet Models for Enhanced AI-Driven Demand Prediction in E-Commerce. *European Advanced AI Journal*, *15*(3), 567–589. DOI: 10.1000/eaaij.aidemandprediction2021

Shende, A. D., Tekade, S. A., Deshmukh, A. A., Tembhurkar, S. P., & Selvakumar, P. (2025). Water Quality Monitoring and Control With AI. In Ruiz-Vanoye, J., & Díaz-Parra, O. (Eds.), *Smart Water Technology for Sustainable Management in Modern Cities* (pp. 51–80). IGI Global Scientific Publishing., DOI: 10.4018/979-8-3693-8074-1.ch003

Shobhana, N. (2024). AI-Powered Supply Chains Towards Greater Efficiency. *AI Dynamics and Interactions in Management, 15*(3), 567–589. DOI: 10.1000/igi-global.aisupplychain2024

Singh, N., & Adhikari, D. (2023). AI and IoT: A Future Perspective on Inventory Management. *International Journal for Research in Applied Science and Engineering Technology, 15*(3), 567–589. DOI: 10.22214/ijraset.2023.57200

Sodiya, E. O., Umoga, U. J., & Amoo, O. O. (2024). AI-Driven Warehouse Automation: A Comprehensive Review of Systems. *Advanced Research in AI and Logistics, 15*(3), 567–589. DOI: 10.1000/researchgate.aiwarehouseautomation2024

Sreenivasulu, A., & Manimozhi, T. Satheesh kumar P, Selvakumar P. and Singh P, (2022). "The Model of Smart Sensing Device For Sensitive Nanoclusters Modification in Sensing Properties," 2022 2nd International Conference on Advance Computing and Innovative Technologies in Engineering (ICACITE), Greater Noida, India, 2022, pp. 1043-1047, DOI: 10.1109/ICACITE53722.2022.9823846

Sugumaran K M D, Pandiyarajan V Y B S, Selvakumar P. and M. S, (2024). "Analysis of Electromyography Signals for Control of Mechanical Prosthesis using Machine Learning Techniques," 2024 2nd International Conference on Intelligent Data Communication Technologies and Internet of Things (IDCIoT), Bengaluru, India, 2024, pp. 1059-1067, .DOI: 10.1109/IDCIoT59759.2024.10467707

Tyler, V. (2025). How Event-Driven Architectures Are Powering Predictive Decision Intelligence. [DOI]. *ResearchGate Preprint, 15*(3), 567–589.

Verma, P. (2024). Transforming Supply Chains Through AI: Demand Forecasting, Inventory Management, and Dynamic Optimization. *Integrated Journal of Science and Technology, 15*(3), 567–589. DOI: 10.1000/researchgate.aisupplychainoptimization2024

Vijayakumar, G. (2024). Muniyandi Rajkumar, Rajiv Chandar N, Selvakumar P and Ramesh Duraisamy, (2024). Environmentally friendly TDS removal from wastewater by electrochemical ion exchange batch-type recirculation (EIR) technique. *Environmental Science. Water Research & Technology, 10*, 826–835. DOI: 10.1039/D3EW00793F

Yazdi, M. (2024). Maintenance Strategies and Optimization Techniques. *Computational Mathematics for Industrial Systems*, *15*(3), 567–589. DOI: 10.1000/springer.aimaintenanceoptimization2024

Chapter 5
Optimizing Demand Forecasting and Inventory Management With AI-Driven Solutions

K. Vijayakumar
https://orcid.org/0009-0005-6245-3165
Panimalar Engineering College, India

L. A. Anto Gracious
https://orcid.org/0000-0003-2386-9182
R.M.K. College of Engineering and Technology, India

ABSTRACT

This chapter explores the transformative role of Artificial Intelligence (AI) in optimizing demand forecasting and inventory management across various industries. Traditional methods, while useful, often fail to meet the demands of modern, dynamic supply chains, leading to inefficiencies such as stockouts and overstocking. AI-driven solutions, leveraging machine learning, deep learning, and predictive analytics, offer enhanced accuracy, efficiency, and cost reductions. By automating key processes and integrating with advanced technologies like IoT and blockchain, AI enables real-time decision-making and smarter resource allocation. The chapter also examines challenges in data quality, system integration, and workforce adaptation, while highlighting future research directions in AI adoption, optimization, and the integration of emerging technologies.

DOI: 10.4018/979-8-3373-3176-8.ch005

1. INTRODUCTION

1.1 Background

Demand forecasting & inventory are essential elements of the supply chain that is used in the retail industry, manufacturing companies, and logistics. In any competitive business environment, demand forecasting & inventory management are important factors that would enable organizations to meet their customers' needs while at the same time avoiding the vice versa. They can be closely associated with the company's performance in terms of sales, profits, and customer satisfaction. Demand forecasting involves identifying the likely demand for products in the future, which depends on the previous records, trends, seasonality, economic factors, and promotions. Demand forecasting is a critical aspect of any business since it helps in the determination of the demand for specific products by customers in the future in order to avoid oversupply or lack of stock. While on the other hand, inventory management entails the handling of the stock, control and replenishment of the same. Inventory management is the proper procurement and storage of products to meet the demand that is at the right place, right time with the right quality & quantity as well as the right cost. For many years, the forecast techniques which were in use included the moving average, Exponential Smoothing, Rule of Thumb etc for demand forecasting and inventory management. Although these approaches offered some level of insight, they were not able to consider other factors that influence demand and supply such as changes in the market, changes in the behavior of the customer, and any other shock to the economy such as changes in the political environment or a natural disaster. However, traditional approaches were constrained by the ability to handle massive and diverse data, which are characteristic of the current business world. These have led to some disadvantages like wrong forecasting, over stocking, customer loss, and high operating costs.

1.2 Objective

The aim of this research is to explore how AI-based solutions are disrupting the demand forecasting and inventory management and how businesses can benefit from it. This chapter explores the ability of using machine learning and deep learning models to improve the accuracy of the forecasts by minimizing the time and effort spent on inventory management(Ramesh P et al., 2024). It is especially convenient that AI systems are capable to analyze the large datasets within a short period of time and discover hidden tendencies which are difficult to recognize for a human specialist. These systems have the ability to learn from new data and enhance the predictive and decision making ability of the systems. There are several advantages that have been associated with the use of AI for both demand forecasting and inventory management. The first advantage is cost cutting which is one of the most important factors that are considered when it comes to business. They also reduce the likelihood of stockouts, whereby businesses end up with inadequate stock, which would mean loss of sales and customer dissatisfaction, and overstocking, which puts down capital in unsold products. Also, AI can control the stock levels and help businesses to always stock the right products to avoid high costs of storage and risks of obsolete stocks. The other advantage of using AI in demand forecasting and inventory management is the enhancement of accuracy. All types of AI models especially the ML and DL models are capable of analyzing large volumes of historical sales data, market data, and even social media data to provide very accurate forecasts. This is especially common in industries with fluctuating demand or emerging markets, this includes fashion industries, technology and communication industries among others. Moreover, AI models have the advantage of flexibility to update the forecasts in real-time to help the business adapt to new conditions in the market, customers' preferences, or supply chain disruptions. The third significant advantage, therefore, is the enhancement of decision-making through the use of AI solutions. It is worth acknowledging the fact that AI systems can offer businesses a range of insights that can be valuable in the decision-making process. For instance, the AI algorithms can help identify the possible patterns of demanded products replenishments, detect possible vulnerabilities in the supply chain, and adjust the pricing policies for specific intervals. When implementing the use of AI in a business, it becomes easy to make decisions based on the data collected instead of using guesswork. In addition, it should be noted that the use of AI solutions provides flexibility and adaptability. In addition, the AI systems can be trained again if there are changes in the market and the consumers so that the forecasts and the recommendations can be changed as well. This makes it possible for business entities to be flexible in a dynamic market environment by adapting to changes in demand as well as any other occurrences in the market. Further, AI systems can be easily interfaced with the current ERP & supply chain management systems that organizations use to improve the functioning of the business.

Over the past few years, the use of AI technologies in demand forecasting & inventory management has been on the rise in different industries. Retailers, e-tailers and manufacturers are also using AI for supply chain solutions. The combination of AI with other advanced technologies like IoT and blockchain improves the demand forecasting and inventory management systems to another level of efficiency and transparency. Nevertheless, like in any other method, there are challenges associated with the use of AI solutions in demand forecasting & inventory management. Potential challenges may include the quality of the data required for learning as well as the intricacy of the models that may be hard to implement by non-professionals.

Also, the organisations may encounter some challenges in implementing AI-based systems especially for those industries that have already set their norms and tend to use conventional techniques. However, with the advancement in AI technology and the availability of technology for application in demand forecasting and inventory management, it is believed that more businesses will incorporate the technology to gain competitive advantages. The scope of this chapter is to discuss AI-based solutions in demand forecasting & inventory management, to describe the existing techniques and models, the difficulties faced by the companies, as well as the opportunities for further development. In the case studies and future trends, the chapter will discuss how AI has revolutionalise these key business activities and the supply chain environment.

2. DEMAND FORECASTING

2.1 Traditional Methods of Demand Forecasting

Businesses of all industries have been using demand forecasting for several decades as an essential activity. In the past, the demand forecasting techniques mainly employed statistical models that relied on past data to predict the demand in the future. The traditional techniques that can be applied are moving averages and exponential smoothing. The simplest technique is the moving average technique, in which the demand for the future periods is estimated as the average of the past period's demand over the number of periods that are specified. This method is most suitable when demand is constant and does not fluctuate with the changing of the season. But it is not effective when the demand is unpredictable or has cycles such as trends or seasonal patterns as seen here. Exponential smoothing is another form of the moving average and it is another popular approach to be used; it attributes exponentially decreasing weights to period data. Here, more importance is given to recent data points and with time the earlier data points are gradually eliminated. This method is appropriate for demand patterns that have some level of seasonality or trend. However, exponential smoothing also has its pitfalls and one of them is that it does not take into account any other factors that may affect the demand in the future other than the previous demand. Though these methods of analysis have been useful, they also have their demerits. This is because the models are incapable of modeling complex and non linear forms of interactions between variables. For example, it is difficult to consider some things such as promotion, weather or macroeconomic factors which can significantly affect demand. Further, such models are generally of the reactive type rather than being proactive. They rely on past trends and can therefore be slow to adapt to new trends, market shifts, or any other changes that may occur such as disruption of supply, or adverse economic conditions. When there are many businesses and a large volume of data, simple methods of forecasting are not suitable for capturing all the features of demand(Nagaraj, S., & Kumar, K. V. R 2023).

2.2 AI-Driven Demand Forecasting

It can be therefore described as a revolution in the way that demand is forecasted and managed by businesses. Using AI-based methods, a large volume of data can be analyzed, and relationships in the data can be identified, and predictability can be accurate, precise and up-to-date(Amosu et al 2024). Among the most popular methods used in AI-based demand forecasting, there are ML, DL, and NN. Most of the ML algorithms like regression & decision trees can work with non-linear data and can learn and adapt from the new patterns that are being observed in the data(Boualam et al 2025)(Chakraborty, M., & Parida, P. 2024). These algorithms are the ones that are trained from the previous data and are capable of evolving overtime. For instance, in the case of regression models, the algorithm can determine which variables (products, their attributes, price, and promotional activities) have a great influence on the demand levels. The second is the decision trees which enable one to segment the data depending on certain conditions and come up with the forecasts of different segments of customers or categories of products. Deep learning is a subset of ML that applies a special and more advanced set of algorithms called NN to even enhance demand forecasting by analyzing numerous variables and their interactions in large volumes of data. A well-known deep learning model for demand forecasting includes the use of the LSTM, a type of RNN that is suitable to work with time series. LSTMs are very suitable for products whose demand is influenced by seasonal or cyclic factors, as they can model both short & long-term dependencies and use all of them in making its predictions. There is another deep learning algorithm in demand forecasting known as Prophet, is used for forecasting the time series data with daily or seasonal trends. One of the advantages of using Prophet is that it is able to deal with both missing data and outliers, as well as holidays hence suitable for forecasting demand in different environments(Suresh et al 2024).

However, there are also other models that are still under development which are the integration of the traditional models with AI techniques. For instance, short-term forecasting of stock can be done using exponential mechanism while for long-term trends and arrivals variables that include events and other conditions within the market, then long term machine learning models can be applied(Girija P et al., 2023). These combined models are more useful in demand forecasting than the traditional models and AI models since they combine the best features of both the models(Subramanian et al 2023).

Figure 1. **Demand Forecasting**

2.3 Applications of AI in Demand Forecasting

The use of AI in demand forecasting has emerged as a new way of managing supply chain & inventory in organizations. It is also known as real-time demand forecasting where business organizations can make changes to their forecast depending on the latest input data(Annor Antwi, A., & Al-Dherasi, A. A. M. 2019). This is especially beneficial in the sectors where demand is volatile, for instance, the retail and online businesses. Online data allows the AI system to update its predictions based on the new customer data, external changes or market fluctuations. It is also useful in predicting the demand for such products that are seasonal, promotional offers, and new products in the market(Walter, A et al 2025). For instance, AI can help determine the periods of high demand based on the historical data, and thus help businesses prepare for the high demand periods. Likewise, it can analyze the effect of marketing campaigns or promotional prices, and make relevant changes to the expected sales. In new product, for instance, the existing products or the segments of consumers of existing products can be used to estimate the potential demand that will be associated with the new product before it is out in the market. In addition, AI-based forecasting helps the business to make informed decisions that will help in managing the inventory and the risks of having stock out or overstocking. Forecasting helps in determining the probable demand for a particular product in the future hence helping the business to make the right decision on the supply side. It also serves customers better by availing the products and at the same time avoids situations where the organization has to dispose of products which have not been sold due to various reasons like expiry dates among others.

3. INVENTORY MANAGEMENT

3.1 Traditional Inventory Management Approaches

Inventory management is a very important factor in the SCM because it helps a business to ensure that it has the right stock to meet the consumers' needs without incurring unnecessary costs on excess stocks. This chapter aims to discuss about the traditional methods of inventory control that are Manual Systems, Just-In-Time (JIT) and Economic Order Quantity (EOQ). However, these methods have proven to be effective for businesses but they have also brought about some challenges that affect the implementation of the methods in the current world of business that is characterized by fast and massive production of data.

Figure 2. **Inventory Management**

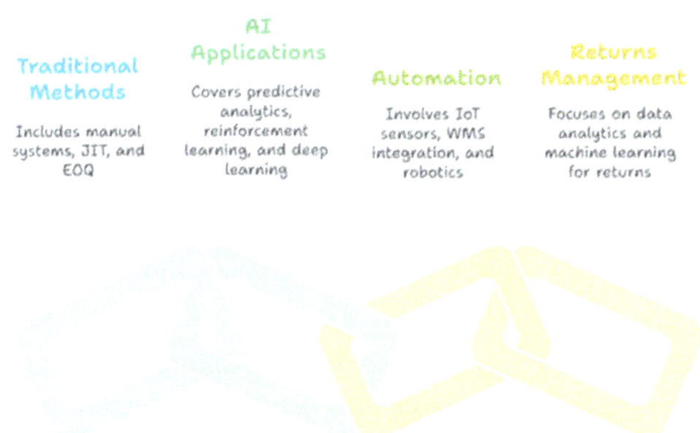

Manual systems of inventory control have been used for years, where inventory level updates are made manually and are usually recorded in a spreadsheet or by counting the stock physically. Although this method is quite easy to implement, it is quite vulnerable to human factors and time-consuming particularly when large volumes are involved. manual systems also do not offer real time visibility hence one cannot be able to know whether to restock or not or even the number of stocks available immediately. The JIT inventory system, which was adopted by Toyota and many other industries, involves receiving inventory when it is required in the production line. JIT reduces storage costs and also eliminates the risk of stockpiling. However, JIT systems are very vulnerable to supply chain disruptions since they rely on a tight schedule and coordination of supply chain partners. This makes it very dangerous to have a single delay in supply because it is likely to lead to production stops and stockouts. The Economic Order Quantity (EOQ) model is one of the traditional models that is used in the determination of the most appropriate order quantity that will help to reduce the total cost of ordering and holding inventory. EOQ is based on certain assumptions such as constant demand rate and fixed ordering costs which are not very realistic in today's unpredictable markets. Nevertheless, the EOQ model does not consider changes in demand or market conditions and results in either stock out or overstocking of the inventory. While the above strategies are quite useful, they have some drawbacks in the modern environment that is characterized by high volatility. Some of the problems that are likely to be encountered include stockouts, overstocking, and lack of real time information on stock positions. The instability in the volume of business activities and capacity of these traditional systems to accommodate increased complexities are some of the major factors that make it necessary to employ more intelligent and improved systems.

3.2 AI in Inventory Management

AI has become influential in the management of inventories through the automation of processes, rationalization of decisions and information delivery. With the help of AI, big data is processed and patterns are recognized which are very difficult to identify using conventional means, thus providing great benefits for the management of inventories to businesses (Singh, N., & Adhikari, D. 2023). Of the various applications of AI in inventory management, the most common one is predictive analytics. To forecast future sales, AI models take into consideration the past sales records, external factors like weather, promotion, and other economic factors and the current market trends. These predictions help the business organization to make better decisions on stocking up inventory as they avoid situations where they have too much or too little stock. Also, predictive analytics can predict the sales effects of campaigns, new products, or seasonal trends and adapt the stock levels. Another machine learning technique which is being applied in inventory optimization is reinforcement learning (RL)(Anusha et al., 2024). In RL, an agent has the capability to learn the decision making process by interacting with the environment and getting the feedback of the action taken by it. In terms of inventory control, RL algorithms can be trained to improve the stock levels by constantly updating the replenishment policies depending on the demand and inventory parameters. RL can be used to estimate the inventory policies that can change with the current conditions in the environment to provide the best stock levels in a dynamically changing environment. Another one of the powerful AI methods, deep learning, is applied for the real-time demand prediction and stock restocking. Neural networks especially the ones that are intended for time series forecasting can capture several patterns in the historical demand data and make accurate predictions of future demand. That is why deep learning can effectively integrate sales data in real-time as well as supply data to ensure that stock is real-time replenished to match the current market demand(Albayrak Ünal et al 2023).

3.3 Applications of AI in Inventory Management

AI has a significant role in inventory management through automation of activities & improvement of the supply chain processes. The latest innovation is the use of AI in tracking of inventories with the help of automation. AI systems can also be applied in tracking inventory by adopting IoT sensors to track the inventory status, location, and more. Sensors connected to the products or placed on the shelves can be able to sense when the product stock is low and therefore initiate the purchase process thus eliminating the chances of stockouts. It is also applied effectively in the operations of Warehouse Management Systems (WMS). This makes it possible for the businesses to automate the process of picking, sorting and packing of products through integration of WMS and AI. Worthy of note is the fact that robotics and autonomous vehicles that are driven by artificial intelligence are today widely adopted in warehouses for the purpose of moving products in an optimal manner hence cutting on the number of cases where humans are needed for movement. These advanced systems of warehouses not only enhance the performance of the warehouse but also help in avoiding human mistakes and decreasing the expenses of workers.

In addition, AI is enhancing the process of returns and reverse logistics which is a challenging and expensive process in managing inventory. Using data analytics and machine learning on returned products, businesses can be in a position to forecast the number of returns and stock appropriately(Siva Subramanian, R et al., 2023. This in a way minimizes the cost of returns and makes sure that there is proper match between stock and demand.

4. INTEGRATION OF DEMAND FORECASTING AND INVENTORY MANAGEMENT

4.1 Benefits of Integration

This is one of the most revolutionary areas where AI technologies are applied integration of demand forecasting & inventory management. Historically, demand forecasting & inventory management were two different functions which had different systems and processes. However, integration of these two areas into one using AI has benefits which are beneficial to businesses as highlighted below. Supply chain management can be easily optimized and costs can be saved through the use of AI in integrating demand forecasting with inventory management. Another advantage of demand forecasting in integration with inventory is the reduction in the lead time. Earlier, companies used static inventory models and fragmented techniques of forecasting, thus leading to long cycle times. AI can also help in managing supply chain through demand forecasting, this means that instead of waiting for long periods of time to restock or overstock products since it will make the required number of products known in real-time. This results in quicker product release, reduced customers' waiting time and an overall improved supply chain responsiveness. The other advantage of integration is that it is able to improve service levels. When demand forecasts are in congruency with inventory control, it becomes easier to stock products or material that will meet the demand of the consumers.

This eliminates the likelihood of stock out and back orders through proper stock management to fit the demand of the customers. This does not only lead to increase in satisfaction among consumers but also ensures that the consumers keep coming back for more. In addition, an integrated system which is driven by AI can learn the changing demands and make sure that the companies are ready to meet the requirements at any given time, be it a result of a promotional event or a high sales season. Demand forecasting is also integrated with inventory management to reduce waste as well. The conventional methods of inventory management are characterized by over stocking that causes congestion of the inventory with items that may take long to sell or may not sell at all. On the other hand, understocking poses severe risks that include loss-making since some stocks do not sell as expected. When demand is forecasted and the inventory level is adjusted, it ensures that the right stock is ordered and available hence minimizing on storage costs. This not only advances the financial aspect of the business but also contributes to the sustainability as it eliminates wastage, products that have expired or are likely to expire or become valueless in the market are sold.

4.2 Techniques for Integrated Optimization

For the purpose of integration, the demand forecasting is integrated with inventory control using AI-driven platforms and advanced models. These systems employ AI and processing of large data to offer real-time data on demand on inventory as well as monitoring of supply chain strategies. Hence, AI models are capable of processing and analyzing large sets of data, the internal data inclusive of the sales history and the external factors, for instance, promotions, weather conditions, and market undertakings so as to arrive at accurate demands. These forecasts are then used directly in the inventory control systems, which are capable of altering the buying pattern, stock holding, and replenishment of the inventories. One of the often-used integrated optimization processes is AI integrated forecasting and inventory management system that are executing in a single system. These platforms use demand forecasting as a way of creating demand, which is then used to inform replenishment of stock. For instance, it can suggest the right quantities and time to replenish the inventory in order to meet the forecasted demand and reduce chances of having no stock or having excess stocks. Further, these integrated platforms can learn from real time data and this can help businesses to change their forecast and inventory management on the go. The ability of AI to learn implies that it can easily adapt to changes in the market demand, consumer preferences or any other factors; this makes the supply chain to be more responsive. Integrated inventory management models that extend into the realms of forecasting, replenishment as well as production planning are also being implemented in the current world. These models try to enhance the SCM by coordinating the demand management with the production and purchasing plans. For example, in demand forecast, AI systems can combine demand and production tools for planning to mean that production lines are configured in readiness for a particular demand without having to formulate an excess of stocks. Likewise, these models can correlate demand with procurement which means that businesses can order raw materials or finished products in a right proportion to the anticipated demand. This leads to the enhancement of the supply chain efficiency by cutting on the costs and time taken to deliver the goods and services while ensuring high service delivery. It is also crucial to note that with the help of RL, AI-based inventory management systems can learn inventory decisions in the long run. In RL, an AI agent works with an environment with which it has to deal with for instance, an inventory system and gets a result that informs its decision making in the future. In SCM this technique can be adopted by training an AI model to have a best knowledge of stock holding's, reorder points and supplier performance. In this way, RL can improve inventory management as it adapts to new demand patterns and disruptions of supply chains by using real-life data.

4.3 Challenges and Limitations

Although the integration of demand forecasting and inventory management through the aid of AI has many advantages, there are also several risks and disadvantages that organizations should consider to achieve these benefits. This is due to challenges such as data quality and issues arising from integration of data from different sources. To be able to make accurate forecasts and to be able to manage inventories efficiently, AI systems need clean and large volumes of data. Nevertheless, there are problems with data in terms of its isolation and poor formats as well as complex IT architectures that hinder the integration of data across functional areas. AI models cannot make intelligent decisions if they are not provided with adequate and detailed data from both demand forecasting & inventory management to yield the optimum results. The other problem is to provide adequate inventory forecasts that reflect realistic values while at the same time being feasible. Although the forecasts generated by AI are usually very accurate, they do not always take into account the actual conditions of working with inventory. For instance, the forecast may indicate that the demand for a particular product will be high, the organisation has various constraints that include space, time it takes to purchase raw materials, and its production capacity. To achieve this, AI models have to be integrated with these operational constraints to enable the supply chain to meet demand as forecasted yet within the operational constraints. This needs complex models that include issues to do with stock management and the available capacity in the warehouses, lead times from the suppliers, and the manufacturing capacity of the company. Another important problem is how to manage change resistance in traditional organizations. Some companies have not yet adopted the AI mechanisms for demand forecasting and inventory management since their current processes are built on legacy systems and paperwork, which can be a hindrance to change; some organizations have worries about either the complexity of the developments or the costs required to implement such technologies; some simply may not know how AI works. Also, the employees may not have confidence in the AI systems over conventional methods, as they may lose their job or decision-making power. To counter this, management needs to provide adequate training, change management and communication initiatives that would educate employees on the benefits of using AI and the fact that it is meant to complement, not replace, human employees in the organization. Nevertheless, the application of demand forecasting & inventory management with the help of AI is a very effective tool that opens up vast opportunities for increasing efficiency, minimizing costs, and improving customer satisfaction. Future advancements in the field of AI and further investments in advanced data management and change management will make it possible for more organizations to obtain near-optimal supply chains further ahead, which will open new opportunities for companies to remain competitive in the constantly evolving market.

5. AI MODELS AND TECHNIQUES FOR OPTIMIZATION

5.1 Machine Learning Models

Demand forecasting & inventory management are two areas where the application of ML is of great importance. Machine learning models are categorized in a way that they can analyze data and make a predictive analysis using data from the past. The two most common categories of ML that can be applied to demand forecasting and inventory optimization are supervised and unsupervised learning. Some of the widely used techniques of demand forecasting models are regression and decision trees which are part of supervised learning algorithms(Indu V T et al., 2026). In supervised learning, based on a definite set of input data like past sales data, prices, features, and the likes, there are corresponding output values that can be predicted by the model. For instance, regression models estimate continuous values such as, the number of sales expected for a certain product in the next one month. However, decision trees offer more interpretability since it splits the data into segments based on the feature values, enabling business organizations to gain insights on factors that affect demand and inventory.

These models are particularly appropriate in the situations where the mapping between the input features and the output predictions is linear in nature and can be easily defined.

In contrast, unsupervised learning algorithms are applied for the identification of structures or patterns in the datasets without having the outcome labels. For instance, clustering can categorize products if they share similarities in terms of selling patterns or seasonality. This can be especially useful for organizations that have hundreds, if not thousands of products which need to be sorted for easier tracking. In this way, similar products are gathered together and make it easier to determine the right inventory replenishment points, as well as the right shelf space. Another common class of machine learning is reinforcement learning (RL) which is also applied in dynamic inventory systems. Unlike supervised and unsupervised learning, RL is a form of learning where an agent operates in an environment and, through experience, aims for some goal. In inventory management, this could be as simple as understanding how to change the order quantities and the stock level to fit into the overall framework of minimizing holding cost, ordering cost, stockout cost, and at the same time maximize the service level. RL is especially valuable in scenarios where demand is uncertain or volatile because it can be applied in real-time and therefore the inventory decisions are constantly being improved.

5.2 Deep Learning and Neural Networks

There is no doubt that demand forecasting and inventory optimization have been significantly boosted by the use DL, which is a sub-field of ML. Deep learning models, for instance, LSTM network is specifically developed to learn temporal dependencies in the data which makes it ideal for time series forecasting, an important part of demand forecasting. LSTM stands for Long Short Term Memory, and it is a kind of RNN to handle long term dependencies in sequences. In the context of demand forecasting, LSTM networks are capable of identifying the past sales data and make a suitable forecast of the future demand. For instance, an LSTM model will be able to capture patterns in sales, seasonality, trends, and other factors that may take months or even years to manifest. This capability enables the business to estimate the demand for its products with more precision, especially when the data has some noise or unevenness. LSTM models have become especially beneficial when it comes to industries that have unpredictable and fluctuating demand, like fashion, electronics, and consumer goods. Another deep learning technique that can be applied for inventory optimization is Convolutional Neural Networks (CNNs), which is mainly used for image recognition but can be used in the context of supply chain and inventory management as well. In inventory management, CNNs can be applied to analyze data coming from smart shelves, barcodes, RFID systems to monitor inventory in real-time. CNNs are capable of learning patterns from the data, for instance, fluctuations in the inventory of products in various regions or stores, and can be used to improve the ways of restocking products. Autoencoders and the GANs have also been used in cases of anomaly detection as well as demand simulation and are quickly becoming popular tools due to the fact that they enable the discovery of unauthorized patterns or supply chain disruptions in business systems. Self-learning is a benefit of deep learning where there is no need to make changes to the forecasting system or inventory system manually. This technique improves the performance of DL algorithms by the use of layers of neural nets in that it can identify abstract features in data and relationships that will not be detected by traditional models thus making it easier for businesses to get accurate and efficient results of forecasting and optimization. However, deep learning models need a large amount of data for training & their explainability is an issue of concern for many organizations that need to understand how and why the AI model made certain decisions.

5.3 Hybrid Models

It is stimulated that in quite a few occasions, integration of AI together with other optimization algorithms can bring about improved investigations as compared to relying on the singular technique. They combine the advantages of using artificial intelligence methods and traditional techniques of optimization – is an effective weapon when facing the inventory and demand forecasting challenges. For instance, two popular types of optimization methods are linear programming and genetic algorithms that have been combined with machine learning. Linear programming is an analytical technique that seeks to achieve a given goal or objective, for instance, minimize inventory cost or maximize service level, within the confines of certain constraints for instance inventory capacity, lead time or production schedule. If applied in conjunction with the use of machine learning models, LP can be used to select the best solution that is informed by machine learning models such as demand forecasting. This means that businesses can use the AI capabilities of the system to solve some of the most difficult supply chain issues while maintaining the ability to control the outcomes. Another type of the inventory optimization problem can be solved with such metaheuristic technique as genetic algorithms which are based on the natural selection strategy. GAs operate by creating many solutions at a time and optimising them in regard to the fitness function.

When it is combined with machine learning models, GAs can be used to optimize the replenishment strategies or production schedule that matches the demand forecast by the AI systems. It means that by incorporating these traditional optimization methods in combination with machine learning, one can enhance the reliability of the decision-making procedures of the enterprises. These models have the benefits of using the AI and optimization with high effectiveness in the multi-variant tasks of demand forecasting and inventory management.

5.4 AI-Driven Analytics Tools

Artificial intelligence-based analytics tools have brought an innovative shift in the decision support systems in forecasting as well as inventory planning. These tools employ big data methods, artificial intelligence, and techniques in the presentation of insights that tend to inform business activities. For instance, there is a possibility to use artificial intelligence analytics tools and systems to collect information from sources like past sales and purchases, the behavior of customers, and general trends in the market and present this information in a form where it can be easily interpreted and analysed in terms of risks, as well as inventory management. One of the most common use of Intelligent auto analysis of resources is in the creation of dynamic cockpits of inventories, sales and supply chain reports. These can be linked to demand forecasting models to enable the users to check the accuracy of the forecasts and the inventory management policies. For instance, a manager could use a dashboard to monitor the current sales and demand forecasts, and consequently, decide on restocking, changing promotion tactics or redistributing stock between stores. Also, the AI decision-support systems can include other optimization methods like predictive maintenance and risk assessment to enhance the effectiveness of the inventory management systems. Predictive analytical tools for instance, borrow the expertise of artificial intelligence to determine when a certain equipment for instance the cloned automobiles; the warehouse robots or the conveyor belt is expected to fail. This makes it possible for companies to have their machines serviced before they break down hence reducing on the idle time. In general, AI-based analytical solutions are critical to supporting management decision-making processes, helping to efficiently manage both the demand and inventory levels. These tools are revolutionizing traditional decision making by giving better, quicker and useful information, thus enhancing efficient and effective flow of supply chain.

6. CHALLENGES IN IMPLEMENTING AI IN DEMAND FORECASTING AND INVENTORY MANAGEMENT

6.1 Data Challenges

Another area of difficulty in using AI for demand forecasting & inventory management is data requirements that have to be accurate, structured, and clean. AI models depend on big data and use them in learning to make accurate predictions concerning a given event or phenomenon. However, it should be noted that clean and structured data is not always easy to obtain, and many organizations encounter rather serious challenges in this regard. Lack of quality data can result in bad forecasts and poor decisions which may reduce the value that can be provided by AI in demand forecasting & inventory management. Data quality is a critical issue since the model's performance is dependent on the quality of the data used in training. This is because inaccurate, incomplete, or outdated data may lead to wrong predictions that can lead to overstocking or understocking of products. For instance, historical sales data that lack some information or have some data entry errors may cause wrong demand forecasts, subsequently affecting the inventory management. Moreover, the data used for creating AI models can be obtained from various sources that include the customers, sales, market trends, and suppliers among others. The collected data must be preprocessed to make them standardized and of high quality. Unclean data implies that there is duplicated, missing or even contradictory information on one system to another hence it hinders the performance of AI algorithms in learning from the data and deriving meaningful insights. Also, the data privacy issues are getting paramount, especially after the GDPR regulation has been implemented in the EU. All organizations must implement the necessary measures to ensure that they are compliant with these regulations especially when it comes to customer data. Privacy challenges are a problem that can hinder the integration of multiple sources of data and limit the ability of an organization to use their data for AI-based demand forecasting & inventory management.

6.2 Computational Complexity

One of the problems of AI application in demand forecasting & inventory management is the computational cost in training and operating the AI models. Deep learning and reinforcement learning, for instance, take a lot of time and computational power to process the large data and perform the calculations for precise forecast and optimization. The problem of scalability is a critical factor when it comes to using AI models in supply chain digitalization, mainly due to the large volumes of real-time data produced by today's supply chains. It means that as the size of the operation increases, the amount of information that has to be analyzed also increases. Conventional forms of AI could potentially underperform in dealing with a great number of transactions as it is the case with customer transactions, unit sales, timely inventory changes, and the likes. This may lead to increased time taken in processing, slow decision making and general inefficiency. To address this issue, organizations require acquiring systems with powerful computations, such as servers or cloud services to support the models' expansion according to the organization's needs. However, training DL models may be expensive especially to the smaller firms or organizations or even when using some complex models. Deep learning models allow the system to build worldly and intricate representations while in actuality, deep learning for NLP entails enormous data and compute resources. This results in long training times, for instance, when training large models and high energy costs which not only are costly but also pose environmental concerns of running such power-hungry models.

Figure 3. **Challenges in Implementing AI in Demand Forecasting and Inventory Management**

6.3 Integration with Existing Systems

Thus, for AI to be effective in demand forecasting & inventory management, it has to be integrated with the current ERP, CRM, and the supply chain management systems. As a result, there are still numerous organizations that still use systems that were not developed to accommodate the current AI technologies. Such systems do not possess the necessary flexibility, scalability, and compatibility needed for integration with the new AI-based systems. However, the problem is in integrating the AI capabilities with these systems that are already in existence. These include data silos, old-fashioned databases, low computational capacity, and incompatible formats of data integration. For example, the demand forecast generated from the real-time sales data using an AI model may need to be integrated with an ERP system that keeps records of stock. If the two are not integrated, the AI model may not be able to retrieve data or it has to transfer data manually, which would lead to time wastage and errors. AI can be integrated with various systems such as supply chain management systems, CRM tools, etc., and the integration process may be a lengthy and expensive affair.

6.4 Human Factors and Adoption

Another more hidden, yet very important, barrier to the adoption of AI in demand forecasting & inventory management is the resistance to change. Employees may also not have faith in the new systems as many organizations have been relying on conventional approaches to demand forecasting & inventory management for a long time. This is because such a transition may be due to ignorance of how artificial intelligence operates, fear of automation, or staying with what has always been productive for a while. Some of the employees may feel that the use of AI is taking away some of the responsibilities that were in their discretion to perform. For instance, inventory managers may be concerned that the development of AI models will lead to automation of decision-making processes thus leading to joblessness for the managers. Such feelings can cause employees to resist the use of AI tools and be skeptical of the results provided by these tools. To this effect, there is the need for organizations to engage the employees in the uptake of AI and assure them that AI is there to assist the employees and not replace them. AI can assist the employees in making better decisions based on data and can also optimize the processes instead of replacing workers. Education of the employees is another crucial factor that is relevant to the successful implementation of artificial intelligence. People have to understand how it operates and how it can be applied at the workplace to increase efficiency. It may mean offering talks or seminars, demonstrations, or actual practice in the utilization of such tools. Organisations should also encourage the employees to embrace lifelong learning and have freedom to test different AI tools and practices in the workplace.

7. FUTURE TRENDS AND EMERGING TECHNOLOGIES

7.1 AI and Blockchain for Supply Chain Optimization

AI when combined with the blockchain technology will help in bringing major improvements in the aspects of demand forecasting & inventory management. Therefore, applying AI in these domains can be complimented by blockchain, a decentralized digital record to offers greater control, provenance, and security to the supply chain activities. Real time tracking of the transactions and movements of the goods is made possible by blockchain which makes it easy to manage inventory and accurately forecast demand. In the context of demand forecasting, the use of blockchain would make it possible to maintain a record of all the orders and shipments made, as well as the current level of inventory. This makes it possible to train the AI models using accurate and current data to avoid cases of errors that could have originated from the discrepancies or fraud. For example, using the blockchain-based data in AI models, the demand forecasts can be more accurate due to the increased certainty that the data cannot be altered, for example, in industries with multiple links in the supply chain or multiple parties involved. The decentralized nature also implies that the information is available to the supplier, retailers, and every other party in the chain in real-time enhancing the cooperation. Additionally, there is an enhanced tracking of inventories through the use of blockchain because it has an open record of products from manufacturers to consumers. It is especially important for fields where the origin of the goods is rather vital like pharmacy or food production. The integration of blockchain data in AI models ensures that they make the right choices on stock replenishment, thus minimizing on overstocking or stockout situations and improving on the stock. Thus, integrating AI and blockchain technologies makes it possible to achieve better efficiency and enhance the level of trust in supply chain management.

7.2 Explainable AI (XAI)

As AI enhances its usage in the demand forecast and inventory management, there is an increasing awareness in the organizations on the need for the operationalizing of AI models. XAI refers to the methods used to increase the level of interpretability of artificial intelligence models that have been developed. When it comes to supply chain management, XAI is instrumental in not only providing correct predictions but also justifying those predictions to the decision-makers in case the decisions are bound to have a major impact in operations. Some of the traditional approaches of artificial intelligence, for example deep learning networks, have a closed form where the decision-making process is not easily understandable by users. However, these models may have high accuracy that is not easy to interpret hence making it hard for the supply chain managers to believe in or even check the results. This is where XAI comes in since it offers explanations on why particular decisions were made, how factors affected predictions as well as the data utilized by the model. This is especially important in industries where there is need to show accountability and compliance to the law. For instance, in demand forecasting, XAI can assist the supply chain managers comprehend why an AI model is forecasting high demand of a certain product in a certain period. If a model recommends increase stock for a certain product, then XAI can explain what factors (historical data, seasonality, promotions or external environment) led to such decision. This makes it possible for the decision makers to certify the recommendation given by the AI, make changes to it and ensure that the forecast made is in line with their business goals.

7.3 Autonomous Supply Chain

The idea of the autonomous supply chain is one of the most potential ideas in the future of the use of AI and inventory. An autonomous supply chain is one that has adopted artificial intelligence and robotics as well as other advanced technologies such as self-driving vehicles, drones among others that function independently of human input. This vision of the supply chain of the future will eliminate most paper and human intervention while making the supply chain to be more of a system that is controlled and operated by artificial intelligence with little or no human interferences in activities such as inventory, orders, and transportation. Robotics, for instance, are being used in various industries particularly in performing simple activities such as sorting, picking as well as packing. These systems when integrated with real-time demand forecasting models will be able to facilitate the right products for the right time without the need for human intervention. Self-driving cars and drones will also be involved in the transportation element of supply chain where transport will be done by machines and not human beings. These vehicles will incorporate artificial intelligence solutions in real time to solve delivery routes depending on the traffic situation, changes in demand or other possible aspects that can cause loss of time and money in the traditional methods of logistics. This is one of the greatest advantages of autonomous supply chains since it eliminates human mistakes and drawbacks. Since the routine activities like inventory monitoring, order processing, and supply chain management can be automated, businesses can get more accuracy and efficiency. This will help in making a smoother and much faster supply chain that is capable to adapt new changes in demand or in supply system. Nevertheless, the implementation of ASC will be a difficult process because it will involve changes in infrastructure and regulation. The major requirement in this aspect will be for the businesses to adopt new technologies like autonomous vehicles, drones, and AI robotics in the supply chain. Furthermore, there will be a need to establish legal and safety standards for the operation of these systems to allow for proper and socially acceptable use of the technologies. However, the idea and opportunity of having AI autonomous supply chain, which is still in its early stages of development, has the capability of transforming industries such as retail, manufacturing and logistics.

7.4 Quantum Computing

Another example of such technologies is quantum computing, which can help have a significant influence on demand forecasting & inventory management in the future. In contrast to the modern conventional computers that use binary digits (0s and 1s), quantum computers work on principles of quantum mechanics to solve problems at speeds and volume that are beyond the ability of the regular computers. Quantum computers are capable of performing optimization problem solutions better than current systems and this means that AI solutions in supply chain management could benefit from it. In demand forecasting, quantum computing could help in the processing of large data in a shorter time and hence enable organizations to make accurate predictions. For instance, quantum algorithms can be applied in modeling and solving supply chain problems by considering a number of factors such as lead time, production capacity, market factors and shocks. This would help businesses to use the inventory in a more efficient way with less holding cost and variability even in conditions of high uncertainty or variability of demand. Quantum computing is also useful in solving complex problems like the stock level that is required in a network of warehouses or distribution centers. Most of the current optimization techniques including linear programming is faced with the challenge of solving large scale problems with many variables and constraints. In contrast, quantum algorithms may be able to solve such problems much more efficiently which will lead to more efficient resources use and cost reduction. As of the moment, quantum computing is still in its infancy, and actual uses for supply chain optimization are not yet fully explored but the possibilities are endless. Quantum computing in the future can offer the computational resources needed to solve some of the most difficult problems of demand forecasting, as well as inventory, which will significantly change the supply chain. Therefore, the future of AI in demand forecasting and inventory management is dependent on the future of these technologies blockchain, Explainable AI, autonomous systems, and quantum computing. They will help companies manage their supply chain in a way that has never been possible before, making the future of business look forward to.

8. CONCLUSION

The advanced technologies in the demand forecasting and inventory management have made it easy for businesses to enhance their supply chain management. Conventional approaches that were estimating the incidents using past data and easy models are slowly giving way to newly developed and highly sophisticated AI, ML, and DL, and Predictive Analytics (PA). These techniques have helped in improving the accuracy of demand forecasts hence helping the business to better respond to such changes in demand. This is because through AI, there will be increased operational efficiency, decreased costs, and better decision making hence improving customers' satisfaction and also optimizing the inventory. AI has also played a great impact in inventory management as well as in other aspects of businesses. The accurate tracking, analysis and automation of stock management enables businessmen to avoid stock out situations while at the same time avoiding congestion of excess stock. AI in connection with IoT and robotics has also improved warehouse management and order fulfilment, which are more intelligent agreements in the supply chain. However, like any innovation, the integration of AI in the aspect of demand forecasting & inventory management has its limitations. Some of the challenges include the quality of the data being fed to an AI model, the compatibility of new AI systems with the already established ones in an environment and computational capabilities. However, there are other issues which are people related such as resistance to change and need for training to address in order to implement. As the next steps, there should be more investigations on how to make AI models more scalable and interpretable so that businesses could benefit from the AI systems without being restricted by data or system constraints. One can presume that further development of innovative technologies, including blockchain, quantum computing, and XAI, will allow for making improvements to the current approach to demand forecasting and inventory management. Further development and investigation in these areas will help address current limitations and bring the best of the ability of AI to the supply chain.

REFERENCES

Albayrak Ünal, Ö., Erkayman, B., & Usanmaz, B. (2023). Applications of artificial intelligence in inventory management: A systematic review of the literature. *Archives of Computational Methods in Engineering*, *30*(4), 2605–2625. DOI: 10.1007/s11831-022-09879-5

Amosu, O. R., Kumar, P., Ogunsuji, Y. M., Oni, S., & Faworaja, O. (2024). AI-driven demand forecasting: Enhancing inventory management and customer satisfaction. *World Journal of Advanced Research and Reviews*, *23*(2), 100–110.

Annor Antwi, A., & Al-Dherasi, A. A. M. (2019). Application of artificial intelligence in forecasting: A systematic review. *Available at SSRN* 3483313.

Anusha, S., Nithyanandhan, R., Anish, T. P., Girija, P., & Nalini, M. (2024, October). Deep Reinforcement Learning: Key Algorithms, Applications, and Future Challenges. In *2024 2nd International Conference on Self Sustainable Artificial Intelligence Systems (ICSSAS)* (pp. 565-571). IEEE.

Boualam, M., El Farouk, I. I., & Jawab, F. (2025). Revolutionizing the Demand-Driven Supply Chain: AI and Machine Learning Applications. In *Supply Chain Transformation Through Generative AI and Machine Learning* (pp. 347-378). IGI Global Scientific Publishing.

Chakraborty, M., & Parida, P. (2024, January). Comprehensive Study of AI-Driven Market Forecasting Models and Their Applicability. In *2024 International Conference on Advancements in Smart, Secure and Intelligent Computing (ASSIC)* (pp. 1-5). IEEE. DOI: 10.1109/ASSIC60049.2024.10508015

Girija, P., Aunradha, M., Dinesh, M. G., Aswini, J., & Divya, P. (2023). Heterogeneous ensemble variable selection to improve customer prediction using naive bayes model. *International Journal on Recent and Innovation Trends in Computing and Communication*, *11*(5s), 64–71.

Indu, V. T., Elumalai, M., & Gracious, L. A. (2026). Optimization Algorithms for Machine Learning Model Enhancement in Healthcare. In *AI Model Design and Data Management for Disease Prediction* (pp. 159–192). IGI Global Scientific Publishing.

Nagaraj, S., & Kumar, K. V. R. (Eds.). (2023). *AI-driven intelligent models for business excellence*. IGI Global.

Ramesh, P., Nithyanandhan, R., Faizal, M. M., Nivakumar, V., & Nalini, M. (2024, December). Machine Learning Strategies for Customer Churn Prediction in Competitive Enterprises. In *2024 International Conference on Sustainable Communication Networks and Application (ICSCNA)* (pp. 653-660). IEEE. DOI: 10.1109/ICSCNA63714.2024.10864143

Singh, N., & Adhikari, D. (2023). AI in inventory management: Applications, Challenges, and opportunities. *International Journal for Research in Applied Science and Engineering Technology*, *11*(11), 2049–2053. DOI: 10.22214/ijraset.2023.57010

Siva Subramanian, R., Maheswari, B., Nikkath Bushra, S., Nirmala, G., & Anita, M. (2023). Enhancing Customer Prediction Using Machine Learning with Feature Selection Approaches. In *Inventive Computation and Information Technologies* [Singapore: Springer Nature Singapore.]. *Proceedings of ICICIT, 2022*, 45–57.

Subramanian, M., Cho, J., Veerappampalayam Easwaramoorthy, S., Murugesan, A., & Chinnasamy, R. (2023). Enhancing sustainable transportation: AI-Driven bike demand forecasting in smart cities. *Sustainability (Basel)*, *15*(18), 13840. DOI: 10.3390/su151813840

Suresh, C., Nyemeesha, V., Prasath, R., Lokeshwaran, K., Raju, K. R., & Boopathi, S. (2024). AI-Driven Energy Forecasting, Optimization, and Demand Side Management for Consumer Engagement. In *Intelligent Solutions for Sustainable Power Grids* (pp. 112–131). IGI Global. DOI: 10.4018/979-8-3693-3735-6.ch006

Walter, A., Ahsan, K., & Rahman, S. (2025). Application of artificial intelligence in demand planning for supply chains: A systematic literature review. *International Journal of Logistics Management*, *36*(3), 672–719. DOI: 10.1108/IJLM-02-2024-0120

Chapter 6
AI and IoT-Enabled Automation for Energy Optimization and Efficiency

Vishnu Gandhi V.
Easwari Engineering College, India

M. Ezhilvendan
https://orcid.org/0009-0003-3213-1304
Panimalar Engineering College, India

ABSTRACT

This chapter explores the integration of Artificial Intelligence (AI) and Internet of Things (IoT) technologies for energy optimization and efficiency across various sectors. It highlights fundamental concepts, key enabling technologies, and applications including smart grids, buildings, industrial automation, and renewable energy management. The synergy between AI's advanced algorithms and IoT's pervasive sensing enables real-time monitoring, predictive analytics, and automated control, driving significant reductions in energy consumption and environmental impact. The chapter also discusses critical challenges such as scalability, data quality, security, and deployment barriers, while outlining promising future trends like explainable AI, federated learning, and 5G connectivity. Emphasizing sustainability and cross-domain integration, this chapter provides a comprehensive overview of how AI and IoT empower smarter, more efficient energy systems for a sustainable future.

DOI: 10.4018/979-8-3373-3176-8.ch006

1. INTRODUCTION

1.1 Background and Motivation

Economic growth, new technology and the development of people have always depended on energy. Currently, energy plays a bigger role than before, especially since we are tackling the needs for economic growth and looking after the environment. Energy consumption on a global scale has gone up significantly in recent years due to increasing cities, more manufacturing and service businesses and the spread of technology that requires a lot of energy(Kavitha et al., 2023). The combination of modern cities, innovative infrastructure and industrial sites makes it necessary to supply a lot of energy for their lighting, heating or cooling, movement, data work and various manufacturing activities. At the same time, much of this energy is lost since there are inefficiencies, old systems and no real-time monitoring. There are additional effects of energy inefficiency, not only financial ones. Overconsumption results in using more fossil fuels, using up valuable resources and emitting a lot of greenhouse gases which are main reasons for climate change. In addition, when energy systems are overloaded during high demand, there may be less reliability, grid issues and power failures. In this situation, minimizing energy use has become a serious issue for the world that affects the economy, the environment and energy security. Therefore, researchers, engineers and policymakers are focusing more on managing energy in an intelligent way. In the past, focusing on energy efficiency required people to watch and record energy use, perform scheduled maintenance and look at consumption after it happened. Even though these approaches give important results, they cannot react fast to current changes or predict the future accurately. For this reason, people are now using smart systems that can adapt energy use with little help from people. Automation is making a big difference in dealing with issues related to energy. With the help of automation, it is possible to keep an eye on energy systems, control them and optimize them which helps reduce wastage, improve performance and allow for better maintenance. Advanced automation systems can change lighting, temperature control, ventilation and air conditioning depending on who is in the building, the weather and the prices of energy. Automated control in industry helps machines to be used more efficiently, less time is spent idle and the workload is distributed properly. Because of these improvements, there is less money spent on operations, machines last for a longer time and the environment is less affected. Also, automation is in line with the increasing focus on environmental sustainability and fighting climate change. Automated systems make it possible to use energy more efficiently and support the goals set by different countries and organizations for better energy efficiency. They also play a role in using renewable energy by helping energy systems adjust faster and, in turn, respond well to changes in solar and wind production. All in all, the study is motivated by the immediate need to find and use modern technologies that can deal with energy inefficiency from many angles. Since the world experiences high demand, less available resources and climate issues, intelligent automation can help create energy systems that are strong, less harmful to the environment and economical. The importance of AI and IoT in this change points out how data-backed, connected technologies can help bring about energy saving and sustainability.

1.2 Overview of AI and IoT Technologies

Because of AI and the IoT, we are now seeing new levels of automation, quick data analysis and smart decision-making. AI and IoT make it possible for smart systems to understand, learn from and deal with different conditions in real time(Kumar et al., 2023; Sathya U et al., 2024). Using these technologies in energy management is very attractive, as they help monitor things closely, analyze future trends and control various energy-using areas automatically(Arumugam et al., 2024). AI means that machines and software can carry out activities such as learning, thinking, spotting patterns and making decisions, just as humans do(Prasad et al., 2023). AI is usually used in the energy industry by applying machine learning (ML), deep learning (DL), reinforcement learning (RL) and expert systems(Ramesh P et al., 2024). They allow energy systems to use old information, spot any issues in their workings, know what to expect in the future and make live adjustments for better results. For instance, with the help of AI, grid operators can look at past usage trends to expect future highs in demand and respond ahead of time(Narayan et al., 2023). Reinforcement learning which focuses on making decisions sequentially, is now used more often to develop better control strategies for energy systems. Such a system enables systems to find the best actions by experimenting, adjusting their controls when new inputs like weather, the number of people or production requirements appear. For this reason, this is helpful in managing HVAC systems, industrial processes and battery storage systems, since conditions in these systems are always changing(Radhamani et al., 2024). On the other hand, IoT includes physical devices that have sensors, actuators and communication tools to collect data from the environment in real time. In an energy system, the IoT devices are like sensors that monitor energy usage, temperature, humidity, performance of equipment and other details. These data streams are moved over communication networks to centralized or edge computers where they are studied and explained. Some examples of IoT in energy management are smart meters that watch household energy consumption, IoT sensors in industries that check for machine problems and weather stations that guide the use of solar panels(Ponmalar A et al., 2024). IoT devices in building management systems can find out if an area is occupied and adjust the lighting or HVAC to cut down on extra energy use. Their true value can be seen when AI and IoT are connected in a single framework. IoT gives high-speed data which is used to track system activities and AI takes care of analyzing this data and spotting any patterns. As a result, this system can receive environmental information, understand it and take actions to control itself. They are able to learn from the past and keep improving their functions as time goes on. Besides, this combination makes it possible to use various vital functions aimed at saving energy. Predictive maintenance uses information from IoT and AI to predict any equipment problems before they take place which cuts down on downtime and makes the equipment more reliable. They can change the amount of energy people use according to what's going on in the grid or pricing which helps make the system more stable and saves consumers money. In addition, AI-IoT technology makes it possible to easily link renewable energy by anticipating changes in availability and controlling how energy is used. AI-IoT systems in distributed energy allow micro-grids to be controlled, energy storage to be improved and communication between producers and consumers to be carried out in a decentralized manner. Due to this, the energy infrastructure becomes stronger and can adjust well to changes in supply, demand and external factors.

All in all, bringing AI and IoT together has greatly changed the way energy systems are built, controlled and looked after. Turning old, hand-controlled systems into smart and self-controlled networks allows these technologies to build the basis for future energy management solutions that are efficient, flexible and eco-friendly(Syamala et al 2023).

1.3 Scope and Objectives of the Survey

The purpose of this survey is to assess in detail how AI and IoT-based automation is changing the way energy is used and managed. Because the need for intelligent and sustainable energy is increasing, we need to learn about the technology, current trends and future possibilities in this area. The goal of this chapter is to join research and practice by bringing together the latest studies, innovations and examples of use. This survey includes many types of AI and IoT technologies and how they are used in energy management systems found in various domains. They consist of, among others, smart buildings, industrial automation, smart grids, transportation systems and managing renewable energy. In all of these industries, the survey looks into how automation can save energy, make operations more efficient and support both the environment and the economy. A main aim of the survey is to outline the several AI approaches, for example, supervised and unsupervised learning, neural networks, genetic algorithms and fuzzy logic, that have tackled problems in the energy field. The techniques are examined in the chapter in terms of their applications for forecasting, optimization, control and fault detection in energy systems. The role of IoT structures such as sensor networks, communication protocols and edge computing, is also studied to see how they make real-time data collection and control possible. It is also necessary to evaluate what benefits and difficulties come with using AI-IoT systems for energy management. Still, the use of these technologies is usually delayed due to data privacy, problems linking different systems, their ability to work on a large scale and cybersecurity. The survey mentions these barriers and suggests approaches to deal with them such as federated learning, blockchain and strong encryption methods. Besides, the chapter points out some new trends and topics that researchers should pay attention to. Such things are needed as standardized ways to combine AI with IoT, better ways to model energy use and algorithms that can adapt to changing and unknown conditions. The survey also investigates how these technologies can help with global sustainability, for example, by cutting carbon emissions, increasing energy for everyone and encouraging the use of green energy. In the end, the goal of this survey is to be useful for researchers, people practicing in industry, system designers and policymakers. To inspire new innovations and good decision-making, the chapter gives a complete picture of technology, its use and the latest research in intelligent energy systems. All in all, this survey marks the beginning of a closer look at the joint efforts of AI, IoT and automation aimed at saving energy. It points out the major changes these technologies can bring to energy production, distribution and use and lays out a plan for making use of them to create a greener future.

2. FUNDAMENTALS AND CONCEPTS

2.1 Energy Optimization and Efficiency: Definitions and Metrics

Sustainable, cost-effective and intelligent energy management depends on energy optimization and efficiency as its main supports. In short, energy optimization means making changes to how systems, devices or processes use energy to get the best balance between their efficiency and energy use. This idea includes several actions, for example, managing schedules, improving equipment and updating designs so that energy is used efficiently and comfort or quality are not affected. When optimizing, the main goal is to find the best energy use that matches economic, environmental and practical needs in different locations. On the other hand, measuring energy efficiency means seeing how much input energy is changed into useful energy. Usually, efficiency is measured by the ratio of useful energy to the total energy used. A highly efficient system can help cut energy use and cost while still giving the same or better service which is good for the environment. As an example, an energy-efficient motor would turn most of the electricity into motion with only a little heat or noise produced. It is necessary to use defined metrics and indicators to assess and measure energy performance. Total energy consumption which is often shown in kilowatt-hours (kWh) or megajoules (MJ), is the simplest way to measure how much energy a country uses during a certain period. Even so, the total amount of energy used does not show if the energy is used efficiently. This is the reason why energy efficiency ratios (EERs) and similar indices matter a lot. EUI is usually used to measure energy use in buildings. It stands for how much energy is used up in a certain space, usually shown as kWh per square meter per year. Less energy used shows that the building is more energy efficient. In industrial processes, people often use Specific Energy Consumption (SEC) which means the amount of energy required to make a unit of product such as MJ per kilogram. SEC helps companies review their energy performance by comparing it to the best standards in the industry. We can also look at carbon footprint which measures the effects of energy use on the environment by working out the quantity of greenhouse gases, mainly CO_2. By measuring their carbon footprint, organizations and governments work towards climate targets and follow the rules. Electricity produced from coal results in a bigger carbon footprint than electricity made from renewable sources. In environments where things are always changing such as smart buildings or industrial systems, it is important to have real-time performance indicators. Some of these are load factor which measures the ratio of average to highest energy use and demand response indicators which check the ability of systems to react to changes in energy supply or price. PUE is used in data centers to determine how much energy is consumed by the computing part of the center as opposed to the rest of the infrastructure. A better efficiency rating is shown by a PUE value that is as close to 1.0 as possible. All these metrics, when combined, offer a clear way to check and boost the performance of energy systems. They help those in charge find ways to save energy, decide which areas to focus on and check their achievements over the years. In addition, using these metrics makes it easier to bring AI and IoT into energy systems, as they give the data needed to train AI, set goals for control and check the results of automation. Apart from measuring, companies are now focusing on making energy use more efficient to become both resilient and sustainable. Such systems are able to continue working well even with peak demand, problems with equipment or changes in renewable energy. By having efficient systems, businesses can cut their operating fees and reduce stress on the environment. For these reasons, energy efficiency is often called "the first fuel," since it is always available, inexpensive and friendly to the environment. Thus, more and more companies and governments are using AI and IoT to take energy management from reacting to events to planning and managing it automatically.

2.2 AI Techniques for Energy Management

AI has become a key player in today's energy systems by learning from information, noticing inefficiencies, anticipating future situations and managing complicated systems on its own. As more data becomes available in energy systems, AI becomes very important for making energy management more intelligent, adaptable and automatic. With the help of AI, tools and methods have been made to manage energy use in various sectors, working well for both forecasting, control, finding faults and designing energy systems(Boopathi, S., & Kanike, U. K 2023). Machine Learning (ML) is used very often in AI applications for energy management. ML models are made to find patterns in old data and use those patterns to decide or predict outcomes. In energy applications, people often rely on linear regression, decision trees, support vector machines and neural networks to forecast demand, identify problems with equipment or sort out various energy consumption patterns. When trained on previous energy use, a supervised learning system can forecast the power requirements of a building based on the weather, how many people are there or the time of day. That allows for timely changes that save costs and keep energy demand under control(Balaji et al., 2024). With unsupervised learning, there are no labels on the data and it helps to spot hidden problems or unusual activities in energy systems. Using clustering algorithms, you can group similar behaviors, spot equipment that is not working well and separate users into groups for individual energy services. Such insights are very useful for managers of utilities and buildings, as they guide them in choosing energy-saving measures that fit consumers' habits and the equipment they use. Another important type of AI is Deep Learning (DL), since it can use networks with multiple layers to model complex relationships. Deep learning is especially good at working with large amounts of data and is used in predicting energy levels, spotting faults in energy systems and checking energy facilities with drone-taken images. An example is that CNNs can analyze thermal images of solar panels to spot any issues and RNNs are helpful for forecasting the future electricity demand in a reliable manner. Because RL works best in changing and step-by-step decision-making settings, it is very effective in controlling energy in real time. An agent in RL tries different actions, receives feedback as rewards or penalties and learns which actions are best. This way of controlling is best for HVAC systems, energy in microgrids, charging/discharging batteries and control operations that face continuous and unpredictable changes. Because of RL, an RL-based controller can control building temperature in a way that maximizes comfort for the people inside and, at the same time, saves energy, adapting as needed as people's routines change. Apart from standard AI, fuzzy logic is used in energy management when there is some uncertainty, ambiguity or imprecision. Fuzzy systems work like the human mind by using maybe instead of only yes or no. This is why they are used for controlling heating and cooling systems, lighting and power grids, since the conditions might not be clear-cut. So, a fuzzy controller alters the fan's operation depending on conditions other than temperature, for example, when it feels moderately warm or slightly humid. Algorithms used for optimization are especially important for solving issues that have several goals, rules and factors to deal with. These algorithms support the identification of the right configurations for energy scheduling, load distribution, network planning and resource allocation. These techniques assist in industries by figuring out the best order of tasks and selecting suitable machines so that energy is used more efficiently and power bills are lowered.

AI methods are not used by themselves anymore; instead, they are being combined with other approaches like learning, reasoning and optimization. As a result, an energy management system can forecast demand, quickly change control approaches and optimize schedules. They make it possible to have strong, adaptable and flexible energy solutions for many different conditions. One of the most important things for AI in energy is access to good and sufficient data. Models need data from IoT sensors, smart meters and old records to become accurate and reliable. It is important to focus on how clear, reliable and fair AI solutions are, mainly when they are involved in decisions that matter for infrastructure or customers. All in all, AI helps in managing the challenges of energy that are constantly changing and vary in complexity. With AI, it is now possible to use data, predictions and automated systems which help make energy strategies more proactive. When they are added to smart grids, buildings, industrial systems and renewable energy platforms, they can help us save energy, money and protect the environment(Dayal et al., 2024).

2.3 IoT Architecture and Components in Energy Systems

The Internet of Things is what makes automated energy management possible through physical and communication components. The main part of IoT are sensors and actuators that are set up on energy assets. All the time, sensors check temperature, humidity, power use, voltage and equipment status and transfer the results as real-time data. Within a smart home, actuators are told by the control system to change the environment for instance, dim the lights, change the temperature or set the motor's speed. They connect with each other via different protocols and Wi-Fi, Zigbee, Bluetooth Low Energy and LPWAN are a few that are selected for their range, bandwidth, power requirements and the place they are used. Most of the time, IoT data goes through nested systems that use edge computing to handle the first analysis and filtering which lowers the demand for both latency and bandwidth. Whereas the edge focuses on local processing, fog computing adds more computing resources close to the network and cloud computing provides extra storage and advanced processing for detailed analysis and storing data for a long time. All these elements make energy systems responsive and allow AI-based decision-making and automation.

2.4 Automation Systems for Energy Control

With AI and IoT data, automation systems give control commands to maximize the use of energy. Examples of these systems are simple programmable logic controllers in machines, advanced building systems and control platforms for the smart grid. Some automation is guided by set rules and others are flexible, where AI learns from data and changes the controls accordingly. Automation in energy efficiency matters a lot as it allows immediate and exact solutions to shifts in the number of people, weather or how equipment functions. For instance, an automated HVAC system can control heating or cooling depending on the people in the area which helps save energy. In industries, automation helps to organize work schedules and machine operations to reduce the highest levels of energy demand. Because of continuous surveillance, intelligent decisions and self-control, automation systems decrease the need for human involvement and save a lot of energy in different sectors.

3. AI AND IOT-ENABLED ENERGY OPTIMIZATION APPLICATIONS

3.1 Smart Grids and Energy Distribution

Smart grids are a major change from traditional power grids because they use AI and IoT to boost energy distribution, dependability, and efficiency. AI is mainly applied in demand response, which uses real-time data taken from IoT sensors in residential, commercial, and industrial areas. Such systems help predict when there will be the greatest demand and adjust the supply of energy or motivate users to use less at those times, so the load is balanced and the grid does not overload. Load forecasting is also aided by machine learning models and it enables utilities to make optimal decisions regarding electricity generation and reduce their costs. The assistance of smart meters, predictions, and previous energy history, AI-based load prediction stabilization maintains the grid and conserves energy. The uniformity of the grid is improved by the continuous surveillance provided by the IoT devices in its various components. These kinds of sensors can sense any difference or variation in voltage and frequency that occurs in real time. Based on the gathered data, AI models can forecast the possibility of problems and address these problems automatically, rerouting or turning on backup resources. As a result, AI and IoT improve the strength, efficiency, and capacity of smart grids to work with different sources of energy.

3.2 Smart Buildings and Homes

AI and IoT cooperate in smart buildings and homes to ensure energy use is efficient and people are still comfortable. Since heating, ventilation, and air conditioning (HVAC) systems use a lot of energy, smart control of them is very important. Sensors in the IoT check the temperature, humidity, presence of people, and air quality and give this information to AI, which adjusts HVAC settings according to what is happening in real time, the weather, and how the building is being used. As a result, you avoid wasting energy and can save a lot on your utility bills. Automation also improves how lighting is controlled in various areas. Sensors notice if a room is occupied and how much natural light is present, so AI can control the artificial lights and turn them down or off if no one is there. As a result of this automation, electricity is saved and occupants feel more comfortable due to the right amount of light. In addition, noticing people in the building with motion sensors, cameras, or Wi-Fi gives the management system a clear picture of energy usage at any given time. Such devices and appliances help save energy by adjusting their schedules based on people's habits and how much power is available.

3.3 Industrial Automation for Energy Efficiency

It is well known that industries use a lot of energy, and automation greatly helps in making manufacturing and processing plants more efficient. The use of AI and IoT data allows processes to be optimized since equipment and environmental data are monitored all the time and changes are made to operations on the spot. As an illustration, AI can manage the speeds of motors, the temperature, and the materials used so that less energy is consumed without affecting the quality of the product. Predictive maintenance is very important, and in this case, IoT sensors keep an eye on vibration, temperature, and any wear in the machines. AI is used to examine the data and predict when equipment might fail, so maintenance can be done ahead of time to avoid losing energy from machines that are not working properly. Also, energy-aware scheduling uses AI to organize production when energy prices are low or when there is an abundance of renewable energy. This way, both costs and the load on the grid during peak hours are lowered, which improves the whole energy system.

3.4 Renewable Energy Management

Adding solar and wind energy to the power grid is difficult because they are not always available. Because of AI and IoT, it is easier to manage renewable energy since predictions of energy output are based on weather information and past trends. Because of this, those who manage the grid can handle changes in supply and demand more easily. Storing energy in batteries is very important in dealing with renewable energy. The algorithm establishes charging and discharging plans based on the forecasts of the energy consumption, thus even greater energy can be saved with less fossil-fuel energy being consumed. The IoT devices check the state of the storage and the system at regular intervals to ensure that it is reliable. Due to these methods, renewables can easier enter the energy supply, the grid can be more stable, and power systems can be more sustainable.

3.5 Transportation and Electric Vehicles

The introduction of electric cars and technologies that communicate cars and other vehicles changes the transportation industry significantly. The use of AI and IoT applications in this area is important to optimize energy. To assist EVs, such algorithms examine road concerns, traffic, and energy consumption and suggest more energy-efficient and shorter routes. The development of the way charging stations are constructed and operated is also of high importance. IoT sensors monitor chargers and their usage and power consumption at the grid, and AI models analyze it to predict the future demand and establish the most suitable time to charge. Availability of renewable energy is also considered in such strategies and therefore charging rates are adjusted to maximize them. With the help of AI and IoT, V2G technologies enable EVs to provide additional energy to the grid at the time when it is needed the most, enhancing the flexibility of the grid and making energy use more effective.

4. KEY ENABLING TECHNOLOGIES AND FRAMEWORKS

4.1 AI Algorithms and Models for Energy Optimization

The core technology of intelligent energy optimization is made of AI algorithms. Among the key categories are supervised learning, unsupervised learning and reinforcement learning and most of them have various roles to play as far as managing energy is concerned. Training models with data sets containing labels that are regarded as the correct ones is the process of supervised learning. They play a role in energy systems by forecasting energy consumption, identify any problems and categorize equipment by using past sensor data. By doing so, regression models are applied to predict the consumption of electricity, whereas classification algorithms are applied to identify unusual events. It is a good and correct method but only when you possess a sufficient amount of annotated data. In contrast, unsupervised learning examines unlabeled data to determine potential patterns or structures which are not specified in advance. Clustering can be used to sort out similar patterns of energy consumption, and it can also be used to identify anything out of the ordinary, with the dimensionality reduction used to identify significant features in large datasets. Using these approaches, it is more straightforward to identify abnormal behavior and categorize users or devices based on their use, which can be used to tailor energy-saving approaches. RL makes it possible for systems to learn the best way to control themselves by trying different approaches and seeing the results. In contrast to supervised models, RL does not need labels but gains experience by reacting to the results of its actions. This is why it is well-suited for handling tasks that need to be handled in real time, such as managing HVAC systems or meeting demand response, as the environment keeps changing and several factors, like comfort, budget, and efficiency, must be considered. Apart from this, other AI tools including fuzzy logic and optimization methods (genetic algorithms and particle swarm optimization) are popular for dealing with uncertainty or hard scheduling and resource allocation problems. Because of these algorithms, automation adjusts how energy is used, doing so in a timely and efficient manner.

4.2 IoT Platforms and Middleware

It is important that AI-driven energy optimization depends on strong IoT systems and software to smoothly gather, send, and integrate data. These platforms organize the way thousands of sensors and actuators are set up and how data is transferred in the energy sector. At this stage, IoT devices keep collecting multiple parameters, including temperature, voltage, occupancy, and equipment status, which gives AI systems the data they require. Ensuring that data is captured accurately and reliably is important, and usually this is done by edge devices that process data to make it more manageable and quicker. Connecting these devices to each other depends on using certain communication standards and protocols. Using Wi-Fi, Zigbee, Bluetooth Low Energy, and Low-Power Wide-Area Networks (LPWAN), it is possible to get scalable, low-power, and long-range connectivity for each deployment type. When different devices and systems follow the same standards, it becomes easy for them to connect and communicate with each other. They hide the differences between devices and communication, giving programmers a standard way to work with data and tools for management. Middleware technology assists in discovering devices, securing data, dealing with failures, and handling a lot of data. Bringing together IoT platforms and middleware supports the easy flow of data from sensors to AI, making it possible for energy automation to work well.

4.3 Edge and Cloud Computing in Energy Systems

Since energy systems are getting larger and more complicated, a hybrid method of computing using edge and cloud technology is needed to ensure they perform well and are scalable(Maheshwaran, T et al., 2024). Edge computing helps by processing data next to IoT devices so that instant decisions can be made with very little delay. On the edge, these nodes can check and remove unnecessary data, find unusual events, and take control measures without needing to contact remote servers. Thanks to this, bandwidth is saved, replies are faster, and the system is more dependable, most importantly for quick tasks like detecting faults or responding to high demand. At the same time, cloud computing enables users to use a lot of computational resources and storage for detailed analytics, long-term data saving, and training AI models. Cloud platforms allow the gathering of a lot of energy data, make predictions easier, and help optimize different sectors. They offer support for user interfaces, remote control, and integration with different third-party services. Through the combination of edge and cloud computing, energy systems are able to respond quickly and at the same time use extensive intelligence for better energy management in many different situations(Anusha, S et al., 2024).

4.4 Cybersecurity and Privacy Considerations

Because AI and IoT are being used more in energy systems, security and privacy are now very important. A disruption in energy systems can cause economic issues, possible hazards, or a break in the service that people rely on. As IoT devices are not always powerful or secure, they provide chances for cyber threats to take place, such as unauthorized entry, tampering with data, or denying the system its service. You should use several security measures, for example, authentication of devices, encryption of communication, secure booting, and updated firmware. Using AI, cybersecurity can be aided by finding signs of intrusion and unusual activities. In addition, keeping data privacy is very important, as energy consumption data may disclose private details about users' habits or how often they are at home. Companies are more and more required to design their solutions with privacy in mind and to handle data openly. People are studying data anonymization, differential privacy, and federated learning to help solve privacy issues. It is essential to have solid cybersecurity and privacy to earn users' trust and protect the reliability and security of AI and IoT-based energy optimization systems.

5. CHALLENGES AND OPEN ISSUES

5.1 Scalability and Interoperability

A major difficulty in this field is making sure AI and IoT can be used at a large scale. Growing and more complex systems of energy find it hard to support the increasing number of linked devices, sensors, and control points. It is important for systems to handle a huge amount of data from different areas without affecting their performance or dependability. Therefore, it calls for architectures that are able to process data fast and transfer it efficiently over networks made up of various types of devices. Another issue that is related is interoperability. In an energy ecosystem, devices and platforms made by different producers usually apply unique ways to communicate, format data, and control things. As there is no standardization, it becomes hard to join systems and share information or responses automatically. Using the same standards and middleware hides differences between hardware and software, letting systems communicate and making it easier to develop and add more features.

5.2 Data Quality and Management

The performance of AI algorithms increases with the better and larger amount of data from IoT devices. Still, many real-world deployments deal with problems such as missing data, sensor errors, noise, and irregular data formats. If data is not of high quality, it may result in wrong predictions, less effective decisions, and fewer energy savings. Also, managing data raises many difficulties. Because IoT networks provide a huge amount of data all the time, it is important to use efficient techniques for collecting, cleaning, storing, and retrieving the data. Ensuring that the data is safe and available at all times for both current and past use is a very difficult task. In addition, it is necessary to unify and check the meaning of data from various sources to generate useful insights using AI.

5.3 Real-time Processing and Latency

Energy systems are designed to process information in real time or close to real time so they can react fast to changes in demand, problems, or the amount of energy coming from renewable sources. Lowering latency in data handling, moving data, and making decisions by AI is very tricky, mainly in deployments that cover many places or a large area. It is important to split the work between edge and cloud devices to decrease the time for tasks. Edge computing is able to manage urgent jobs locally, but its processing power is not very high. On the other hand, cloud computing gives access to many resources, though it experiences delays because of the network. Allocating different parts of a task to the right parts of a hybrid architecture is still a tough task to meet tight real-time demands.

Figure 1. **Challenges and Open Issues**

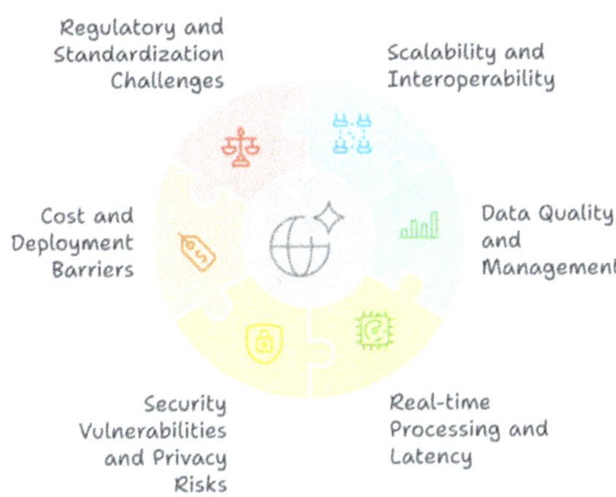

5.4 Security Vulnerabilities and Privacy Risks

AI and IoT have made it possible for threats to security to grow. Since IoT devices are not very secure, they can be easily targeted by cyberattacks that allow hackers to get in, copy data, or control the devices. A breakdown in the energy system may cause services to stop, financial problems, or endanger the public. There is a privacy concern since energy information can disclose details about a user's habits at home and when they are around. If an organization's data is used without permission, it may cause both ethical and legal issues. These risks can only be handled by using cybersecurity methods that rely on encryption, authentication, detecting intrusions, and safeguarding privacy. Even so, it is still very difficult to set up strong security features without affecting the system's performance or how easily users can use it.

5.5 Cost and Deployment Barriers

It can cost a lot to set up AI and IoT-based systems for energy optimization. Examples are the expenses needed for sensors, communication systems, computer equipment, and bringing all the parts together. For most small and medium enterprises as well as residential users, the high cost of these investments may not be justified unless they see fast financial rewards. Besides, the fact that these systems are complex and require specialized skills makes it hard to deploy them. The use of different technologies, setting up AI models, and managing the systems cannot always be done by just anyone, since these tasks usually require skilled workers. Overcoming the related costs and operations is important for more people to take advantage of energy-efficient automation.

5.6 Regulatory and Standardization Challenges

There are many obstacles in the way of AI and IoT development in energy systems due to rules and standards. Introducing new technology in the energy sectors is restricted by following strict rules for safety, compatibility, and data protection. In a number of places, the regulations have not yet covered everything that is needed for AI-driven automation and IoT connectivity. Since there are no worldwide standards for communication, data exchange, and security, progress and deployment are delayed. Besides, the rules in each country are not always the same, which adds difficulty to creating solutions that can be used worldwide. It is still a major and unresolved issue to set standards and policies that ensure innovation and also protect users and the infrastructure.

6. FUTURE TRENDS AND RESEARCH DIRECTIONS

6.1 Integration of AI Advancements

AI and IoT are set to shape energy optimization in the future when they add more transparency, privacy, and learning together. One area that is gaining attention is explainable AI (XAI), as it tries to make AI decisions clear for human users. Since energy systems are becoming more independent and sophisticated, people involved need to trust actions taken by AI. XAI will provide a clear view of the energy optimization process, which will support better verification, debugging, and following regulations. Federated learning is another significant advance, since it enables AI models to be developed by several connected IoT devices or edge nodes, without them sharing their individual data. Thanks to this method, sensitive information is stored locally and people still gain from learning that happens across the network. Since federated learning cuts down on communication and makes data safer, it is perfect for energy systems that require cooperation among many parties and regions.

Besides, researchers are working on better reinforcement learning algorithms and AI models that mix symbolic reasoning with data learning, which may improve the flexibility and efficiency of systems in dynamic energy conditions. They will make energy automation more intelligent and flexible, so it can continue to work even when things are uncertain and conditions keep changing.

6.2 Advances in IoT Hardware and Communication

The fast development of IoT hardware and communication technologies is helping energy optimization systems become more advanced. Because 5G networks are being deployed everywhere, they now offer much faster speed, more bandwidth, and can connect many devices, all of which will improve real-time data handling in the energy sector. It allows better and quicker communication among sensors, edge devices, and cloud platforms, so energy can be controlled more precisely and quickly. At the same time, LoRaWAN and NB-IoT, known as LPWAN, are making it easier for IoT devices used in energy systems to reach more places and last longer on batteries, especially in places that are far from the network or used in big industries. With these technologies, monitoring solar panels, wind turbines, and smart meters takes less time and cost, since they require very little upkeep. With better sensors, devices are now more precise, more energy-saving, and can serve several needs. With solar-powered sensors, energy harvesting makes it possible to reduce the number of batteries used in IoT setups. These changes will help AI and IoT better manage and improve the use of energy in many places.

6.3 Cross-Domain Energy Optimization

A valuable future development is coming up with methods that manage energy across different areas such as smart cities, industries, agriculture, and transportation in a unified way. It is important to look at energy optimization across the whole company, since different parts of the business are interconnected. This means in smart cities, energy systems need to link buildings, public transport, lights on the streets, and renewable resources to reach their sustainability goals. Such AI models are able to simultaneously manage different objectives, such as efficiency, expenses, emissions, and what users feel comfortable with. In the same way, agricultural activities can gain from energy management that helps optimize water pumps, control climate in greenhouses, and make use of renewable energy. Using data and controlling different domains together, cross-domain optimization helps save more energy, boost productivity, and protect the environment. Creating systems and AI solutions that work across different industries is a big research task, but it could greatly help people and the environment.

6.4 Sustainability and Circular Economy Considerations

Future studies in energy optimization will more and more center on following principles of sustainability and the circular economy. Also, systems will work to lessen carbon emissions, encourage the use of renewable energy, and support effective use of resources in managing equipment.

AI and IoT make it possible to predict maintenance and watch over assets, which saves energy infrastructure and uses less waste. It is also possible for them to use recycled resources more efficiently and back energy recovery initiatives. Besides, sustainability-based policies will require companies to be open and responsible in their energy and emission practices. AI will supply detailed environmental studies that will enable organizations and consumers to decide wisely. In order to bring circular economy into AI and IoT systems, it will be important to develop innovative algorithms that take into account reusing resources, recycling them, and the energy inside materials, so that materials and energy do not go to waste.

7. CONCLUSION

This study examined how AI and IoT are key to making energy use and efficiency better in different industries. The study points out that AI and IoT working together make it possible to save energy, boost the efficiency of operations, and promote sustainable use of resources. Using AI and IoT, it is possible to closely monitor, manage, and decide on energy use in real time to help reduce wastage and lessen the effects on the environment. Such technologies make a big difference, helping to solve the problems of more people needing energy and climate change. With both AI algorithms and IoT platforms, it is possible to do predictive analytics, detect anomalies, and apply optimized control methods, as the platforms are always acquiring and sharing data. Because of them, energy systems can react better, withstand shocks, and focus on users. Still, there are several problems, such as being able to handle large data, checking data accuracy, ensuring security, and paying for deployment, that need more study and new solutions. In the near future, explainable AI, federated learning, 5G, and optimizing energy in several sectors are expected to make AI and IoT in energy management more effective and popular. To get long-term positive results for the environment and the economy, it will be necessary to include sustainability and circular economy ideas in these frameworks. All things considered, the introduction of AI and IoT automation is a key step toward saving energy. Progress in different fields and technology will play a key role in making the world's energy systems greener and more sustainable.

REFERENCES

Anusha, S., Ezhilvandan, M., Gayathri, S., Sivaraman, V., & TP, A. (2024). Emerging trends in edge AI for Industry 4.0 and 5.0: Technologies, applications, and challenges. *AI Innovations in Service and Tourism Marketing*, 211-228.

Arumugam, T., Arun, R., Natarajan, S., Thoti, K. K., Shanthi, P., & Kommuri, U. K. (2024). Unlocking the power of artificial intelligence and machine learning in transforming marketing as we know it. In *Data-Driven Intelligent Business Sustainability* (pp. 60–74). IGI Global Scientific Publishing.

Balaji, S., Bharathi, S. P., Balaji, T. S., Hariharan, R., & Subin, P. G. (2024, November). Enhancing Predictive Accuracy with Machine Learning and IoT Integration in Energy Management. In *2024 Second International Conference Computational and Characterization Techniques in Engineering & Sciences (IC3TES)* (pp. 1-5). IEEE. DOI: 10.1109/IC3TES62412.2024.10877551

Boopathi, S., & Kanike, U. K. (2023). Applications of artificial intelligent and machine learning techniques in image processing. In *Handbook of Research on Thrust Technologies' Effect on Image Processing* (pp. 151–173). IGI Global. DOI: 10.4018/978-1-6684-8618-4.ch010

Dayal, G., Verma, P., & Sehgal, S. (2024). A comprehensive review on the integration of artificial intelligence in the field of education. *Leveraging AI and Emotional Intelligence in Contemporary Business Organizations*, 331-349.

Kavitha, C. R., Varalatchoumy, M., Mithuna, H. R., Bharathi, K., Geethalakshmi, N. M., & Boopathi, S. (2023). Energy Monitoring and Control in the Smart Grid: Integrated Intelligent IoT and ANFIS. In *Applications of Synthetic Biology in Health, Energy, and Environment* (pp. 290-316). IGI Global.

Kumar, M., Kumar, K., Sasikala, P., Sampath, B., Gopi, B., & Sundaram, S. (2023). Sustainable Green Energy Generation From Waste Water: IoT and ML Integration. In *Sustainable Science and Intelligent Technologies for Societal Development* (pp. 440-463). IGI Global.

Maheshwaran, T., Muthumarilakshmi, S., Suganya, K., Maheswari, B., & Girija, P. (2024). Securing E-Commerce Strategies With Cloud, Blockchain, AI, and ML. In Strategies for E-Commerce Data Security: Cloud, Blockchain, AI, and Machine Learning (pp. 470-500). IGI Global.

Narayan, Y., Selvaraj, S., Suryawanshi, M. A., & Ramya, S. (2023, October). Machine Learning Model used Efficiently to Manage the Smart Grid and Renewable Energy. In *2023 International Conference on New Frontiers in Communication, Automation, Management and Security (ICCAMS)* (Vol. 1, pp. 1-3). IEEE. DOI: 10.1109/ICCAMS60113.2023.10526077

Ponmalar, A., Vijayakumar, K., Lakshmipriya, C., Karthikeyan, M., Gracelin Sheena, B., Beslin Pajila, P. J., & Siva Subramanian, R. (2024). Meta-Heuristics and Machine Learning Applications in Complex Systems. In *Metaheuristic and Machine Learning Optimization Strategies for Complex Systems* (pp. 257–275). IGI Global Scientific Publishing.

Prasad, A., Kumar, A. S., Sharma, P., & Irawati, I. D. DV, C., Musirin, I. B., & Abdullah, H. M. A. (2023). Artificial intelligence in computer science: An overview of current trends and future directions. *Advances in artificial and human intelligence in the modern era*, 43-60.

Radhamani, R., Karthick, S., Kumar, S. K., & Gokulraj, M. (2024, March). Deployment of an IoT-Integrated Home Energy Management System Employing Deep Reinforcement Learning. In *2024 2nd International Conference on Artificial Intelligence and Machine Learning Applications Theme: Healthcare and Internet of Things (AIMLA)* (pp. 1-4). IEEE. DOI: 10.1109/AIMLA59606.2024.10531519

Ramesh, P., Nithyanandhan, R., Faizal, M. M., Nivakumar, V., & Nalini, M. (2024, December). Machine Learning Strategies for Customer Churn Prediction in Competitive Enterprises. In *2024 International Conference on Sustainable Communication Networks and Application (ICSCNA)* (pp. 653-660). IEEE. DOI: 10.1109/ICSCNA63714.2024.10864143

Sathya, U., Hashini, S., Lakshmipriya, C., Varun, C. M., & Nalini, M. (2024, December). Emerging Trends and Future Prospects in Internet of Things (IoT) Technology. In *2024 International Conference on Sustainable Communication Networks and Application (ICSCNA)* (pp. 79-85). IEEE.

Syamala, M., Komala, C. R., Pramila, P. V., Dash, S., Meenakshi, S., & Boopathi, S. (2023). Machine learning-integrated IoT-based smart home energy management system. In *Handbook of Research on Deep Learning Techniques for Cloud-Based Industrial IoT* (pp. 219–235). IGI Global. DOI: 10.4018/978-1-6684-8098-4.ch013

Chapter 7
Smart Warehousing Technologies for Energy Efficiency:
AI-Driven Optimization and Sustainability

Beslin Pajila
https://orcid.org/0000-0002-3905-2460
Anna University, India

ABSTRACT

This chapter explores the role of Artificial Intelligence (AI) in enhancing energy efficiency and sustainability in smart warehousing. AI technologies such as predictive maintenance, smart inventory management, dynamic load balancing, and automation have been shown to significantly reduce energy consumption, optimize warehouse operations, and minimize environmental impacts. AI's integration with renewable energy sources, like solar and wind, further supports sustainability goals by optimizing energy usage and reducing reliance on non renewable power. The future potential of AI in revolutionizing warehouse energy efficiency is immense, with advancements in deep learning, edge computing, and real-time analytics. This chapter also highlights areas for future research, particularly in renewable energy integration, AI algorithm development, and scalability for small and medium-sized enterprises.

DOI: 10.4018/979-8-3373-3176-8.ch007

1. INTRODUCTION

1.1 Overview of Smart Warehousing

Smart warehousing involves integrating the most current technology such as automation, data analytics, robotics and Internet of Things (IoT) with the objective of enhancing efficiency, accuracy and speed of the warehouse(Batarlienė, N., & Jarašūnienė, A 2024). It is a breakthrough of the traditional manual-driven warehouse management systems and offers new answers to the emerging need of faster delivery and high efficiency of operations. The necessity to have increased accuracy, flexibility and productivity in the modern supply chains has contributed to the development of intelligent warehouse technologies. In the very early days, warehouses were operating with the basic inventory management facility and manual processing of goods. However, as e-commerce increases, the world becomes global and consumer demands increase, there is a need to transport goods more efficiently and faster thus necessitating the use of more advanced technology integration(Van Geest, M et al., 2021).

The operations of a warehouse have changed due to smart technologies comprising robotics, AI, and IoT. The technologies allow the warehouses to work with the lowest number of people in them thus reducing the error margin and speeding up the operating time. Robotics and automated guided vehicles (AGVs) are now part of the standard at warehouses where they enable the inventory to be retrieved, orders to be picked, and materials to be transported. Moreover, the observance of the condition of the goods and the environmental factors, such as the temperature and the humidity can be tracked in real-time because of the IoT-enabled devices and sensors and it is highly valuable when it comes to delicate goods, such as pharmaceuticals and perishables.

The evolutions of the technologies have transformed the smart warehousing not only as a tool of optimization but as a must in the management of the increasingly complex needs of the contemporary supply chain. As companies are keen on becoming competitive, streamlining business processes and improving the efficiency of their warehouses, smart warehousing is a feature of business success.

1.2 Energy Efficiency in Warehousing

The use of energy in warehouses is an interesting issue, both on operational and environmental level. Warehouses usually use a lot of energy in heating, cooling, lighting as well as in running of machinery. Energy consumption in a conventional warehouse is usually ineffective because it is determined by antiquated infrastructure, ineffective insulation and inability to automate the process of controlling energy consumption in real-time. Not only does this result in an augmented cost of operations but also causes carbon emission, which impacts the sustainability of supply chain operations.

The increasing energy use in warehouses is the side effect of the growth of e-commerce and the increased demand of warehousing to serve the global supply chains. Recent research indicated that warehouses contribute a significant percentage to the overall energy use in industries and this is likely to be on the rise since more industries are demanding 24 hours services. In most situations, the situation is even worsened by energy-intensive activities like lighting all the time, heating or cooling that is not necessary. The increasing demand of quicker and more efficient deliveries put pressure on the companies to deliver and the energy consumption may not be controlled and this results into wastages and inefficiencies(MAY.P et al., 2025).

The realm of sustainable practices in the sphere of warehouse activities is urgently required in order to decrease the amount of consumed energy and limit the adverse influence on the environment. With the world struggling to manage climate change and its impacts of excessive carbon emission, industries are coming under pressure to use energy efficient solutions. This has seen tremendous increase in activities geared towards integrating energy-efficient technologies in warehousing, streamlining the use of resources and adopting sustainable practices in warehousing which in addition to cost-cutting in operations also help in achieving environmental objectives like minimizing carbon footprints(Domagała, J. 2022).

1.3 Role of AI in Optimization and Sustainability

Artificial Intelligence (AI) will play the most important role in ensuring that warehouses are energy-efficient and sustainable. The technologies based on artificial intelligence cause enormous gains associated with the optimization of the work of the warehouse, energy consumption, and sustainability. The technologies have the potential of assisting the warehouses to maximize on their energy use, predict and prevent energy wastage and automate energy saving. The AI applications can cover a wide variety of areas of a warehouse, such as robotics, energy management, and much more, helping the managers of the warehouses make more efficient and sustainable data-driven decisions.

The initial steps of the optimization of energy use by AI are predictive analytics and machine learning algorithms. The technologies allow AI systems to analyze huge volumes of warehouse activities and environmental monitoring systems to identify trends and patterns of energy use. AI can help to schedule the operations at off-peak hours, lighting systems and HVAC systems based on real time occupancy or temperature data and provide dynamic allocation of energy resources by being aware of peak energy demands. Taking an example, AI-based systems can adjust the warehouse lighting to the one outside of the daylight, reducing the number of artificial light sources and, consequently, energy consumption(Jayasutha, D et al., 2024; Bhuvanya, R et al., 2025).

Moreover, AI is also quite significant in predictive maintenance, which is the practice that could save significant energy amounts that are spent unnecessarily. Using data analysis of machines and equipment, AI systems are able to predict the likelihood of the equipment failure or maintenance and remove downtime and ensure the highest efficiency of the equipment operation. It is not only an efficient use of energy but it also makes the warehouse property last longer since it does not need repairs or replacement that consume energy.

The middle of the optimization of the layout of the warehouses and the inventory managing is also being carried out with the help of AI. Warehouses will be in a position to arrange their storage in the most optimal manner, reduce the number of movements of unwanted transportation and store the goods in the most energy-saving manner, through AI algorithms to predict the demand of the products. Another use of AI is to optimize the motion of autonomous mobile robots (AMRs) within the warehouse and ensure that the goods are moved in the most efficient way, without unnecessary movements or wait times, wasting energy.

One of the most promising ways of AI implementation in the smart warehousing is the possibility of using renewable sources of energy in the work of a warehouse. The utilization of renewable energy (solar or wind) can be also optimized by AI systems, depending on the estimation of the energy level and correlating it with the activities occurring in the warehouse. An example would be when the production of solar energy is high, AI will be in a position to alter operations at the warehouse to use more energy and vise versa.

1.4 Purpose of the Chapter

The goal of this chapter is to provide an in-depth overview of AI-based technology in smart warehousing and specifically how they are used to enhance energy efficiency and sustainability. The necessity to find ways to minimize energy consumption in warehouses has never been so urgent as the energy consumption is steadily increasing in the warehouses all over the world. In the current chapter, the author aims to cover the different AI-powered technologies which are currently being applied to enhance energy efficiency in warehouse operations, including predictive maintenance, dynamic load management, smart inventory systems, and energy-efficient robotics.

This survey will help us to outline the possibilities of AI to make drastic changes in the energy efficiency of warehouses and operation cost reduction and the contribution to sustainability principles. Using the available trends and the emerging technologies, this chapter will give a better insight into how AI will affect the energy landscape of modern warehouses. Also, the chapter will address the obstacles and hindrances to the wide usage of AI-driven solutions, such as technological constraints, integration problems, and cost-related consequences of implementation.

Finally, the chapter aims at providing informative details to the researchers, practitioners, and policymakers about logistics and supply chain management. With the clear picture of the role of AI in enhancing energy-efficient warehouse processes, it will add to the existing body of knowledge on how technology may be used in the future to establish more sustainable and efficient supply chains.

2. BACKGROUND AND KEY CONCEPTS

2.1 Smart Warehousing Technologies

The introduction of smart warehousing technologies is a revolutionary change in the functioning of warehouses as it was previously based on manual work and old systems, whereas nowadays a warehouse is highly automated, smart, and data-driven. These are robotics, Internet of Things (IoT), and Artificial Intelligence (AI) technologies to enhance the process of managing the warehouse and make it more efficient, accurate, and speedy(Gopalan, B et al., 2025). The key element of this transformation is automation that is about using machines, systems, and processes to perform tasks with a minimal input by a human being. The warehouse will be able to save on labor as well as increase the pace of operations and reduce human error, which makes the process more efficient.

Some of the major technologies that are used in smart warehouses include robotics and automated guided vehicles (AGVs) that are meant to transport goods and materials within the warehouse without any human assistance. Warehousing robotics are used in order picking, sorting, packing, and retrieval of goods hence speeding and accuracy of the operations. Besides, autonomous mobile robots (AMRs) and AGVs will contribute to the optimization of the movement of goods in the warehouse because workers will have to walk less and less congestion will occur in the areas with the highest traffic.

The other crucial technology is that of the Internet of Things (IoT) where sensors and other connected devices are used to gather real-time data regarding the warehouse activities. This information may comprise temperature and humidity of storage facilities, equipment condition and location of stock and flow of commodities in the warehouse. IoT technology allows operator to have a clearer picture of the whole supply chain and have a more informed decision. IoT devices also assist in inventory tracking, which means that the warehouse will work with the current information about the quantity of goods, orders, and delivery.

AI is the key to the combination of these technologies as it allows warehouses to move beyond simple automation. The AI-powered technology is capable of processing massive amounts of data that IoT sensors and robotics can collect and enhance predictions, inventory, and scheduling. The example is machine learning algorithms that allow predicting the demand of products and helping to optimize storage space so that the products could be on the places where the energy consumption is minimal and the whole warehouse works more efficiently. In addition, AI has the ability to study past data to improve the route that robots and AGVs will follow, making the delivery of goods faster and more economical in terms of energy consumption.

Smart warehouses can be of two types fully automated and partially automated. Most operations in fully automated warehouses are completed using automated systems such as the inventory management, picking, packing, and shipping. Such warehouses do not demand much human interaction, and the process of functioning depends on AI and robots significantly. Partially automated warehouses on the other hand incorporate human work and automation. Although a lot of work is automated including inventory management and sorting the products, the human labor is still important in such processes as order picking, packing, and quality control. Both the models have their own merits, which are dependent on the size of the warehouse and the requirements of the operations.

2.2 Energy Efficiency in Warehousing

Warehouses are major users of energy thus energy efficiency is one of the major factors to consider. Warehouses are energy-intensive areas; they have lighting, HVAC systems, and the machinery that is used in everyday activities. Energy optimization can help to minimize the cost of operations but also to meet sustainability objectives through a reduction in carbon emissions and the environmental impact of the activities of the warehouse. Energy management is very important especially as there are increased concerns of climate change and the high cost of energy.

The major energy consuming sources in warehouses are lighting systems, heating ventilation and air conditioning (HVAC) systems and the use of machinery. The conventional lighting systems like incandescent bulbs use much energy and usually they are left on at all times, without respect to the status of the warehouse. Conversely, the energy efficient lighting systems of today such as LED lights and smart lighting systems assist in cutting down the consumption of energy by regulating the amount of light in real time. The smart lighting systems can be programmed to switch off in the areas of the warehouse when no one is working there or can be programmed to dim down according to the availability of natural light thereby reducing the excessive use of energy.

The other source of high energy consumption in warehouses is the HVAC systems especially in warehouses that handle sensitive goods that need to be maintained under strict temperature and humidity conditions. HVAC systems may also be inefficient because of lack of proper insulation, obsolete equipment and inefficient temperature control. Warehouses can use the HVAC operations to their advantage by using advanced control systems and relying on real-time data provided by the IoT sensors, to adjust the temperature and airflow depending on the requirements of the space.

Besides lighting and HVAC, another energy consuming machinery in warehouses includes conveyor belts, automated sorting systems, and robotic arms that are used in the operation of the warehouse. Inefficient machines do not only consume energy but may also add costs to the operations, because of the frequent maintenance and shutdown. Predictive maintenance is one of the most powerful methods that can be used to reduce energy wastages in the machinery of the warehouse as it allows to prevent disruptions with the help of real-time data provided by sensors and AI algorithms that can determine when the equipment will need maintenance or replacement. The strategy will make sure that the machines are functioning optimally and hence, reduce wastage of energy.

Warehouses can use energy audits to determine the energy efficiency of a facility, energy audit is a structured evaluation of energy consumption in a facility. Auditing of energy is done to determine the places where energy is wasted and how to make energy efficient. They usually entail an in-depth examination of the structure of the building, the illumination, the climatic systems, the equipment and anything that can cause energy consumption. Once the audit has been conducted, there is certain information that can be acted upon, such as the necessity to switch to energy efficient lighting, improved temperature control system or new equipment. Through regular energy audits, warehouses will be able to identify and correct their energy usage thus leading to low operating cost and ecological resource footprint.

2.3 Artificial Intelligence (AI) in Warehousing

Artificial Intelligence (AI) is one of the new technologies that have transformed the way warehouses are being run and AI is offering powerful ways to maximize the running of warehouses, its efficiency and make it sustainable(Ravindran et al., 2025). The tools and approaches that AI offers to warehouses are numerous and include machine learning, predictive analytics, and optimization algorithms that can help the warehouse to control the energy it is using, streamline processes and reduce operation cost. Warehouses will be in a position to make decisions based on data that will result in high productivity, reduction of wastes and overall enhancement of the energy efficiency with the assistance of AI(Pandian, A. P. 2019).

Some of the significant AI tools being used in warehouses include machine learning whereby the system can learn using the past data and then make an inference or a decision based on trend and patterns. With regard to an example, the machine learning algorithms are able to predict the demand of certain products and suggest the most appropriate storage locations where the transportation of the goods will consume less power. Similarly, one can forecast the volumes of orders using machine learning and also adjust the number of workers and use of equipment in warehouses. This prediction capability eliminates the wastage of resources that can lead to the wastage of energy and ensures that there is a proximity between the energy consumption and the operations that are needed.

The next significant AI technique that is used to optimize the operations of the warehouse is predictive analytics. Based on the data that will be offered by various sources, including IoT sensors, inventory systems, and historical trends, AI will be able to predict the issues that can emerge without necessarily manifesting themselves. To illustrate this, it can be said that predictive analytics can be used to determine when a particular equipment is going to break down so that it can be fixed before it breaks down and thus reduce the downtimes and the machine can also run without wasting any energy. Moreover, AI can be also used in the energy consumption of different zones in the warehouse or equipment, hence giving managers a chance to discover the means to reduce the energy consumption or optimise it.

Speaking of the functioning of a warehouse with the assistance of AI, optimization algorithms are one of the most important roles to be played, and in particular, in the work with energy resources. These algorithms make use of real-time data gathered by IoT sensors and previous patterns to optimize resources, such as energy, equipment and labor. To give an example, the HVAC system can be optimized by AI dynamically adjusting the temperature levels based on the occupancy data currently available so that energy is used where and when it is needed. Similarly, AI will also be in a position to optimize the route of autonomous robots and any other automated system in order to reduce the wasteful consumption of energy through inefficient routes.

The energy efficiency is particularly critical in the case of AI in forecasting, load balancing and real-time adjustments. In the operation, AI systems have the ability to forecast the necessity of energy based on the pattern, as well as external factors, e.g. weather conditions or changes in the prices of electricity and control the consumption of energy. Moreover, with the help of AI-based systems, in real-time, one will be able to adjust the energy consumption to its maximum, e.g., turning off the lighting or reducing the power of HVAC during low traffic, further enhancing energy savings.

3. AI-DRIVEN TECHNOLOGIES FOR ENERGY OPTIMIZATION

3.1 AI and Predictive Maintenance

Predictive maintenance is also one of the most promising technologies driven by AI in energy efficiency pursuit in warehouses. Conventionally, maintenance was either reactive or periodic with predetermined time intervals that in most cases results in inefficient utilization of resources and unwarranted downtimes. Conversely, predictive maintenance takes the data of several sensors fitted on equipment, machinery, and infrastructures in warehouses and applies AI algorithms in analyzing the information. Such sensors will check things such as vibration, temperature, pressure and other parameters of operation. This data is then fed to the AI to find patterns and predict such failures and maintenance needs of the equipment.

With the ability to forecast when equipment requires servicing, AI can only schedule maintenance when it is required, hence preventing the use of expensive energy-consuming repair or replacement. The maintenance schedules can be streamlined to make sure that machines perform optimally without any wastage of energy because of malfunctioning or under performing machines. An example would be a conveyor belt not working at its optimum speed because of a worn-out motor and could consume more energy than necessary. This problem can also be identified in advance and preventive measures taken in time with the help of the predictive ability of AI, as a result of which the system will not break down at all and the loss of time and unnecessary energy expenditures will also be minimized.

Moreover, predictive maintenance with the help of AI will be able to increase the service life of warehouse equipment considerably. Machineries that are maintained properly work efficiently using less energy and have less chance of making unnecessary repairs that can further use more energy. In case of warehouses where the automatized systems are in use (e.g. sorting machines or AGVs), it is important to make sure that the systems are in the best possible condition so that they can be energy efficient throughout the years.

3.2 Smart Inventory Management

AI is also important in streamlining of inventory and this has direct impact on energy efficiency of warehouses. The inventory management systems are charged with the responsibility of tracking the products, their movement and the places they are stored in the warehouse. The conventional inventory management techniques usually entail manual storage of products in the storage facilities without the care of the overall performance of the product movements. This may result in poor routes, unproductive movements and wastage of energy.

Inventory management can also be improved with the help of AI since it can be done based on algorithms that analyze the patterns in the demand of products, the space in the warehouse, and the conditions of storing the products. The machine learning models can forecast the demand changes and change the storage points accordingly, so that the products which are in high demand are stored in the places that are easily accessible and the workers or robots do not have to travel a long distance to get them. Such an optimized positioning minimizes the number of redundant movements and that automatically saves energy.

As an illustration, AI will make sure that frequently chosen products are placed near the area of dispatch so that less time will be spent on transportation of the goods. Likewise, when there is a low demand, AI has the potential to propose the rearrangement of inventory to make more efficient use of energy. The AI can also prevent re-handling and relocation of products by predicting the most efficient storage places based on product turnover, which leads to fewer energy costs and relieves the human and robotic systems of operational pressure.

Inventory management systems using AI can also generate real-time information about the amount of stock in hand, so that the excess stock is not kept in the warehouses unnecessarily, which not only occupies the valuable space but also uses more energy. Warehouses are also able to minimize the energy requirements on cooling and heating of the huge storage rooms, by keeping the optimal inventory levels.

3.3 Dynamic Load Management

The optimization of dynamic load management, which is the main feature of AI, is the key to increasing the efficiency of energy consumption in modern warehouses. Dynamic load management is a way of balancing energy use of different systems and equipment's in the warehouse to prevent sharp energy demands that result in inefficiency and higher energy costs. With the use of AI-driven systems, energy consumption of different types of equipment can be monitored and the operation can be dynamically adjusted to ensure the equal use of energy during the day.

Among the major advantages of dynamic load management, one must mention the possibility to cut peak energy consumption. The energy demand in warehouses is usually variable and some activities within a warehouse like automated handling of materials or machine work may lead to an instant increase in energy consumption. AI systems can forecast when such peaks are apt to take place and make the functioning of non-critical systems change. As an example, when the energy demand is high, AI will be able to reduce the energy consumption of some systems that are not that important, e.g., lighting or air conditioning, but not affect the overall warehouse operation.

Besides the energy balancing, AI can help to optimize the warehouse energy consumption regarding the external parameters, i.e., the time-of-day electricity prices or weather conditions. At off-peak hours when energy is less expensive, AI systems can program energy-intensive actions to be done to use the cheaper energy and this will make the running of the warehouse less costly. AI-driven load management allows controlling the costs of energy and maintaining the warehouse operations without any issues by optimizing energy consumption during the day.

Besides, AI is able to incorporate renewable energy sources, i.e., solar power or wind power, into the load management system. With the help of AI, one will be able to regulate the energy consumption of the warehouse by predicting when these sources will be most abundant. This will minimize the use of conventional sources of energy, which will be another factor towards energy efficiency and sustainability of the warehouse in general.

3.4 AI in HVAC and Lighting Systems

Two of the greatest consumers of energy in a warehouse are the HVAC (heating, ventilation and air conditioning) systems, as well as lighting systems. The enhancement of such systems with the help of AI technologies can play a powerful role in energy savings. The HVAC AI-based control system is developed to control the temperature and airflow in the warehouse depending on the actual data, including the level of occupancy, weather conditions, and requirements of a particular zone inside the warehouse.

Conventionally, HVAC units are programmed to operate under a specific temperature, irrespective of the way the space is utilized. AI systems however will set the settings depending on live data and thus energy is not wasted. As an example, in case some parts of the warehouse are not used, AI can cut the heating or cooling of the parts and save energy without affecting comfort or the conditions under which the products are stored. AI is also able to learn historical data and be able to anticipate the best temperatures to set in different conditions and thus use energy better over time.

Likewise, the smart lighting systems with the help of AI can maximize the lighting usage in the warehouse. Conventional lighting systems tend to be left on more time than is necessary despite whether the place is occupied or not. Nonetheless, with the help of AI-based lighting systems, the brightness can be changed automatically depending on the time of the day, the amount of natural light etc. as well as the number of workers in a particular location. As an example, in case one part of the warehouse is vacant, the lights in that part can be dimmed or switched off, and the amount of energy used will be greatly reduced. Moreover, AI will make sure that only the lighting that is required is turned on, therefore, keeping the operations efficient but reducing the use of energy that is not required.

HVAC systems and lighting systems are two of the systems that can use IoT devices and sensors to give real-time information to AI systems. Through this data, AI will be able to make informed choices when and how to change energy-consuming systems to further optimize energy management in warehouses.

3.5 AI in Robotics and Automation

The energy efficiency of warehouse operation is based on robotics and automation, and AI is essential to the optimization of robots. Robotic and automated systems are programmed to perform repetitive activities including sorting of inventory, handling of materials as well as picking orders. Such systems can allow to minimize manual work, speed up the operations, and decrease the number of errors, at the same time optimizing energy use.

AI-based robots have the ability to perform with minimum consumption of energy since they learn the most efficient routes and actions. As an example, AI programs can be used to optimize a path that a robot will use to pick up items at various positions within the warehouse and use the minimum amount of energy in the process. Besides, AI enables robots to change the pace and energy consumption relative to the items they are carrying, depending on the weight and the size of the items to be carried, and the energy consumption is minimized.

AI is also relevant in maximizing charging and battery life of autonomous robots. AI can make the charging of a robot as efficient as possible by predicting the time when a robot will need to be charged, which will not disrupt the working processes. Based on working data, AI can also decide when it is most efficient to recharge robots, so that they are always ready to use at the necessary moment, but not spending much energy at the times when they are not in use.

Also, AI assists in improving interaction between robots and human employees. AI can be used in a warehouse that has both robots and human workers to do the same thing but in a manner that minimizes the redundancy of the two, minimizing the energy expenditure of assigning the tasks to the wrong parties.

4. SUSTAINABILITY AND ENVIRONMENTAL IMPACTS

4.1 Carbon Footprint Reduction

The world is also experiencing an increasing number of environmental challenges and thus the industries are increasingly being pressured to embrace sustainable practices to minimize their carbon footprints. As an important component of the international supply chain, warehousing has a significant influence on energy use and emissions. But, with the application of smart warehousing technologies using AI, there has been a significant promise to reduce this environmental footprint. Automation, the use of advanced technologies, and the optimization of energy consumption are the reasons why AI can make a great contribution to cutting carbon emissions in the operation of warehouses.

The AI-driven systems lead to the process of efficient consumption of resources, so there is no need to spend much energy and, therefore, to emit carbon dioxide. To give an example, predictive algorithms may examine the data about the warehouse processes to optimize the HVAC systems, the light, and the machines so that the energy is used only when it is necessary and in the place where it is necessary. This specific energy consumption does not only reduce expenses but also avoids excessive use of energy, which is the main source of unnecessary $CO2$ emissions. In large scale refrigeration or climate control warehouses, the AI can even dynamically adjust the temperatures according to real-time occupancy and external weather conditions to reduce emissions by further avoiding energy wastage.

Besides, AI can also help in cutting down carbon emission in the transportation sector of the warehousing. A lot of vehicle movements are generally involved in logistics and material handling, and this may be a significant source of carbon emission. Route planning of autonomous mobile robots (AMRs) and Automated Guided Vehicles (AGVs) can be optimized by AI, which can decrease the distance and time it takes to move goods and, therefore, the number of trips required and the carbon footprint of transportation in the warehouse. AI can also be used in managing the fleet through the monitoring of the vehicle performance, as well as, proposing the best route to take to avoid traffic and wastage of fuel.

With the help of AI-driven demand forecasting, the warehouses will be able to more precisely estimate the inventory needs, in turn, leading to fewer emergency deliveries and the surplus of the products. This reduces the wastage of movements of goods thus reducing the carbon footprint of logistics processes(Krishnan, R et al., 2024). Moreover, with the sorting, storage, and picking processes automated, AI will result in minimal energy wastage since goods will be stored in the most optimal positions and will not need as much energy to access in the future.

4.2 Renewable Energy Integration

One of the most vital aspects of sustainable warehousing is the introduction of the renewable energy sources in the warehouse operations, and AI can be of much help in facilitating its utilization(Malinowska, M. 2022). As the world gets to know about climate change, most warehouses are turning to renewable energy such as solar and wind energy to reduce the need of using fossil fuel and further reduce the effects on the environment. However, the process of adding renewable energy to the warehouse is a complicated one, at least in case of renewable energy variability and intermittency. In their part, these problems can be addressed with the help of AI, as it will allow integrating and utilizing renewable energy in a more efficient way.

The renewable energy can be handled in the most optimal way using AI-based systems to forecast energy generation based on weather, time of day, and other environmental factors.

As an example, in solar-powered warehouses, AI algorithms can forecast the amount of solar energy that will be produced using the weather data and time of the year to enable the warehouse to make decisions on the amount of energy that it consumes. AI will be able to optimize on energy intensive activities like robotics operations or automated sorting when there is abundant solar power and hence energy will be utilized in the most efficient manner without the grid. In the same way, in case of insufficient solar power, AI can switch to energy storage or limit energy consumption.

AI can also be used in prediction and management of energy in wind powered warehouses. The AI algorithms are able to process data on real-time wind and predict the energy generation, which will enable the warehouse to match its energy consumption with the accessible wind energy. Also, AI can be employed to make energy storage systems (like batteries) optimal, so that when the renewable energy is produced a lot, the surplus is stored, and it can be used when there is a minimal production. Such incorporation of renewable energy sources decreases the level of reliance of the warehouse on the non-renewable energy sources thus making it less harmful with regard to the environment, and thus contributes to green future.

AI can be also applicable in the management of hybrid energy system where both renewable energy and traditional grid power are used. AI allows predicting the changes in the production and consumption of renewable energy and, therefore, the warehouse will use more renewable energy when it is abundant and only switch to grid power when there is no other option. This dynamic energy management strategy enables warehouses to make maximum utilization of renewable energy, which has a long-term effect on the carbon footprint of warehouses and the cost associated with energy consumption.

In addition to that, AI will enable integration of microgrids that will enable warehouses to be independent of the primary energy grid. Microgrids are regularly equipped with a mixture of renewable energy resources, energy storage capacities and backup generators. AI algorithms can be used to optimize the work of such microgrids by controlling the flows of energy, deciding when it is the best to use renewable energy, and when it is better to use the stored one. This degree of control does not only raise the level of energy independence, but also facilitates the shift towards a more sustainable energy infrastructure.

4.3 Waste Minimization

Another important element of sustainability in warehousing is the waste minimization, and AI technologies are essential in maximizing the use of resources to minimize wastes. The warehouses tend to be victims of different types of waste such as material waste, energy waste and food or product spoilage (particularly in food storage industries). These waste streams can be reduced significantly by the capabilities of AI to streamline processes, predict demand and inventory management.

Regarding material waste, AI can be used to eliminate the necessity of unnecessary packaging, which is the typical side effect of poor inventory control and handling procedures. The optimalisation of the number of products in the stocks, along with the ability to predict the demand of the products with the assistance of AI-based systems, will ensure that the warehouses will not overstock the products and, consequently, will not use as much packaging material. Besides, AI assists in managing wastage by overproducing or spoiling the products by improving the inventory management and product spoilage accuracy(Masudin, I et al., 2021).

Another way of using AI to enhance waste management in warehouses involves optimization of the recycling processes. To illustrate, AI algorithms have the ability to classify and segregate recyclable waste products and non-recyclable products and automate the waste sorting procedure. Smart sorting systems with the aid of AI can better separate various materials such as plastic, metal and chapter and therefore recycle them more effectively. This will reduce the burden of the waste in the landfills and contribute to reuse of materials, which will assist in a circular economy.

Use of AI in minimisation of wastes is also used in energy wastes. AI ensures that the energy is not being wasted in the areas of the warehouse that are not functioning by making the maximum use of energy through the predictive analytics and real-time changes. To give an example, AI can automatically adjust the lighting or HVAC state of operations in the areas of the warehouse which are not used, thereby ensuring that energy is only used when it is needed. This smart way of managing energy can not only reduce the cost but also reduce the negative effect on the environment by the wastage of energy by using too much of the energy.

Additionally, AI would allow minimizing waste by eliminating the amount of superfluous resources or energy and streamlining robotic and automation systems. This makes such systems programmable and this implies that they can be programmed to operate to the most efficient level and this will save the energy that is required to be used in accomplishing the tasks in a warehouse and reduce the carbon footprint of a warehouse. The use of AI eliminates the waste of energy and material resources because robots and other automated systems only perform the necessary work and at the most efficient rates.

5. CHALLENGES AND BARRIERS

5.1 Technological and Integration Challenges

The fact that the introduction of Artificial Intelligence (AI) in the work of warehouses encounters significant challenges is indisputable. The lack of possibility to combine the new AI technologies with the ones which are already implemented in the warehouses is one of the primary challenges as these systems may not be modified to use the advanced technologies. The majority of warehouses are on the legacy systems that were developed long before the creation of AI, robotics, and IoT technologies. These legacy systems tend to be founded on other data formats, protocols, and infrastructure and integration of AI-based solutions is not straightforward without massive changes or upgrades. The process of AI implementation is associated with a complex reorganization of both hardware and software, which can be time-consuming, costly and disruptive to the business under development(Nandhini et al., 2025).

Moreover, there are cases when integration process can be linked with alignment of a great variety of different technologies in a warehouse environment. The one that is likely to use different technologies such as robotics, AI, and IoT is smart warehouse(Tiwari, S. 2023). All these technologies are motivated by different sets of protocols and architectures and therefore this may be an issue of compatibility. As an example, one kind of the protocols could be utilised in the robotic systems, and another kind in the IoT sensors. In the instance where such systems cannot effectively talk to each other, then there is the possibility of inefficiencies, delays or even system failures. The question of the seamless integration of robotics, AI algorithms, and IoT devices appears even more urgent, the more complex the operations in the warehouse are.

Other than the compatibility challenges, AI in the warehouses requires a strong data infrastructure. All the incoming information in different forms, e.g., inventory management systems, robotic operations, and environmental sensors, must be integrated into a single system that can subsequently be processed and analyzed through the AI algorithms.

This is usually a process that needs advanced data pipelines and cloud computing systems, which do not necessarily exist in a warehouse environment that is older. The amount and the number of different types of data produced by IoT sensors, robots, and AI-driven decision-making systems are too high and diverse to be managed by simple data management systems and they must ensure that the operational performance remains high regardless of the real-time processing.

These issues of integration are costly, time consuming and also need specific skills. The migration to the AI-driven solution might be a nightmare to the warehouses using older systems, and it is necessary to plan this carefully and introduce it in phases so that it does not interfere with the operations.

5.2 Data Privacy and Security

With more and more warehouses embracing the AI and IoT technologies, the quantity of data generated and processed becomes exponentially larger. This information normally contains sensitive data of inventory, customer orders, shipping and operational performance. Moreover, a vast number of warehouses handle personally identifiable information (PII), particularly in such industries as e-commerce where the personal data of customers and their shopping history is collected and managed. The mass adoption of AI-based solutions and IoT devices in the smart warehouses creates a serious issue of data security and privacy.

The greatest issue that revolves around the privacy of data is that of breaches and unauthorized access. The AI-based warehouses involve an endless stream of sensitive data exchange between IoT devices, robotic systems, cloud servers, and AI algorithms. Unless such systems are secured accordingly, they can be hacked, information leaked, and attacked by other parties. The potential impact of a data breach might entail a substantial loss of money and reputation and legal consequences, in particular when the breached information concerns the PII of customers or sensitive business data.

Moreover, the fact that AI systems will increasingly rely on machine learning to make predictions and optimize the work of the warehouses leaves a possibility that such systems will reveal or misuse sensitive information by mistake. In some instances machine learning algorithms trained on large data sets can learn patterns or correlations that can jeopardize data privacy. To illustrate, the AI systems can inadvertently match some of the customer data with the action of the operations and this information can be utilised in the event that the data management processes of the system are not sound enough.

The issue of data privacy and security also makes management quite complex given that different technologies are implemented in the warehouse. The IoT devices with the tendency to be connected to the internet increase the scope of attack of the warehouse network. All these devices accumulate much information and as they do not have the required security measures such as encryption and authentication, they give the opportunity to be attacked by cyberattacks. In addition, there might be external risks in the management of warehouses owing to the utilization of the third-party AIs and IoT vendors. The external systems may compromise the security of the warehouse in case the external providers do not embrace appropriate security.

To eliminate these risks, a comprehensive data security strategy in warehouses needs to be adopted. This entails application of advanced encryption, multi factor authentication, scheduled security audit and real time monitoring network to detect any kind of suspect activity. The warehouses are also expected to comply with the data protection regulations such as the General Data Protection Regulation (GDPR) enacted in Europe to ensure that the data of the customers are being processed in a fit and safe manner.

5.3 Cost of Implementation

Implementation cost is among the biggest barriers to the popularization of AI and smart technologies in warehouses. Though the long-term benefits of AI-based solutions are huge (improving efficiency, energy savings, and making better decisions), the cost of initial investment into such solutions might be too high, particularly in the case of small and medium-sized enterprises (SMEs). Introducing AI technologies is not free and the cost is hardware infrastructure, software development, system integration and training.

One of them is that the cost of funding robotics and automated systems in a warehouse can be very high in terms of capital investment on robot units, sensors and other hardware components. Also, the AI-powered systems require efficient data storage systems, high-performance computing system, and cloud services to process and handle the huge data volumes produced by the IoT devices. These infrastructural needs may be expensive, especially to the SMEs which cannot afford big-scale technological investments.

The expenses do not only lie in buying hardware and software. Incorporation and adaptation of AI solutions to the individual requirements of a warehouse may also become quite costly. Most AI solutions also need personalization to guarantee its compatibility with the current systems, maintenance, and updates. Furthermore, the process of implementation may also lead to the downtimes in the operations of the warehouse since it will have to adjust itself to new systems, which can cause some short-term losses in productivity.

The second financial problem of SMEs is the absence of in-house knowledge necessary to introduce and streamline the AI technologies. Whereas big companies can afford to hire teams of engineers, data scientists, and AI experts, smaller corporations might not be able to do so and must hire consultants or vendors to address the installation and configuration of AI systems. Such dependence on the third party expertise escalates the implementation cost even further.

To a lot of SMEs, the economic cost of implementing AI and smart technologies may be overwhelming. Some of these are however mitigated with the increasing availability of cloud based solutions and the emergence of AI-as-a-Service (AIaaS). Through cloud-based platforms, SMEs can use AI tools and resources without having to invest heavily on infrastructures. Also, cloud solutions are scalable, which means that SMEs can gradually introduce AI technologies as their needs change, which eliminates the risk of overinvestment.

5.4 Skills Gap

AI-powered systems in warehouses need very sophisticated skills, and such skills are usually limited. Lack of highly qualified specialists is one of the key challenges that cannot allow warehouses to make full use of AI and smart technologies. AI, machine learning, and robotics are very advanced subjects, which involve the knowledge of programming, data analysis, and optimization of systems.

The companies that want to implement AI technologies in their warehouses should have professionals who can design, implement, and support complex AI algorithms, handle big data, and guarantee smooth interaction between robotics and other intelligent technologies. Nevertheless, there is a worldwide shortage of skilled employees in such areas, and the skills gap is especially acute in such industries as warehousing, where AI and automation are still at their beginning.

The shortage of such professionals forms the major adoption barrier to most organizations, which are unable to hire or develop the right talent to run and fine-tune AI-based systems. Also, in case the companies could find skilled professionals, it may be difficult to retain them. The speed at which technology is developing implies that AI and robotics professionals should constantly monitor the changes in their knowledge and skills to keep up with the changing technologies. This will need constant training and workforce upgrading which can also be a burden to the organization.

6. FUTURE DIRECTIONS

6.1 Emerging AI Technologies

As we continue to introduce AI-based solutions into the work of warehouses, some of the emerging AI technologies have a potential to further increase energy efficiency and optimization of processes. Deep learning and edge computing are some of the most promising developments that can transform the way warehouses can operate their operations and energy consumption.

Deep learning is a subfield of machine learning concerned with the creation of artificial neural networks. When it comes to smart warehousing, deep learning algorithms may be used to better forecast demand, to manage stock more efficiently and even to predict energy use patterns. Deep learning has the potential to discover nuance relationships and trends in historic and real-time data that more traditional techniques may overlook. As an example, deep learning models might be deployed to forecast the change in the customer demand with a greater level of precision so that the warehouses could change the staffing levels and energy consumption accordingly. The systems can also be implemented to maximize the pathways and timings of the robots and automated systems so that energy is used efficiently and all the unnecessary movements are avoided and overall efficiency of the operations is increased.

The other emerging technology is edge computing which is the processing of data near its origin instead of transmitting it to a centralized cloud server to be analyzed. Edge computing may be an important factor in smart warehouses to decrease latency, response time, and the amount of bandwidth needed to handle large quantities of information produced by IoT devices, robots, and AI. In energy management Edge computing enables real-time tracking and optimization of energy-using systems such as lighting, HVAC and equipment, to make sure that such systems are running effectively even in high-demand situations. Processing data on the edge allows the warehouses to adjust to the changing conditions immediately, which means that the warehouses do not need to rely on the external data processing services and that they will not have to wait to make important decisions. This will not only raise the speed of operations but also minimize the energy consumption since it will be able to react to changes in demand or environmental conditions almost immediately.

6.2 Smart Cities and Smart Logistics

The task of smart warehouses lies not only in the individual operations, but it also is closely connected to the concept of smart cities and smart logistic systems as a whole. With the constantly increasing urbanization and the growing urgency to streamline the processes of supply chain operations, the implementation of the smart warehouses into the greater scheme of smart cities is becoming a major priority.

Within the framework of smart cities, smart warehouses are regarded as one of the players that would design an intelligent and sustainable urban environment. Warehouses can play a part in ensuring the maximization of resources within a city by connecting to other elements of a smart city, including energy grids, transport systems, waste management systems, etc. An example of this is that a smart warehouse which connects with a smart energy grid may dynamically shift its energy consumption in real-time depending on the information provided by the grid, using less energy during peak times and more renewable energy when it is available. The relationship assists in levelling the supply and demand of energy within the city, making it sustainable and lessen the burden in the conventional power infrastructure.

On the same note, intelligent logistic systems are important towards smooth flow of goods within cities. The use of AI-powered smart warehouses within such systems can enhance the supply chain through the optimization of routes both of the trucks and autonomous vehicles, which will guarantee the delivery of goods promptly and effectively with minimum fuel usage and emission. The warehouse real-time data regarding inventory levels and order fulfillment can be exchanged with transportation networks so that the coordination between the warehouses and the delivery systems can be improved. This does not only enhance the efficiency of the operations, but also minimizes the carbon-footprint of the logistics operations.

Predictive maintenance and optimization of the supply chain also belong to smart logistics, which is also based on AI technologies. Within the framework of a smart city, the information gathered in different warehouses, transport vehicles, and environmental monitors can be processed to anticipate possible disruptions, optimize the transportation timetables, and make sure that goods are delivered in the most energy-saving way possible. Even the transportation routes and schedules with the help of AI-driving optimization tools can be optimized according to the real-time traffic information, weather, or the availability of renewable energy and thus help to make logistics in cities more sustainable(Munuhwa, S. 2023).

6.3 Evolution of Energy Management Systems

Energy management in warehouses will be defined by the innovations of the AI-driven technologies dedicated to the real-time optimization, predictive analytics, and integration of renewable energy. With energy efficiency becoming one of the primary objectives of the warehouse operations, AI will keep on developing in terms of how it can manage its energy consumption in a more accurate and controllable way.

The future of AI-based energy management systems (EMS) in the near future is most likely to become the more autonomous ones that predict energy use and make adjustments based on real-time data without the need of manual control. Such systems will be able to check the usage of energy on individual level thus being able to detect which processes consume most energy and how to make them more efficient. To cite an example, AI can be used to regulate temperature in a warehouse by training it to determine which parts of the structure needs heating or cooling at various times of the day or in relation to outside weather conditions. EMS AI will also be able to incorporate renewable energy sources into the energy management strategy of the warehouse and make its work more flexible and environmentally friendly.

Among the most significant future trends of energy management, there will be optimization of energy storage systems with the help of AI. As more renewable sources of energy (solar and wind) are used, warehouses will require effective systems to store surplus energy during off-peak times to be used during high demand times. The utilization of these energy storage systems will be critical in the ability of AI to decide when to store energy and when to unload the energy depending on the demand patterns. Such combination of AI and energy storage technologies will decrease the dependency on the traditional grid energy and will make warehouses more independent and sustainable.

The other important trend in the progress of the energy management systems is the emergence of the decentralized energy systems. Warehouse operations are becoming increasingly popular with the use of microgrids that are a combination of local generation of renewable energy (e.g. solar panels or wind turbines) and energy storage systems. AI will make these microgrids to be autonomous and they will be managing energy flows, and will make sure that energy is consumed efficiently. With the help of AI, warehouses can develop self-sufficient and energy-efficient systems that will reduce their dependence on external energy sources and help to decrease the general amount of consumed energy.

The future energy management in warehouses will be marked by even more intelligent and autonomous systems, which are even more energy-saving, sustainable, and less harmful to the environment since the AI technologies continue to evolve. The additional development of the AI-based energy management will be the decisive move to help the warehouses to reach their energy efficiency and sustainability goals.

7. CONCLUSION

Artificial Intelligence (AI) in warehouses has witnessed a high degree of energy efficiency and sustainability. Some of the AI-enabled technologies that have been found to optimise warehouse operations and reduce energy consumption include predictive maintenance, intelligent inventory management, dynamic load balancing and robotics with energy efficiency. Because AI automates routine operations and predicts energy demand and dynamically changes energy demand based on real-time information, it can reduce the costs of operations and environmental impact of warehouse operations to the minimum. The use of AI in HVAC, lighting, and robotics management ensures that the energy is not consumed more than it is necessary, and that is why it makes the use of energy much more efficient and decreases the carbon footprint. In addition, AI contributes to the realisation of sustainability as it is easy to introduce renewable sources of energy in the work of warehouses such as wind and solar energy. The AI systems can predict these sources of energy in order to minimize the consumption of energy within a warehouse and maximize on the renewable energy. This does not only reduce the reliance on the non-renewable energy but also the sustainability of the warehousing activities in the long run. Lastly, the AI technologies are useful in transforming the conventional warehouses into energy saving and environmental friendly activities. The potential of the warehousing-related AI-powered technologies is enormous in the sense of changing the energy efficiency. AI technologies, such as deep learning, edge computing, and real-time analytics, are developing and making warehouses smarter, more responsive, and energy-efficient. The technologies will be applied to result in the reduction of the wastage of energy, incorporation of renewable sources of energy, and the utilization of warehouses to their full capacity with no impact on sustainability. One of the best directions to the creation of more efficient and sustainable global supply chains is the popularization of AI-based solutions in warehousing. Although much has been already discussed, one can say that several points should be investigated to make the best out of AI technologies regarding energy sustainability in warehousing. It requires further research on the combination of the renewable energy sources with the AI systems both in terms of energy storage and real-time energy optimization. In addition to that, even more progressive AI algorithms to control the energy usage and predictive analytics may allow controlling the energy usage even more accurately. In addition, one should carry out a study on the scalability of AI-powered solutions to small and medium-sized enterprises (SMEs) to ensure that the technology may be accessible to any business size. The further advancement of these lines of research will clear the path to new opportunities of making the warehousing operations more sustainable and energy-efficient in the future.

REFERENCES

Batarlienė, N., & Jarašūnienė, A. (2024). Improving the Quality of Warehousing Processes in the Context of the Logistics Sector. *Sustainability (Basel)*, *16*(6), 2595. DOI: 10.3390/su16062595

Bhuvanya, R., Kujani, T., Matheswaran, P., Yamini, B., Yuvasri, B., & Sathya, V. (2025). AI and the Future of Economic Systems: Analyzing Technological, Ethical, and Governance Dimensions. In *Reshaping the Economy With AI* (pp. 111-136). IGI Global Scientific Publishing.

Domagała, J. (2022). Green warehousing: a state-of-the-art literature review. *Sustainable Logistics*, 43-71.

Gopalan, B., Vijaya, G. S., & Tiwary, A. (2025). Exploration of IoT Applications, Lean Management, and Green Practices for Sustainable Urban Logistics. In *Bridging Technology and Development for Sustainable Innovation and Geopolitical Dynamics* (pp. 225–252). IGI Global Scientific Publishing. DOI: 10.4018/979-8-3693-9072-6.ch009

Jayasutha, D., Prabhu, R. V., Sassirekha, S. M., Maithili, K., Girija, P., & Subramanian, R. S. (2024). Exploring the Intersection of Sustainable Energy, AI, and Ethical Considerations. In *AI Applications for Clean Energy and Sustainability* (pp. 139–160). IGI Global. DOI: 10.4018/979-8-3693-6567-0.ch008

Krishnan, R., Perumal, E., Govindaraj, M., & Kandasamy, L. (2024). Enhancing Logistics Operations Through Technological Advancements for Superior Service Efficiency. In *Innovative Technologies for Increasing Service Productivity* (pp. 61–82). IGI Global. DOI: 10.4018/979-8-3693-2019-8.ch004

Malinowska, M. (2022). The concept of decision support system to increase the level of sustainable warehousing. *Procedia Computer Science*, *207*, 4046–4054. DOI: 10.1016/j.procs.2022.09.467

Masudin, I., Sumah, B., Zulfikarijah, F., & Restuputri, D. P. (2021). Effect of information technology on warehousing and inventory management for competitive advantage: a theoretical framework. In *Handbook of Research on Innovation and Development of E-Commerce and E-Business in ASEAN* (pp. 570–593). IGI Global.

MAY. P. M. A., Savitha, V., Gayathri, S., & Mahalakshmi, S. (2025, April). Exploring Sustainable Computing: IoT, Big Data, and Energy-Efficient Solutions. In *2025 5th International Conference on Trends in Material Science and Inventive Materials (ICTMIM)* (pp. 1218-1223). IEEE.

Munuhwa, S. (2023). Sustainable logistics and competitive positioning. In *Integrating Intelligence and Sustainability in Supply Chains* (pp. 198–220). IGI Global. DOI: 10.4018/979-8-3693-0225-5.ch011

Nandhini, S., Yamini, B., Anurekha, G., Thilagam, T., & Pajila, P. B. (2025). The Role of Risk Management in Modern Business and Technology. In *AI Methods for Environmental Protection and Resource Conservation* (pp. 217–244). IGI Global Scientific Publishing. DOI: 10.4018/979-8-3373-3246-8.ch009

Pandian, A. P. (2019). Artificial intelligence application in smart warehousing environment for automated logistics. *Journal of Artificial Intelligence and Capsule Networks*, *1*(2), 63–72. DOI: 10.36548/jaicn.2019.2.002

Ravindran, D., Revathi, S., Sowndharya, V., Farzhana, I., Swetha, S., Vinoth, R., & Subramanian, R. S. (2025). Environmental Monitoring for Sustainability: Technologies, Trends, and Future Directions. *Leveraging Urban Computing for Sustainable Urban Development*, 323-340.

Tiwari, S. (2023). Smart warehouse: A bibliometric analysis and future research direction. *Sustainable Manufacturing and Service Economics*, *2*, 100014. DOI: 10.1016/j.smse.2023.100014

Van Geest, M., Tekinerdogan, B., & Catal, C. (2021). Smart warehouses: Rationale, challenges and solution directions. *Applied Sciences (Basel, Switzerland)*, *12*(1), 219. DOI: 10.3390/app12010219

Chapter 8
Smart Warehousing With Predictive Water Usage Insights Using Deep Learning Models

Usharani Bhimavarapu
https://orcid.org/0000-0002-0246-1420
Department of Computer Science and Engineering, Koneru Lakshmaiah Education Foundation, Vaddeswara, India

ABSTRACT

Water conservation in warehousing operations has become a critical concern due to increasing water scarcity and rising operational demands. Warehouses consume water for a variety of purposes including sanitation, cooling systems, fire safety, and landscape maintenance. This study explores the application of intelligent water conservation techniques supported by advanced deep learning models, specifically the Bi-Stacked Gated Recurrent Unit (Bi-Stacked GRU), to forecast water usage patterns and identify optimization opportunities. A dataset comprising 1,195 household-level responses was collected from eight districts and preprocessed for analysis. Feature selection was performed using Particle Swarm Optimization (PSO) to isolate key variables influencing water usage. The Bi-Stacked GRU model enabled accurate prediction of water consumption trends and detection of inefficiencies, allowing for proactive water-saving strategies in warehousing environments

DOI: 10.4018/979-8-3373-3176-8.ch008

INTRODUCTION

Water is draining increasing amounts due to global warming, urbanization, and growing populations. Warehousing activities usually consume significant water in cooling systems, lawn maintenance, washing, and fire protection. Functioning on water-saving settings helps warehouses make a tremendous impact in reducing stresses in local water supplies. This is most significant where deserts or water-poor habitats exist. Water conservation enables the plant to function even during local bans. Water conservation ensures warehousing operation is within sustainability goals. A water-resilient warehousing operation will be insensitive to perturbations. Water-dependent externally is reduced and enhances self-sufficiency. Water conservation also ensures the vision of corporate social responsibility in the entire corporation.

Water conservation in warehousing ensures overall environmental sustainability. Warehouses are saving water too and hence their green footprints. Water-saving campaigns trigger more prudent use of natural resources. More efficient water practices eased the load on city wastewater treatment plants. Treated or reused water wherever possible is suitable for circular economy application. Low-flow unit sets or rainwater harvesting water-saving sets may be installed in warehouses. It saves freshwater without reducing the working requirements on a daily basis. Green warehousing is being sought increasingly by investors and stakeholders such as themselves. Water used applied means practice warehousing the environment. It is left with positive environmental impact with cost-saving benefit in sustainable operation.

Water used in practice saves money. Water used in conservation equipment saves water and, as a result, reduced utility bills. Sealing leaks and maintenance keeping the infrastructure open save water loss and repair money. Water recycling plant and greywater reuse systems pay back in the long term. Savings filter down to the bottom line of the warehousing business. Metering and controlling the consumption of water also has the additional value benefit of greater resource stewardship. The savings on these programs fund the underlying supply chain and logistics efficiency. Water-conservation equipment saves energy, and that carries through to other cost savings. Low-cost economies make the warehouse economically viable for a low-margin company. Save water, in general, is economic and environmental effectiveness.

Water preservation keeps the warehouse on par with current environmental regulations. Governments are imposing tighter restraints on water consumption, outflow, and sustainability. Non-compliance will result in penalty, shut-down, or bad reputations. Warehouses can achieve, and even surpass, the regulation requirement through preservation. This minimizes the risk of business and legal. It also makes the company socially responsible corporate leadership. Water plants that are used to meet regulation risk nothing in water waste and contamination. Water-saving warehouses operate full capacities to address inspections and audits. Readiness to react to water supply disasters is one of the measures through which the risk is mitigated. Proper management of water is that which ensures long-term business continuity.

Water is equally one of the most vital elements of fire protection equipment system of warehouses like hydrants and sprinklers. Vast quantities of water are wasted during fire drill or testing, though. Technologies conserving water ensure maximum use of water in fire protection systems without any compromise on their security whatsoever. Intelligent systems, for example, can potentially recognize leaks and control pressure. Rain water gathered or grey water can be used for practice and practice exercises. i.e., tanks of water storage get filled and fresh water gets preserved. Water conservation and fire safety in combination ensure compliance and flexibility. Green water systems also lead to LEED certification. The warehouses thus become capable of attaining safety levels without harming the environment. Water-conserving fire systems help to save the building and the environment.

Water-saving warehouses describe the extent of engagement with CSR. Firms in the current era need to be responsible and sustainable. Water-conserve in warehousing is eco-friendly and society-friendly. It is an excellent means of gaining more customers' confidence, investors' confidence, and local society's confidence. Employee and community conservation drives are CSR. Water management in the region is also conducted through collaboration with the local government or NGOs by warehouses. Green responsibility enables the creation of brand equity and customer trust. Water disclosure makes reporting valid for CSR. Water disclosure brings the firm on par with the world's expectation level for sustainability. Conserving water is given social license to operate.

Conserving water in warehouses provides end-to-end supply chain-wide greening. Warehouses are nodes of high centers where material and product are imported, sorted, and exported. Green activity in the nodes is an effect multiplier across the chain of logistics. Water-saving customers are served by water-saving carriers and warehouses. This legacy provides room for accompaniment of green certification requirements in the work logistics contract. Greenness is achieved by partnership between retailers, carriers, and manufacturers. Transparency and two-way water partnership guarantee responsibility. Water savings and supply chain KPI guarantee long-term sustainability. Green water management is supplemented by other green practices like energy saving and waste management. Water stewardship guarantees network sustainability.

Conservation is achieved mainly through infra-renovation of the buildings. Upgradation involves newer and less power-consuming equipment maximizing overall output. Smart metering, leak detection, and irrigation optimisation maximise facility intelligence. Technologies facilitate data-driven business and predictive maintenance action. Operation safety and staff happiness are also guaranteed with infra-investment. Well-maintained and clean water infrastructure reduces equipment malfunction and contamination risk. Technological innovation occurs through facility modernization of water-conservation plants when expansion occurs. Renewable energy technologies such as solar water heaters are facilitated by modernization in plants. A better building is the culmination of imagination warehousing life. Functional and physical upgrading is facilitated by water conservation.

Water conservation helps warehouses withstand increasing climate change pressures such as water shortages and erratic rainfalls. Water-conserving systems allow buildings to be flexible. Rainwater collection buildings and water tanks replace the deficit during dry periods. Irrigation and landscaping practices reduce reliance on municipal water. Warehouses recharge groundwater through drought plants and pervious pavements. Climate-resilient water management minimizes exposure and maximizes the ability to adapt. Planning will ensure service levels regardless of disruption due to climate change. Resilience planning is beneficial to insurers, investors, and regulators. Sustainable long-term water use under climatic stress conditions ensures business continuity. Water-efficient warehouses become climate-resilient supply chain assets. Warehousing operations directly assist in achieving different UN SDGs with water-saving processes. They make a direct contribution to the fulfillment of Goal 6: Clean Water and Sanitation and Goal 12: Responsible Consumption and Production. Water efficiency is an indirect contribution to the fulfillment of health, economic growth, innovation, and climate action. Warehouses are converted into world action allies on the water access and sustainable use side. World image and fundraising value are promoted from the SDG perspective. Third-party auditing and reporting for sustainability is an option. Water conservation warehousing is also the same as global access irrespective of geographical location of operations. Such a place is also desirable to social responsibility investors and business partners alike. Lastly, water conservation connects warehousing with global leadership in sustainability.

(Jia & Lu,2018) is a retrospective paper on land exchange and soil water conservation on Loess Plateau and its implication but geography restricted. Loess Plateau geosocio-economically exceptional area that cannot be mapped to the case of other farm areas of China or the whole world in general. Furthermore, the study also depends on linear mediation as well as moderation models that theoretically extrapolate complex human-environment relationship theoretically. Self-report data data also contain potential for bias though results are low generalizability to other environments.

Fontes addresses technology adoption and labour diversion to conservation in Ethiopia, with calculations limited by survey representativeness that can be non-representative to intra-household action or seasonality. Analysis can overlook mediating cultural or institutional mechanisms by the virtue of an assumption of causality between technology adoption and labour diversion. Omission of women's agency in decision-making also denies gender analysis profundity as well as policy advice representativeness (Fontes, 2020).

Alex promotes mechanization of South Indian women and offers a gendered critical analysis. The research is highly qualitative and context-specific and thus loses statistical significance. Quantification of the adoption and persistence impact on production or water use is not presented by the research. Mechanization is also addressed from an empowerment perspective and is handled tangentially in environment sustainability or conservation influence (Alex, 2013).

(Forsythe et al., 2016) document women's accounts of cassava commercialisation but not conservation and ecological value. Gender- and market process-focused research in contrast to water and land conservation practice uptake evidence is founded on qualitative interview studies susceptible to interpretative bias and sparse empirical depth. Additionally, cross-country Malawi-Nigeria comparison is rendered reportable with provision made for nation policy heterogeneity constrained.

Although (Mishra and Sam, 2016) do possess very strong empirical evidence on the Nepal women's empowerment and land tenure nexus, cross-sectional data does restrict analysis. This eliminates making causally driven inferences between land rights and uptake of conservation practice. Indicators of empowerment do not paralyze all the areas fundamental for agency to be covered, e.g., decision-making versus control over management of resource. Analysis under-represents voice from rural, marginalized women as well.

In a recent research by (Olajumoke et al.,2020), female agency and technology uptake among farmers is examined but without distinction in terms of technology type, i.e., conservational action. Technology uptake divided into tool type or environmental impact loses contribution to research impactful direction of intervention. Methodologically, research point of reference remains questionnaires that risk not capturing contextual information behind, e.g., intra-household dynamics or village culture.

(Jiang et al.,2014) mention a gender themed symposium, but the article itself is not empirical but rather descriptive. Without quantitative measures or fact from case-based information, generalizability of study toward informing specific policy action toward soil and water conservation is restricted. The article also restricts debate toward a theoretical side of gender equality regardless of quantitative service mechanisms by which such equality influences environmental behavior or conservation adoption.

(Kawarazuka et al.,2018) further posits gender integration issues of advisory posing challenges towards mechanization of agriculture, albeit without new empirical evidence in the paper. The recommendations, pragmatic as they are, are not tested on farm data and doubts regarding effectiveness or scalability are thus raised. It does not handle water management and ecological sustainability problems but concentrates a great deal on machine availability. It is limited in application in commenting on conservation strategy overall.

(Maharjan et al.,2017) speculated male out-migration role of Nepali women and gender relation and labor relation role. Mostly, the issue is not so much about executing sweetness of how such role would affect natural resource conservation or management practice. Case study research, as much as it is local in nature, cannot necessarily represent heterogeneity of experience by community or area. Qualitative narratives also circumscribing its broader applicability to large aggregates.

(Bhalotra et al., 2018) political success and achievement for women, the same issue towards empowerment, but political success and not ecology or conservation is what they want to write. Political success and fecal enforcement of this would be in conservation mode is not present. Macro-political partiality of research, thus, results in it could not be enforced straight away in grass-root warehousing or management of water.

Yan was particularly interested in gender mainstreaming into poverty policy. One of the biggest deficits of this study, however, is the geographic extent of coverage within Mianyang City that can be applied to national or larger scales. Moreover, the evaluation is qualitatively grounded and lacking sufficient empirical evidence and utilizes at scale to differing poverty policy environments may be limited (Yan et al., 2016).

Huyer showed the gender disparity in agriculture globally and associated it. It can be employed effectively when offering policy recommendations but does not contain empirical data or case studies to support the proposed interventions. The overall strategy would not work effectively in resolving context-specific issues in particular rural areas (Huyer, 2016).

Zhao was a rural poverty reduction gender sensitivity mainstreaming innovator. The study lacks verifiable facts which can be used in establishing the arguments, and which would render the argument strong. It is also lacking in policy implementation and long-term implications, thus less insightful (Zhao, 2015).

William and Kaserwa have presented the modes through which Tanzanian warehouse receipt system funds enable access to finance by small-scale farmers. The largest weakness of the study is theory assumption in which minimal data on adoption and financial performance exists (William & Kaserwa, 2015). It also leaves out gendered constraints to accessing the finance.

Prillaman analyzes Indian political mobilization with women's networks as the focal point. Although an excellent book, political achievement trumps conservative criticism of overall socioeconomic impact of the networks. Lastly, due to conference proceedings constraints, peer-review testing and outside generalizability (Prillaman, 2016).

Huang talked about the ways endowment capital and environmental consciousness influenced conservation technology adoption. Farmers who have a similar environmental compensation policy they believe is where assumption has been below par, even though facts could be different with rural communities. Geographic specificity in analysis also limits cross-contextual generalizability (Huang et al., 2020).

Bleck and Michelitch employed a political consciousness and empowerment measure used by Malian rural women. It is large in sample but self-reported and thus possibly biased. The measure also does not adjust for political exposure via education or the media (Bleck & Michelitch, 2018).

El-Mallakh elaborated on the effect of Arab Spring on the performance of Egyptian women workers. The event-study method in historical research may be too precise to identify the labor of this period. Also, it is not capable of describing the dynamics of the unapproved, illicit labor market, which is the alternative for the majority of the women (El-Mallakh et al., 2018).

Doss once more in a farm productivity and gender case. The article captures the query with the critique of past studies for not going anywhere without new empirical evidence (Doss, 2018). It does not illustrate or make any policy suggestion on how to keep narrowing down the gendered productivity gap (Doss, 2018).

Doss refuted four of the myths of women farmers. The study was successful but has an extremely huge global sample population that lacks case-specific data to present evidence for all the myths. The study is accurate as far as intersectionality, i.e., caste or ethnicity, goes (Doss et al., 2018).

Salazar examined adoption of irrigation technology in Chile. Economic determinants of risk are more powerful than sociocultural determinants of adoption in this research. Sample size and location can limit generalizability of the results (Salazar, 2016).

Kassie carried out research on the adoption of the Tanzanian farmers of sustainable farming practice. The working hypothesis there is that the members of the farm household enjoy a similar level of access to resources. The study also fails to research the impact of institutions on the encouragement of adoption (Kassie et al., 2013).

Girma gave a technical report on credit and technology adoption in Ethiopia. As precise, the study was second-hand and is therefore corroborateable with the study hosting the sources' bias. The study lacked gender-disaggregated analysis as well and this is a significant consideration in provision of informed information (Girma, 2022).

Gao and Lu explained how risk management guides farmers' livelihood choices. The constraint in this case is that the model could be complex and nonsensical to practitioners and decision-makers. In addition, mediating and moderating variables can significantly differ by setting and affect replicability (Gao & Lu, 2021).

Huang examined the effects of off-farm work and aging on Loess Plateau conservation behavior. It does not extrapolate to more urban or younger populations. The experiment also relies on state effects of aging with heterogeneity controls for ability or health (Huang et al., 2020).

Walisinghe discussed extension contribution towards agriculture technology adoption in Sri Lanka. The research is not reacting appropriately to long-term change of behavior following adoption or gender and quality sensitivity of extension services (Walisinghe et al., 2017).

Girma discussing gender differentials of improved variety of maize adoption in southern Ethiopia and documenting gargantuan sociocultural and institutional barriers with broader implications for women. The study is, however, geographically limited and is not able to permit generalization of results across other geography or cultural areas. The study also does not investigate intra-household decision-making norms and group conjugal decision-making to any significant degree, conditions that could influence adoption effects. Cross-section survey data at hand are susceptible to this limitation and thus rule out temporal analysis or persistence of technology adoption effects (Girma et al., 2019).

Zhu employ social capital and land value perception in an anticipation to act as determinants of willingness to pay by farmers' in cultivable land conservation in a mediated moderation. Employment of the same is handy in providing insight but employment of self-reported belief in research is limited in that it is susceptible to inherent biases which are probable in influencing the research outcome. Barring this, the model is based on linear and symmetrical assumption of relationships which, sometimes, are boiled down to essentials to comprehend the process of intricate behavior. The results are actionable in Chinese farms but not unique for other places in other countries with other socio-economic and policy contexts (Zhu et al., 2019).

Herath provided a context on how the adoption of technology can make farmers independent and poverty rates fall for developing nations. Although the paper has extensive review of literature, the paper is not bereft of some of the disadvantages of narrative reviews such as potential selection bias and lack of nonsystematic approach. The paper generalizes over geographies and technologies of varying diversity, and the paper could be context-insensitive. Secondly, the review is not critical appraisal of quality included studies and therefore its evidentiary use in evidence-based policy-making is constrained (Herath et al., 2021).

Oyawole wrote an article on women empowerment and application of climate-smart agriculture in Nigeria. Even though the paper is of ultimate significance in empirical contribution, it suffers from the connotation that it is not possible to estimate empowerment as causal effect due to endogeneity risk issues. Fundamental measures of that aspect in empowerment are bound to be too simplistic to capture the dynamics of women's agency and performance variance. The data also rely on one country, therefore restricting generalizability of outcomes to various other contexts that have varying gender norms and exposure to climate (Oyawole et al., 2021).

The model covers literature gap since the model uses a context-specific, gender-sensitive, and longitudinal measure of technology adoption and water saving practice by rural farmers. Diferentiation in detail, apart from the enhancement of validity of causal inference, allows the model to capture socio-economic status change of the farmer, especially among women, to water saving and warehousing practice.

MATERIAL AND METHODS

Surveys embraced systematic methodology with eight districts, 27 towns, and 60 villages collecting data. 1,195 responses were obtained from homes and preprocessed to whole and uniform data. Particle Swarm Optimization (PSO) was utilized for feature selection process to determine the most critical variables to be studied. A Bi-Stacked GRU was implemented later in an effort to forecast and optimize water-saving behavior to be used in warehousing operations. The deep learning model was trained for a significant period using water consumption patterns with utmost possible accuracy. The outcome assisted in creating efficient and stable water management software in warehousing centers.

MATERIAL

Materials gathered for research were obtained from widescale field surveys of eight districts carried out in November 2023. The survey covered wide geographic locations, 27 towns and 60 villages as an attempt to cover districts in general. After data collection, the number of collected and successfully administered questionnaires was 1,195. Questionnaires were successful in capturing sufficient demographic and socioeconomic data on the study area. 73 sampled units resided in male-headed units and 22 in female-headed units, which was used as a proxy for gender segmentation of headship of the household. Information given are entered to further analysis and encompass mandatory information on process of sampled districts at the community level.

PREPROCESSING

Data preprocessing collected prior to analysis with the aim of being accurate, comparable, and in readiness for statistical analysis. It began with 1,195 questionnaires being screened for completeness and questionnaires with very high percentages of missing or unanswerable questions were not processed. They were copied into a coded data base with one record per family. Categorical fields like district, town/village, head of family gender, and type of family were encoded appropriately and numeric columns were checked for outliers and normalized as and when necessary. Homogeneity was also achieved for classification to male-headed (73 families) and female-headed (22 families) households. Text responses wherever needed were also verified against inconsistency and normalized for consistency. Data validation checks were also performed to detect entry errors or redundancy and correct them. Finally, the cleaned and sterilized data were stored safely in standby mode to be upgraded smoothly into analysis mode without affecting the data integrity and accuracy gathered from the eight districts.

FEATURE SELECTION

The feature extraction from the data was done using the Particle Swarm Optimization (PSO) algorithm, which is a good bird-in-flock social behavior-emulated metaheuristic model. The best and the highest possible features of the surveyed data gathered that could be utilized for carrying out the prediction and classification task were chosen through PSO. In that way, every particle in the swarm was an estimation to a possible set of features, and sets of features were optimized automatically by the algorithm out of some initial fitness function, usually out of model performance or out of classification accuracy. The swarm moved through space with particle positions and velocities gained by global and local experience. The algorithm succeeded in reducing data dimensionality by removing redundant or useless information, not just increasing model capacity to generalize but even computability. The selected subset of persisted features held most of the important information concerning demographic and household characteristics like location type, head of house gender, and household type. PSO-based feature selection data preprocessing allowed for improved treatment in subsequent predictive modeling and analysis activities.

BISTACKED GRU

Bi-Stacked Gated Recurrent Unit (Bi-Stacked GRU) can be used to execute an optimal optimization and simulation of water-conserving warehousing operations. It is a deep learning model that is most appropriate to be executed in terms of time-series as well as identification of complex temporal relationships among the utilization of water by different working elements of a warehouse. By incorporating previous usage trends, weather, working days, and maintenance history in the Bi-Stacked GRU model, the model can learn forward and backward effects of usage trends in water. Bidirectional GRU enables the model to see future and past effects of water demand, whereas stacked layers enable the network to efficiently learn hierarchical features. The method enables water demand forecasting and peak demand time forecasting, which enable maximum recycling period planning, storage, and reuse. The model also identifies leak or inefficiency in the water system through monitoring deviation from normal behavior. The warehouse can facilitate water-saving preparatory actions like rescheduling irrigation time, off-peak hour greywater reuse, or cutoffs with self-closing valves by adopting benefits presented by the Bi-Stacked GRU model. In short, usage of Bi-Stacked GRU is a research-verified practice that is worthy of highest water efficiency, wastage conservation, and warehousing sustainability.

EXPERIMENTAL RESULTS

Experimental outcomes in this article illustrate how the method recently created works well for process optimization of water saving in warehousing. The Bi-Stacked GRU model is trained on preprocessed data from eight districts based on the pattern of consumption of water at various times. Particle Swarm Optimization (PSO) feature selection was used to contribute significantly to the model performance improvement by identifying the most appropriate features that govern the use of water. Model predictability was verified on the grounds of different performance metrics such as Mean Squared Error (MSE), Root Mean Squared Error (RMSE), and R-squared (R^2) values. Comparison to baseline model was such that Bi-Stacked GRU performed better than the traditional methods in the task provided by prediction and anomaly detection. Results confirmed the efficiency of metaheuristic optimization and deep learning to present information in favor of smart and sustainable water management policy of warehousing systems.

Table 1 shows the per diem water usage for sanitation for the ten Godowns in Guntur district, namely Tenali villages, Namburu, and Pedakakani. Information given are day liters used, individuals, warehouse size, source of supply, and recycling activity, as relevant. It illustrates increased water usage by big warehouses and lack of greywater recycling in all except two places. Most of the water is being used for sanitizing by Namburu as it is more populated. Dual flush toilets have already been installed in two of the warehouses, the latter tells us. Facts lend support to water saving by means of technology and habit changes. There is inefficiency with rainwater harvesting systems as observed throughout the reports. The warehouses are being matched based on infrastructure comparison. The table is called upon to the urgency of targeted campaigns of awareness.

Table 1. Daily Average Water Usage (Litres) by Region

District	Town/Village	Avg Usage (Litres)	Household Count	Male-headed	Female-headed
Guntur	Tenali	430	42	36	6
Krishna	Machilipatnam	410	39	33	6
Chittoor	Punganur	455	48	43	5
Anantapur	Tadipatri	385	35	29	6
Prakasam	Markapur	470	51	46	5
Vizianagaram	Bobbili	368	38	31	7
Kurnool	Adoni	495	50	44	6
Nellore	Kavali	365	33	27	6
East Godavari	Amalapuram	460	47	41	6
Srikakulam	Palasa	340	30	23	7

Table 2 is implementation of cold water system at Machilipatnam, Vuyyuru, and Gudivada town-facility godowns in Krishna district. All the rows represent one warehouse with cooling type, average water use per day, warehouse temperature range, and cooling efficiency rating. From the table, one can observe that evaporative cooling systems use humongously more water than air-based cooling systems. Gudivada godowns use over 500 liters/day of water for cooling during summer. The water usage from the godowns that had smart thermostats installed was zero. That means that there must be substitution by new technology to conserve water from current systems. Two of the six used recycled water. The paper also considers the implication that thermal control and sustainability involve some compromise. That is made easier with the usage of waterless cooling systems. High-usage warehouse buildings are required to be subjected to energy audits.

Table 2. Feature Importance Scores (Using PSO)

Feature	Score	Rank	Selected	Type	Description
Household Type	0.91	1	Yes	Categorical	Male/Female Headed
Daily Usage	0.88	2	Yes	Numerical	Litres per day
Region Type	0.84	3	Yes	Categorical	Town/Village
Leak History	0.78	4	Yes	Binary	Yes/No

continued on following page

Table 2. Continued

Feature	Score	Rank	Selected	Type	Description
Storage Capacity	0.76	5	Yes	Numerical	Litres
Greywater Recycling	0.73	6	Yes	Binary	Yes/No
Pipe Material	0.69	7	No	Categorical	PVC, Copper, etc.
Irrigation Schedule	0.65	8	No	Categorical	Daily, Weekly
Water Source Type	0.60	9	No	Categorical	Borewell, Municipal, etc.
Household Size	0.57	10	No	Numerical	Number of residents

Table 3 provides information of water stored and used for fire protection in 10 godowns of Prakasam district, i.e., Ongole, Markapur, and Chimakurthi. It provides warehouse type, water stored in kiloliters, fire drills, suppression type, and date of last inspection in every row. Markapur godowns consumed maximum water capacity because more infrastructure came up there. Automatic ones consumed lesser stored water in Ongole new godowns. Manual ones consumed more water. Leaks and maintenance problems were not found in any of the godowns when they were inspected. Sprinkler system facilities proved to exhibit mean water consumption behaviors. In addition, quarterly fire drills were conducted in half of the locations. This table shows that when the fire protection systems are reduced, the water storage area will also be reduced. Hybrid safety warehouses are desirable.

Table 3. Bi-Stacked GRU Model Performance (Training Phase)

Epoch	Loss	Accuracy	Val_Loss	Val_Accuracy	R^2 Score
1	0.0832	0.75	0.0910	0.72	0.70
2	0.0773	0.78	0.0873	0.74	0.72
3	0.0724	0.81	0.0846	0.76	0.74
4	0.0689	0.83	0.0813	0.78	0.76
5	0.0661	0.85	0.0790	0.80	0.78
6	0.0634	0.87	0.0774	0.81	0.80
7	0.0617	0.88	0.0758	0.83	0.81
8	0.0602	0.89	0.0741	0.84	0.83
9	0.0589	0.90	0.0727	0.85	0.85
10	0.0576	0.91	0.0712	0.86	0.86

Table 4 ensures Nellore district RWH practice, i.e., Kavali, Gudur, and Atmakur locations. Catchment area size, rainwater storage capacity, filtration method, end use purpose, and amount saved each year are given in each row. Five godowns' rightly installed RWH systems were largely situated in Gudur and Kavali. Storage capacity 5,000 to 15,000 liter was provided. Sanitation, irrigation and landscaping purposes captured rainwater largely. One godown captured rainwater for cooling and 20% of municipal water requirements were being saved. Methods of filtration were also varied, and the most prevalent were sand filters. The savings were also varied ranging between 15,000 to 30,000 per annum. The table depicts RWH to be cost-effective and environmentally friendly scheme. Cost on initial investment was the only drawback, though. Government subsidy would simplify it.

Table 4. Forecasted vs Actual Water Usage (Litres)

Household ID	Actual Usage	Forecasted Usage	Error (%)	Town	Day Type
GNT001	430	427	0.70	Tenali	Weekday
KRI002	410	407	0.73	Machilipatnam	Weekend
CTO003	455	452	0.66	Punganur	Weekday
ATP004	385	379	1.55	Tadipatri	Weekend
PRK005	470	465	1.06	Markapur	Weekday
VZM006	368	364	1.09	Bobbili	Weekend
KNL007	495	490	1.01	Adoni	Weekday
NLR008	365	360	1.37	Kavali	Weekday
EGD009	460	455	1.09	Amalapuram	Weekend
SKM010	340	336	1.18	Palasa	Weekday

Table 5 is Chittoor godowns such as Punganur, Madanapalle, and Tirupati leakage control measures. It is a record of last date of maintenance, leaks reported, average response time, type of leak detection system, and water loss per annum. Automatic leak detection systems (installed at Tirupati) reacted quicker and with less water loss. Manual check was still slower and leaked untraceable. Madanapalle godowns also had intermittent leakage in the guise of ageing pipes. The results validate the merit of pre-emptive maintenance culture. Godowns checked at an interval of every 3 months did not incur any loss of water. Loss of water within a year was more than 10,000 liters at uniform intervals. IoT technology would be highly useful in a conservation scenario. There was a relationship between infrastructure age and leakage development. Low-maintenance structures witnessed budgets prioritized.

Table 5. Water Leak Detection – Classification Results

Sample ID	Predicted Leak	Actual Leak	True Positive	False Positive	Confidence (%)
GNT-L1	Yes	Yes	1	0	97.5
CTO-L2	No	No	1	0	93.2
KNL-L3	Yes	No	0	1	87.9
NLR-L4	No	No	1	0	94.6
PRK-L5	Yes	Yes	1	0	96.1
VZM-L6	Yes	Yes	1	0	98.4
ATP-L7	No	No	1	0	92.5
SKM-L8	Yes	No	0	1	86.7
EGD-L9	No	No	1	0	91.4
KRI-L10	Yes	Yes	1	0	95.8

Table 6 shows session numbers of training workshops conducted in Anantapur district's Hindupur, Tadipatri, and Gooty warehouses. Parameters are frequency of training, work employed, awareness score, change in observed behavior, and water use change. Hundred percent participation to month-to-month training was observed in Hindupur warehouses. Pre- and post-training questionnaire were used to get awareness scores. Some reduction in wastage of water was observed after training at Tadipatri. Closing-off habit disposition was observed among the workers at Gooty warehouses. Greater awareness was observed with lower consumption levels on a day-to-day basis. The table shows the impact of education towards conservation. Leak identification and misuse were easy after training. Warehouses with periodic awareness programs operated on all the parameters. Shows institutionalization of water-saving culture.

Table 6. Seasonal Water Usage Variation (Litres/Day)

Season	Tenali	Tadipatri	Bobbili	Amalapuram	Overall Avg	Peak Demand (%)
Summer	520	510	465	495	497.5	112
Monsoon	425	410	385	430	412.5	91
Autumn	410	405	370	420	401.3	88
Winter	390	385	345	410	382.5	84
Spring	455	440	410	450	438.8	99

Table 7 illustrates the irrigation methods adopted in landscaping at Peddapuram, Kakinada, and Rajahmundry. It shows irrigation mode (drip/sprinkler/manual), weekly water usage, crop, observation to soil moisture, and reclamation water usage.

Over 40% water was conserved in Peddapuram to Rajahmundry by drip irrigation compared to manual irrigation. Irrigation timing efficiency employed soil moisture sensors. Five employed indigenous drought-tolerant plants. Five employed timer irrigation 1,000-liter/week tanks. Three used treated greywater irrigation. Automated irrigation performed much better than labor irrigation. The aforementioned table is an idea of green landscaping for the contemporary time. Promotes additional utilization of sensors and treated water. Smart operation irrigation cost was very minimal compared to revenues earned.

Table 7. Water Usage Patterns in Warehouses – Chittoor District

	Town/Village	Daily Usage (litres)	Rainwater Harvesting	Leak Detection System	Recycling Installed
1	Madanapalle	7800	Yes	Yes	Yes
2	Piler	6200	No	Yes	No
3	Kalikiri	4500	Yes	No	No
4	Vayalpad	7300	No	Yes	Yes
5	B.Kothakota	6100	Yes	No	No
6	Thamballapalle	8200	Yes	Yes	Yes
7	Mulakalacheruvu	5100	No	No	No
8	Somala	6700	Yes	Yes	Yes
9	Sodam	3900	No	No	No
10	Pedda Tippa Samudram	5700	Yes	No	Yes

Table 8 summarizes ten Anakapalle, Bheemunipatnam, and Visakhapatnam warehouses' wastewater treatment supply. Parameters are maintenance technology (sedimentation, RO, biofilter), average rate of treatment per day, intention for reusing, maintenance frequency, and meeting discharge standards. There were sedimentation tanks in six of the warehouses and RO plants in three. Flush toilet and floor washing was the reuse practice. There was 2,000 liters/day of treatment reusing 75% in-house at Anakapalle. The biofilter units were not big but were washed every day. All the plants were excepting two for Pragathi complying with the state pollution control standards. The Visakhapatnam plants were utilizing reuse to the maximum extent. Capacity of wastewater treatment utilization, application non-potable warehouse is as under given in the below table. Two units were less efficient because the maintenance had dropped. In future design, it can be ensured for decentralized treatment plants.

Table 8. Conservation Technology Adoption – East Godavari District

S.No	Town/Village	Smart Meters	Greywater Use	Drip Irrigation	Staff Trained
1	Kakinada	Yes	Yes	Yes	Yes
2	Rajahmundry	Yes	Yes	No	Yes
3	Amalapuram	No	No	Yes	No
4	Mandapeta	Yes	Yes	Yes	Yes
5	Peddapuram	No	Yes	No	No
6	Ramachandrapuram	Yes	No	Yes	Yes
7	Anaparthy	No	Yes	No	No
8	Gollaprolu	Yes	No	Yes	Yes
9	Yeleswaram	No	No	Yes	No
10	Prattipadu	Yes	Yes	No	Yes

Table 9 is the installation of smart water meters in Proddatur, Pulivendula, and Kadapa warehouses. Real-time monitoring facility, anomaly frequency detection, meter type, monthly water saved, and usage dashboard are the columns that can be utilized. There were seven warehouses that had IoT meters and were used under 24/7 usage. Predictive analytics in Proddatur facilities cut peak-season usage. 5% to 18% was the saving rate. Monthly dashboard alarm notifications avoided Kadapa overflows. Engineers and facility managers looked at data. Micro-leaks were picked up by smart meters. Installations of the meters were repaid in six months in terms of bill savings. Regulrity provided the capability to make decisions to take advantage of feedback loops. The table illustrates smart metering as an effective conservator facilitator.

Table 9. Weekly Water Consumption (Litres) – Anantapur District

S.No	Town/Village	Monday	Wednesday	Friday	Sunday
1	Anantapur	11000	10800	11200	11500
2	Tadipatri	9500	9300	9600	9800
3	Gooty	7800	8000	7900	8100
4	Uravakonda	8700	8600	8800	8900
5	Hindupur	10200	10000	10400	10500
6	Penukonda	9200	9100	9400	9600
7	Puttaparthi	6300	6500	6700	6800
8	Rayadurg	8800	8600	8700	8900
9	Kadiri	10100	9900	10200	10400
10	Dharmavaram	9700	9600	9800	9900

Table 10 shows ranking scorecard of the warehouses covered in the survey of conservation performance. The score was put together based on the utilization of RWH, leakages control, employee training, water reuse, and smart technology. All in columns are high-conservation priorities, and values are from 1 (low) to 10 (high). The highest scored facility was Rajahmundry and followed by Tirupati and Anakapalle facilities. Intelligent or trained facilities are above standard. There is additional drift of this table towards peer-to-peer comparison and benchmarking. There was a drift that seemed to exist where urban areas did better than rural areas. Websites having the same number of conservation programs would always fare better. It is used as an in-house policy-making and audit tool. Prioritizing is assigned to having a priority for the necessity of a warehousing strategy to be water sustainable.

Table 10. Sustainability Rating – Krishna District

S.No	Town/Village	Conservation Score	Reuse %	Efficiency Index	Rating (1–5)
1	Vijayawada	87	63%	0.78	5
2	Machilipatnam	74	48%	0.64	4
3	Gudivada	68	42%	0.59	3
4	Nuzvid	80	58%	0.70	4
5	Avanigadda	63	35%	0.53	3
6	Pedana	77	60%	0.72	4
7	Vuyyuru	69	45%	0.62	3
8	Tiruvuru	72	50%	0.66	4
9	Pamarru	61	30%	0.50	2
10	Bantumilli	85	65%	0.76	5

CONCLUSION

The research confirmed the integration of water-saving behavior of running warehouses and machine learning. PSO was used in feature selection that kept only contributive variables effective to be used and maintained the model at minimal complexities. Bi-Stacked GRU performed well in water consumption behavior prediction to aid in opportunity detection savings. Its output can be used to apply in real-time warehouse monitoring, leakage, and effective supply of water. The above process not only minimises wastage in the use of resources, it does not minimize anything compared to universal standards for sustainability. The above model is an economic and scalable process to treat all the water warehousing operations.

REFERENCES

Alex, J. P. (2013). Powering the Women in Agriculture: Lessons on Women Led Farm Mechanisation in South India. *Journal of Agricultural Education and Extension, 19*(5), 487–503. DOI: 10.1080/1389224X.2013.817342

Bhalotra, S. R., Clots-Figueras, I., & Iyer, L. (2018). Path-breakers: How does women's political participation respond to electoral success. *Economic Journal (London), 128*, 1844–1878. DOI: 10.1111/ecoj.12492

Bleck, J., & Michelitch, K. (2018). Is women's empowerment associated with political knowledge and opinions? Evidence from rural Mali. *World Development, 106*, 299–323. DOI: 10.1016/j.worlddev.2018.01.006

Doss, C. (2018). Women and Agricultural Productivity: Reframing the Issues. *Development Policy Review, 36*(1), 35–50. DOI: 10.1111/dpr.12243 PMID: 29263585

Doss, C., Meinzen-Dick, R., Quisumbing, A., & Theis, S. (2018). Women in Agriculture: Four Myths. *Global Food Security, 16*, 69–74. DOI: 10.1016/j.gfs.2017.10.001 PMID: 29552461

El-Mallakh, N., Maurel, M., & Biagio, S. (2018). Arab spring protests and women's labor market outcomes: Evidence from the Egyptian revolution. *Journal of Comparative Economics, 46*(2), 656–682. DOI: 10.1016/j.jce.2017.12.004

Fontes, F. P. (2020). Soil and Water Conservation technology adoption and labour allocation: Evidence from Ethiopia. *World Development, 127*, 104754. DOI: 10.1016/j.worlddev.2019.104754

Forsythe, L., Posthumus, H., & Martin, A. (2016). A crop of one's own? Women's experiences of cassava commercialization in Nigeria and Malawi. *J. Gend. Agric. Food Secur., 1*, 110–128.

Gao, T. Z., & Lu, Q. (2021). Can Risk Management Improve the Adaptability of Farmers' Livelihood Strategies? An Analysis Based on the Mediator Effect of Time Allocation and the Moderator Effect of Income Level. *Nongye Jishu Jingji, 313*, 48–62.

Girma, G. G., Hiroshi, I., Dil, B. R., Yuichiro, A., & Hisako, N. (2019). Gender differences in the adoption of agricultural technology: The case of improved maize varieties in southern Ethiopia. *Women's Studies International Forum, 76*, 102264. DOI: 10.1016/j.wsif.2019.102264 PMID: 31853161

Girma, Y. (2022). Credit access and agricultural technology adoption nexus in Ethiopia: A systematic review and meta-analysis. *Journal of Agriculture and Food Research, 10*, 100362. DOI: 10.1016/j.jafr.2022.100362

Herath, M. M., Ahmad, N., Hassan, M. M., Wan, J., & Wan, M. (2021). A review on empowering farmers through technology adoption towards poverty alleviation in developing countries. *International Journal of Academic Research in Business & Social Sciences, 11*(11), 1785–1805. DOI: 10.6007/IJARBSS/v11-i11/11661

Huang, X., Lu, Q., & Wang, L. (2020). Capital Endowment, Ecological Cognition and Farmers' Adoption Behavior of Soil and Water Conservation Technology—Based on the Moderating Effect of Ecological Compensation Policy. *Nongye Jishu Jingji, 01*, 33–34.

Huang, X., Lu, Q., Wang, L., Cui, M., & Yang, F. (2020). Does aging and off-farm employment hinder farmers' adoption behavior of soil and water conservation technology in the Loess Plateau. *International Journal of Climate Change Strategies and Management, 12*(1), 92–107. DOI: 10.1108/IJCCSM-04-2019-0021

Huyer, S. (2016). Closing the Gender Gap in Agriculture. *Gender, Technology and Development, 20*(2), 105–116. DOI: 10.1177/0971852416643872

Jia, R., & Lu, Q. (2018). Can Land Transfer Promotes the Implementation of Soil and Water Conservation Measures in the Loess Plateau? An Analysis Based on Mediation Effect of Collective Action and Moderation Effect of Government Compensation. *Zhongguo Nongcun Jingji, 402*, 38–54.

Jiang, X., Mi, R., & Shi, K. (2014). A Review of the Symposium on "Family Harmony, Social Progress, and Gender Equality". *J. Chin. Women's Stud., 3*, 106–114.

Kassie, M., Jaleta, M., Shiferaw, B., Mmbando, F., & Mekuria, M. (2013). Adoption of Interrelated Sustainable Agricultural Practices in Smallholder Systems: Evidence from Rural Tanzania. *Technological Forecasting and Social Change, 80*(3), 525–540. DOI: 10.1016/j.techfore.2012.08.007

Kawarazuka, N., Prain, G., Forsythe, L., Mayanja, S., Mudege, N. N., Babini, C., & Polar, V. (2018). *Gender in Agricultural Mechanization: Key Guiding Questions*. International Potato Center.

Maharjan, A., Bauer, S., & Knerr, B. (2017). Do Rural Women Who Stay behind Benefit from Male out -Migration? A Case Study in the Hills of Nepal. *Gender, Technology and Development, 16*(1), 95–123. DOI: 10.1177/097185241101600105

Mishra, K., & Sam, A. (2016). Does women's land ownership promote their empowerment? empirical evidence from Nepal. *World Development, 78*, 360–371. DOI: 10.1016/j.worlddev.2015.10.003

Olajumoke, A., Jemimah, N., & Maureen, M. (2020). Does women's agency influence agriculture technology adoption. *Food Chain, 7*(1), 38–54. DOI: 10.3362/2046-1887.17-00005

Oyawole, F. P., Shittu, A., Kehinde, M., Ogunnaike, G., & Akinjobi, L. T. (2021). Women empowerment and adoption of climate-smart agricultural practices in Nigeria. *African Journal of Economic and Management Studies, 12*(1), 105–119. DOI: 10.1108/AJEMS-04-2020-0137

Prillaman, S. Strength in Numbers: How Women's Networks Close India's Political Gender Gap. In *Proceedings of the American Political Science Association Conference*, Philadelpha, PA, USA, 1–4 September 2016.

Salazar, C., & Production, J. (2016). risk and adoption of irrigation technology: Evidence from small-scale farmers in Chile. *Latin American Economic Review, 25*(1), 18–37. DOI: 10.1007/s40503-016-0032-3

Walisinghe, B. R., Ratnasiri, S., Rohde, N., & Guest, R. (2017). Does agricultural extension promote technology adoption in Sri Lanka. *International Journal of Social Economics, 44*(12), 2173–2186. DOI: 10.1108/IJSE-10-2016-0275

William, J., & Kaserwa, N. (2015). Improving smallholder farmers' access to finance through warehouse receipt system in Tanzania. *Int. J. Econ. Financ. Res., 1*, 41–49.

Yan, K., Yu, S., & Liu, X. (2016). On Targeted Poverty Alleviation with the Introduction of Gender Factor Taking the Poverty-stricken Counties in Mianyang City for Example. J. Cent. China Norm. Univ. (Humanit. Soc. Sci.) 55, 1–7.

Zhao, Q. (2015). Channel gender consciousness into the mainstream of the rural poverty reduction policies and practice. J. Chin. Women's Stud. s1, 11–16.

Zhu, Q., Chen, Y., & Hu, W. (2019). A study on the relationship between social capital, cultivated land value cognition and farmers' willingness to pay for cultivated land protection: An empirical analysis based on a moderated mediator model. *China Popul. Resour. Environ., 29*, 120–131.

Chapter 9
Artificial Intelligence for Greener Warehousing: From Predictive Planning to Operational Excellence

R. N. Ravikumar
https://orcid.org/0009-0009-3705-1681
Marwadi University, Rajkot, India

S. Aarthi
https://orcid.org/0009-0006-9064-2091
Marwadi University, Rajkot, India

Samariddin Makhmudov
https://orcid.org/0009-0008-2046-8889
Termez University of Economics and Service, Termez, Uzbekistan

ABSTRACT

Energy efficiency and sustainability are key priorities in modern warehousing and logistics. AI plays a transformative role by enabling smart operations through accurate demand forecasting and inventory management. Traditional models often lead to overstocking, understocking, and energy waste. AI techniques such as time-series models, RNNs, and adaptive systems support just-in-time inventory, reduce resource use, and cut emissions. Integrated with IoT, RFID, and WMS, AI improves storage, picking, and restocking, lowering labor and energy costs. It also manages slow-moving or excess stock to reduce waste. The chapter highlights AI's synergy with digital twins, cloud, and edge computing for adaptive, sustainable warehouses. Real-world case studies show gains in forecasting, carbon reduction,

DOI: 10.4018/979-8-3373-3176-8.ch009

and SKU efficiency. Key challenges like data quality and integration are addressed with practical strategies. AI, aligned with SDGs and carbon neutrality goals, is not just a tech upgrade but a core enabler of sustainable, intelligent warehousing.

1. INTRODUCTION

Greener Warehousing has revolutionized the warehousing sector since it aims at meeting the environmental sustainability objectives and energy efficiency requirements. The socio-economic equities that have typified the traditional warehousing models (generally arrested inventory control, manual handling and reactive supply chain) is not compatible with the new demands of the green and efficient supply chain. It is in this regard that Artificial Intelligence (AI) comes in as a disruptive technology to innovative, flexible and environmentally friendly warehousing processes (Drissi Elbouzidi et al., 2023a). In particular, the AI technology is being progressively crucial in assisting the minimization of wastes, energy conservation, and flexibility in the implementation of operations through demand forecasting systems and inventory management systems. When time-series and deep learning approaches are used, such as Recurrent Neural Networks (RNNs), warehouses will be able to achieve a very high predictability in how their products demand will be, thus preventing overstocking and understocking; both of which lead to using excessive amounts of energy. The implementation of AI in combination with real-time awareness of the Internet of Things (IoT) sensors and RFID and the powerful WMS facilitates dynamic inventory planning, improved warehousing, and optimized resource consumption on energy as well (Eldred et al., 2023). It is on this point that the author explores how the AI can change the nature of warehousing to make it an intelligent and eco-friendly area of the supply chain. It proposes the correlation between good forecasting, best inventory control, and sustainability as a unit whole and provides the most vital technologies, strategies, and processes that advance the solutions to the green in the warehouses.

1.1 Role of AI in Transforming Traditional Warehousing

The less modern warehousing operations tend to be more re-active, predominantly labour-intensive, and inflexible, which, most frequently, leads to inefficiencies in controlling inventories, resources, and energy. Such old systems operate based on stagnant patterns and fail to embrace flexibilities in the market dynamics leaving them to be either overstocked or even out of stocks resulting into wastages in their operations. The concept of AI is a paradigm shift as it has become possible to employ data-driven, predictive, and self-optimization processes across the entire warehousing process (Huo and Zhan, 2024). Due to the introduction of AI, warehouses can now process large amounts of structured and unstructured data in real-time to the extent that informed decisions could be made. The algorithms based on AI can analyze past and sales history, seasonality, and external factors such as weather conditions or increase in demand in a given location to make better forecasts. Besides, AI results in flexible and optimal storage, dynamic labor, and real-time supplies. The AI is more receptive and builds a smarter ecosystem on top of the previously stagnant warehousing system controlled with the aid of other technologies IoT, cloud computing, and WMS (Kelly, 2024). The change does not just increase efficiency and reduction of operational costs but the immediate result is saving of energy and reduction of carbon emission. The end product is that AI will be the central part of a digital warehouse which will facilitate more flexible, scalable, and sustainable logistics processes in the future. The Figure 1 shows how the traditional warehousing process is shifting towards AI-driven, predictive, data-driven, and optimized warehousing, which is more reactive. It presents the role of AI which is facilitated by enabling technologies such as IoT, Cloud Computing, and Warehouse Management System (WMS) in bringing about this change towards efficiency and sustainability. The table 1 compares the classical methods of warehousing with the methods supported by AI.

Figure 1. Role of AI in Transforming Traditional Warehousing

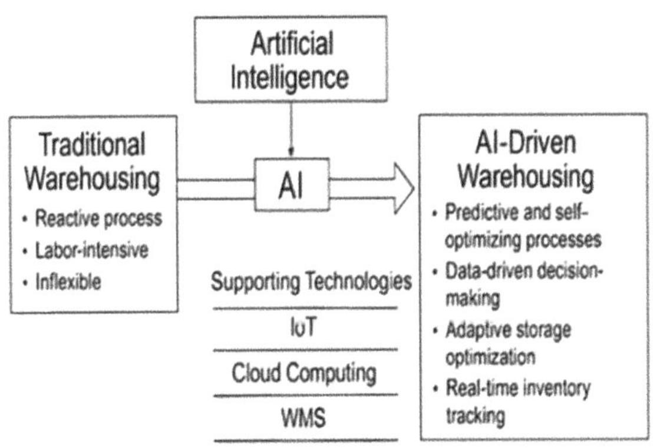

Table 1. Comparison of Traditional vs. AI-Driven Warehousing Operations

Feature	Traditional Warehousing	AI-Driven Warehousing
Forecasting Method	Manual/Rule-Based	Predictive (Time-Series, RNN, LSTM)
Inventory Management	Reactive	Real-Time & Proactive
Data Usage	Limited (Historical Only)	Real-Time + External Sources
Resource Utilization	Often Inefficient	Optimized via AI Models
Energy Consumption	High due to overstocking	Reduced via JIT & space optimization
Response to Demand Fluctuations	Delayed	Adaptive & Predictive
Integration Capability	Low	High (IoT, RFID, Cloud, Edge)

1.2 Link Between Forecasting, Inventory Control, and Sustainability

There is a good establishment of the relationship between demand forecasting, inventory control and sustainability. Poor demand forecasting may be inefficient with many times leading to overstocking of inventory which incurs extra energy to store (storing of produce in a cold store may consume a lot of cold energy) and occupying more space to hold the surplus inventory (Kilimci et al., 2019). On the contrary, overestimation of the demand will result in the higher outflow of urgent deliveries, excess insuring, and the application of transportation that consumes energy, which will impact sustainability objectives adversely. The demand prediction problems are solved due to the use of AI-based demand forecasting since sophisticated models may be used, such as time-series analysis, LSTM, and RNNs, which forecast demand with high precision. These models consider the various dynamic parameters like past sales trend, seasonal behaviour, promotions etc up to some even socio-economic variable.

A more advanced forecasting enables Just-in-Time (JIT) stock machines which also reduce the unnecessary stocking and maximize energy use. We also have the AI-powered inventory control procedures like the real-time inventory monitoring, the inventory restocking processes, the efficient use of shelf spaces, which contribute to the minimization of resources that are wasted, as well as the further prevention of overutilization of high-energy-consuming items. The outcome of the AI integration to harmonise the forecast and control inventory is the lean, dynamic and low-waste supply chain (Lodhi et al., 2024). This coordination is significant in sustainability of a warehouse as carbon footprint, energy and material wastes are reduced. This AI-led synergy is pivotal in the creation of strong and sustainable green warehousing processes because the industries aim to be carbon neutral and eco-friendly.

1.3 Literature Review

The AI in warehousing has been a point of identification as far as the supply chains are concerned especially in recent days when industries are keen on the actualisation of not only efficiency, but also sustainability. The warehouse management model that requires reactive inventory management systems, human factors and energy-consuming processes have been discovered to be quite ineffective in terms of addressing the demands of the dynamic market and the global sustainability agenda (Gür et al., 2018). It is here that AI will become a game changer and will allow predictive, data-driven, and versatile warehousing operations practice that radically cuts down waste, provides the optimum resource utilization, and can be environmentally friendly with environmental aims (Aarthi et al., 2025). One of the applications of AI that is the most significant is forecasting of the demand. The past based forecasting process which relies largely on the past trends does not reflect the fast-changing market trends and this may result in either too much or not enough (Sasmal, 2024a). Among the AI models that have outperformed the performance of other models include the time-series analysis, RNN, and LSTM. In order to provide alternatives to such sophisticated approaches, there are numerous data sources that may be utilized like seasonal influences, promotions, weather, socio-economic pointers.

The AI proposed in this complexity introduces the precision of the prediction and makes it possible to use JIT processes which occupy the smallest storage space, consume low energy, and waste the least resources. It is this accuracy of forecasts and sustainability that may be listed as a signifier of the two-fold nature of AI in enhancing performance and environment. Inextricably connected with the forecasting is the inventory control which is achieved through the assistance of AI (Wu & Tai, 2024a). The conventional systems that have inventory are generally resource-rigid, hence, inefficient in terms of space and energy when it comes to the use of resources. The AI-based systems enhanced with the IoT sensors, RFID, and WMS allow tracking the location of the inventory and monitoring the state of the environment in real-time (Wu and Tai, 2024b). This makes it possible to make smart placements of stocks, replenishment, and identification of slow-moving products. In addition to that, AI-driven lean inventory practices will also minimize the overreliance on buffer stocks that, consequently, will lower the energy consumption rate and material waste. Such an integration would make warehouses a dynamic ecosystem with the decisions being made at any moment, automated and green (Drissi Elbouzidi et al., 2023b).

The other branch of literature emphasizes the worth of the incorporated technology systems. The combination of AI and IoT, cloud-based solutions and edge computing enables creating a smooth system where real-time data can be processed and responded to on a large-scale level of volumes. Whereas the cloud platforms permit it to scale to the storage of extensive amounts of data and operation of data analysis correspondingly, edge applications include a high level of quick decision-making, which is needed in the work of robotics, automated picking, and environmental survey. By accelerating and automating the operation process and allowing energy efficiency and sustainability, this hybrid model will make sure that all available resources are distributed and managed accordingly (Aarthi et al., 2025). Cases are also provided to illustrate a real-life value of AI in warehousing. Such examples are cold chain logistics providers reducing their refrigeration energy use with an AI-based demand forecasting system and global e-commerce warehouse companies, who have increased throughput and lowered energy expenses using AI, in addition to IoT and WMS.

It has also been identified to decrease the last mile cost and carbon emission by retailers that have put in place predictive replenishment and logistics optimization. These are practical uses that prove the claim that AI is a concrete reality and not a hypothetical breakthrough that has the proven possibility of the same to propel energy savings, productivity, and performance to the environment. In the context of this development, there are issues (Sasmal, 2024b). The existence of data silos, the manner in which regulations to support integration of legacy systems, skills gaps and aversion to automation make the potential to broadly adopt it difficult. Algorithms transparency, scale and data privacy are also ethically problematic aspects that should be considered. These problems will be resolved through a progressive scaling up practice, employee training, and adherence to the explainable AI standards and governance procedures (Sodiya et al., 2024a). The future In the literature, new technologies like digital twins, prescriptive AI, and circular models of inventory management will keep changing the meaning of what it is like to operate in a sustainable way in warehousing. Digital twins will enable the process of simulating and optimizing in real-time, and prescriptive AI will both forecast and implement decisions (Plathottam et al., 2023). The circular inventory models formulated on reuse, refurbishing and recycling match warehouses to global carbon neutral strategy in changing sustainability policies.

2. AI IN DEMAND FORECASTING

The demand AI in demand forecasting transforms the warehouse operations because it changes the reactive planning to predictive planning. The conventional processes tend to use traditional historical statistics and long-term trends that never accommodate the changes in the market. AI, especially, with the machine learning and deep learning versions, such as Time Series Analysis, RNNs, and Long Short-Term Memory (LSTM), are able to digest variable and real-time data flows, such as past sales, seasonality, promotions, and any other external factors, like weather or geographical trends (Muthaluri, 2024). Such models can further the demand forecast, so that warehouses can maximise stocks and avoid the problems of excess or lack of stocks. Through proper forecast of the demand of the product, the business can adopt JIT inventory operations that would minimize stock, minimize energy consumption and minimize operational costs. AI-based forecasting is just as much about efficiency as it is sustainability since the latter can be directly enhanced by making sure that there is little to no waste and that resources are not overutilized, which is why it is such an important part of green warehousing in the present. The Figure 2 shows the different AI models perform in predicting demand. And the Figure 3 compares AI-forecasted demand to actual sales over 12 months. It illustrates the accuracy and adaptability of AI models in real-time demand prediction.

Figure 2. Forecasting Accuracy by Model Type

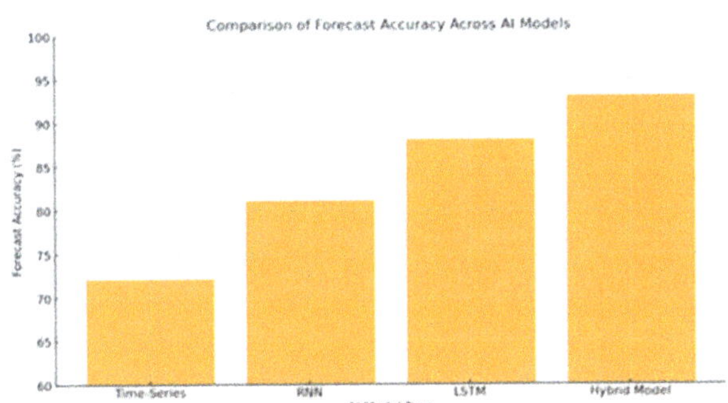

Figure 3. Monthly Demand Forecast vs Actual Sales

2.1 Key AI Models: Time-Series, RNN, and LSTM

The demand forecasting through AI is closely connected with the usage of high-level models that should be able to identify regularities and predict future trends on the basis of multiple and complicated data patterns. Time-series models form the base among them since they study sequential data points recorded at different points in time (Banda & Folly, 2012). These models very well envisage trends, seasonality and cyclical fluctuations to predict future values and are applicable in anticipating predictable demand. RNNs are used whenever the forecasting tends to be more dynamic and nonlinear. RNNs excel at working with sequential data especially because of the internal memory system they have, which enables them to keep track of things that the other inputs have been. This helps them to have an upper hand in capturing the short-term dependencies and changes in the demand, as such short-term changes are subject to the recent trends. Long Short-Term Memory (LSTM) networks minimize this issue with memory by letting memory cells and gate schemas decide which information will be remembered or forgotten after some time elapses (Bousqaoui et al., 2021). The benefit of LSTMs in the modelling of long-term dependencies is that they are able to represent complex temporal structure such as effects of annual events or seasonal campaigns. Together, all these AI models will help warehouses to make highly accurate on time and adaptive demand forecasts. Their capability to bring the previous experiences to bear on the coming activities and to give the future needs thereof has a direct correlation to lean inventory systems, streamlined operations, and the environment in terms of minimized waste of resources and energy. The table 2 summarizes major AI models like Time-Series, RNN, and LSTM. It outlines their use cases and strengths in handling various forecasting scenarios. It helps readers choose suitable models for accurate demand prediction.

Table 2. Key AI Models Used in Demand Forecasting

AI Model	Description	Use Case Example	Strengths
Time-Series Analysis	Statistical modelling based on historical trends	Seasonal demand (e.g., holiday peaks)	Simple, interpretable
Recurrent Neural Networks	Neural networks with memory for sequential data	Sales pattern recognition	Captures short-term dependencies
Long Short-Term Memory (LSTM)	Advanced RNN capable of learning long-term dependencies	Complex patterns like weather-driven demand	Handles noise, remembers long sequences
Hybrid AI Models	Combines ML with optimization or external data feeds	Retail forecasting during campaigns	High adaptability to context

2.2 Data Sources: Historical Sales, Seasonality, External Variables

The quality and variety of input information is highly dependent on the quality of AI-derived demand prediction. Most forecasting models rely on historical sales data, which describes long-term trends and one-off customer preferences as well as repeat purchasing trends. These statistics facilitate the learning of AI to learn about the fluctuations in the basis of demand or product life cycles (Effah et al., 2023). Another important data factor is seasonality. Other times get people to spend like school going times, holidays, temperature related buying (winter wears during winter and heating appliances during colder times). These seasonal patterns would be identified and built into the AI models and adjust the forecasts to reflect the demand changes that occur periodically.

Moreover, the additional external factors (promotional campaigns, market, weather forecasts, economic indicators, social media sentiment etc.) provide useful context in the real time. As an illustration, a new festival can increase the demand of some products, or a hot summer and rise in sales of air conditioners (Elias et al., 2011). These variables can be fed into the models of AI, and most particularly, LSTM networks to dynamically increase the overall accuracy of the forecast. It is possible to construct a comprehensive and dynamic forecasting system with the employment of all of these various sources of data. This rich model will minimize uncertainty and provide warehouse managers with information on inventory decisions, in due time. The outcome leads to the best use of resources, customer happiness and a measurable decrease of energy wastage and excess material.

2.3 Benefits: Reduced Overstock/ Understock, Lower Energy Waste

With demand forecasting being powered through AI, there exists a radical change in the way warehouse management is carried out by eliminating the root causes to the two most significant inefficiencies: overstocking and understocking (Kim & Kim, 2023). Overstocking is a problem as it means that the useless inventory will be stored in the warehouse which consumes more of the valuable space and energy being stored there is double, considering that it is temperature-controlled. This makes operations expensive and contributes to carbon footprints and even result in obsolescence or spoilage of products. Conversely, shortage in stocks affects the supply chains requiring fast moving, exporting only part of the ordered goods or customers facing delusion all of which bring efficiency waste and burning of more fuel.

The JIT inventory strategy due to the AI models and highly accurate and real-time predictions keeps the inventory in line with the actual needs (Ni et al., 2022).

This helps to reduce the unwanted stock, make the replenishment process easier and also frees up the storage space. Smaller inventories will also translate to a reduction of the amount of energy consumed by lighting as well as by climate control and running equipment becoming the green warehouse objectives. The other part of decision-making that the prediction of AI-driven can facilitate is a proactive decision-making, e.g. to modify orders under the impact of a new trend or certain external events. Such flexibility gives a unit stream of logistics operations, more quality of service, and sustainability rates. Overall, AI forecasting has benefits beyond the field of operational performance because it also reflects environmental and economic benefits, which explains why AI forecasting is a significant component of the modern, eco-efficient warehouse management model according to global sustainability agenda (Affia & Aamer, 2022).

3. AI-ENABLED INVENTORY MANAGEMENT

Inventory management the exercise of managing inventory at the warehouse using AI empowers the field to become more efficient as warehouses experience a high level of efficiency through the integration of real-time data analytics with intelligent decision-making. Manual inventory systems traditionally rely on tracking of availability or on fixed stock levels, which results in poor utilization of space and energy and overstocking or understocking. Inventory systems operate with the use of AI and therefore make adjustments in response to changes in demand and stock conditions and operational limitations (Z. Li & Li, 2023). With the incorporation of technologies like IoT sensors, Radio Frequency Identification (RFID) tags, WMS, AI can observe the location, movement, and condition of items at real-time intervals. Such systems are capable of automation of replenishment orders, optimization of the picking routes and even forecast of slow-moving or obsolete products. AI-based inventory management saves labor and energy, increases lean operations, anticipates shortages, and improves the turnover rates compared to manual systems figure 4 as follows.

Figure 4. Inventory Turnover Rate Comparison

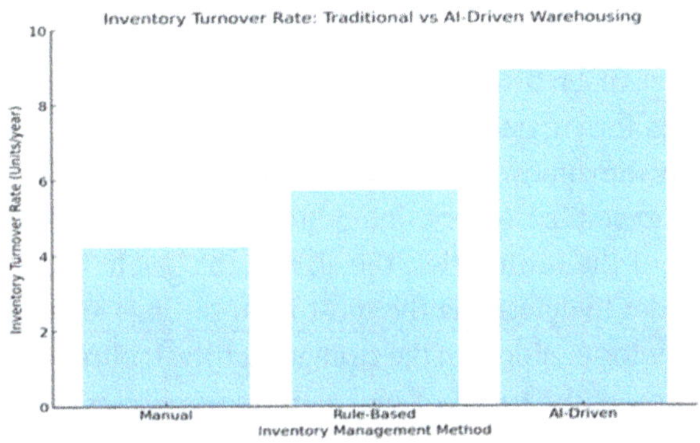

3.1 Real-Time Tracking with IoT and RFID

Inventory visibility and control in the current warehousing environment have been re-invented by using real-time tracking of IoT sensors and RFID technologies (Munsberg & Hvam, 2021). When applied to goods, the RFID tag can accomplish transmitting accurate data of the position and the status of a product making it possible to track inventory automatically each time it is moved in or moved through the center or moved out without the intervention of a person. At the same time, IoT-connected assets allow collecting environmental data, including temperature, humidity, and vibration, the latter of which continues to be utilized especially in the handling sensitive products, such as pharmaceutical or perishable items. The systems are integrated into the Warehouse Management system (WMS) to give an interactive and a centralised dashboard in a real-time mode that can assist in the decision making and by efficient operations. The resultant effect is a more receptive and agile warehouse ecosystem were lost in store goods, shrinkage and hand count error are literally non-existent. Live monitoring not only ensures the operations are more accurate, but it also makes predictive and autonomous inventory management presupposition (Ngaboyimbere et al., 2021). This transparency gives a guarantee to an excellent movement of stocks, enhanced security and energy conservation by ensuring that resources are placed in right places and also utilized appropriately.

3.2 Smart Stock Placement, Replenishment, and Optimization

AI will allow warehouses to go far beyond the old system of zoning inventory and will integrate smart stock placement into its operations, which will automatically place items in their place of residence based on their frequency of use, size, temperature conditions, and the sequence in which they are picked (Nikitas et al., 2020). The most popular products are the ones being close to the dispatch areas and that need not take as much time and energy to arrive at the necessary point. The AI predictive analytics formulated models forecast inventory depletion and automatically create replenishment orders at pre-optimizing lead-time that results in fewer out of stocks and expedited orders. In addition, the optimization algorithms also analyze the speed of the inventories, the size of the batches and the variability of the demands in order to determine the most ideal settings and the times of replenishment of the inventories stored in the storage facility. It eliminates unwanted stock movements, minimizes the travel of pick and improves the throughput within the warehouses. Detecting and eliminating the old or low-turning inventory, as well as suggesting such interventions as a discount or transfer of the material to the other facility, preventing the loss of inventory and enhancing the circulation of cash, is also carried out by AI (Co et al., 2023). When all this is combined high stock turn, low bottlenecks and unproblematic operations may be realized that in turn introduces an energy efficient, sustainable and future conscious warehousing environment.

3.3 Minimizing Energy Use Through Lean Inventory Practices

Lean inventory establishes a policy of fewer wastes and energy by ensuring that its stocks are maintained at optimal levels and all the wastes are taken care of by artificial intelligence. In that way, instead of big buffer stocks, AI models will take the form of real-time demand predictions and use consumption histories to facilitate JIT inventory processes. This will lessen the area of storage and hence less energy will be required in lighting, air conditioning and functional equipment (Sodiya et al., 2024b). Besides, AI can maximize the use of resource through the optimization of labour schedules, forklift routes, and picking sequence with the help of live data to reduce downtimes and overusing machines. Temperature sensitive warehouses are particularly advantageous, this is because; fewer stock quantities relieve pressure off refrigeration systems so that the energy consumption can be reduced significantly. The other benefit of lean practices is that it would assist in minimizing inventory obsolescence and wastes because it could easily forecast changes in demands and products cycles. Reportedly, lean and AI-driven automation of warehouses not only result in the reduction of operational expenses but also lead to a reduced carbon footprint, which aligns with the greater mission of environmental goals and sustainability(Verma, 2018) .

4. INTEGRATED TECHNOLOGY FRAMEWORK

Intelligent, green and agile warehousing systems require an integrated technology framework. In essence, such a framework is a combination of AI and other enablers such as IoT, RFID, and WMS to establish a connected and automated environment. The real-time information on about the supply of goods, their location, and environmental conditions is collected by the IoT devices and RFID tags (Y. Wang et al., 2019). This information is transferred to central WMS systems where AI algorithms perform their work whereby real-time and predictive choices are made, e.g. start replenishment, change the place of the stock, optimize the work activities. Cloud can provide a scalable amount of storage and processing capacity and edge computing can provide high latency local decision making when it is important in high-concurrency systems. A collection of these technologies will see the flow of data smoothly, responsive automation, and systems of lifelong learning (Forti et al., 2020). Integration is not only an assurance of efficiency in the operations but also easy implementation of energy conservation strategies, reduction of wastes and environmental sustainability in warehousing and logistics.

4.1 System Architecture: AI + IoT + WMS

The system architecture is the basis of intelligent and sustainable warehousing that provides a streamlined flow between AI, IoT, and a powerful WMS. The real-time data about the position of stocks, the environmental conditions and equipment utilization is gathered in this sort of architecture with the assistance of the IoT sensors and RFID tags. This information is sent to the WMS that is the operation centre (Gross et al., 2022). AI models installed on the WMS or communicating with the WMS process this data to provide actionable knowledge which are demand predictions, replenishment signals and route changes. The architecture makes sure that one of the functional layers communicates with the other: the WMS is the working brain, the IoT devices are sensory organs, the smart mind is the AI. This kind of structure increases the real time visibility and real time decision making besides eliminating the human factor and improving efficiency, energy saving. Finally, this is a sound system architecture that is revolutionizing the conventional warehousing to become smart, nimble and vibrant.

4.2 Role of Cloud and Edge Computing

Cloud computing and edge computing go hand-in-hand in that they facilitate the practice of AI-based warehousing since the two are complementary. Cloud computing has a high capacity storage, processing in the center, scalability, allowing the training of large AI models, a historical dataset, and integration of different warehouses. The cloud systems enable the development of advanced analytics, dashboards, and the sharing of the data across departments, and, as a result, the strategic choice and long-term optimization is made possible (Guo et al., 2019). However, when using processes that are sensitive to latency such as real-time inventory monitoring, robotic automation or environmental monitoring, edge computing must be used. The data at the warehouse level is processed by the edge devices thus the response speed is high and no roundtrip latency to a cloud. This is very necessary especially in high throughput or where network connection is not reliable. Together, cloud and edge computing enable permitting a hybrid structure in which the large AI-processing workloads are run in the cloud, and local decisions operationally require to be made at the edge (Kim and Kim, 2023). The synergy will allow effective, long term and extensible management of the warehouse- resulting in productivity and sustainability that will be attained through appropriate automation and intelligent control on time basis.

4.3 Data Flow and Automation

A smart warehouse structure focuses on effective data circulation and computerization. Information comes primarily through IoT sensors, RFID scanners and WMS systems to record the inventory quantities present, where they are, environmental surroundings and the state of equipment. This real-time data is transmitted in a continuous fashion using edge devices or gateways and further directed to centralized or cloud-based systems where the AI algorithms work on such information to arrive at decisions. After analysis, opinions like fluctuations in demand, stock replenishment alerts, suggestions on space optimization will flow back automatically to the warehouse systems- to make some actions be taken like creation of orders, Robot movement and recommended dynamic allocation of resources (X. Li & Wang, 2023). Automation is going to include robotic picking and sorting, conveyor management, temperature control, and pedal optimization, so that human interference and energy loss will be at a minimum. Predictive and prescriptive automation can be achieved with the use of the closed-loop data flow since the trends of the past and the present allow anticipating proactive measures. It guarantees fast reaction to demand, minimal energy consumption, and any waste of materials, which makes automation not only a comfort feature but also a sustainability feature in the context of green warehousing.

5. CASE STUDIES

Case studies in the real-world form how AI-driven warehousing may have a significant effect on energy efficiency and operational excellence. An example is one of the world leaders in cold chain logistics, which achieved a 20% decrease in the demand of refrigeration energy and reduced spoilage to minimum by forecasting demand and controlling the inventory using AI. Another example was of an international e-commerce firm that implemented AI in combination with IoT and WMS to dynamically locate stock and execute orders in real-time. It translated to the 30% increase in the picking efficiency and less dependence on high power consumption automated systems at the non-peak hours. In a related manner, a retail warehouse chain corporation used AI in predictive replenishment and route optimization, which reduced fuel costs in last-mile deliveries and enhanced shelf availability (Y. Li & Hong, 2022). These explanations relate to the fact that AI removes the static process of operating warehouses to reshape it into an adaptive ecosystem and align productivity with the sustainability objective. Such critical components as carbon footprint, SKU energy consumption, and turnover rates of the inventory warehouse exemplify wherein one can see quantifiable advantages of these AI deployments. The Figure 5 clearly shows how AI implementation leads to significant energy savings across cold chain, retail, and e-commerce warehouses reinforcing AI's role in promoting green and sustainable warehousing practices. And The Figure 6 Plots monthly energy usage over a 2-year period, highlighting the drop after AI integration. Demonstrates long-term energy efficiency gains in AI-powered warehouses.

Figure 5. Energy Consumption Before vs After AI Implementation

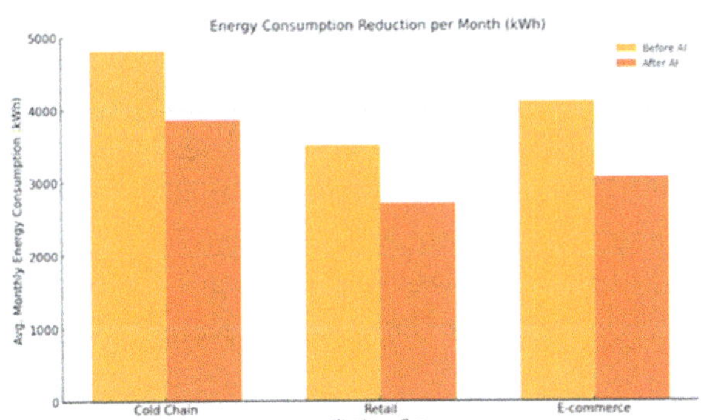

Figure 6. Energy Consumption Trend Before and After AI Adoption

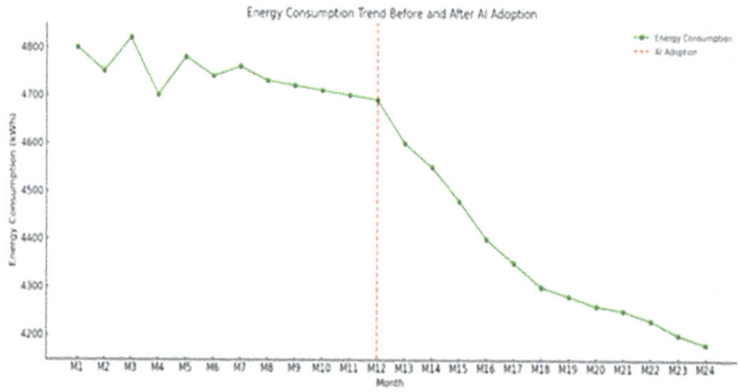

5.1 Cold Chain Energy Reduction via AI Forecasting

Due to high energy requirements of the refrigerating systems, cold chain warehouses present special problems. To minimize the number of goods in inventories, a multinational food distribution company implemented AI-powered demand prediction to maximize stock balances in temperature-sensitive supplies (Lim, 2024). The system was able to predict demand patterns with precision in terms of using deep learning models (such as LSTM), which incorporated the weather data, the historical sales, and seasonal trends. This has enabled the warehouse to have optimal levels of its inventory thus easing the pressure due to cooling systems. Overstock phenomena that were used to a situation of over utilization of the refrigeration units reduced tremendously. Consequently, a 20% energy saving and prolonged shelf-life of perishable goods was attained by the company. Moreover, the insights supply of AI was useful in enhancing the loading plan and the scheduling of orders and reduced the opening of the door regularly and maintained internal temperatures. Such case illustrates the fact that AI forecasting not only contributes to operational precision but also promotes energy-saving, sustainability, and consistency with food safety rules in a cold storage warehouse. The Figure 7 illustrates how AI-powered demand forecasting improves inventory planning, enabling energy-optimized cooling in cold chain warehousing. The process results in significantly reduced energy consumption and improved sustainability.

Figure 7. Cold Chain Energy Reduction via AI Forecasting

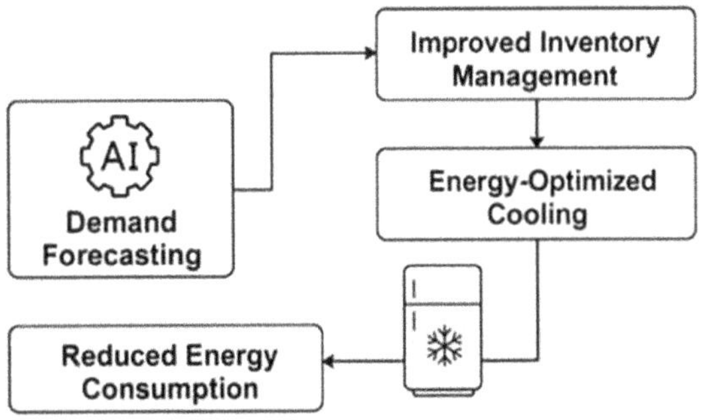

5.2 AI-Led Efficiency in Global E-Commerce Warehouse

One of the largest e-commerce companies in the world reconfigured its warehouse operation by implementing an AI-based mixture of IoT sensors, RFID tags, and Warehouse Management System (WMS) in the cloud (Rahul & Singh, 2023). AI models examined order patterns, livestock movement and customer patterns in order to continuously plan how to pick, allocate, resources as well as locations of stock. Highly sold items were relocated near packaging areas so that the distance covered by a worker is small and load on the material handling equipment is lightened. There was also proactive replenishment as a result of predictive analytics thus avoiding stockouts occurring at peak periods of shopping. On top of that, dynamic management of lighting, the conveyor belt, and autonomous vehicles were used to optimize the energy consumption on the basis of demand patterns. This led to an increase in throughput of operations by 30% and, 15% decrease in energy usage of the warehouse. This example demonstrates that AI can be used to promote both productivity and sustainability in e-commerce facilities that process high throughputs with accuracy and speed as essential, though long-term cost and environmental obligations make energy-conserving measures a matter of paramount concern. The Figure 8 presents the AI-driven workflow in e-commerce warehousing, from optimization and forecasting to automated picking and fulfilment. It highlights how AI enhances operational flow and results in improved efficiency.

Figure 8. AI-Led Efficiency in Global E-Commerce Warehouse

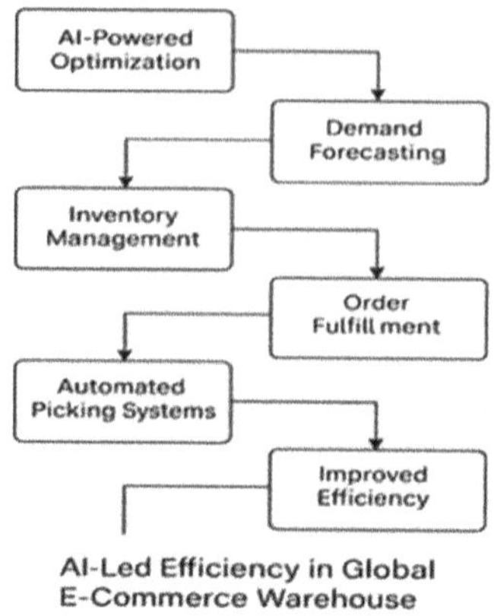

5.3 Predictive Replenishment and Logistics Optimization

One of many examples of national retail chains takes advantage of predictive replenishment enabled by AI to streamline inventory allocation between various regional warehouses and stores. AI algorithms were able to predict replenishment requirements at very high accuracy levels, by considering historical sales, events in the area, running promotional campaigns as well as limitations on inbound transportation (Sasmal, 2024c). The system would create purchase orders automatically and release the stocks depending on the demand predictions and the transportation facilities. At the same time, AI-based logistics optimization was used in order to design the most efficient deliveries routes avoiding any additional vehicle wanderings thus minimizing fuel costs. The integrated system enabled active reconfiguration of paths in line with the current time traffic or weather requirements. They resulted in a 25%decrease in last-mile delivery expenses, an 80% decrease in carbon establishment, and a 15% rise in the availability of products on the shelves in retail (Taleb & Mohamed, 2020). The specified case illustrates the role of AI in balancing the warehousing and logistics processes and in rendering such operations sustainable at the same time ensuring that the customer satisfaction with the adequate, timely, and efficient supply of goods. The Figure 9 visualizes the overall end-to-end process of forecasting the demand, managing inventory, operations in the warehouse, coordinating supply, and optimization of routes. It points out that AI will result in smarter, co-ordinated logistics choices in order to make replenishment more efficient.

Figure 9. Predictive Replenishment and Logistics Optimization

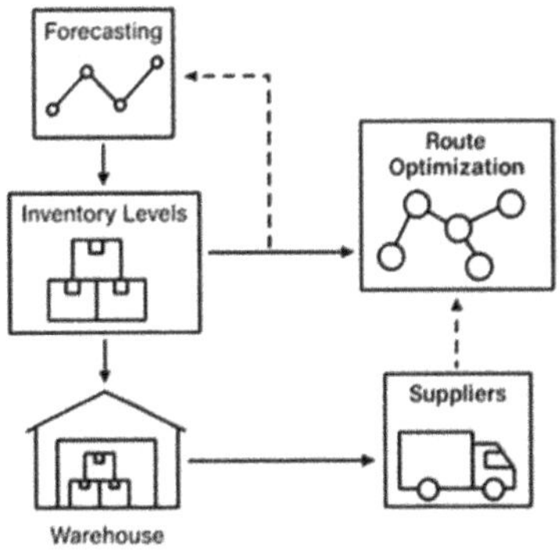

6. CHALLENGES AND SOLUTIONS

Despite the change in the paradigm of AI benefits of warehousing, there are several challenges that come with its use. The actual problem of data integration is actually one large hurdle given that warehouses have their own respective legacy systems which barely communicate with each other at times. The major infrastructure comes with data standardization in the process of integrating AI with these systems (Varghese et al., 2018). The adoption may also be hampered due to the skills deficiency among the personnel in the warehouse and the IT departments as the operation of AI systems demands the skills of data science, analytics, and machine learning. The conventional operations teams also create resistance to change, especially in cases where the automation alters job role.

Furthermore, the regulatory and ethical challenges may emerge because of the problems with transparency, biasness, and privacy of data employed in the algorithmic systems. Overcoming such challenges, it is recommended that organizational change-agents should use processes in stages, start upskilling employees, and concentrate on explainable AI models. The integration may be significantly extensive with the assistance of scaled cloud-based solutions and ready AI modules. Collaboration with technology providers, adherence to ethical codes and good leadership is the key to the full potential of AI in the achievement of smart and sustainable warehouses. The table 3 lists common implementation challenges such as data silos and skill gaps. It pairs each challenge with practical solutions and strategic recommendations.

It serves as a quick-reference guide for overcoming AI adoption barriers.

Table 3. Challenges and Solutions in AI-Driven Warehousing

Challenge	Impact	Recommended Solution
Data Integration Issues	Inconsistent and siloed data impedes model performance	Use APIs, data lakes, and standardization protocols
Skill Gaps in Workforce	Poor handling of AI tools and analytics	Upskill staff through targeted AI training programs
Resistance to Automation	Cultural pushback from traditional operations teams	Change management, stakeholder involvement
Lack of Algorithm Transparency	Difficulty in trust and regulatory compliance	Adopt explainable AI (XAI) frameworks
Scalability of AI Models	Inflexibility across different warehouse types	Build modular, cloud-native AI systems
Ethical and Legal Compliance	Risk of biased or unfair automation	Establish AI governance and auditing mechanisms

6.1 Data Integration, Skill Gaps, and Change Resistance

Data integration, skill gaps and change resistance are three main operational impediments to the adoption of AI in warehousing. In the majority of warehouses, assorted legacy solutions are employed which do not support smooth information exchange (N. Wang et al., 2019). To combine AI with the existing ERP, WMS, and IoT infrastructure, it is necessary to align the data formats and APIs data, which is quite expensive and time-consuming. At the same time, technical illiteracy of the warehouse staff is an obstacle, since AI-powered systems require proficiency in the field of data mining, machine learning, and cloud computing. Unless well-trained, the personnel cannot operate or rely on AI suggestions. Additionally, it is normal to have change resistance in a scenario where automation is seen to take away the job positions. In order to mitigate such issues, firms ought to implement an incremental application of AI, focus on personnel training and cross departmental collaboration, and develop a sense of the productivity and environmental sustainability improvements AI makes to the organization - making it an aid and not a substitute.

6.2 Mitigating Risks: Transparency, Scalability, Ethics

Ethics is another important area of concern given the question of AI-driven warehousing and the issue of transparency, scalability, and ethical application of such technology. Most deep learning systems may be called black box, and no one is even able to comprehend how the warehouse managers come to the decisions (Wei and Ma, 2024). This kind of transparency in the algorithms may undermine the trust and make it very difficult to comply with the industrial regulation. Another issue is scalability; as an AI solution that has been implemented with success in a single facility, may not easily apply to different types of facilities, having different inventory, layout and even operation processes. Also, the ethical aspect of such AI recommendations in the case of biased or partially-incomplete information is also present, as such AI recommendations can create an inefficient distribution of resources, or even be unjust. Explainable AI models, auditable pipelines, and continuous validation methods have been suggested to reduce such risks, in the organizations. It is not fixed so that AI solution development platforms are modular and scale (Yang et al., 2022). Governance plans of AI should include ethical rules and checks and balances so that it can use AI responsibly, accountably and fairly in the high-impact environment like warehousing.

7. DEPLOYMENT FRAMEWORK AND USE CASE EXPANSION

AI in warehousing space requires a blueprint which links technology, operations and greenhouse emissions. The IoT devices, RFID tags and smart meters can track real-time data within a warehouse on the very bottom and with scalability and low-latency processing features, offered by cloud and edge computing, vendors are able to store more data and transmit it to clients with an incredibly small time (Forti et al., 2020). WMS are then installed with the AIs to enable the predictive analytic, demand forecasting, and automated decision-making systems. The privacy and regulatory compliance are handled by a layer of security and compliance; the workforce is trained to adopt it and form human-AI relationship. The expansion use case uses the expansion to elaborate how AI can be scaled in other inventory tracking. Predictive maintenance is reducing downtimes, computer vision is improving quality and safety and digital twins are assisting in optimizing layout and energy usage. In frozen logistics, AI promotes stable temperature operation, and the majority of adaptive hybrids and tailor-made logistics improve sustainable logistics and content with their consumers. The combination of the framework and different use cases introduces the possibilities of AI in the world of warehousing as a powerful and radical facilitator of intelligent and sustainable warehouses.

7.1 Deployment Framework for AI in Warehousing

A proper deployment structure is also a requirement so as to integrate AI technologies in the warehousing process. The initial step of the process should be the introduction of an effective data infrastructure, and the IoT devices, RFID sensors, and smart cameras should be introduced to obtain continuous data about inventory, equipment and any alteration of the environment (Shahroudnejad et al., 2024). This information is processed within the cloud and edge computing systems to ensure that the data is scalable, secure and decisions are made based on the low-latency data. Then, AI algorithms are deployed to the WMS and Enterprise Resource Planning (ERP) systems, which enables conducting predictive analytics, demand planning, and intelligent resource planning. The other important level of this framework is the data security and regulatory compliance which ensures confidence and privacy in the decision making based on data. The final step is workforce preparedness and change management during which the employees are trained to operate with AI and adjust to the alterations in workflow. This is a well-planned model of AI deployment that assists organisations to attain maximum efficiency, sustainability, and scalability in the application of AI in warehousing operations.

7.2 Use Case Expansion in Industry

The area of AI in warehousing is quickly expanding beyond the traditional inventory management apps and creating a new dynamism in the whole logistics ecosystem. The predictive maintenance, which is enabled by machine learning, may also be used to monitor the equipment in a manner that issues are detected and fixed before they escalate and lead to the changes in the system as a whole. The computer vision systems not only make the working environment safer, but also preserve the quality assurance by detecting the operational hazard and detection of product defect on real-time basis. The models of digital twins will be used to imitate conditions in the warehouse, in order to simulate performance, optimal warehouse layout, and capability to integrate green power solutions. Cold-chain logistics is applying AI to make sure that the temperatures of drugs and other perishable goods are maintained at an optimal point, which significantly decreases wastage (Židek et al., 2020). Flexibility of AI is also highlighted through customized logistics to enable them to package, route, and deliver products in a customer-centered way. The improvement strategies that are already adopted by the industry giants such as Amazon, or DHL, are observed to be rather promising in terms of magnitude, and the beneficial results are measurable. Alongside these developments, AI in warehousing has ceased to be a test technology and has become a market transforming force that is constructing a sustainable, intelligent, and resilient supply chain (Aarthi et al., 2025).

8. FUTURE OUTLOOK

The future of AI in warehousing lies in having the creation of autonomous, adaptive and sustainable systems that are not predictive analytics in the first place. The key part will consist of digital twins of warehouse settings to allow simulating and streamlining operations in real-time and thereby develop an ongoing process of enhancement and faster reaction to disturbances. Prescriptive AI will enhance our existing decision-making abilities by not only forecasting what transpires next but also include a suggestion and may even execute the suggested action (Zhuang and Jiang, 2022). Besides, the carbon-neutral and zero-waste programs will be supported by the circular inventory models where the ideal results under the conditions of returns, refurbishing, and redistribution are the best possible output with the help of AI. The combination of AI and blockchain, robotic automation and 5G/edge computing will make the interaction of AI and blockchain easier in terms of traceability, speed of the operations and data protection. Policy systems that promote the application of green logistics and rewards given on the application of AI will also promote the transition to smart warehousing. Ultimately, AI will change the perception of the warehouse management as a rather stationary resource-consuming process to a smart self-optimizing environment in accordance with the country-specific global sustainability and digitalization goals (Aarthi et al., 2025). Figure 10 depicts a downward pattern in the quantity of CO 2 emitted by the warehouse activities in 5 years. Represents the potential impact of scaling AI on logistics networks on the environment. And the table 4 shows quantifiable KPIs to measure the effect of AI on warehousing. Measures are accuracy of the forecasting, energy consumption, and carbon reduction. It helps in tracking sustainability and efficiency performance based on evidence.

Figure 10. Projected Carbon Emission Reduction with AI Implementation

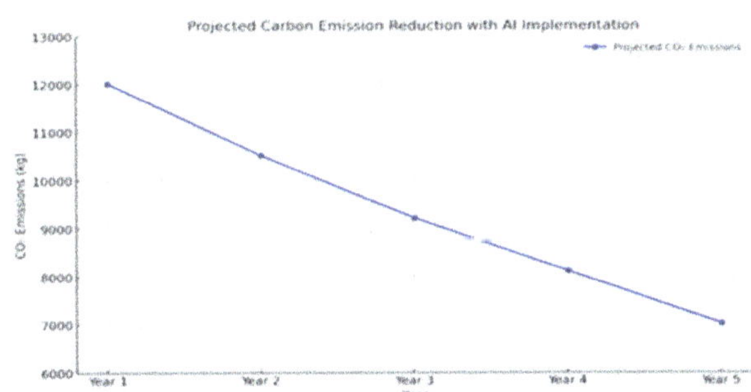

Table 4. Key Performance Indicators (KPIs) for AI-Based Green Warehousing

KPI	Description	Measurement Method
Forecast Accuracy	Precision of demand predictions	Mean Absolute Percentage Error (MAPE), RMSE
Inventory Turnover Rate	How quickly inventory is sold and replaced	Cost of Goods Sold / Average Inventory
Energy Consumption per SKU	Energy used for storing each unit of inventory	kWh/SKU (via energy monitoring systems)
On-Time Replenishment Rate	Efficiency of restocking aligned with demand	Timely Orders / Total Orders
Carbon Emission Reduction (%)	Improvement in environmental performance	CO_2 output before vs after AI implementation
Waste Reduction Rate	Decrease in expired or obsolete stock	(Old Waste – New Waste) / Old Waste × 100
Automation ROI	Financial return from AI-driven automation	(Savings – Investment) / Investment × 100

8.1 Digital Twins, Prescriptive AI, Circular Inventory Models

The next-generation of AI-enabled warehousing will include the use of digital twins, prescriptive AI, and the circular model of inventory. Digital twins simulate the behavior of a physical warehouse to apply to the virtual environment and thereby allow real-time simulation of operations, layout optimization and energy flows modeling. These twins can be used to gather the decision-making and optimization in advance without affecting the physical processes (Drissi Elbouzidi et al., 2023a). Prescriptive AI will further position it to making the best decisions possible and acting on a truly autonomous level even in areas such as when to replenish supplies, rerouting, or altering the labour schedules. Sustainability will be pegged on the shift to circular forms of inventory in the meantime. The AI will support returns management, refurbishment, recycling and redistribution through analyzing product lifetimes and consumer behaviour. The practice reduces wastage, reutilization and makes warehousing green as conceptualized in green supply chain. All these developments will be the transformation towards autonomous and sustainable ecosystems that are not only responsive but also regenerative providing the immediate long-term resilience to both, the environmental and the business (Aarthi et al., 2025).

8.2 Policy Support and Evolving Sustainability Standards

Future effectiveness of AI usage in warehousing is also predetermined by effective policy systems and sustainability norms. Green logistics is also being encouraged by governments and international organisations by fuelling carbon neutrality limits, energy efficiency requirements and supporting the digital transformation (Huriye, 2023). To promote the use of AI, it is possible to introduce the policies which will support the use of intelligent systems by SMEs, including tax relief, technology grants, and cheap loans. Adherence to the arising international standards such as the ISO 14001 (environmental management) and conformity to the UN Sustainable Development Goals (SDGs) is becoming a strategic issue to warehouse operators. Additionally, data ethics, AI transparency, and worker rights guidelines frameworks will have to develop in order to guarantee responsible deployment (Ogundipe & Abaku, 2024). Energy reporting standards, carbon footprint per SKU and autonomous audit systems will likely become obligatory. All these advancements will demand warehouses not only aim to optimize their operations but also to comply with environmental regulations with quantifiable KPIs. Policy and AI will be responsible together in creating sustainable, smart logistics ecosystems in the next few years (Aarthi et al., 2025).

9. CONCLUSION

Warehousing business is undergoing a paradigm shift as it tries to find the right balance between operational efficiency and sustainability. With the global supply chains getting complex, and sustainability becoming an urgent issue, the implementation of AI has turned out to be an innovation that will enable the actual green and energy-saving warehousing. The past warehousing systems that tend to be resources-intensive and reactive are failing to respond to the demand in a timely fashion, agility and sustainability. The issues of overstocking, wastage of space and the overstocking of excess handling not only raise the operation cost but also add to a huge quantity of carbon emission. AI is a proactive, data-based framework, using allowing technologies, such as the IoT, RFID, WMS and cloud-edge computing, that provides operational excellence and sustainability.

Advanced models (time-series, RNNs, Long Short-Term Memory (LSTM) models) of the Ai of demand forecasting provide the accurate forecasts of the product demand including the past sales, seasonal fluctuations, and market forces. This precision minimizes its overstock and inventories, therefore, unnecessary storing that uses storage energy and resources. The inventory management is premised on AI with the support of IoT and RFID databases, and thus assists in the efficient tracking, replenishment, and distribution of inventory, consuming the least amount of energy when the lighting, HVAC systems, and equipment are in operation, and makes the work of labor and delivery easier. Cloud computing to train AI on a scalable basis and edge computing to react in real-time enhance end-to-end visibility and agility in the work of the warehouse to stimulate automation and flowing data with no interruptions.

Another application of AI is also spread to more specific areas, such as cold chain logistics, online shopping, and retail, where energy and space savings are of high importance. AI changes innovation to measurable sustainability outputs through improving SKU accuracy, forecasting accuracy and throughput efficiency. However, the limitations to the widespread adoption are data integration, skills gap in the workforce, interoperability, and resistance to change. Ethical concerns, transparency, and compliance with the emerging AI rules are also present. The obstacles should be overcome through creating explainable AI systems, training employees, and creating scaling, interoperable systems, and strong governance.

The warehousing industry will embrace better AI capabilities in the future such as digital twins, prescriptive analytics and autonomous systems. They will enable the existence of predictive maintenance and zero-waste operations and regenerative supply chains, which will contribute to sustainability additional. Policy support and the evolving international standards will be used to improve this change. And lastly, AI is not among the tools to streamline warehousing but a tactical help of sustainable change. By introducing AI into the work principles of the warehouses, they will turn into an intelligent and environmentally friendly component of the global logistics system, which is willing to join the carbon neutrality and sustainable development agenda in the future.

REFERENCES

Aarthi, S., Ravikumar, R. N., Sobirova, K., & Mamadiyarov, Z. (2025). Empowering a sustainable future through digital learning and the SDGs. In *Harnessing E-Learning to Create a Sustainable Future* (pp. 145–172). IGI Global Scientific Publishing. DOI: 10.4018/979-8-3373-0112-9.ch006

Affia, I., & Aamer, A. (2022). An internet of things-based smart warehouse infrastructure: Design and application. *J. Sci. Technol. Policy Manag.*, *13*(1), 90–109. DOI: 10.1108/JSTPM-08-2020-0117

Banda, E., & Folly, K. A. (2012). Day ahead load forecasting using an artificial neural network & Elman recurrent network. 760: 761: 762: 763: The IASTED 2012 African Conferences.

Bousqaoui, H., Slimani, I., & Achchab, S. (2021). Comparative analysis of short-term demand predicting models using ARIMA and deep learning. [IJECE]. *International Journal of Electrical and Computer Engineering*, *11*(4), 3319. DOI: 10.11591/ijece.v11i4.pp3319-3328

Co, M., Sham, R., Ahmad, N., & Acesor, S. (2023). Warehouse automation implementation environment: Case of central warehouse management in Mali, West Africa. Environ.-. *Behav. Proc. J.*, *8*(26), 59–66. DOI: 10.21834/e-bpj.v8i26.5127

Drissi Elbouzidi, A., Ait El Cadi, A., Pellerin, R., Lamouri, S., Tobon Valencia, E., & Bélanger, M.-J. (2023a). The role of AI in warehouse digital twins: Literature review. *Applied Sciences (Basel, Switzerland)*, *13*(11), 6746. DOI: 10.3390/app13116746

Drissi Elbouzidi, A., Ait El Cadi, A., Pellerin, R., Lamouri, S., Tobon Valencia, E., & Bélanger, M.-J. (2023b). The role of AI in warehouse digital twins: Literature review. *Applied Sciences (Basel, Switzerland)*, *13*(11), 6746. DOI: 10.3390/app13116746

Effah, F. B., Kwegyir, D., Opoku, D., Asigri, P., & Frimpong, E. A. (2023). Short-term EV charging demand forecast with feedforward artificial neural network. J. Nas. Tek. Elektro.

Eldred, M. E., Thatcher, J., Rehman, A., Gee, I., & Suboyin, A. (2023, January). Leveraging AI for inventory management and accurate forecast – an industrial field study. Day 1 Tue, January 17, 2023.

Elias, R. S., Fang, L., & Wahab, M. I. M. (2011, June). Electricity load forecasting based on weather variables and seasonalities: A neural network approach. ICSSSM11.

Forti, S., Ferrari, G.-L., & Brogi, A. (2020). Secure cloud-edge deployments, with trust. *Future Generation Computer Systems*, *102*, 775–788. DOI: 10.1016/j.future.2019.08.020

Gross, J., & Törngren, M., Dán György and Broman, D., Herzog, E., Leite, I., Ramakrishna, R., Stower, R., & Thompson, H. (2022). TECoSA – trends, drivers, and strategic directions for trustworthy edge computing in industrial Applications. *Insight (American Society of Ophthalmic Registered Nurses)*, *25*(4), 29–34.

Guo, M., Li, L., & Guan, Q. (2019). Energy-efficient and delay-guaranteed workload allocation in IoT-edge-cloud computing systems. *IEEE Access: Practical Innovations, Open Solutions*, *7*, 78685–78697. DOI: 10.1109/ACCESS.2019.2922992

Gür, N., Pedersen, T. B., Zimányi, E., & Hose, K. (2018). A foundation for spatial data warehouses on the Semantic Web. *Semantic Web*, *9*(5), 557–587. DOI: 10.3233/SW-170281

Huo, Z., & Zhan, X. (2024). Revolutionizing inventory management: The role of data mining in industry 4.0. Advances in Economics. *Management and Political Sciences*, *109*(1), 67–72.

Huriye, A. Z. (2023). The ethics of artificial intelligence: Examining the ethical considerations surrounding the development and use of AI. *American Journal of Technology*, *2*(1), 37–45. DOI: 10.58425/ajt.v2i1.142

Kelly, A. (2024). Impact of Artificial Intelligence on supply chain optimization. *Journal of Technology and Systems*, *6*(6), 15–27. DOI: 10.47941/jts.2153

Kilimci, Z. H., Akyuz, A. O., Uysal, M., Akyokus, S., Uysal, M. O., Atak Bulbul, B., & Ekmis, M. A. (2019). An improved demand forecasting model using deep learning approach and proposed decision integration strategy for supply chain. *Complexity*, *2019*(1), 1–15. DOI: 10.1155/2019/9067367

Kim, S.-H., & Kim, T. (2023). Local scheduling in KubeEdge-based edge computing environment. *Sensors (Basel)*, *23*(3), 1522. DOI: 10.3390/s23031522 PMID: 36772562

Li, X., & Wang, Z. (2023). A cloud-edge computing method for integrated electricity-gas system dispatch. *Processes (Basel, Switzerland)*, *11*(8), 2299. DOI: 10.3390/pr11082299

Li, Y., & Hong, Y. (2022). Prediction of football match results based on edge computing and machine learning technology. *International Journal of Mobile Computing and Multimedia Communications*, *13*(2), 1–10. DOI: 10.4018/IJMCMC.293749

Li, Z., & Li, M. (2023). Design and implementation of automated stereoscopic warehouse based on artificial intelligence. *Advances in Computer and Communication, 4*(5), 318–323. DOI: 10.26855/acc.2023.10.010

Lim, J. (2024). Adaptive scheduling based on intelligent agents in edge-cloud computing environments. *Wangji Wanglu Jishu Xuekan, 25*(4), 609–617. DOI: 10.70003/160792642024072504011

Lodhi, S. K., Gill, A. Y., & Hussain, I. (2024). AI-powered innovations in contemporary manufacturing procedures: An extensive analysis. *Ijmdsa, 3*(4), 15–25. DOI: 10.47709/ijmdsa.v3i4.4616

Munsberg, T. M., & Hvam, L. (2021, December). Potential of streamlining warehouse processes and IT integration to increase implementation of automation. *2021 IEEE International Conference on Industrial and Engineering Management (IEEM)*. DOI: 10.1109/IEEM50564.2021.9672816

Muthaluri, J. (2024). Optimizing supply chain management and marketing strategies: AI and ML integration for competitive advantage. Educational Administration: Theory and Practice, 8990–8997.

Ngaboyimbere, F., Leo, J., Michael, K., & Muteteshe, D. (2021). Development of RFID based automatic warehouse management system: A case study of ROK industries limited Kenya. *Int. J. Adv. Sci. Res. Eng., 07*(08), 112–126. DOI: 10.31695/IJASRE.2021.34076

Ni, S., Peng, Y., & Liu, Z. (2022). Logistics demand forecast of fresh food E-commerce based on bi-LSTM model. *J. Comput. Commun., 10*(09), 51–65. DOI: 10.4236/jcc.2022.109004

Nikitas, A., Michalakopoulou, K., Njoya, E. T., & Karampatzakis, D. (2020). Artificial intelligence, transport and the smart city: Definitions and dimensions of a new mobility era. *Sustainability (Basel), 12*(7), 2789. DOI: 10.3390/su12072789

Ogundipe, D. O., & Abaku, E. A.Damilola Oluwaseun OgundipeEmmanuel Adeyemi Abaku. (2024). Theoretical insights into AI product launch strategies for start-ups: Navigating market challenges. *Int. J. Front. Sci. Technol. Res., 6*(1), 62–72. DOI: 10.53294/ijfstr.2024.6.1.0032

Plathottam, S. J., Rzonca, A., Lakhnori, R., & Iloeje, C. O. (2023). A review of artificial intelligence applications in manufacturing operations. *Journal of Advanced Manufacturing and Processing, 5*(3), e10159. DOI: 10.1002/amp2.10159

Rahul, P., & Singh, A. J. (2023). A study of Mobile Edge Computing for IOT. *Advances in Science and Technology (Owerri, Nigeria), 124*, 856–863. DOI: 10.4028/p-2u34v7

Sasmal, S. (2024a). Data warehousing revolution: AI-driven solutions. *INTERNATIONAL RESEARCH JOURNAL OF ENGINEERING & APPLIED*, *12*(1), 1–6.

Sasmal, S. (2024b). Data warehousing revolution: AI-driven solutions. *INTERNATIONAL RESEARCH JOURNAL OF ENGINEERING & APPLIED*, *12*(1), 1–6.

Sasmal, S. (2024c). Data warehousing revolution: AI-driven solutions. *INTERNATIONAL RESEARCH JOURNAL OF ENGINEERING & APPLIED*, *12*(1), 1–6.

Shahroudnejad, A., Mousavi, P., Perepelytsia, O., Sahir, , Staszak, D., Taylor, M. E., & Bawel, B. (2024). A novel framework for automated warehouse layout generation. *Frontiers in Artificial Intelligence*, *7*, 1465186. DOI: 10.3389/frai.2024.1465186 PMID: 39474600

Sodiya, E. O., Umoga, U. J., Amoo, O. O., & Atadoga, A. (2024a). AI-driven warehouse automation: A comprehensive review of systems. *GSC Adv. Res. Rev.*, *18*(2), 272–282. DOI: 10.30574/gscarr.2024.18.2.0063

Sodiya, E. O., Umoga, U. J., Amoo, O. O., & Atadoga, A. (2024b). AI-driven warehouse automation: A comprehensive review of systems. *GSC Adv. Res. Rev.*, *18*(2), 272–282. DOI: 10.30574/gscarr.2024.18.2.0063

Taleb, N., & Mohamed, E. A. (2020). Cloud computing trends: A literature review. *Acad. J. Interdiscip. Stud.*, *9*(1), 91. DOI: 10.36941/ajis-2020-0008

Varghese, B., Reano, C., & Silla, F. (2018). Accelerator virtualization in fog computing: Moving from the cloud to the edge. *IEEE Cloud Computing*, *5*(6), 28–37. DOI: 10.1109/MCC.2018.064181118

Verma, D. (2018). Analysis of smart manufacturing technologies for industry using AI methods. [TURCOMAT]. *Turk. J. Comput. Math. Educ.*, *9*(2), 529–540. DOI: 10.17762/turcomat.v9i2.13857

Wang, N., Varghese, B., Matthaiou, M., & Nikolopoulos, D. S. (2019). ENORM: A framework for edge NOde resource management. *IEEE Transactions on Services Computing*, •••, 1. DOI: 10.1109/TSC.2017.2753775

Wang, Y., Pan, J., Zhang, P., & Tesfamariam Semere, D. (2019). Application on automated warehouse simulation system in location optimization. *Proceedings of The First International Symposium on Management and Social Sciences (ISMSS 2019)*. DOI: 10.2991/ismss-19.2019.20

Wei, Z., & Ma, Y. (2024). Research on integration management of big data application precision portrait marketing resources based on cloud and edge collaborative computing. Proceedings of the First International Conference on Science, Engineering and Technology Practices for Sustainable Development, ICSETPSD 2023, 17th-18th 2023, Coimbatore, Tamilnadu, India. DOI: 10.4108/eai.17-11-2023.2342626

Wu, P.-J., & Tai, Y.-C. (2024a). Artificial intelligence-based food-quality and warehousing management for food banks' inbound logistics. *Journal of Enterprise Information Management*, *37*(1), 307–325. DOI: 10.1108/JEIM-10-2022-0398

Wu, P.-J., & Tai, Y.-C. (2024b). Artificial intelligence-based food-quality and warehousing management for food banks' inbound logistics. *Journal of Enterprise Information Management*, *37*(1), 307–325. DOI: 10.1108/JEIM-10-2022-0398

Yang, J., Lee, T.-Y., Lee, W.-T., & Xu, L. (2022). A design and application of municipal service platform based on cloud-edge collaboration for smart cities. *Sensors (Basel)*, *22*(22), 8784. DOI: 10.3390/s22228784 PMID: 36433381

Zhuang, Y., & Jiang, N. (2022). Progressive privacy-preserving batch retrieval of lung CT image sequences based on edge-cloud collaborative computation. *PLoS One*, *17*(9), e0274507. DOI: 10.1371/journal.pone.0274507 PMID: 36107827

Židek, K., Piteľ, J., Adámek, M., Lazorík, P., & Hošovský, A. (2020). Digital twin of experimental smart manufacturing assembly system for industry 4.0 concept. *Sustainability (Basel)*, *12*(9), 3658. DOI: 10.3390/su12093658

Key Terms and Definitions

Artificial Intelligence (AI) The use of machine learning: deep learning, and algorithmic models to enable warehouses to make predictive and autonomous decisions.

Demand Forecasting The process of predicting future product demand using AI models such as Time-Series: RNN, and LSTM to reduce overstocking and understocking.

Inventory Management The monitoring: placement, and replenishment of goods using AI, IoT, and RFID to optimize space, reduce waste, and minimize energy use.

Just-In-Time (JIT) An inventory strategy enabled by AI forecasting to keep stock levels minimal while meeting real-time demand: thus conserving energy and resources.

Internet of Things (IoT) A network of connected sensors and devices that provide real-time data on inventory: environment, and equipment for AI-driven decisions.

Radio Frequency Identification (RFID) A tracking technology using tags and scanners to monitor the movement: status, and condition of goods in real time.

Warehouse Management System (WMS) A digital platform that integrates AI: IoT, and RFID for centralized monitoring, inventory optimization, and process automation.

Edge Computing Localized data processing close to the warehouse floor to enable rapid: low-latency AI decisions.

Cloud Computing Centralized storage and processing systems that allow scalable AI model training: data analysis, and integration across multiple warehouses.

Digital Twin A virtual replica of a warehouse environment used for simulating: optimizing, and testing operational strategies in real-time.

Prescriptive AI An advanced AI approach that not only predicts outcomes but also recommends and executes the best actions for warehouse operations.:

Circular Inventory Model A sustainable approach in warehousing where products are reused: refurbished, recycled, and redistributed to minimize waste.

Chapter 10
Sustainable Innovations in Energy-Efficient Warehousing

R. Vasanthi
Panimalar Engineering College, India

Siva Subramanian R.
https://orcid.org/0000-0002-7509-9223
SRM Institute of Science and Technology, India

ABSTRACT

This chapter explores innovations in green and energy-efficient warehousing, focusing on sustainable technologies that minimize environmental impact and optimize energy consumption. Key advancements, including energy-efficient lighting, automated systems, renewable energy integration, and smart warehouse designs, are explored for their role in reducing carbon footprints and improving operational efficiency. The chapter also discusses emerging technologies like AI, IoT, and robotics that enhance warehouse operations and contribute to sustainability goals. Challenges such as high initial investment costs and regulatory compliance are addressed, along with future trends like carbon neutrality and zero-energy warehouses. The chapter highlights the importance of adopting green technologies for long-term environmental and economic benefits.

DOI: 10.4018/979-8-3373-3176-8.ch010

1. INTRODUCTION

1.1 Overview of Warehousing Industry and Its Role in Logistics

Warehousing plays an important role in the logistics and supply chain management, as it offers storage services for various products. A warehouse is a central facility where stock, goods, and products are stored, sorted and relayed to other centers, retailers, wholesalers or customers(Masudin et al 2021). As the pressure for time is constantly mounting, the warehouses have become more and more sophisticated and play a central role in supply chain management. The industry has also experienced enhancement in technology, automation and digitalization in order to enhance and speed up the process of fulfilling the orders. However, as the warehousing sector expands, the concerns related to the sector increase in terms of energy consumption and carbon footprint(Batarlienė, N., & Jarašūnienė, A. 2024).

1.2 Importance of Sustainability and Energy Efficiency in Warehousing

It is now a popular concept in industries across the globe, and the warehousing industry is no exception. As the effects of climate change and the depletion of resources become more and more familiar, its importance to implement energy-saving ecological measures(Munuhwa, S. 2023). As a result, energy consumption in warehouses is relatively high due to the characteristics of this type of facilities. Lighting, heating, ventilation, cooling systems along with heavy machineries used in material handling are the major source of energy consumption. Thus, sustainability and energy efficiency have become non-negotiable for warehouse management to sustain the business and abide by the rules. The trend of shrinking environmental impacts has been set in most industries in the recent past. This has given rise to a shift in focus where firms are gradually beginning to see that there are advantages of adopting green practices. These practices not only help in the conservation of the environment but also reduce the operational costs, increase company reputation and makes the company more competitive in the market that is inclined towards sustainability. Energy efficiency in warehousing practice is, therefore, tied to both environmental and economic performance, therefore, opening a window to the possibility of innovation in order to meet the increasing global focus on sustainability in business.

1.3 Environmental Impact of Traditional Warehousing Practices

Conventional methods of managing and storing goods and products while still functional from the operational perspective are known to have negative impacts on the environment. Another of them is the high energy consumption of lighting, heating, and cooling systems. Some of the warehouses are still operating under conventional systems where there is a lot of wastage of energy due to outdated structures, excessive heating or bad lighting system. These inefficiencies are a source of carbon footprints, which are commonly known to cause global warming and air pollution. In addition, the use of fossil fuel in transportation, machinery, and the equipment used in the warehouse also has these adverse impacts on the environment. Forklifts, conveyor systems and trucks used for transportation are usually operated on diesel or gasoline which causes emission of greenhouse gases and exhaustion of natural resources. Also, the warehouses themselves are not well insulated or insulated with non-ecological materials and therefore require a lot of energy to warm or cool. The effects of these practices on the environment have been felt and this has led to pressure from environmental authorities, shareholders and consumers who are looking for green solutions.

1.4 Objectives and Structure of the Chapter

To this end, this chapter aims at discussing the new technologies and approaches in the field of warehousing that will lead to sustainability of energy. This chapter will therefore discuss how warehouses can ensure that they minimize their negative effects on the environment through the use of other practical measures such as renewable energy sources, automation, efficient building designs and other green technology. It will also discuss the problems associated with the change in the warehousing and manufacturing industry toward sustainability, such as financial, technological, and legal issues. This chapter is organized as follows: Section 2 will define the principles of sustainability and energy efficiency in relation to warehousing in order to establish the basic knowledge in the topic. Section 3 will focus on individual green technologies currently being adopted in the industry that include lighting systems and renewable energy. Section 4 will discuss the use of automation and robotics in the aspect of energy efficiency of the warehouse, and Section 5 will be on the concept of sustainable warehouse layout. Section 6 will expand on sustainable packaging and material handling that also contribute to the reduction of the total amount of energy used. Section 7 will discuss the issues likely to be encountered by the industry in the implementation of these innovations while Section 8 shall compare different trends that may be adopted in green warehousing in the future. Last, the chapter will give a brief conclusion of the major findings and a suggestion of the future research. Therefore, this chapter seeks to present a review of green and energy-efficient warehousing, as well as present recommendations on how the industry can improve its sustainability.

2. SUSTAINABILITY AND ENERGY-EFFICIENCY IN WAREHOUSING

2.1 Definition of Green Warehousing and Energy Efficiency

Green warehousing is defined as the use of sustainable practices and technology in the managing of the warehouse with the aim of reducing the impact on the environment(Nicoletti, B., & Appolloni, A. 2024). This concept is all about the reduction of energy usage, waste and optimization of resources. Green warehousing mainly covers the use of natural energy sources, energy conserving systems and environmentally friendly products in the construction of the warehouse and the running of the building. The objective is to establish a condition that will enable the efficient use of resources, reduction of waste in the warehouse and a considerable reduction of the carbon footprint(Tiwari, S. 2023). Energy efficiency in warehousing, therefore, relates to the level at which a warehouse can operate with least amount of energy that is needed to perform activities such as lighting, heating, cooling and powering equipment. The implementation of energy efficient measures would mean that warehouses can attain the same or even higher performance at the same time using less energy hence meaning lower costs and reduced negative effects on the environment. Energy conservation is important not just from the economic aspect of saving money on electricity bills but also from the aspect of environmentally friendly practices that are aimed at meeting the global goal of cutting down on carbon emissions that have been deemed to contribute to climate change(Yamini et al., 2025).

2.2 Key Principles of Sustainable Warehousing

Sustainable warehousing is based on a number of principles that should be followed in the process of sustainable development of the warehouse and its activities(Malinowska, M. 2022). These principles are crucial on the way towards obtaining a higher level of energy saving and environmental friendliness. The other important principle of sustainable warehousing is the use of resources. This can be done by using better technologies and practising methods that require less raw materials, energy and water. For instance, the use of efficient lightings for instance light emitting diodes (LEDs) can greatly cut down on electricity usage. Also, the design of the warehouse can be made in such a way that transportation within the facility requires minimal movement thus saving energy in transportation by the machines and vehicles. Waste Reduction: The other principle is the reduction of waste as much as is possible. Sustainable warehouses aim at minimizing the use of materials, reusing and recycling the materials used in the construction of the warehouses. It can be done by providing proper waste disposal techniques, using eco-friendly packing materials and inventory control. Such practices implemented by the warehouses not only help in environmental conservation but also assist in cutting down on costs of waste management. Renewable Energy: This is the use of renewable energy sources like the solar, wind, and geothermal energy in the green warehousing system. Take for instance, installation of photovoltaic solar systems on warehouses which can be used to power lighting systems among other utilities thereby minimizing on the conventional sources of power. Renewable energy is a very significant way of reducing greenhouse gas emissions that are present in the operation of the warehouses and moving towards carbon-neutral status. Sustainable warehousing also involves the use of green building materials in the construction of the warehouses. These are the health of the workers and people in the community in case of accidents, use of natural and green products in construction, and proper insulation and windows to keep the cost of heating and cooling down. Programs that certify environmentally friendly construction and design like the LEED are widely used by companies that seek to express their concern for the environment.

2.3 Benefits of Implementing Green Technologies in Warehouses

The integration of green technologies in warehouses gives a number of benefits that are not only concerned with environmental conservation (Domagała, J. 2022). These are not only beneficial for the sustenance of warehouses but also have actual financial and legal benefits.

Cost Saving: The first and rather obvious advantage of implementing energy efficient and sustainable measures is that of saving on costs. For instance, facilities like the use of LED lighting or automation in controlling temperatures in warehouses help cut down on electricity usage. Likewise, the installation of solar power system can reduce or even eliminate electricity costs in the long run and in some cases, offer an added stream of income through energy credits or selling back power to the utility (Bhattacharjee, A., & Jahanshahi, A. A. (2024).

Regulatory Compliance: With various governments placing increased focus on the use of energy as well as the emission of carbon, the more the warehouse implements green practices the better placed he is in meeting the required environmental standards and laws. This may refer to following legal requirements of the local, national, or international energy efficiency codes, carbon reduction targets and waste management standards. Also, green warehouses may be privileged to claim tax credits, subsidy, or rebates from government for embracing green technologies.

Environmental Conservation: Green technologies used in warehousing help in minimizing the effect of the company's operations on the environment. There are measures that warehouses can take to ensure that they reduce their emissions of greenhouse gases to the atmosphere and some of them are; This is not only a positive impact on the environment but also an improvement to the image of the business organizations as environmentally friendly to the customers and investors(subramanian R et al., 2026).

Better Brand Image and Market Value: Corporate goals that embrace sustainability tend to be favored by consumers and the market in general. This is because more consumers are willing to purchase products from companies that are environmentally conscious hence green warehousing helps the business gain market share and customer loyalty. Also, the implementation of green practices can also set the business apart from its competitors whereby it can form partnerships with other green organizations and companies as well as investors who need to fund green businesses(Lija A et al., 2025).

3. INNOVATIVE GREEN TECHNOLOGIES IN WAREHOUSING

3.1 Energy-Efficient Lighting

Among the most effective and easily applied green technologies in warehousing, energy efficient lighting is on the top of the list. Compact fluorescent lamps, incandescent lamps and other conventional lamps are energy intensive and their life expectancy is also short. On the other hand, LED lighting system is a very sustainable system that is energy efficient and has a longer lifespan. LED lights are more efficient by up to 80% as compared to incandescent lighting and can help lower the electricity bills through optimal usage in large warehouses which are open for long hours. Apart from the energy-efficient bulbs, the use of automatic lighting system has also been incorporated in warehouses. These systems are capable of sensing movement or fluctuations in natural lighting and hence turning on the lights only when necessary. For instance, it is possible to control lighting to switch off in areas that are not occupied or reduce brightness when there is adequate natural light. This further enhances the energy consumption and makes sure that energy in the form of electricity is not wasted. Daylighting is another effective technique that can be used to light up indoor spaces using natural light. The use of skylights, light shelves, and windows that allow sunlight to get into the rooms can help to decrease the usage of artificial lighting. The best way to achieve both natural and artificial lighting at the warehouses would be to incorporate the two, thus saving energy and costs. Another energy conservation technique commonly used in the warehouses is the use of motion sensors. These sensors are capable of sensing when the areas are not in use and thus switch off the lights. This minimizes wastage of energy in areas that are rarely used such as storage, or the loading bay where lights are commonly left on for long(Gopalan et al 2025).

Figure 1. Green technologies in warehousing

3.2 HVAC and Climate Control Systems

Another important aspect in the energy conservation in warehousing is the Heating, Ventilation, and Air Conditioning (HVAC) system. Conventional HVAC systems are highly energy consuming and one of the biggest challenges is to heat or cool large warehouse spaces. On the other hand, energy-efficient HVAC solutions include; VRF systems that control the cooling and heating based on the amount of energy needed. This makes the technology to be efficient as compared to the fixed-speed systems since there is no wastage of energy when heating or cooling is not required. For the purposes of improving energy conservation, there is an increased use of smart sensors for temperature and humidity in warehouses. These sensors maintain the environmental conditions and regulate the HVAC system to maintain the set standards of comfort with less energy usage. For example, sensors are able to measure temperature and only switch on the cooling or heating mechanism when required and not otherwise. In the same regard, the sensors can be interfaced with the WMS to adapt the use of energy in the warehouse depending on the occupancy and operations timetable. Also, there are ideas to implement the geothermal heating and cooling as the energy efficient solution. It controls the indoor climate by harnessing the temperature of the earth thus minimizing the use of energy for conventional heating and cooling.

3.3 Renewable Energy Sources

Since the demand for clean energy is growing, the warehouses now opt for renewable energy sources to minimize their impacts on the environment. Solar energy is among the most common renewable sources of energy that is being implemented in the warehouses. Thus, solar power systems on the roof or any other open area of the warehouse can be effectively utilized to generate a good amount of electricity for lighting, HVAC, and other necessities. This not only has the advantage of lowering the operational costs but also the environmental cost through the reduction of the use of non-renewable energy sources. Another renewable energy source that is also available for use in the operation of warehouses is the wind energy. Wind generators can also be installed on-site or in other part of the site to produce electricity. While large wind turbines are used for generating electricity, small wind turbines are now being considered for use in industrial buildings such as warehouses. Due to the unpredictable nature of renewables, there is a need to store energy for the times when the resources are not available. Battery or energy management systems enable warehousing of excess energy produced during certain times of the day or year such as when there is an abundance of solar energy production. This helps to cut down the dependence on the grid and ensure that the energy is used in a constant and reliable manner all year round.

3.4 Building Materials and Construction

In the same manner, the design and construction of warehouses contribute to the overall process of making them more sustainable. Some of the common green building certification tools are LEED (Leadership in Energy and Environmental Design) and BREEAM (Building Research Establishment Environmental Assessment Method). These covenants promote the application of green building material, proper utilization of energy and environmentally sound construction methods.

The other factor that needs to be incorporated in constructing green warehouses is the use of sustainable materials. Recycled steel, bamboo, or low-emission paints are some of the materials, which can be used to reduce the effects on the environment. Thirdly, the use of green insulation materials makes it easy to control the temperature and therefore, there is no need to use a lot of energy to either heat or cool the building. Some of the common organic materials used include cork or cellulose as they are more efficient in use of energy and environmentally friendly than the synthetic materials. They also include green roofs that help to save heat from radiating during the night, hence helps to counter the heat island effect experienced especially in urban areas. In addition, rainwater harvesting systems can be installed in the roofs of the warehouse to collect rainwater for use in irrigation, washing or for toilet use, thus cutting down on the facility's environmental footprint(Krishnan et al 2024; Manjunathan et al., 2025).

4. AUTOMATION AND ROBOTICS FOR ENERGY EFFICIENCY

4.1 Role of Automated Systems (AGVs, Robots) in Reducing Energy Consumption

Automations and robotics are now part of the integral processes in the warehousing industry and they also help in energy conservation. AGVs are one of the most effective ways of automation that are currently being used in ware houses. These are self-driven vehicles that are required to move the stock within the warehouse, thus eliminating the need of human energy and effort. AGVs are fitted with some form of navigation technology like sensors and cameras to enable them to navigate the warehouse on their own, thereby conserving power and covering as little distance as possible. The main source of energy efficiency in AGVs is the fact that the vehicles are automatically controlled and do not need the human input, which otherwise results in energy wastage in form of time spent in manual handling of loads. In addition, AGVs can be programmed to follow the path provided by the WMS thus guaranteeing that the products are transported in the shortest possible distance. Since AGVs do not require forklifts and other equipment for their operations, they contribute to the reduced energy consumption and the overall environmental impact of the warehouse.

Besides AGVs, other robots are now employed in pick and place, sorting and packing in warehouses. They include the robots with speed and precision to perform the tasks that take much energy to be repeated. For instance, the robotic arms can help in picking items from the shelves more efficiently than human beings, reducing on time spent on searching for and moving the goods. Through performing the operational processes, robots take lesser time as compared to human beings and hence, uses less energy. In addition, most robots are equipped with other technologies like machine learning and artificial intelligence that help the robots to learn new and changing conditions in the warehouse. This continues to optimize their movements and eliminate unnecessary time, force or path that may lead to wastage of energy in their operations.

Figure 2. **Automation and Robotics for Energy Efficiency**

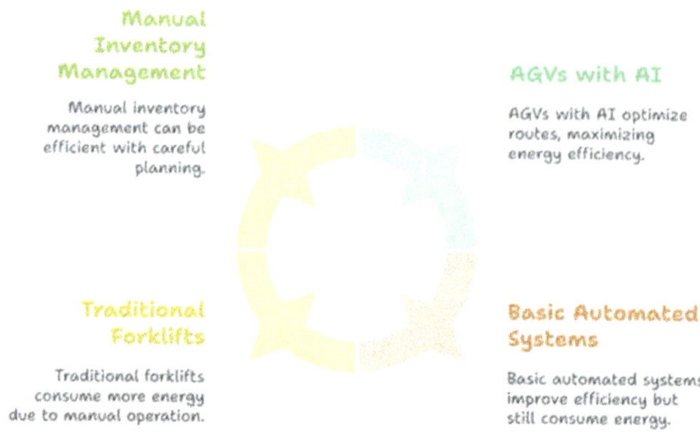

4.2 Efficiency of Automated Inventory Systems in Optimizing Warehouse Operations

Computerized inventory is a key factor in managing the warehouse and helps to save a lot of energy. The conventional systems of inventory management are cumbersome and a lot of time is spent on it as it is mostly done manually. Mechanized systems however, reduce the flow of inventory by employing technology to manage, record and access the stock without much interference from human beings. These systems use later generation barcode scan and RFID tags along with real time data feed for keeping the inventory up to date and products identified and localized in real time. Automated inventory systems are advantageous to organizations because they help minimize mistakes such as overstocking and misplacement of products, which leads to wastage of energy. For instance, if there is no automation then workers may have to look for the products all over the warehouse and this is going to lead to wastage of energy as forklifts or workers move up and down the warehouse. Automated inventory systems, however, will ensure that the products are placed in the right place so that the robotic systems or AGVs can easily access the products. This reduction in search time has the effect of reducing energy use as people do not move around as much as before. Also, real-time tracking of stocks in automated systems helps in determining the status of orders and inventory. This leads to better demand forecasting eliminating the cases of overstock and understock in the stores. The problem of overproduction and, as a result, overstocking of inventories is one of the major culprits as they waste energy and space. This way, through automation, it is easy to match supply with demand whereby energy is used to store what is necessary in the warehouse. Also, there is a reduction in transportation needs as most of the demand-driven supply chains do not require unnecessary shipment. This also means that there is less human interference in the running of the inventory systems which makes the environment more suitable for energy saving technologies. For instance, automation can control the lights and the heating, ventilation, and air conditioning (HVAC) systems to regulate the energy usage depending on the data collected from the warehouse area that is not in use at that particular time.

5. ENERGY-EFFICIENT WAREHOUSE LAYOUT AND DESIGN

5.1 Optimal Warehouse Layout for Energy Efficiency (Space Utilization, Ventilation)

An effective planning of the warehouse involves the use of energy efficient designs to reduce energy usage while increasing functionality. It is important to point out that one of the most critical factors that determine the energy efficiency of the warehousing process concerns the space management. If the layout of the floors is planned well, then much less energy will be needed for heating, cooling, and lighting the warehouse's space. Organizing layout in a proper way helps to keep stocks in areas where they are likely to be used much, so that there is little movement of stocks. This has the effect of saving time and efforts used in moving products from one part of the warehouse to another. Furthermore, there is a need to recommend some principles of organization of energy-efficient warehouses such as natural ventilation and air movement. Ventilation is critical to have a proper supply of fresh air in the building especially in large rooms such as those found in warehouses with high ceilings. It is possible to design the ventilation systems in such a way that they promote natural air circulation and thus, minimize the use of the air conditioning systems. Some windows or vents can be opened at strategic areas to allow fresh air to circulate freely within the warehouse since such buildings normally have a large area. This design strategy ensures that temperature fluctuations are kept to a minimum hence no need to use the air conditioning units during hot weather or the heater during cold weather. Another consideration of the layout of the warehouse is the zoning systems that divide the warehouse into different regions depending on the type of products being stored and their susceptibility to the climate. For example, perishable products can be stored in the areas with temperature control while the products that do not require such conditions, can be placed in the other areas. This zoning makes it possible to avoid the wastage of energy with regard to heating and cooling since some parts of the building may not require these services.

5.2 Design Strategies for Minimizing Energy Consumption (Smart Storage Systems, High Ceilings for Air Circulation)

The layout of the warehouse greatly determines its energy consumption, and the following methods can be used to reduce the energy consumption levels. This is probably one of the best strategies, which involves the management of storage equipment to minimize energy use in the process of storing and finding the products. These systems use equipment like robotic arms, conveyors and automated guided vehicle (AGVs) for storage and picking of items. Through automation of the inventory, there are several ways that it will help to avoid much physical exertion and speed up the activities of a warehouse, which in turn will help to use less energy to move around the warehouse. However, there are some other advantages of the presented design, such as high ceilings, which increase energy efficiency. Large spaces with high ceiling height are effective in allowing air to circulate, and this will mean that the temperature of the building will not fluctuate often and this will not require a lot of energy. High ceiling helps in the circulation of warm air to the top of the building and therefore, makes it easier to regulate temperatures in the warehouse. This design also ensures free air circulation hence no need for artificial sources of ventilation such as fans. From the above description, it is clear that proper air flow management can help reduce the need for HVAC systems, which are usually the biggest energy consumers in a warehouse. The design and location of the aisles and racking systems also contribute in a big way to reduction of energy use. In energy-efficient warehouses, positions of the racking systems must be in such a way that would enable workers to access them with ease without having to move around a lot, which would require a lot of energy. Through the narrow aisles and proper spacing of the aisles to fit the particular automated systems, the warehouses can enhance the storage density without much energy needed for forklifts or manual handling.

5.3 Use of Thermal Insulation and Natural Lighting for Energy Savings

In this case, thermal insulation is a critical factor that needs to be incorporated in the design of an efficient warehouse. Proper insulation ensures that the temperature inside the warehouse is regulated to the desired levels as it minimizes the heat exchange with the external environment. It is especially true for large warehouses with high heating or cooling load requirements. Walls, roofs and floor can be insulated using spray foam, fiberglass, or polystyrene to avoid wastage of energy. Heating and cooling are the most important aspect that ensures the temperatures within the facility are well regulated within the required standards and thus the use of thermal insulation is very useful in cutting down on energy usage. Therefore, other than insulation, the use of natural lighting also contributes to the reduction of artificial lighting during the day. This helps in the incorporation of the natural light especially in the form of large windows or skylights to allow natural light into the warehouse. This also helps in saving electricity used in lighting systems and in addition, enhances the working conditions by providing adequate lighting and a favourable physical atmosphere. Daylight control systems that regulate artificial lighting in relation to the amount of natural light admitted in the building also enhance the utilization of energy. Through proper placement of windows, skylights and light shelves, it is possible for the warehouses to allow as much light as possible to penetrate the interior hence saving on energy and being environmentally friendly.

In addition, the use of light coloured roofing material or application of reflective coat on the exterior of the warehouse reduce the heat gain. These materials also ensure that the inside temperature of the warehouse does not rise high in the summer to cause a lot of heat. This cuts down on the use of air condition since this is a major factor of energy use in large warehouses. Hence, the energy and layout of a warehouse should be effectively done to reduce the amount of energy used. This means that the warehouses can reduce energy wastage by increasing space utilization, efficient air circulation, and smart storage solutions. The high ceiling and large openings for better air circulation, proper insulation techniques and natural lightings are also effective in energy saving. Therefore, through application of these design strategies, warehouses can be made to be more efficient, less costly to run, as well as environmentally friendly than those currently in the logistics and warehousing industry.

6. SUSTAINABLE PACKAGING AND MATERIAL HANDLING

6.1 Green Packaging Solutions (Recyclable Materials, Reduction in Packaging Waste)

As for sustainable packaging, it remains one of the key factors that can help minimize the negative impact on the environment in warehouses and logistics centers. Plastic and non-recyclable cardboard packaging materials also pose a great threat to the environment as they increase the rate of waste production. Green packaging, on the other hand, is the packaging that concerns the use of recyclable, biodegradable or reusable materials in the packaging process. The most often used one is the use of reusable packaging, which is a material that can be recycled and reused, for example, cardboard, paper or plastic. There is also a growing use of biodegradable packaging material derived from plant raw materials as this has been found to minimize on waste and biodegrade easily in the landfill. In order to cut down on packaging waste, a lot of organizations are following the policy of minimizing packaging. Packaging optimization means the process of redesigning the packages with the aim of using as little material as possible while at the same time protecting the contents from damage or harm. This can be done through the use of small packaging or redesigning packaging to enable more space to be created for packaging and storage. One of the approaches that are regularly adopted include the downsizing of packaging to limit the use of non-recyclable products like bubble wrap by replacing them with fabrics or even air bags that can be reused. One such strategies is right sizing where the packing is made to fit the size and shape of the product. Thus, companies are able to minimize the use of fillers and additional packaging materials thus minimizing on packaging waste and the harm done to the environment. Besides, the other minor packaging features like the removal of plastic films or labels on the packaging adds to the reduction of waste that is produced in the entire supply chain. In addition, there are some organizations that are interested in packaging recycling programs to guarantee that packaging that has already been used is recycled or reused. Through partnering with recycling companies and offering incentives to customers with the recyclable materials, organizations are in a position to make a difference in minimizing their impact on the environment through recycling instead of disposal.

6.2 Energy-Efficient Material Handling Systems (Palletizers, Conveyors)

Conveyors are considered as material handling systems that are required in any warehouse or a distribution center and have a direct influence on the energy utilization of the operation. Energy conservation equipment like palletizers, conveyors and automatic sorting equipment contribute to lower energy consumption because they increase the rate, effectiveness and precision of the movement of products within the depot. For example, there exist palletizers which are machines that palletize products for storage or shipment. New palletizers consume less energy than the older models due to the incorporation of technologies such as variable frequency drives and modern motors that enhance the use of energy. These machines only work when needed and are used in conjunction with automation systems for the warehouse that make certain that the machines consume as little energy as possible. Conveyor systems are another facet of material handling systems that is crucial to the success of an organization. These transport vehicles move the goods across the warehouse from the entry points of receipt to the storage sections, and from the storage points to the shipping areas. Conventional conveyor systems are energy consuming but new generation conveyor systems have energy efficiency features such as variable frequency drives, energy return systems and low friction rollers. For instance, gravity conveyors that rely on the force of gravity to transport products are cheaper than the powered conveyor systems. Also, the introduction of AGVs can be used to reduce the energy consumption by doing away with the common forklift trucks that are known to consume a lot of fuel and require manpower. Another innovative idea in material handling is the use of smart sensors and IoT in the material handling systems. These systems utilize sensors to identify the existence and the flow of the products, and control the energy usage according to the flow of goods in the warehouse. For example, the conveyor systems can reduce their speed or even halt during periods when there are no products to transport for instance during the night. Likewise, intelligent control system can also determine the path of the AGVs or robots in order to reduce energy waste by avoiding unnecessary movement and keeping the distance of the travelled path as short as possible.

6.3 Integration of Circular Economy Principles in Packaging and Handling

The concept of adopting circular economy within packaging and material handling is a revolutionary concept that seeks to alter the current supply chain management. The circular economy is a business model that optimises the use of resources by using, repairing, refurbishing and recycling products and materials in order to reduce the waste and use of the earth's limited resources. In the context of warehousing, this means redefining the traditional linear approach of using resources, creating goods, using them and then disposing them, and replacing it with a model where resources are kept in use for as long as possible. Looking at packaging, circular economy involves the reusing or recycling options. For instance, firms are using returnable packaging items like pallets, bins, and crates, which are reusable in other logistic chains. This has the effect of cutting down the use of packaging material that is used once and also saves the company a lot of money that would have been used to purchase the material. It is especially important for businesses that have a high rate of shipping and transportation, for instance, retail and e-commerce businesses. Also, the packaging systems are being implemented in closed-loop systems where the packaging is returned, washed, and circulated again, thus minimizing the use of new packaging and the negative effects on the environment. In the material handling aspect, circular economy can be implemented through the remanufacturing and recycling of materials and equipment. For instance, conveyors, rollers and palletizers are used items that can be recycled and maintained for a longer use rather than replacing them. Also, material handling systems may be made to be modifiable for the purpose of recycling the components in case they are no longer useful. Through the application of these principles, warehouses help to prevent wastage at the same time as they support a general sustainability cause through the conservation of raw materials. Therefore, it is crucial to incorporate the use of green packaging and energy-efficient material handling systems in a warehouse in order to make it sustainable and productive. This chapter identifies ways through which warehouses can reduce the environmental impact, including the use of recyclable and biodegradable packaging, the use of appropriate size of the packaging, and the use of energy efficient technologies when handling materials. Furthermore, the use of circular economy in packaging and handling ensures that logistics is driven to a more sustainable supply chain, therefore embracing the protection of resources, cutting on cost, and the attainment of sustainable environment in the future.

7. DATA-DRIVEN OPTIMIZATION IN GREEN WAREHOUSING

7.1 Role of IoT and Data Analytics in Energy Monitoring and Optimization

The use of IoT and data analysis in the management of energy consumption is now a key aspect in managing warehouses. IoT allows for the integration of numerous devices and sensors all over the warehouse to form a data stream that provides information about energy consumption. These sensors can track other factors such as temperature, humidity, levels of light, and usage of equipment which will help the managers of the warehouse to assess energy usage. In this way, the data can be gathered to help warehouses go beyond the simple energy audit and become more proactive in their energy use.

Figure 3. Data-Driven Optimization in Green Warehousing

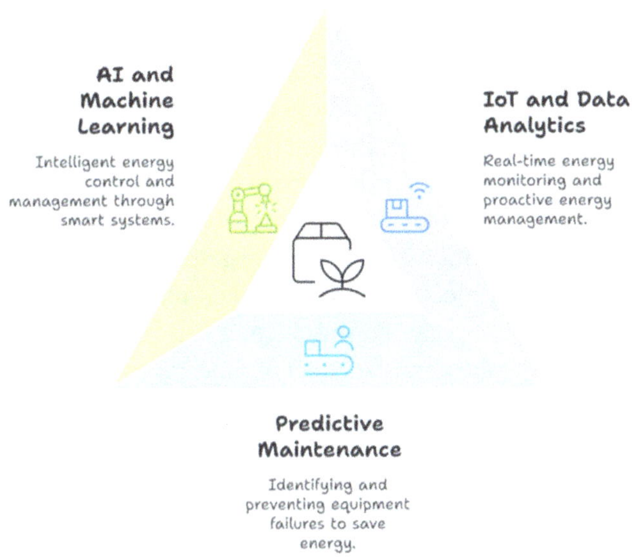

For instance, smart lighting systems that have been integrated with IoT sensors have the capability of changing the lighting intensity depending on the people presence and lighting condition. These systems regulate the amount of natural light coming into the building, and therefore minimize the use of artificial lighting to the most extreme of circumstances. Likewise, IoT-based systems for heating, ventilation, and air conditioning in warehouses can control temperature and air flow depending on the real-time information, which makes the working environment comfortable for the employees while enhancing energy utilization. Data analytics is also useful in energy management by using the data collected from the IoT sensors to analyze the patterns, inefficiencies and possible improvements. Sophisticated analysis can identify pre-heating, over-lighting, over-cooling, or over-use of appliances among other things. From such patterns, the warehouse operators can then be able to develop appropriate strategies that would reduce energy consumption. For instance, there can be lighting schedules set and equipment use can be regulated in a way that there is little energy wastage during idle periods. In addition, analytics can be used to predict energy requirements and thereby enhance the management of resources with a view of expected high energy demand. Besides, data analytics also help the planning of energy consumption for future use since it involves the analysis of past data. By this way, it is possible to apply energy saving measures or to invest in new energy efficient technologies at the right time, when their application brings most benefit for the warehouse.

7.2 Predictive Maintenance and Energy Management

One of the other important aspects of optimization through data in green warehousing is the predictive maintenance. This is an approach that involves using data from IoT sensors to identify the current status of the crucial assets and the likelihood of their failure. Thus, through real-time analysis of this data, the warehouses can alert them on any problems with machines such as HVAC systems, conveyor belts, and refrigeration units before they cause any failure or inefficiency. This approach is effective in avoiding situations when an emergency repair is needed or when a breakdown occurs due to neglecting the systems, equipment, and machinery, as well as in minimizing energy wastage. For instance, if it is discovered that a conveyor system is dragging due to the abrasive effect of time, then the conveyor system can be serviced or even changed to operate as efficiently as possible without the use of much power.

In the same way, the application of predictive maintenance is also effective for the energy management system. EMS can be linked to other predictive maintenance tools so as to get confirmation that all the equipment is operating optimally. Where there is any problem in an energy intensive system, one can draw its attention before the problem results to wastage of energy. For example, if a motor is identified to draw more current than necessary, predictive analysis can raise a service request for the same. This not only saves energy but also helps in increasing the lifespan of some important equipment that are in use thus increasing the efficiency of the operation. However, it is also possible to apply the concept of predictive maintenance to the overall structure of the warehouse. In this way, by accumulating and analyzing data of different systems and processes, the warehouses are able to gain a clear view of the energy efficiency and identify the areas that require maintenance, upgrades or changes in the processes. This makes it possible for the warehouse management to manage the energy consumption of different systems at the same time thus making a positive impact in the sustainability of the warehouse.

7.3 Smart Warehouses: The Integration of AI and Machine Learning to Optimize Energy Usage

The advancement of AI and ML in recent years has brought more changes in the optimization of energy usage in warehouses. Smart warehouses can be defined as the warehouses in which artificial intelligence and machine learning algorithms analyze real-time data received from sensors and machines, as well as from external sources. Smart warehouses are also able to implement real-time learning from patterns in energy utilization hence adjusting energies automatically over machine learning. For instance, the smart systems on energy management can also be applied to monitor and control the lighting, heating, cooling, and use of equipment in the warehouse. These systems can therefore estimate the energy demand of the warehouse, this can be due to time of the day, the weather or the different activities taking place in the warehouse . These systems are able to cut down on energy consumption by a considerable degree while at the same time sustaining the efficiency of the operational parameters. Machine learning models can also be used in predicting energy demand so that the warehouses can be able to make informed decisions regarding energy purchase. With the help of the historical data and the current consumption it is possible to predict the future needs in energy which can be purchased at the most beneficial time or use demand-response programs and avoid using energy during the most expensive time. Such a change can lead to major cost savings and will contribute to the minimization of carbon footprint within the warehouse environment. Furthermore, AI can be incorporated with the warehouse automation systems to increase energy efficiency. Robotic arms and automated guided vehicles (AGVs) are some of the systems that employ AI in that they try to minimize energy consumption in their movements to increase operational throughput. Such systems can learn from the changes in the environment like traffic pattern or the stock available in the warehouse and alter the path accordingly to reduce wastage of energy. When integrated with WMS, AI and ML are used in allocating workers, storage, and material handling, which all leads to the efficient use of energy in a warehouse. Therefore, data-driven optimization is a critical factor in the green warehousing in terms of energy efficiency. The integration of IoT and big data also helps in the real-time analysis of energy consumption in warehouses and making of informed decisions that will lead to the reduction in the energy consumption. Predictive maintenance makes it possible to ensure that some of the energy-intensive equipment is effectively used without any wastage of energy. Last but not the least, the application of AI and machine learning in smart warehouse systems the smart warehouse results in intelligent energy control and management. With the advancement of the technology, the issues of efficient application of data analysis for making the warehouses smart and energy-efficient will become more significant.

8. ENERGY-EFFICIENT TRANSPORTATION AND LOGISTICS IN WAREHOUSING

8.1 Electrification of Vehicle Fleets (Electric Trucks, Forklifts)

One of the major changes that has been seen in warehousing industry with regards to energy efficiency is the conversion of vehicle fleets. Conventional fleets used in warehouses include forklifts, trucks and delivery trucks which are predominantly diesel operated hence consuming a lot of energy and emitting carbon. The use of electric trucks and electric forklifts is one of the ways through which the environmental effects of transportation in the warehouse can be addressed. EVs are more efficient than ICE vehicles since they utilize energy from electricity directly without converting it through other forms such as mechanical energy. Electric trucks are slowly being adopted in warehouse operations because they help in the reduction of emissions and operating costs. These vehicles are suitable for short distance transport from one warehouse to the other or from a warehouse to delivery points in towns. Electric trucks mean that warehouses can reduce their emissions and at the same time save on fuel costs and expenses on maintenance. However, with the growth of the electric vehicle technology where charging points are becoming more common, the use of electric trucks in warehouse companies becomes more feasible. Likewise, electric forklifts are used in most of the material handling activities that take place in the warehouse. These electric models of internal combustion engine forklifts are less noisy, less pollutive, and require less energy to operate making them suitable for indoor use especially where air quality is a concern. Electric forklifts also have low emission levels which are beneficial to the working environment of the employees. However, they consume less energy, which makes the energy cost of running a warehouse to be low. Electric forklifts are also integrated with smart charging systems so that the charging time of the forklifts can be synchronized with the off-peak electricity tariffs.

8.2 Use of Alternative Fuels in Warehouse Operations

Besides the conversion of vehicles to electric ones, the use of other fuels like hydrogen and biofuels are also a good strategy in the reduction of energy consumption and overall effects on the environment during the transportation of goods in the warehouse. Hydrogen fuel cells can be used for forklifts, trucks and other material handling equipment used in the warehouses. These fuel cells produce electricity through the chemical combination of hydrogen and oxygen, the only emission from which is water. This make hydrogen vehicles particularly suitable for commercial use especially where the vehicle is used for many hours at a go and requires quick refuelling. While it is true that electric vehicles might take a long time to charge, hydrogen fuel cell vehicles can be refuelled in a short time and therefore suitable for 24/7 warehouse operations. There are other alternatives to diesel and gasoline, which are biofuels produced from renewable biological materials. Bio fuels can be used in I.C engine vehicles and equipment's thus cutting down on the carbon emission from the transport vehicles in the warehouse. For instance, biodiesel can be used in diesel engines without having to make a lot of changes on the engines. This gives an instant solution to emissions reduction that is not costly as it involves a complete shift from a fleet. Biofuels are also environmentally friendly because they can help in reducing greenhouse gases and air pollution hence being better than fossil fuels. The utilization of AFs in warehouse operations is still limited at present but is perceived with vast opportunity to decrease the environmental footprint of supply chain and material handling systems. The increase in technology as well as the infrastructure will see more incorporation of hydrogen and bio fuels in the transportation of warehoused products to enhance sustainability.

8.3 Energy-Efficient Routing and Transportation Optimization

The other factor that is equally important in energy efficient logistics in warehousing is the aspect of transporting the products through efficient routes. Fleet management is also a common practice in warehouses since most of them require large vehicles for transporting goods for short distance or over long distances. This is because; energy consumption can be reduced by identifying the most effective routes to be taken by delivery vehicles. Smart routing solutions rely on current information about the traffic patterns, the weather at the given time, and delivery schedules in order to identify the most appropriate pathways for the vehicles with the least amount of fuel used, and in the shortest possible time. In recent years, route optimization through the use of artificial intelligence and other related technologies is used to reduce the time on the road, and the time spent on areas with heavy traffic thus reducing fuel consumption on unnecessary detours and other related factors that lead to increased emission of greenhouse gases. These systems can also calculate the loads to be taken by the trucks so that the trucks are fully loaded hence minimizing the number of times the trucks will have to travel. For instance, instead of making many trips with partial loads, transport management can assist in loading as many goods as possible to a single trip by rearranging delivery schedules and routes. In addition, dynamic scheduling allows warehouses to optimize the transport routes and timings according to the current information. When a traffic jam is identified or there is a predicted storm, the vehicles may be rerouted to reduce wastage of time and energy. Moreover, by using the fleet management systems that are connected with GPS and telemetry, the drivers can receive information concerning the ways of improving their driving, for instance, driving at a steady speed, avoiding excessive idling, harsh acceleration, and braking.

8.4 Last-Mile Delivery and Its Impact on Energy Consumption

Final delivery, the last step that involves delivering goods directly to the end consumer is a crucial factor that defines the energy intensity of a warehouse. The last-mile refers to the final stage of the delivery process, from the distribution center to the customer's door. This stage is usually the most power consuming one as it implies many individual deliveries and can lead to high fuel consumption due to delivery vehicles traveling in city environment with frequent stops and traffic jams.

To cut down the energy costs incurred in delivery, especially in the last leg of the delivery process, companies are now using electric vehicles or bikes within the warehouse. Electric cars such as electric vans or electric bikes can be a much better solution for short distances delivering goods. It should also be noted that these vehicles emit no emissions and are much cheaper to run in terms of fuel and maintenance than conventional diesel delivery vehicles. Another new strategy is the micro-fulfilment centres or micro-warehouses near the consumer concentration areas. This way, warehousing is broken down and made more efficient in terms of distance and energy needed to deliver within the last mile when these smaller fulfilment centres are established near densely populated cities(Ponmalar A et al., 2025). This in turn eliminates the need for the large trucks in making deliveries which makes use of the small energy efficient vehicles. Furthermore, it provides an insight into ways such as with the help of heterosexual delivery robots and drones in last-mile delivery. These technologies are electrically operated which can bring down energy usage and discharge related to traditional delivery trucks. Such delivery solutions are still emerging but have the potential to expand more in the future with a focus on the urban environment. Therefore, to come up with a sustainable warehouse operation, energy-efficient transport and logistics should be embraced. Interchangeable batteries of electric cars plus utilization of hydrogen and biofuels for movement of the vehicles painting a promising picture of future storage of carbon emission in transport vehicles meant for moving goods within the warehouse. Additional efficiency in routing and delivery also cuts down energy use through improving efficiency of the vehicles to be used, while new concepts such as electric and self-driverless vehicles can significantly transform the last delivery process. All these strategies make the warehousing and logistics system more sustainable, economical and energy efficient.

9. CHALLENGES AND BARRIERS TO ADOPTION

9.1 High Initial Investment Costs for Green Technologies

The major drawback of implementing green technologies in warehousing is the high initial cost associated with the implementation of the green technologies. Most energy-saving and renewable products like solar power, electric cars, LED lighting, and automation systems cost a great deal at the initial stages. Though these technologies have long-term operational benefits in the form of energy and maintenance cost, the initial cost remains a constraint for many organizations, especially the SMEs. For instance, the use of solar energy for instance involves a large capital investment in terms of purchasing of the panels, batteries and other necessary structures. Likewise, the shift to electric forklifts or electric delivery trucks means not only the costs of the vehicles but also the cost of charging equipment and possibly personnel to operate them. The long payback period of such an investment makes companies reluctant, for example, where they are unsure of the financial return on their investment or due to economic constraints. Furthermore, some of the efficient technologies may not be available in some markets hence the need to import or source them from other countries. This makes it even harder for organizations to embrace the green technologies since the cost of installing them is usually high for the benefit that will be realized in the future.

9.2 Technological Limitations and Integration Challenges

The last factor is technological issues related to the integration of green technologies in the warehousing environment and lack of technological readiness for these solutions. Most of the technologies to enhance energy efficiency, including renewable energy sources, automation systems, smart sensors need to be integrated into the warehouse systems. For instance, adopting automated systems in material handling or inventory control means that one has to purchase new hardware and integrate it with the existing systems. Also, the solutions based on the optimization of data and smart technologies, including IoT sensors and AI-based analytics, depend on the timely data and a high level of algorithms. In warehouses that do not have a well-developed data solution in place, the use of these tools may be challenging and take a lot of time. The use of sensors in the HVAC systems or the lighting controls to manage the energy usage is an extensive process that requires alterations in the electrical system of the warehouse. This might make some companies hesitate to change to a new level of energy efficiency and automation. Some of the green technologies are also not well developed, for instance in integration of renewable energy systems and automation. One disadvantage of having no reference solutions is that the warehouses may have problems with guaranteeing that new technologies will integrate well with the current systems. This is because the implementation of these solutions may be associated with risks of technical failures, incompatible systems or systems that do not deliver the expected level of efficiency.

9.3 Regulatory and Compliance Challenges

Another challenge is the regulatory environment of green technologies in warehousing. While some parts of the world have established standards for green building, sustainability rules, and energy efficiency policies, many warehouses are situated in regions where such rules are either weak or almost non-existent. It is challenging to go through the various local, national, and international rules and regulation, especially for a company operating in different countries. Also, some technologies like renewable energy systems, fully automated vehicles, etc, may come under some legal restraints regarding their safety, environmental policies or ordinances or zoning laws. For example, the installation of the solar panels or wind turbines may be prohibited or may need some permit or may be restricted by the local zoning laws. Likewise, electric automobiles, which are increasing in use, are not backed up by proper charge points or subsidies in certain parts of the world, thus limiting their adoption. Also, there are likely to be some issues that companies may encounter in terms of compliance with energy efficiency and environmental requirements. Some of the regulatory standards call for businesses to meet certain threshold energy consumption levels or cut down on emissions of carbon. The use of green technologies entails having to go through a number of compliance procedures and make sure that the technologies are capable of meeting certain levels of performance. It becomes especially challenging for organizations that do not have a compliance department to keep track of such regulations.

10. CONCLUSION

There has been an active advancement in green and energy usage in the warehousing industry to change how its operations are conducted to enhance environmental conservation. Features such as efficient lighting, automation, integration of renewable energy and smart designs in the warehouses have been critical in reducing energy consumption, carbon emissions and enhancing efficiency. Nowadays, AI, IoT, and robotics are the prominent technologies that help in improving the efficiency of energy consumption, inventory, and material handling. Also, the ideas of sustainable packaging, energy conservation through HVAC systems and the use of renewable energy sources have further enhanced the efficiency of the warehouse with reduced wastage and emissions. The future of green warehousing appears bright, as the energy storage technology, automation, and smart technologies are set to improve the prospects of sustainability. The new trends of carbon neutrality and zero-energy warehouses will become the new industry standards due to the following reasons: The current trends of the regulatory requirements and the issues of corporate social responsibility. The increasing importance of sustainability in the business world and among customers will only drive more companies to seek effective green warehousing solutions, thus encouraging more development and spending on energy-saving solutions. The companies in the industry such as the warehouse operators, supply chain managers and the technology providers are encouraged to adopt these innovations and embrace green technologies for sustainability of the environment and for long-term benefits. Sustainability in warehousing, in the long run, helps the businesses cut costs, meet the legal requirements and support the fight against climate change. The future of warehousing is not only the focus on the best performance indicators but also the care about the future of the planet, so it is crucial for the industry to keep developing the solutions that are more sustainable and energy-saving.

REFERENCES

Batarlienė, N., & Jarašūnienė, A. (2024). Improving the Quality of Warehousing Processes in the Context of the Logistics Sector. *Sustainability (Basel)*, *16*(6), 2595. DOI: 10.3390/su16062595

Bhattacharjee, A., & Jahanshahi, A. A. (2024). Simulation of Green Supply Chain Design With Renewable Energy and Green Technology for Intensifying Sustainability After COVID-19. In *Green Supply Chain Management Practice and Principles* (pp. 219–248). IGI Global. DOI: 10.4018/979-8-3693-3486-7.ch011

Domagała, J. (2022). Green warehousing: a state-of-the-art literature review. *Sustainable Logistics*, 43-71.

Gopalan, B., Vijaya, G. S., & Tiwary, A. (2025). Exploration of IoT Applications, Lean Management, and Green Practices for Sustainable Urban Logistics. In *Bridging Technology and Development for Sustainable Innovation and Geopolitical Dynamics* (pp. 225–252). IGI Global Scientific Publishing. DOI: 10.4018/979-8-3693-9072-6.ch009

Krishnan, R., Perumal, E., Govindaraj, M., & Kandasamy, L. (2024). Enhancing Logistics Operations Through Technological Advancements for Superior Service Efficiency. In *Innovative Technologies for Increasing Service Productivity* (pp. 61–82). IGI Global. DOI: 10.4018/979-8-3693-2019-8.ch004

Lija, A. V., Aswathy, M. P., Kavitha, G., Velammal, M. N., Maheswari, B., & Pajila, P. B. (2025). A Review on Sustainable Development and Green Practices: Strategies, Challenges, and Future Directions. *Adaptive Strategies for Green Economy and Sustainability Policies*, 1-16.

Malinowska, M. (2022). The concept of decision support system to increase the level of sustainable warehousing. *Procedia Computer Science*, *207*, 4046–4054. DOI: 10.1016/j.procs.2022.09.467

Manjunathan, N., Ramana, T. V., Rajasekar, A., Vijayakumar, D., Sameswari, V., Gopal, S. N., & Subramanian, R. S. (2025). Sustainable Cloud Computing: Ecofriendly Strategies and Innovations for Green Data Centers. In *Energy Efficient Algorithms and Green Data Centers for Sustainable Computing* (pp. 25-54). IGI Global Scientific Publishing.

Masudin, I., Sumah, B., Zulfikarijah, F., & Restuputri, D. P. (2021). Effect of information technology on warehousing and inventory management for competitive advantage: a theoretical framework. In *Handbook of Research on Innovation and Development of E-Commerce and E-Business in ASEAN* (pp. 570–593). IGI Global.

Munuhwa, S. (2023). Sustainable logistics and competitive positioning. In *Integrating Intelligence and Sustainability in Supply Chains* (pp. 198–220). IGI Global. DOI: 10.4018/979-8-3693-0225-5.ch011

Nicoletti, B., & Appolloni, A. (2024). Green Warehouse 5.0. In *The Future of Industry: Human-Centric Approaches in Digital Transformation* (pp. 303–320). Springer Nature Switzerland. DOI: 10.1007/978-3-031-66801-2_20

Ponmalar, A., Muthuram, A., Therasa, M., Priya, V., Sathya, V., Subha, S., & Subramanian, R. S. (2025). Convergence of Advanced Technologies for Sustainable Development and Smart Cities. In *AI and Green Technology Applications in Society* (pp. 189–218). IGI Global Scientific Publishing.

Subramanian, R. S., Santhi, K., Sasikumar, D., Shobana, S., Elavarasi, J., Gracious, L. A., & Nalini, M. (2026). Leveraging Sentiment Analysis in Customer Service for Enhanced Customer Experience. In *Harnessing Emotion AI for Customer Support and Employee Wellbeing* (pp. 49-72). IGI Global Scientific Publishing.

Tiwari, S. (2023). Smart warehouse: A bibliometric analysis and future research direction. *Sustainable Manufacturing and Service Economics*, *2*, 100014. DOI: 10.1016/j.smse.2023.100014

Yamini, B., Anish, T. P., Maheswari, B., & Praveenkumar, J. (2025). Towards a Greener Future: Integrated Strategies for E-Waste Management. In *Integrated Approaches for Sustainable E-Waste Management* (pp. 133–150). IGI Global Scientific Publishing. DOI: 10.4018/979-8-3693-7383-5.ch005

Chapter 11
Integration of Renewable Energies in Warehouses

Muhammad Waheed Azam
https://orcid.org/0009-0005-3735-7999
Department of Industrial Engineering, Alma Mater Studiorum-University of Bologna, Italy

Ghulam Qadir Chaudhary
Shandong University, Jinan, China

Giovani Simprini
Department of Industrial Engineering, Alma Mater Studiorum-University of Bologna, Italy

Suoying He
School of Nuclear Science, Energy, and Power Engineering, Shandong University, Jinan, China

ABSTRACT

The adoption of renewable energy technologies to meet warehouse energy demand is a crucial step in achieving sustainable development, as there are global energy shortages and environmental challenges. This chapter provides an overview of the application of renewable energy technologies, including solar, wind, and geothermal, for adoption in warehouses. Moreover, it also describes the potential of these technologies in the reduction of fossil fuel utilization, operational cost reduction, and minimizing carbon footprints. Initially, solar energy, through photovoltaic (PV) and photovoltaic/thermal (PV/T) systems, is discussed in detail as an attractive alternative for electricity and thermal energy requirements. Therefore, this chapter highlights

DOI: 10.4018/979-8-3373-3176-8.ch011

the potential role of renewable energy technologies by carefully examining their benefits, associated challenges, and their practical applications as a way forward toward a sustainable future in the warehouse industry.

1. INTRODUCTION

Governments and businesses worldwide agreed to support sustainable growth, considering the world's increasing energy shortages and environmental challenges. Sustainable development aims to establish more efficient and environmentally friendly production methods by maximizing resource use and minimizing the environmental impact of products across their entire life cycle. The increase in carbon emissions and energy consumption over the past two decades has become a significant barrier to long-term sustainable development and growth (Lei, 2024). Conventional storage techniques frequently fail to monitor and optimize energy use, leading to environmental contamination. Hence, it becomes particularly crucial to develop new strategies that improve energy efficiency and achieve sustainable, environmentally friendly development (Boysen & de Koster, 2024).

Warehousing operations are key components of the global supply chain process because they enable businesses to effectively store, organize, and distribute products across industries (Cannava et al., 2024). However, these facilities often consume substantial amounts of energy, resulting in high operational costs and a substantial carbon footprint. The warehouse sector is rapidly growing as the technology continues to advance and the trend of online sales, therefore, more energy and space are required to meet the demand (Oloruntobi et al., 2023). The transportation and warehousing industry uses between 10% and 15% of the world's industrial energy, and is expected to increase emissions by more than 40% by 2030. More than 100 kWh/m^2 could be consumed annually by warehouses, especially those with refrigeration, indicating a significant possibility for efficiency improvements and the incorporation of renewable energy sources. The standard classification of warehouses is shown in Figure 1.

Figure 1. Classification of warehouses. Source: Author's work

Public and private warehouses: Public warehouses are usually managed by the government and provide a range of services, including shipping and packaging, whereas a particular individual, business, or organization manages private warehouses.

Automated Warehouses: Automated warehouses are innovative storage facilities that utilize sensors, robotics, and advanced software. They have several benefits of implementing automation in the warehouse, including the elimination of errors, higher efficiency, a significant reduction in operating expenses, improved process control, and an increase in the speed and precision of information flow (Igor Cosma et al., 2025).

Distribution warehouses: Large warehouses with the capacity to move heavy loads efficiently, with the use of equipment like pallet jacks and forklifts, are known as distribution warehouses (Manning & Jorgensen, 2025).

Climate-controlled warehouses: In these types of warehouses, the environment is controlled for products that are sensitive to humidity or temperature, such as pharmaceuticals, perishable food items, electronics, cosmetic products, and so forth (Chen et al., 2023).

Contract warehouses: Third-party storage facilities known as contract warehouses offer warehousing services with a long-term contract between the client and the warehouse operator.

Fossil fuels (coal, oil, and natural gas) are usually used to meet the energy needs of various warehouse operations (lighting, HVAC, material handling equipment, etc.) (Azam, Chaudhary, et al., 2023). These fuels contribute to climate change and global warming by releasing significant amounts of carbon dioxide (CO_2) and other greenhouse gases. Additionally, businesses can significantly reduce fuel usage by implementing energy-efficient systems and technologies, as well as switching to sustainable energy resources (J. Wang & Azam, 2024). Therefore, integrating renewable energy resources, such as solar, wind, and geothermal systems, into warehouse operations provides a potential solution as organizations worldwide face increasing demands for more environmentally friendly strategies. These renewable energy technologies could enhance energy reliability, reduce energy costs, and support global sustainability initiatives by lowering dependency on traditional energy sources. The utilization of renewable energy in warehouses can be achieved directly through on-site energy generation or indirectly through integration into the present conventional energy supply infrastructure (Ren et al., 2025). The energy production by renewable sources has become less expensive than traditional thermal power plants and fossil fuels, so business organizations are paying more attention to their use (Fikiin et al., 2017). The energy share of the worldwide installed capacity for power and electricity generation by 2050 is depicted in Figure 2. Warehouses are energy-intensive buildings that consume substantial amounts of electricity for various operations, including lighting, equipment handling, heating, and cooling. The energy consumption components in the warehouses are shown in Table 1.

Figure 2. According to IEA projected global share of energy resources for electricity generation by 2050 under stated policies scenario (Future of Solar Photovoltaic : Deployment, Investment, Technology, Grid Integration and Socio-Economic Aspects, 2019)

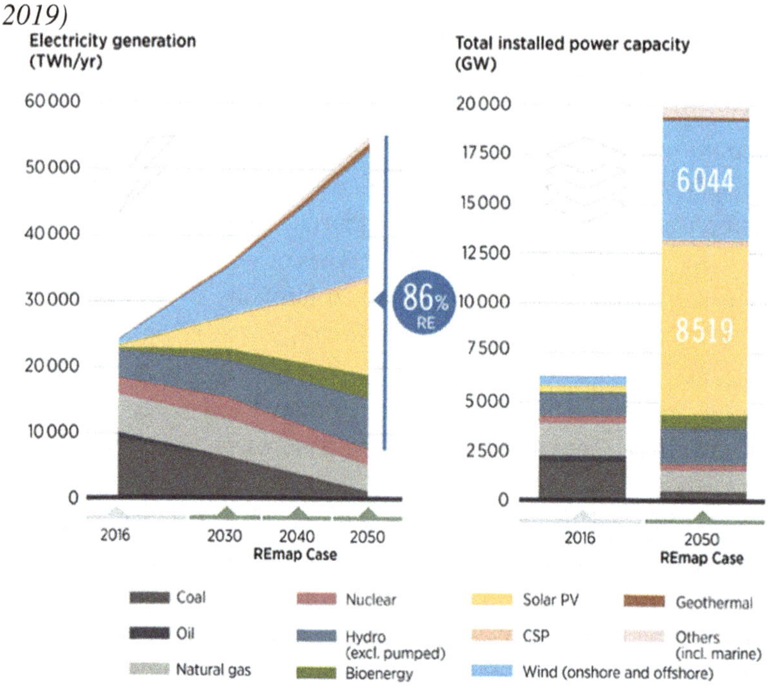

Table 1. Energy consumption components in a warehouse

Energy Consumption Components	Purpose/Use in Warehouse
Lighting	Lighting systems for warehouse operations
Heating, Ventilation, and Air Conditioning (HVAC)	Heating and cooling systems, climate control
Material Handling Equipment	Forklifts, conveyors, pallet jacks, and automated guided vehicles
Industrial Equipment & Production	Machinery for packing, sorting, or processing goods
Lighting for External Areas	Outdoor lighting for loading docks, parking lots, and other areas.
Refrigeration	Temperature control for perishable goods storage (cold storage)
Electric Vehicle Charging (if applicable)	Charging stations for electric trucks or forklifts

The following areas will be covered in this chapter:

- Integrating various renewable energy systems (solar, wind, and geothermal) into warehousing operations
- Energy storage solutions that improve the reliability and efficiency of renewable energy in warehouses
- Examining real-world implementations

2. RENEWABLE ENERGY RESOURCES FOR WAREHOUSING

Renewable energy will play a crucial role in the development of a resilient and environmentally friendly society in the 21st century. It is produced from natural resources from five major groups, including solar, wind, hydro, biomass, and geothermal. It operates under various conditions and has different energy conversion efficiencies.

These resources don't release greenhouse gases or other pollutants that affect the environment, unlike traditional energy production methods. In recent years, renewable energy has been used more and more to generate electricity worldwide, which reduces the dependence on grid energy (Y. Zhang et al., 2026). Moreover, worldwide, there is a significant interest in utilizing renewable energy as a future energy source.

Operating a warehouse has significant environmental impacts, primarily due to high energy consumption and emissions. These emissions are classified into three types based on direct and indirect emissions (Korra & Sadhana V., 2024).

- Emissions originate directly from sources that the facility controls, for example, fuel burning for on-site heating.
- Indirect emissions originate in the production of heat, steam, or electricity that is purchased and used in the warehouse.
- It is challenging to measure and reduce indirect emissions from a broader group, such as those associated with manufacturing and waste management.

Therefore, warehouses can utilize renewable energy sources to meet a significant portion of their energy needs by integrating technologies such as solar, wind, geothermal, and biomass. The benefits, challenges, and applications of renewable energy technology in warehouses are summarized in Table 2. This strategy not only reduces dependency on the grid but also improves energy resilience and reliability, particularly during peak times. Moreover, real-time monitoring and optimization are made feasible by integrating innovative energy management systems, ensuring the effective use of renewable resources. Additionally, the integration of various renewable energy resources into warehouses through hybrid integration is depicted in Figure 3.

Figure 3. Integration of renewable energy systems into warehouses through hybrid integration. Source: Author's work

Table 2. Renewable Energy in Warehouses: Applications, Advantages, and Challenges

Renewable Energy Technology	Application in Warehouses	Benefits	Challenges
Solar Photovoltaic (PV) Panels	Install on the roof and walls of warehouses	Reduces electricity bills Low maintenance Sustainable energy source	High upfront cost Requires sufficient roof space
Solar Thermal Systems	Use to heat water or air for warehouse operations	Reduces heating costs Sustainable source of thermal energy	Efficiency can vary with the weather. Requires a large surface area
Wind Energy	Small-scale turbines on-site for electricity generation	Reduces grid dependency Low carbon footprint	Requires significant wind availability Noise concerns
Geothermal Energy	For heating and cooling the warehouse	Low operating costs after installation Reliable energy source	High initial installation cost Requires suitable ground conditions
Biomass Energy	Use of organic waste for heat generation	Utilizes waste material Can lower energy costs	Needs a supply of organic material Emissions from combustion
Combined Heat and Power (CHP)	Use waste heat from renewable energy sources for cooling or heating	Efficient energy use Reduces overall energy demand	Requires a constant source of waste heat High initial cost

Therefore, the integration of renewable energy resources into warehouse operations provides the following benefits;

- Reduced dependence on fossil fuels
- Minimize carbon footprint
- Lower energy consumption for cost savings
- Compliancc with grccn building standards
- Long-term investment value
- Improved resilience and reliability
- Enhanced corporate image
- Scalability and flexibility
- Incorporation with smart grid energy

2.1 Solar Energy

Among renewable energy sources, solar energy is one of the primary sources of the planet's unlimited free energy and is widely used in the generation of electricity (C. Zhang et al., 2025). Warehouses could significantly reduce their dependence on the grid, normalize energy prices over time, and contribute to global initiatives to address climate change by utilizing solar power. As a result, it is not only an economical choice but also a strategic move toward sustainable and environmentally friendly growth. Moreover, it can be predicted to expand significantly, facilitating the transition to a future with net-zero emissions. The global map of horizontal radiation is depicted in Figure 4. The Sun emits 3.846×10^{26} Watt of energy, this amounts to approximately. 1.214×10^{3} Joules of energy annually. Making financial decisions for the installation of a solar energy plant is helped by an accurate solar evaluation of a location. The following data sets have been considered to be essential for forecasting solar power variability (Esmaeilion et al., 2025). Moreover, a comparative analysis of various solar energy technologies for warehouses is shown in Table 3.

- Measurement data collected on the earth
- Measurement data through satellites
- Numerical weather forecasting data.

Figure 4. Maps of global horizontal irradiation (Solar GIS, Direct Normal Irradiation (GHI); 2024. Available from: Https://Solargis.Com/Resources/Free-Maps-and-Gis-Data, 2024)

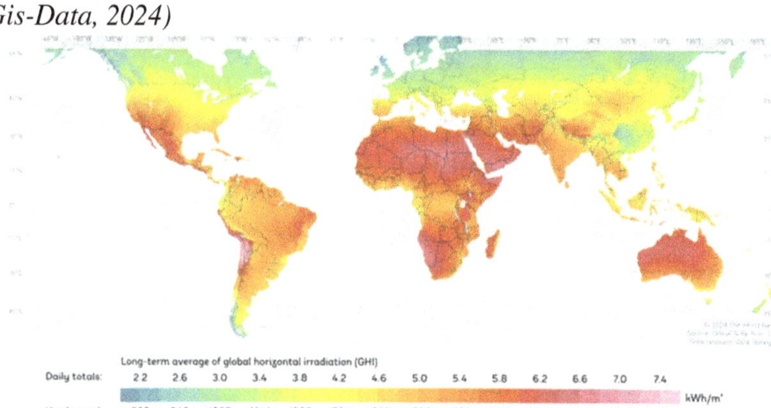

Table 3. Comparative analysis of various solar energy technologies for warehouses

Technology	Output	Efficiency	Key Advantages	Key Challenges & Feasibility
Rooftop PV	Electricity	15-22%	-Mature & Cost-Effective - Utilizes unused rooftop space. - Systems can be sized to match energy demand.	-Roof must support the weight. -Requires storage or grid backup for 24/7 operation. -Energy generation capped by roof area.
BIPV (Building-Integrated PV)	Electricity + Building Material	10-18% (can be lower due to design trade-offs)	- Replaces traditional cladding/roofing, saving material costs. - Can be customised (colour, transparency). - Enhanced Sustainability:	-Significantly higher cost - Requires specialized design and installation. - Potential Efficiency Loss
PV/T (Photovoltaic/ Thermal)	Electricity + Heat	Electrical: 15-18% Thermal: 50-70% (combined >80%)	- Maximizes energy harvest per unit area. - Thermal energy can be used for space or water heating, or solar-assisted cooling.	- More components (heat exchangers, pumps) than PV-only. - More expensive than standard PV systems.
CSP (Concentrated Solar Power)	Heat (High-Temperature) / Electricity	Thermal: 30-40% (Power cycle ~30-35% of this)	-Thermal energy is easier and cheaper to store at scale than electricity. - Can provide heat/ electricity on demand, day or night. - Suitable for industrial processes.	- Not feasible for individual warehouses; designed for utility-scale plants. - Only viable in regions with very high direct solar irradiation.

In recent years, new technologies have been used to harness solar energy and produce power [12]. The most common types of solar energy are shown in Figure 5. Each approach has unique applications, operations, and efficiency in the building sector. Among these technologies, photovoltaic (PV) and thermal photovoltaic (PV/T) are the most promising technologies for warehouse applications. Moreover, advanced building techniques such as thermo-photovoltaic panels (PV/T) are installed on a ventilated façade (which helps cool the building in summer and improve thermal insulation in winter) and connected to a heat pump system. There are several advantages to implementing this approach in warehouses, including generating electricity and hot water, improving the efficiency of heat pumps, and a ventilated façade that reduces internal environmental impact and increases heat pump efficiency in the winter. Figure 5 depicts solar energy collection, utilizing mechanical or electrical technologies to convert heat energy into power energy, as categorized into active and passive solar energy.

Figure 5. Types of solar energy. Source: Author's work

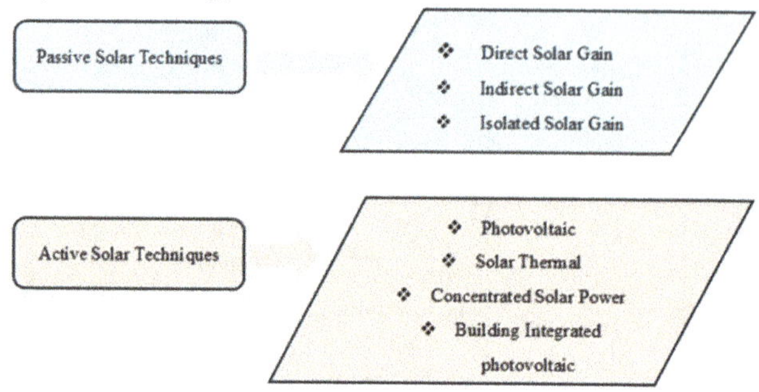

Passive Solar Techniques

The basic concept of passive solar energy is to keep things cool in the summer and warm in the winter. Moreover, when buildings and spaces are constructed to absorb and store solar energy as a heating source, it is known as passive solar energy. Passive solar design reduces dependence on fossil fuels for energy by utilizing solar energy, building materials, and surrounding conditions to produce thermally comfortable and energy-efficient spaces. Regarding this, various strategies can be employed, including shading, spacing, building orientation, window placement, thermal mass utilization, landscaping, and others. These techniques are preferred due to their simplicity and cost-effectiveness, and are crucial for achieving net-zero carbon emissions and decarbonizing buildings (Elaouzy et al., 2024). As warehouses demand significant energy, therefore implementation of passive solar energy could reduce energy consumption. The primary implementation depends on the location, its climatic conditions, and the building materials. The three main categories of passive solar heating systems are sunspace, direct, and indirect gain.

Direct and Indirect Gain

Direct gain systems are the simplest type, allowing sunlight to warm the living area through windows. On the other hand, indirect gain absorbs solar radiation in a separate area, then transfers the heat slowly into the living area. The most common examples of indirect gain are the solar wall and the Trombe wall (Uwa & Pourvahidi, 2024).

Isolated Gain

Figure 6. Passive solar technique: Sun spacing effect. Source: Author's work

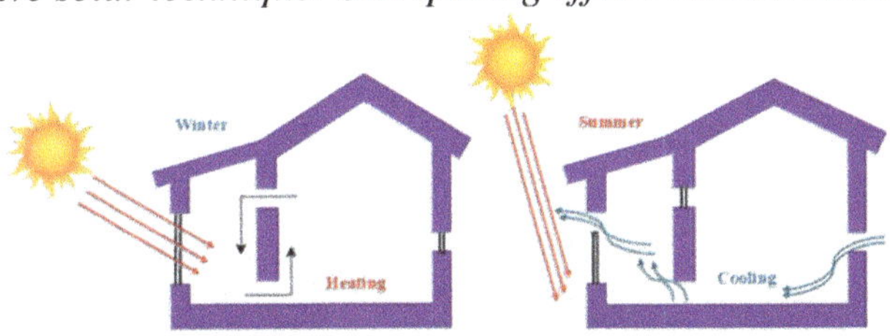

This system utilizes natural heat transfer processes (conduction, convection, or radiation) to move hot air into the living space after collecting solar energy in a separate area. The most popular example of isolated gain is sunspace, a separate sun or solar room on the south side of the house with a large glass area and thermal storage mass, and as a result, it reduces a building's heating load (Gainza-Barrencua et al., 2022). In winter, during the daytime, sunspaces absorb sunlight and store it in a thermal mass, such as the floor or walls, and provide a comfortable place to stay, while at night, it acts as a buffer zone to prevent excessive heat loss. To avoid overheating during the summer, heated air can be kept out of the living area by opening the windows and vents in the sunspace. These effects (winter and summer) are depicted in Figure 6.

2.1.1 Active Solar Techniques

PV and PV/T Solar Systems

Photovoltaic (PV) panels are increasing in demand for several reasons, as their transportation and installation are easier, and their payback period is low compared to other renewable energy resources (Meribout et al., 2023). As photovoltaic technology offers direct conversion of sunlight into electricity, which is a more favourable form of energy in terms of transportation and utilization. Directly convert sunlight into electricity. Furthermore, it is already widely adopted to improve rural living conditions in developing countries, particularly in agriculture, health care, and education. Moreover, such technology has been extensively used and connected with the electric grid in advanced countries (Vázquez & Vázquez, 2023).

As a well-known green energy source, it provides an efficient method to achieve the following key objectives by producing clean electricity locally (N. I. Ahmad et al., 2018).

- Minimizing energy cost

- Reducing environmental impact
- Improving energy independence
- Eco-friendly

Electricity is typically used extensively in warehouse buildings, depending on their size, operation, and degree of automation. It is used for various operations, including climate control, lighting, heating, ventilation, and air conditioning, but the most consistent energy requirement is for lighting and indoor climate control. Energy generation using PV panels in warehouses is more favourable for these types of businesses. The advantages of solar panels vary for each warehouse, depending on its building location, roof type, and energy consumption. Solar PV technology is particularly suitable for warehouses, particularly in areas with high solar radiation and prolonged periods of sunshine (Dada & Popoola, 2023). Warehouses that require air conditioning are well-suited to gain benefit from significant cost savings through the implementation of photovoltaic panels, as peak air-conditioning loads and incident radiation coincide with each other.

The installation of PV panels for electricity generation in warehouses can be accomplished in various ways, as described below and illustrated in Figure 7.

- Roof-mounted PV panels
- Building Integrated PV installation: as materials and products are available that can be used to construct the PV roof
- Wall-mounted PV panels
- PV Parking Lots

Factors to be considered before installing PV panels: The following are various common factors that need to be considered before the installation of PV panels (*INVESTMENT CASE FOR ROOFTOP SOLAR POWER IN WAREHOUSING Introduction & scope of work AUTHORS*, 2022).

- Available roof size & and its condition
- Orientation and angle
- Location
- Climate conditions
- Types of PV panels
- Local regulations
- Maintenance
- Monitoring

Figure 7. Installation of PV panels in warehouses: (a) Roof top, (b) Wall mounted, and (c) Parking lot. Source: Author's work

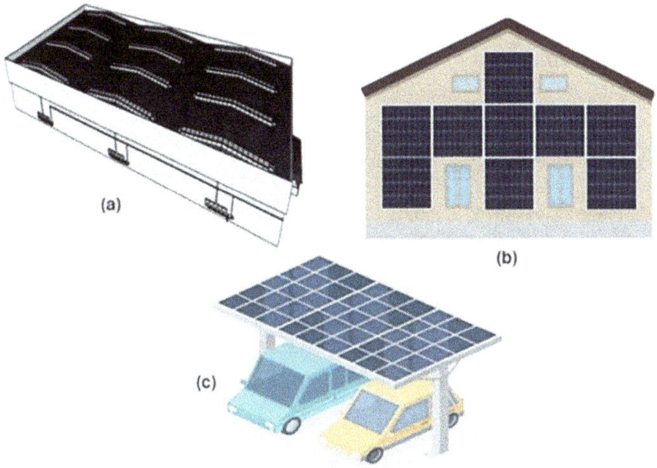

Figure 8 illustrates that solar cells are classified based on their evolution over time, as well as variations in materials and production techniques.

First-generation solar cells: They are built on high-quality crystalline Silicon wafers, which come in polycrystalline and monocrystalline varieties with cell lab efficiencies of up to 20% and 24%, respectively. The silicon crystals in polycrystalline solar cells are not as pure as those in monocrystalline solar cells. Therefore, both have different efficiencies. Moreover, polycrystalline solar cells are less expensive than monocrystalline ones because their manufacturing process is more straight-forward (Sugianto, 2020). Monocrystalline solar cells are particularly well-suited for installations in cloudy areas; however, as temperatures rise, their efficiency decreases. On the other hand, polycrystalline solar cells are more suitable for warm environments, as they are less affected by temperature increases (Pei et al., 2025).

Second-generation solar cells: This generation consists of cells made from thin films that utilize novel techniques for growing and depositing materials with a thickness range of a few micrometers, thereby reducing production costs. When compared to first-generation cells, they are less efficient but more flexible and less sensitive to high temperatures. The creation process of thin-film solar panels involves applying thin films of photovoltaic materials such as copper indium gallium selenide, amorphous silicon, or cadmium telluride onto a substrate. The significant drawbacks of this type of generation are the toxicity of the materials, a shorter lifespan, and the need for more space for installation compared to the first generation. Among all of them, copper indium gallium selenide (CIGS) has the highest efficiencies (22.6%). Moreover, Other benefits of this technology include increased processing simplicity, high radiation resilience, low toxicity, and maximum photon absorption (Casariego et al., 2025).

Figure 8. PV cell technology and its efficiency. Source: Author's work

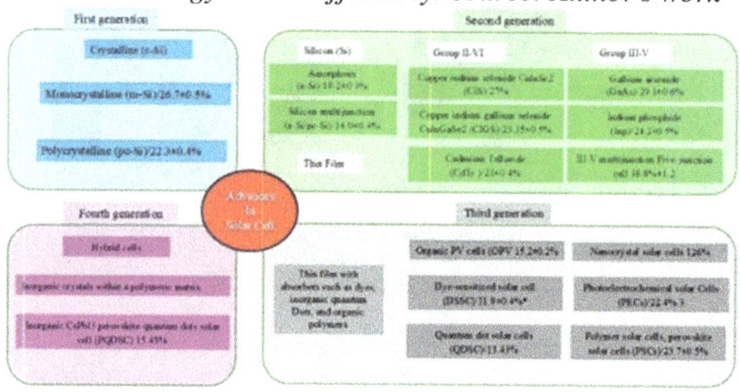

Third-generation solar cells: This generation of organic polymer solar cells combines the best characteristics of the first and second generations, resulting in increased efficiency and lower manufacturing costs using less expensive ingredients. These cells are highly efficient, lightweight, and flexible, and are based on the latest chemical substances. The most relevant third-generation cells are dye-sensitized solar cells (DSSC) and organic photovoltaic (OPV) solar cells. Organic photovoltaic (OPV) solar cells are beneficial to the environment, can be utilized with other technologies, and are appropriate for large-scale production, which makes them interesting for certain photovoltaic applications. However, the implementation of OPV structures has been hindered by instability issues, as research is ongoing to enhance stability and efficiency (Nkinyam et al., 2025). On the other hand, the working principle of dye-sensitized solar cells (DSSCs) is similar to photosynthesis, separating light absorption from charge carrier transfer, allowing their maximum power point to be nearly independent of light levels. Although DSSCs are flexible, low-cost, and efficient energy converters that function well in low light and at wide angles, they present challenges related to temperature stability and the use of toxic substances (Sasikumar et al., 2024).

Fourth-generation solar cells: Fourth-generation solar cells are also known as nano photovoltaics. These cells consist of both organic and inorganic materials combined within a polymer matrix, which is composed of hybrid cells and inorganic crystals with twin structures. Using dye-sensitized porous metal oxide films, both organic-inorganic hybrid solar cells provide improved charge extraction and separation, which could result in increased photovoltaic efficiency. Moreover, hybrid organic-inorganic perovskites are popular optoelectronic materials because they absorb light well and work even with small defects (Tyagi et al., 2013)(Tyagi et al., 2013). A hybrid solar cell with 21% efficiency has been manufactured by Sanyo, one of the largest manufacturers of solar hybrid solar cells in Japan. The combination of an intrinsic thin-film solar cell and a heterojunction is known as HIT.

In photovoltaic solar cells, approximately 20% of sunlight is directly converted into electrical energy, and the remaining 80% is transformed into heat energy, which is either absorbed by PV cells or released into the environment (Togun et al., 2025). When the operating temperature of photovoltaic (PV) panels increases above the standard test temperature of 25°C, the efficiency of producing electricity typically decreases by 0.4% to 0.65% for every degree Celsius. Therefore, photovoltaic/thermal systems (PV/T) utilize the heat extracted from photovoltaic panels by solar thermal systems and use it for various heating applications, such as domestic hot water, rather than releasing the heat to the environment (Qu et al., 2024). The installation of PV/T in warehouses generates both thermal and electrical energy. The recovered heat can be utilized for solar-powered cooling systems, space heating, or water heating, while the electricity generated can power machinery, ventilation systems, lighting, and other operational equipment (Qadar Chaudhary et al., 2024).

The primary characteristics of a photovoltaic/thermal system are:

- Generates electricity and extracts heat for practical applications
- Prolong the life of solar panels by reducing their heat degradation
- PV/T system's overall efficiency is significantly higher than the efficiency of two separate systems
- Less space-consuming than two separate units together.

The following categories of photovoltaic thermal systems (PV/T) can be implemented:

- PV/T based on water cooling
- PV/T based on air cooling
- PV/T based on heat pipes
- PV/T on refrigerant cooling
- PV/T based on nanofluid cooling
- PV/T based on Phase-change material

Concentrated Solar Power:

Another common type of solar energy is concentrated solar power (CSP) systems, which use mirrors or lenses to absorb solar radiation and convert it into electricity using steam turbines or other heat engines. This technique works especially well in large power plants situated in sunny areas. A conventional CSP plant is made up of four main components, including a concentrator, a high-temperature solar receiver, a fluid transport system, and an electricity production block (Goyal et al., 2022). According to the International Energy Agency (IEA), by 2050, it will generate approximately 11% of the world's power generation. It (CSP) has fewer applications in warehouses than photovoltaic (PV) systems because of its particular needs. Still, it is feasible and can offer special advantages in various applications such as industrial process heat, thermal energy storage, and heat and cooling. The following are the primary characteristics of solar power plants that use CSP technology (Saif et al., 2025):

- CSP technology only uses the direct portion of incoming solar radiation, and the reflected and diffused solar radiation is lost into the environment.
- High-temperature input thermodynamic cycles are used in CSP technology to obtain high efficiency.
- CSP technology is not suitable for small-scale solar power projects due to its high initial cost.
- CSP technology requires higher levels of Direct Normal Irradiation.

Various types of CSP technologies, including parabolic troughs, solar towers, and dish engines, are shown in Figure 9, which absorb and transform solar energy in unique ways.

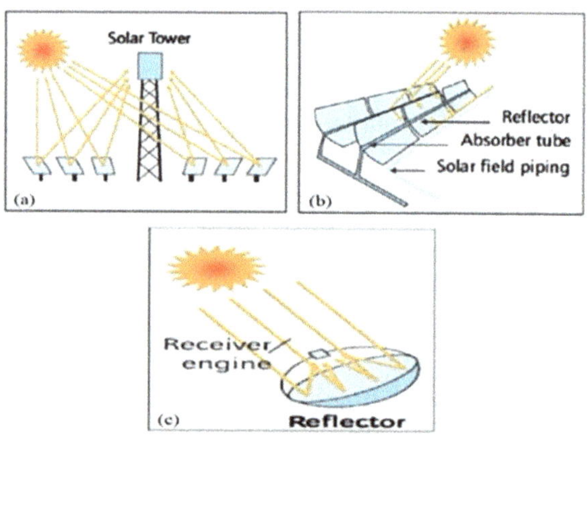

Parabolic Trough Collectors (PTC): The working principle of PTC consists of an absorber tube (filled with a specific fluid) embedded along the center of the mirror with a curved parabolic mirror to optimize the amount of radiation absorbed by the collectors. After that, a connected device uses the heated fluid to boil water to generate steam, which is then sent to a generator to produce electricity (A. Ahmad et al., 2024).

Solar Towers: A solar tower consists of hundreds of sun-tracking reflective mirrors that track solar radiation and concentrate it on a receiver placed at the top of a central tower. The heat generated by this process is used to run a turbine, converting it into electricity. Among CSP systems, they are more efficient, cost-effective, and have higher energy storage capacity (Mohammadi et al., 2025).

Dish engines: This type of CSP system utilizes an array of mirrors shaped like a dish to focus solar radiation onto a receiver positioned at the dish's focal point. These systems transfer the heat they generate to a heat engine, converting it into mechanical energy. After that, this energy drives a generator that generates electricity. Various reflector designs, such as flat, parabolic, and spherical mirror panels, could be utilized because the dish design mainly depends on the reflector shape (Shboul, 2025).

Building Integrated Photovoltaic (BIPV)

Fundamentally, BIPV is an innovative technique for building construction and energy conversion. In recent years, integrating solar technology into building design has become the most popular approach worldwide. This collaboration between solar energy generation and architectural design provides an innovative solution to the challenges associated with conventional energy sources. This indicates that the BIPV method meets the construction and electrical requirements for its application as roofing, façade, or external structure components. Innovation in BIPV modules and systems, adapted to new architectural and construction demands, has been made possible by advancements in photovoltaic technologies, materials, and construction industry technologies. Moreover, the building's energy consumption is influenced by the BIPV modules and systems' performance-related requirements, including optical qualities, thermal insulation level, solar heat gain coefficient, and BIPV electrical performance (Mao & Xiang, 2025). The following are the primary benefits of implementing BIPV:

- Improves the architectural design
- Minimizes the requirement for separate building materials
- Enhances the energy efficiency of buildings
- Supports LEED and other green building certifications.

Building-integrated or applied photovoltaic façades, commonly referred to as BIPV, is an excellent method to harness the vast potential of solar energy globally. About 80% of the BIPV industry is made up of roof-mounted systems, such as skylights, roof panels, and roof-integrated shingles, with the other 20% coming from façade-mounted systems, such as cladding, curtain walls, and shading systems (Shukla et al., 2017). Therefore, integration of BIPV combines electricity production with construction materials to provide a unique approach to improving warehouse energy efficiency. These systems reduce the need for conventional roofing materials by generating sustainable energy and serving as useful components of the building envelope. Moreover, the long-term benefits of BIPV, including energy savings, environmental advantages, and compliance with sustainability standards, make it an excellent choice for modern industrial and logistics facilities. However, the initial costs are considerably higher than those of traditional materials. Figure 10 depicts various examples of building-integrated photovoltaic techniques.

Figure 10. Several examples of building-integrated photovoltaic (BIPV) systems (Martín-Chivelet et al., 2022)

a) BIPV roof in Hägersten, Stockholm (Sweden). Residential building; no architect was involved; BIPV manufacturer: Soltech Energy. © Soltech Energy.

b) BIPV skylight in Málaga (Spain). Institutional building; engineering: Naxfor engineers & infrastructures; supervision: Francisco J. Martín. BIPV manufacturer: Onyx Solar. © Onyx Solar.

c) BIPV coloured ventilated façade in Vacallo (Switzerland). Type of building: residential; architects: Huggenbergerfries architekten; BIPV manufacturer: Solartechnik. © SUPSI.

d) BIPV semitransparent curtain wall in Bratislava (Slovakia). Office building; architects: Siebert+Talas, CEPM, John Robertson Arch.; BIPV manufacturer: Onyx Solar. © Onyx Solar.

e) BIPV double skin façade in Naples (Italy). University building; architects: Agenize di Architettura, Cool Projects S.r.l.; BIPV manufacturer: Sunerg Solar Energy. ©Università degli Studi di Napoli Federico II.

f) BIPV roof and shading devices in Dubai (United Arab Emirates). Expo pavilion; architect: Grimshaw Arch.; BIPV manufacturer: Sunovation. © Sunovation

2.2 Wind Energy

Another promising renewable energy source is wind energy, which is the most efficient, reliable, and cost-effective energy source that can be used in warehouses to generate electricity. Moreover, it is one of the most attractive renewable energy sources due to its widespread availability and efficient applications, which have been utilized worldwide. The working principle of wind energy is based on converting the kinetic energy into electrical energy by a wind turbine (Okonkwo et al., 2023). Wind drives the rotor blades of wind turbines, which convert kinetic energy into rotational energy. A shaft transfers this rotational energy to the generator, which converts it into electrical energy. The schematic representation of a wind energy system is depicted in Figure 11. These turbines have been widely installed worldwide, resulting in a significant reduction in CO_2 emissions. However, it also has some significant drawbacks, including high initial costs, unreliability, threats to wildlife, and noise pollution. Further research is needed to address these challenges and improve wind energy's performance, reduce costs, and minimize its harmful environmental effects (Mahmoud et al., 2023a).

Figure 11. Schematic of a wind energy system (Benbouzid et al., n.d.)

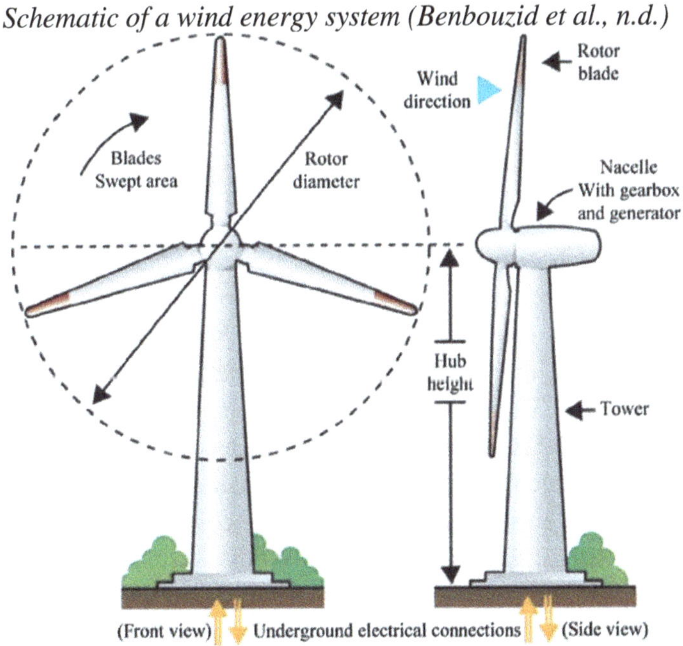

When choosing a wind turbine for a specific location, the following essential criteria should be taken into account (Kalogirou, 2024).

Wind conditions: The most important consideration when choosing a wind turbine is the local wind conditions. To capture more wind, turbines with larger rotor diameters are required in areas with low wind speeds, while turbines with smaller rotors can be used in locations with strong wind. Moreover, the height of the turbine tower should also be considered because altitude can cause wind speeds to increase significantly.

Urban Wind Turbulence: Warehouses are often located in industrial parks or urban areas where surrounding buildings create turbulent, low-speed wind flow. This drastically reduces the efficiency of wind turbines, which are designed for smooth, laminar wind.

Turbine size: Over the past few decades, wind turbine size has increased significantly, as evidenced by the substantial growth of various parameters, including capacity, rotor diameter, hub height, rated wind speed, and cut-in wind speed. Its selection has a direct effect on both its power output and the suitability of its location. Smaller turbines are the best solution for locations with limited space or lower wind speeds; however, larger turbines generate more electricity but need more room and stronger winds. Therefore, the site's characteristics, energy requirements, and the turbine's physical footprint must be considered to determine the ideal size.

Environmental impact: When selecting a wind turbine, environmental factors must be considered to assess its potential effects on local wildlife, particularly birds and bats, and to implement precautions to minimize harm. Moreover, a more sustainable and community-friendly project will also be achieved by taking into account the visual and auditory effects on the surrounding region that support environmental preservation initiatives.

Maintenance requirements: It's critical to know the maintenance requirements of various wind turbines because some have complicated designs that require frequent professional maintenance. In contrast, others are simpler and better suited for routine maintenance. Furthermore, the cost of maintenance services should be considered, as these will impact the financial stability and sustainability of wind energy projects.

Cost consideration: The cost of a wind turbine is an important consideration, including both the initial investment and installation costs as well as regular maintenance expenses (maintenance costs, potential repairs, and the expected lifespan). Therefore, researchers must focus on assessing the cost of wind energy systems to provide comprehensive analyses and identify suitable ways to reduce capital costs.

Structural and Safety Considerations: Mounting a turbine on a warehouse roof requires significant structural reinforcement to handle dynamic loads and vibrations, adding to the cost. Safety concerns regarding ice throw or blade failure in proximity to personnel and property must also be addressed.

Zoning Laws and Permits: Local regulations often impose height restrictions and noise limits. The permitting process for installing a turbine, especially a taller HAWT that accesses better wind resources, can be complex and prohibitive.

Therefore, a thorough professional site assessment is essential before considering wind energy integration. In most cases, wind energy may only be a viable supplement to a broader renewable strategy, primarily in rural, high-wind locations with no height restrictions.

The following are the primary wind turbine solutions suitable for warehouses based on size, design, and suitability for commercial or industrial environments (Jia & Gao, 2025).:

- Vertical axis wind turbines (VAWT)
- Horizontal axis wind turbines (HAWT)
- Building-integrated wind turbines (BIWT)

2.2.1 Horizontal Axis Wind Turbines

These turbines are suitable for warehouses due to their compact size, ability to operate in both windy and low-wind conditions, and simple installation on flat roofs. The most popular type of wind turbine is the horizontal-axis model. In these turbines, the rotor axis is oriented horizontally and could have one, two, or three blades. The turbine blades are designed aerodynamically and rotate because of the aerodynamic lift created by the pressure difference between their upper and lower surfaces. Over the front side of the blade, air flows faster, producing a low-pressure area; on the back side, air moves more slowly, producing a high-pressure area, which produces lift. As a result, air movement from the region of high pressure creates aerodynamic lift that rotates the blades (Mahmoud et al., 2023b).

Pros of HAWT

- A HAWT is known for its high efficiency.
- It is more dependable and rotates faster.
- During storms, the blades can cause the rotor to tilt.
- As the wind shear varies, more wind is captured by the tall tower.

Cons of HAWT

- It requires large machinery to design and manufacture.
- They are challenging to use in distributed generation and are only available in large sizes.

- Compared to other wind turbines, it generates a considerably higher level of noise.
- Moving them from one place to another is difficult due to their size and weight.

2.2.2 Vertical Axis Wind Turbines

A vertical-axis wind turbine (VAWT) is an innovative wind turbine where the rotor's axis of rotation is oriented perpendicular to the wind, effectively harnessing its energy. Maintenance is easy for these turbines due to the position of the generators and gearboxes at the turbine's base. The design does not need particular aerodynamics because the rotor is perpendicular to the wind speed. Savonius and Darrieus are the two primary types of vertical-axis wind turbines (VAWT). Darrieus turbines have an eggbeater-like form with long, wide blades connected at both the top and bottom of the rotor shaft to maximize the swept area, while Savonius turbines have helical blades wound around a vertical shaft, resembling a DNA strand (Kamran, 2023).

Pros of VAWT

- Due to its often smaller size and lower noise level, it is frequently used in distributed generation.
- Maintenance and installation are quite simple compared to HWAT.
- Even at lower wind speeds, it can generate electricity.

Pros of VAWT

- Its efficiency is lower compared to HAWT's.
- It is challenging to replace and maintain elements of a poorly built structure.

2.2.3 Building-Integrated Wind Turbines (BIWT):

The development and demonstration of building-integrated wind turbines have continued, but they have lower popularity than building-integrated photovoltaics (BIPV). The integration of BIWT into the warehouse design is typically located on the roof, sidewalls, or within the ventilation systems. These technologies are referred to as BIWT when installed within urban buildings. They could independently transfer the generated electricity directly to the buildings (at the point of consumption), minimizing transmission power losses and supporting decarbonization objectives, as long-distance grid lines are not involved. Moreover, the development and utilization of BIWT are supported for social, economic, and ecological reasons. Resilience and energy independence are other reasons for providing backup energy sources in areas with unpredictable energy flows. When compared to traditional wind turbines, BIWTs are mounted on facades, rooftops, or other architectural features, converting urban buildings into dynamic sources of energy (Bošnjaković et al., 2025). The implementation of BIWT in the Bahrain World Trade Tower Centre is depicted in Figure 12.

Figure 12. Bahrain World Trade Center: (a) a captivating front view of the wind turbines; and (b) wind speed at 100 meters above ground (Kim et al., 2016)

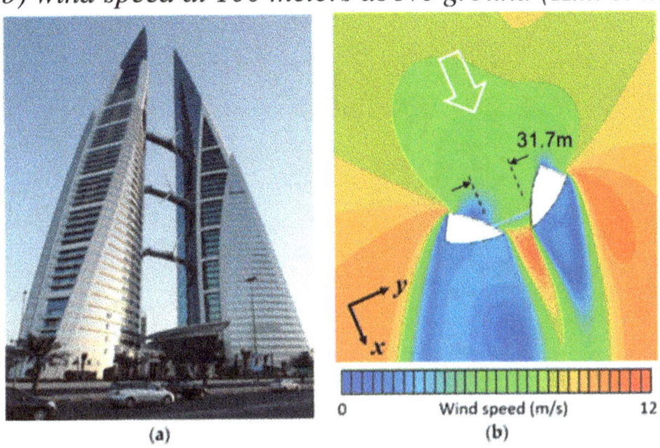

2.3 Geothermal Energy

Geothermal energy (GE) stands out as one of the most reliable and stable sources of renewable energy used in warehouses or the building sector in various applications, including heating, cooling, power generation, and temperature-sensitive goods (pharmaceuticals, food products, or electronics). It is an excellent choice for environmentally friendly warehouse operations due to its long-term cost-effectiveness, low maintenance requirements, and compatibility with other renewable energy systems. Research and application of geothermal energy have advanced significantly, and one of its most significant advantages is that it is a constant and reliable source (Loschetter et al., 2025). Significant advances toward a sustainable energy future include lowering carbon emissions and dependency on fossil fuels while improving energy security. Moreover, without integrating energy storage technologies, it can maintain a steady supply and is not immediately affected by changes in the surrounding air conditions. However, the primary issue with it is its high capital costs, which are greater for deep geothermal energy systems (Mahmoud et al., 2023c). Geothermal energy commercialization is affected by several factors, including as upfront costs, environmental concerns, geological uncertainty, and difficulties integrating with the grid. Typically, heat pumps are utilized to extract heat from the earth and move it indoors for heating and cooling processes. Figure 13 depicts the geothermal energy systems that could be integrated into buildings.

Figure 13. Geothermal integration in a building (Reddy et al., 2024)

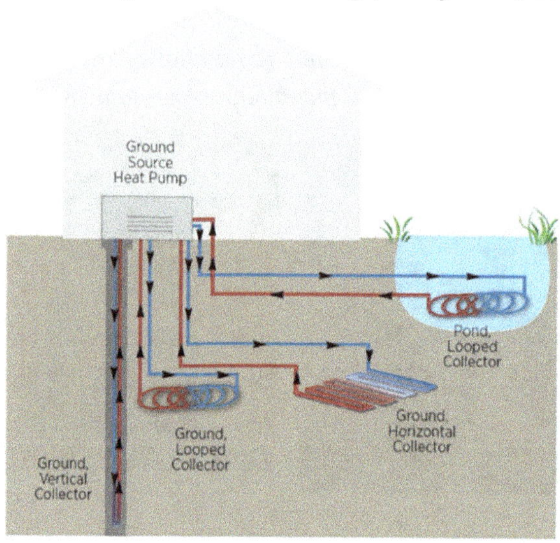

2.3.1 Ground Source Heat Pumps (GSHP)

Geothermal source heat pumps are the most effective and environmentally clean source of sustainable energy in the world; they utilize ground thermal energy for heating (Winter) and cooling (Summer) through a heat exchange process in the building. Therefore, this highly efficient technique can reduce energy use for heating and cooling. Globally, interest in ground-source heat pumps (GSHPs) is growing due to their potential to minimize primary energy consumption and greenhouse gas emissions, particularly in North America and Europe. They are more energy-efficient than air-source heat pumps because they transfer heat to the earth, which has consistent, moderate temperatures compared to the ambient air. Ambient air, which is typically warmer in the summer and colder in the winter than the earth, makes heat transfer less efficient (Lee et al., 2025). The following different geothermal system types exist, each with unique properties and uses:

- Horizontal geothermal systems
- Vertical geothermal systems

- Closed-loop geothermal systems
- Pond loop geothermal systems

Horizontal geothermal systems: Horizontal ground heat exchangers (HGHEs) are the most commonly used type for residential applications and small commercial buildings. It utilizes long, looped horizontal pipes installed in trenches dug at a relatively shallow depth below the earth's surface (typically 1 to 2 meters deep). A dynamic heat transfer fluid that efficiently absorbs and transfers heat to and from the earth, ensuring optimal thermal exchange and energy efficiency. These systems transfer heat from the earth into the building during the winter season, while they distribute heat from the building to the ground throughout the summer season (G. Wang et al., 2025). As shown in Figure 14, the HGHEs are divided into three designs: linear, slinky-coil, and spiral-coil configurations.

Linear-loop – straight pipe laid in trenches (typically 1 to 2 meters deep), requires a large surface area, but it is cost-effective and easy to install.

Slinky-coil – overlapping coils arranged horizontally.

Helical-coil – vertically stacked circular loops in horizontal trenches. It can reduce land requirements while maintaining efficient thermal performance.

Figure 14. Various HGHE designs: (a) linear-loop, (b) slinky-coil, and (c) Helical-coil (Aydin et al., 2015)

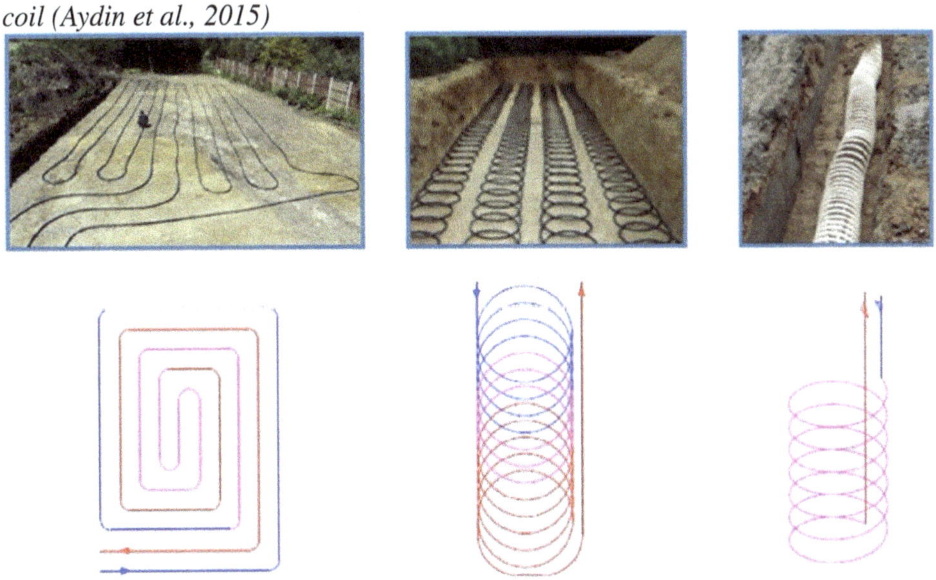

(a) (b) (c)

Vertical geothermal systems: Vertical geothermal systems extract heat from the Earth for energy production by circulating a heat transfer fluid through boreholes (typically ranging from 50 to 150 meters) with inserted pipes. These systems are highly efficient because the ground temperatures increase during the heating season and decrease during the cooling season as depth increases. Moreover, they perform well in areas with limited horizontal space as they require less space. However, these systems are more costly than horizontal geothermal systems due to the higher drilling costs. Figure 15 represents several VGHEs arrangements, which are primarily divided into coaxial and U-shaped tubes (Mahmoud et al., 2023d).

Coaxial Configuration: One or more pipes are positioned inside a larger pipe in a coaxial configuration, allowing fluid to flow in one of two directions through the inner or outer pipe(s).

U-shaped tubes: U-shaped tube configurations have the simplest design, are cost-effective, have high thermal efficiency, and are a widely used type of VGHEs. They are more efficient than coaxial configurations and consist of a single or double U-shaped tube, as shown in Figure 15.

Figure 15. Various vertical ground heat exchangers: (A) simple coaxial, (B) complex coaxial, (C) single U-tube, and (D) double U-tube (Mahmoud et al., 2023d)

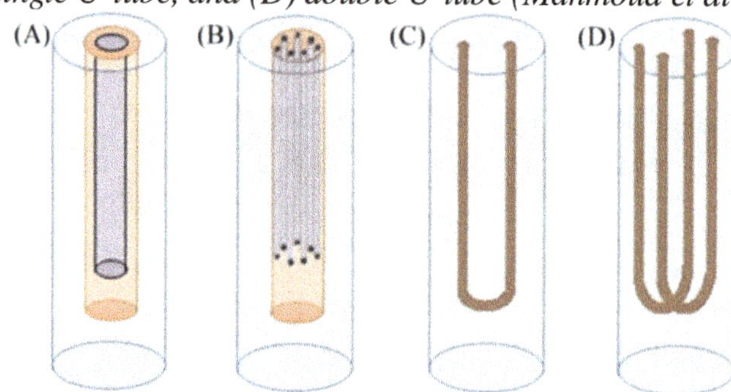

Closed-loop geothermal systems: In closed-loop geothermal systems, a closed network of pipes circulates a heat-transfer fluid, usually a combination of water and antifreeze. It could be installed horizontally or vertically on the ground close to a building. They are frequently utilized in commercial and residential buildings because of their effectiveness and low environmental impact. They are widely used in both commercial and residential buildings because of their effectiveness and low environmental impact (Sharmin et al., 2023).

Pond loop geothermal systems: These systems are also known as lake or surface water loops, as a heat exchange source. Heat is transferred to or from the water by a heat-transfer fluid that circulates through pipes submerged in it. This method is effective when there is a nearby source of water available. However, they depend on the water source's reliability and quality, which must be free from contamination, sufficiently large to prevent drastic temperature changes, and unfrozen in the winter. Moreover, it provides a highly effective and sustainable geothermal heating and cooling solution, especially for rural or semi-rural areas (Fan & Zhang, 2025). The schematic representation of these systems is shown in Figure 14.

3. ENERGY STORAGE TECHNOLOGIES FOR WAREHOUSING FACILITIES

Energy storage technologies are crucial for maintaining a balance between energy supply and demand, especially for intermittent renewable energy sources such as solar, wind, and geothermal energy. Electricity grids and businesses implementing storage technology can save a significant amount of money because of it. These technologies for warehousing play a crucial role in optimizing energy usage, reducing peak demand charges, facilitating the integration of renewable energy, and ensuring backup power for critical operations, such as refrigeration, lighting, and automation systems. Energy storage may be significantly needed to achieve the benefits of local generation and a clean, dependable energy supply, as well as to facilitate the successful integration of renewable energy. It is essential for successfully integrating renewable energy sources, gaining the advantages of local generation, and having a clean, dependable energy source. In recent years, the demand for energy storage has continued to rise as renewable energy production increases. It is crucial for the integration of renewable energy sources and for maintaining a modern energy system that is reliable and stable (Arabkoohsar, 2023). According to their operational lifespan and storage capacity, energy storage solutions are divided into various categories, as shown in Figure 16.

3.3.1 Thermal Energy Storage (TES)

One of the most promising technologies for increasing the effectiveness of renewable energy sources is thermal energy storage. This process involves heating or cooling a medium, such as ice or water, which can then be used to generate power or provide heating and cooling for various applications. For a long time, it has been used for energy efficiency and redistribution, either temporarily or permanently. The application of this in an energy system has several benefits, including increased overall efficiency and dependability, improved economic viability, lower investment and operating costs, and reduced CO_2 emissions and environmental pollution. Moreover, this technology can be utilized in warehouses to store energy for cooling and heating, helping to maintain stable temperatures. Furthermore, this approach also enhances energy efficiency and supports integration with renewable sources. Thermal energy can be stored in various forms, including sensible heat, latent heat, and thermochemical energy storage. Sensible thermal energy storage is the most common technique due to its high specific heat capacity, low cost, long-term stability, and environmentally friendly nature (Azam, Cattani, et al., 2023). On the other hand, latent heat storage systems rely on the changing state of a medium to store energy, rather than maintaining the medium's temperature. There are materials known as "phase change materials" that have the ability to store heat as latent heat within their mass. These materials are frequently used in building materials and solar applications, where they absorb and store excess heat from buildings. Moreover, thermochemical heat storage has the highest energy storage density (per unit mass and volume) and can store energy with minimal losses for long-term use. It works by using reversible reactions that convert heat into chemical potential energy (Yu et al., 2025).

Figure 16. Classification of energy storage systems. Source: Author's work

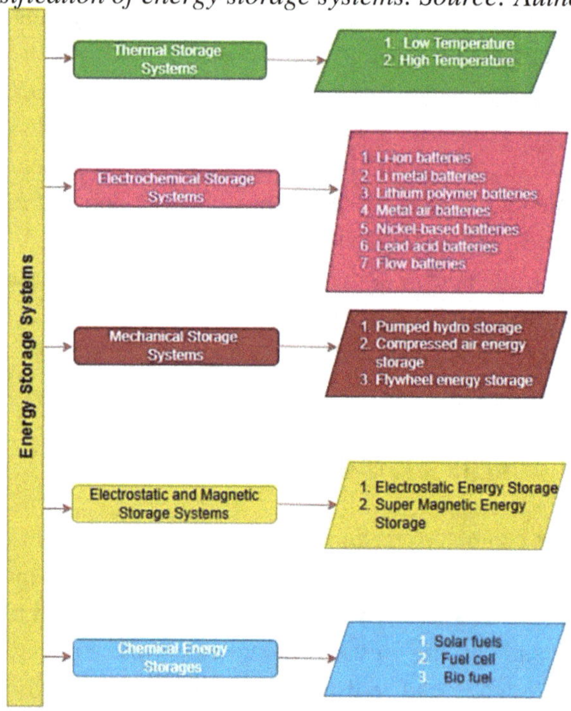

TES is frequently employed with solar photovoltaic (PV) systems, which effectively store solar heat energy. In this way, solar PV systems may ensure the collection of more energy from the sun, which can be utilized at night or during seasonal fluctuations, and generate photovoltaic electricity. Moreover, solar concentrators can also use it to store concentrated heat energy from the sun. Furthermore, they are also widely used in load shifting and electricity generation from heat engines. As an example, ice-based thermal energy storage (TES) is commonly used in refrigerated warehouses. Systems like Calmac's create ice at night using off-peak electricity or solar power, which is then used for cooling during the day, drastically reducing peak demand charges. Based on the range of working temperatures, it can be classified into two groups: high-temperature thermal energy storage (TES) and low-temperature TES. These categories are described below.

Low-Temperature Thermal Energy Storage: Low-temperature thermal energy storage is the process of storing energy at relatively low temperatures, usually less than 100°C, by charging and discharging thermal energy within suitable storage materials (Azam et al., 2024). For example, thermal energy can be stored using water, ice, or phase-change materials in heating and cooling applications. This allows load shifting by storing energy during off-peak hours and utilizing it during peak demand hours (Behzadi et al., 2022).

High-Temperature Thermal Energy Storage: In high-temperature TES systems, sensible and latent heat are two forms of energy storage. Sensible heat is the heat that a body absorbs or releases that causes a change in temperature without affecting its condition (such as pressure or volume). In contrast, latent heat is the heat that a body absorbs or releases, which only changes its condition without affecting its temperature. The quantity of heat energy a material can store depends on its specific heat and overall heat capacity. However, its main drawback is huge space occupation (R. Zhang et al., 2025).

3.3.2 Electrochemical Energy Storage

The electrochemical storage system (ECSS) stores electrical energy as chemical energy through chemical reactions in batteries. It is among the most recognized and ancient technologies in the market. The ECSS is the most extensive electrical energy storage system in the market, with energy densities ranging from 10 Wh/kg to 13 kW/kg. Moreover, they have a high efficiency of 70–80% and emit very few hazardous chemicals. They are an excellent choice for energy storage applications because they require very low maintenance. This technology is being utilized more and more in warehouses to support energy savings, backup energy, Integration of renewable energy and management of peak loads. These storage systems are crucial because electricity must be effectively stored for later use in warehouse operations during demand periods (Dixit et al., 2022). For instance, Amazon has deployed massive Tesla Powerpack battery systems at its fulfilment centers to shift solar energy consumption to peak hours, providing grid services and backup power. The classification of electrochemical energy storage systems (Li-ion batteries, Li metal batteries, Lithium polymer batteries, Metal air batteries, Nickel-based batteries, Lead acid batteries, Flow batteries) are shown in Figure 16, and each type is described below.

Lithium-ion batteries: Lithium-ion batteries have completely changed how we use and store energy, revolutionizing several industries and propelling the transition to a more sustainable future. According to their working principle, lithium ions move from the negative electrode (anode) to the positive electrode (cathode) during discharge and return to the anode during charging. In 1990, commercial Li-ion batteries were first introduced. These rechargeable batteries are suitable for a wide range of applications because of their high energy density, long cycle life, and quick charge-discharge characteristics. While Li-ion batteries offer higher energy density and efficiency, their shorter lifespan (5-15 years) compared to flow batteries (20+ years) must be factored into the total cost of ownership and ROI calculations for warehouse operators. Their efficient energy storage and distribution capabilities have been crucial in promoting the integration of renewable energy sources, lowering greenhouse gas emissions, and improving energy resilience (Hasan et al., 2025).

Lithium Metal Batteries: M. Stanley Whittingham invented lithium metal batteries at Exxon in the 1970s. These primary (non-rechargeable) batteries use metallic lithium as the anode. Li is a highly reactive metal that easily releases an electron to produce an $Li+$ ion, making it ideal as an anode material. Although titanium disulfide was the cathode in the initial design, but the concept was difficult to implement due to safety issues, including explosions. Therefore, to solve this issue, researchers switched to an intercalation design, adding additional materials between lithium layers to improve stability. Since lithium is one of the lightest metals, it has a high charge density and an output voltage between 1.5 and 3.7 V, making it suitable for various applications. However, Li metal batteries had two primary issues: the development of dendrites and their propensity for explosion (Kwon et al., 2023).

Lithium Polymer Batteries: Lithium polymer (Li-Po) batteries are members of the Li-ion battery family and have positive electrodes composed of lithium metal oxides (such as manganese, nickel, or cobalt). It consists of various components such as a positive electrode, a negative electrode, an electrolyte, and a separator. These types of batteries, which were invented in 1978, utilized an electrolyte that consisted of dry solid polymers. The dry polymer electrolyte significantly simplifies the Li-Po production process, but it has poor conductivity and is therefore unsuitable for high current applications (Lennartz et al., 2023).

Metal air batteries: Metal air batteries have attracted significant attention as one of the market's cheapest and most compact batteries. They use metal electrodes (anodes) made of lithium, zinc, magnesium, aluminium, and other metals. Theoretically, lithium-air batteries have an energy density of approximately 3,500 Wh/kg, nearly ten times higher than that of lithium-ion batteries, which typically range from 150 to 250 Wh/kg. These batteries are particularly beneficial in applications where weight and storage capacity are critical, such as drones, electric aircraft, and space travel. However, one of the main drawbacks of metal air batteries is that electrically recharging them is highly challenging and inefficient (Soumya Rai et al., 2025).

Nickel-based batteries: Another type of rechargeable battery known as a nickel-based battery, which has positive electrodes made of nickel oxide. These batteries have been proven durable because of their high stability, extended lifespan, and ability to work dependably at extremely hot or cold temperatures. Ni-MH and Ni-CD batteries are more widely used for vehicle propulsion and power backup applications.

Lead Acid Batteries: Lead-acid batteries (LABs) were invented by Gaston Planté in 1859 and are still very popular because of their high power output, affordability, and dependability, especially in backup power and automotive applications. The most advanced LABs are made of spongy Pb as the anodic material and Pb grid coated with lead oxide (PbO_2) as the cathodic material, which are submerged in sulfuric acid (H_2SO_4) electrolyte. However, this type of battery has certain disadvantages, such as low specific energy and energy density, heavy weight, limited cycle life, cathode oxidative corrosion, water loss, poor low-temperature performance, and electrolytic stratifications. Several approaches have been proposed to improve the electrochemical performance of LABs through modifying the structure and composition of cathodic and anodic materials. However, these approaches could be expensive (Mosallanejad & Akrami, 2024).

Flow Batteries: An innovative aqueous electrolytic energy storage device that combines the characteristics of traditional batteries with fuel cells is known as a flow battery. It stores energy in electrolytes that contain dissolved electroactive species, differentiating it from standard secondary batteries. This kind of electroactive substance transforms chemical energy into electrical energy as it passes through the power core. As a result, flow batteries are made up of active electrolytes and lack self-discharge (Pileri et al., 2025).

3.3.3 Mechanical Energy Storage

One of the most sustainable and effective types of energy storage systems is mechanical energy storage. The main benefits of mechanical energy are its ease of storage, low environmental impact, low cost, and prolonged storage life. There are two types of mechanical energy: kinetic energy, which is held in a body's motion, and potential energy, which is found in compressed springs or pressurized gas. The three primary types of in high-temperature aquifer thermal energy storage (HT-ATES) systems based on coupled thermo-hydro-mechanical (THM) analysis, such as compressed air, pumped hydro, and flywheel, are described below (Mahadevan et al., 2025).

Compressed air energy storage systems: These are a type of mechanical energy storage technology that stores energy by compressed and pressured air in underground caverns, tanks, or other reservoirs. The compressors that pressurize and store air are powered by electrical energy during times when there is a low demand for electricity or when there is excess energy generation (from renewable sources like solar or wind). These methods improve the integration of intermittent renewable energy sources, enhance energy reliability, and help balance grid demand. Moreover, these systems can be used in stationary applications, including load leveling, energy arbitrage, and area and frequency regulation, because of their 35 MW to 300 MW capability.

Pumped hydro energy storage systems: Pumped hydro storage (PHS), which was first used in Switzerland in 1944, is one of the most often used storage methods because of its ease of use and significant storage capacity (1 to 3000 MW). It is a well-established and reliable technology that has a long storage time, an almost infinite life cycle, a low capital cost, a high efficiency of 76–85%, and a very long lifespan of 50 years or more. However, the limitations of this technology include high initial costs, geographic restrictions, long construction times, and dependency on water availability. Their basic idea is to store energy as gravitational potential by pumping water from a lower reservoir to an upper reservoir using excess electricity. After that, turbines are used to generate energy when the water is released back down.

Flywheel energy storage systems: In 1983, John A. Howell first invented the steel flywheel energy storage (FES) for use in military applications. The flywheel energy storage system (FESS) uses a rotating mass to store energy in the form of kinetic energy. Additionally, it has great potential for usage in peak shaving and load leveling applications, where energy is stored during off-peak hours and used during those hours. They are suitable for a variety of energy storage applications in the fields of transportation, satellites, military systems, and power systems, among others.

3.3.4 Electrostatic and Magnetic Energy Storage

Electric and magnetic fields are two ways that energy can be stored through advanced technologies, such as supercapacitors and superconducting magnets, respectively, offering high power density and rapid response for short-duration energy applications. The most common forms of electrostatic energy storage are capacitors and supercapacitors. On the other hand, magnetic energy storage is implemented using superconducting magnetic energy storage (SMES) systems (Guo et al., 2025).

Electrostatic Energy Storage (EES): Electrostatic energy storage capacitors are essential for high-power and renewable energy systems. They combine excellent electrical insulation, mechanical flexibility, low density, affordability, and processability. Therefore, they are not suitable for large-scale storage applications but are highly beneficial for small-scale power control applications.

*Super magnetic Energy Storage (*SMES*):* Energy is stored in the magnetic field generated when a direct current flows through a superconducting coil, using this innovative energy storage technique. The SMES system offers a high power density and a very quick response time (a few milliseconds) when compared to other storage systems. However, the SMES system's major drawbacks are its high cost, material fatigue due to mechanical stress, and low volumetric and gravimetric energy density.

3.3.5 Chemical Energy Storage

Chemical energy storage is the process of storing energy in chemical bonds that chemical reactions can subsequently release. Energy is generated when molecules undergo chemical reactions that involve the transfer of electrons, resulting in reactants and products with completely different compositions. Approximately 100 GWh of energy can be stored using this type of chemical energy storage (Hossain et al., 2020). The three primary types of chemical energy storage systems, such as fuel cells, solar fuels, and biofuels, are described below.

Solar Fuels: Solar fuels, which use solar energy to generate clean alternative fuels by converting sunlight into chemical energy, have attracted significant attention in recent years. They provide clean, storable, and transportable energy, which helps reduce the intermittent nature of solar power and provides a viable path toward decarbonizing industries, transportation, and power generation. The main advantage of this system is its virtual storage capabilities, achieving nearly 100% efficiency and enhancing overall system performance.

Fuel Cell: In 1838, Swiss physicist Christian Friedrich initially proposed the idea of a fuel cell. However, in 1839, Sir William Grove became the first person to successfully develop a fuel cell, which operates by reversing the electrolysis of water. A fuel cell is a device that can directly transform chemical energy from a chemical reaction into electrical energy, with the byproducts being water and hydrogen. The fuel cell has a high energy density and a high rate of charging and discharging. However, the high costs and low round-trip efficiency continue to be significant barriers to the widespread use of fuel cells.

Biofuels: Biofuels are renewable energy sources produced from various biological materials, including algae, plants, and agricultural waste. They can be liquid or gaseous fuels, and utilized as alternatives to fossil fuels for heating, power generation, and transportation. Moreover, it can help in reducing dependence on non-renewable resources and greenhouse gas emissions.

4. REAL-WORLD IMPLEMENTATIONS AND CASE STUDIES

The theoretical potential of renewable energy in warehouses is compelling, but real-world case studies demonstrate its practical viability, economic benefits, and the challenges that must be navigated. This section explores implemented projects across solar, geothermal, and hybrid systems.

4.1 Large-Scale Solar PV: The IKEA and Amazon Model

Case Study: IKEA Distribution Center, Joliet, Illinois, USA: A case study focuses on the Distribution Center located in Joliet, Illinois, USA. The project involved the installation of a 2.8 MW rooftop solar photovoltaic (PV) system on the facility's 1.4 million square foot roof, comprising over 10,000 panels. The system was designed to leverage the vast, unused roof space typical of large distribution centers. IKEA partnered with a third-party developer for financing and installation, a standard model that reduces upfront capital cost. Annually, the system generates approximately 3,500 MWh of electricity, which is equivalent to powering over 300 homes and offsets a significant portion of the facility's base load electricity consumption used for lighting and material handling equipment. This project aligns with IKEA's global commitment to be climate positive by 2030. Key challenges that were overcome included conducting a detailed structural analysis to ensure the roof could support the panels' weight, especially with regional snow load, and navigating the utility's interconnection process to feed excess electricity back to the grid safely.

Case Study: Amazon's Solar Commitment: Another case study, Amazon is deploying solar on a massive scale across its global network of fulfillment and sortation centers; for example, a 4.6 MW system powers its fulfillment center in Mt. Juliet, Tennessee. A key feature of this strategy is the integration of these solar installations with megawatt-scale battery storage systems, such as Tesla Megapacks. The primary purpose of these batteries is load shifting, which involves storing the clean solar energy generated during the day to power the company's highly automated facilities during the evening peak demand periods. This not only reduces reliance on the grid when electricity is most expensive and carbon-intensive but also provides critical backup power to ensure operational resilience.

4.2 Geothermal for Climate Control: The Food Storage Imperative

Case Study: Lineage Logistics Cold Storage, Various Locations: Lineage Logistics, a global leader in temperature-controlled warehousing, has implemented innovative ground-source heat pump (GSHP) systems across multiple cold storage facilities to improve energy efficiency significantly. These facilities have a massive, constant demand for refrigeration, which simultaneously generates large amounts of waste heat. The GSHP system utilizes the earth's stable underground temperature as a heat sink to absorb excess heat from refrigeration condensers during summer and as a heat source to warm office and dock areas in winter. This application has resulted in remarkable efficiency gains of 30-60% compared to traditional HVAC systems, substantially reducing energy consumption and operating costs for these energy-intensive 24/7 operations. The system also enhances reliability by housing mechanical components indoors, protecting them from weather elements and reducing maintenance needs. The company successfully overcame significant challenges, including the high initial capital investment required for ground loop installation, and conducted thorough geological surveys to ensure site suitability, particularly addressing constraints of urban locations with limited space for installation.

4.3 Hybrid Renewable Systems: Maximizing Resilience and ROI

Case Study: FedEx Distribution Hub, Cologne, Germany: A holistic approach to sustainable logistics was implemented at this facility, integrating hybrid solar PV, geothermal systems, and energy-efficient design from the ground up. The hub features a massive 2.4 MW rooftop solar array alongside a geothermal system consisting of 66 boreholes drilled 70 meters deep to provide heating and cooling for office spaces, complemented by LED lighting and skylights for daylight harvesting. This integrated design has earned the building the highest sustainability certifications (DGNB Platinum, BREEAM Outstanding) and reduces primary energy consumption by over 40% compared to conventional facilities. The geothermal system alone meets 100% of heating and cooling demands for administrative areas. This project demonstrates how combining multiple renewable technologies with energy efficiency measures from the initial design phase creates synergistic effects that maximize environmental impact and operational performance.

4.4 Common Implementation Challenges and Solutions

The following are common implementation challenges and their corresponding solutions observed in real-world applications (van Geest et al., 2022).

4.4.1 High Upfront Capital Cost

The biggest challenge is the initial investment for technology and installation, which can be prohibitive. Therefore, the solution is to utilize power purchase agreements (PPAs) or leasing models, where a third-party owns, operates, and maintains the system. The warehouse owner agrees to purchase the electricity generated at a predetermined, often lower, rate. Government incentives, tax credits, and green grants are also critical.

4.4.2 Grid Integration and Interconnection

A significant challenge in deploying renewable energy systems involves navigating the strict protocols utilities enforce for connecting distributed generation to the grid, a process that can often be slow and bureaucratic to ensure overall grid safety and reliability. To overcome these hurdles, developers can employ several key solutions: engaging with the local utility early in the planning process to align expectations and requirements, hiring experienced contractors who are thoroughly familiar with the specific interconnection application and approval procedures, and considering on-site battery storage (Lamas et al., 2025). These batteries can be used to limit the amount of electricity fed back to the grid, which can often simplify the interconnection process and make the project more palatable to the utility.

4.4.3 Maintenance and Expertise

A significant challenge for warehouses adopting advanced renewable energy systems is the potential lack of in-house expertise required to maintain specialized equipment like wind turbines or complex photovoltaic-thermal (PV/T) systems. Without proper maintenance, these high-value assets risk underperformance, frequent breakdowns, and diminished return on investment. To effectively mitigate this, it is crucial to factor long-term Operations & Maintenance (O&M) costs directly into the initial business model and financial planning (Liu et al., 2025). The most practical solution is to secure long-term service contracts with the original installers or specialized third-party O&M providers. These contracts ensure that certified technicians perform regular, expert maintenance, guaranteeing system reliability, maximizing energy production, and transforming a fixed operational expense into predictable budget line items.

4.4.4 Regulatory and Permitting Challenges

Navigating the complex web of regulatory and permitting hurdles presents a major challenge for warehouse energy projects, particularly for systems like geothermal (requiring drilling), Building-Integrated Photovoltaics (BIPV), or wind turbines (requiring new structures). The permit requirements for these specialized technologies can vary dramatically and unpredictably by municipality, creating significant project delays and potential cost overruns. To overcome this, developers must proactively engage with Authorities Having Jurisdiction (AHJs) early in the planning process. The most effective solution is to partner with experienced local consultants and engineering firms who possess deep knowledge of regional codes and, crucially, have established relationships with permit offices. This pre-emptive collaboration helps streamline approvals, clarify requirements, and de-risk the project timeline by transforming a potentially adversarial process into a cooperative one (Pratiwi et al., 2025).

These real-world examples confirm that while barriers exist, they are surmountable. The successful integration of renewables leads to enhanced energy resilience, significant operational cost savings, and tangible progress toward corporate sustainability goals, making it a strategic investment for the modern warehouse industry.

5. CONCLUSION

In conclusion, warehouses can derive operational, financial, and environmental advantages by integrating renewable energy sources into their system. By implementing renewable energy resources, such as solar panels, wind turbines, geothermal energy systems, and energy storage options, warehouses can significantly reduce energy costs, enhance energy security, and support sustainability initiatives. In addition to minimizing the carbon footprint, warehouses can enhance long-term operational resilience and energy efficiency through careful planning and integration. For warehouse operators, a phased approach is recommended: (1) conduct an energy audit, (2) implement efficiency measures (e.g., LED lighting), (3) install rooftop solar PV, (4) integrate energy storage to optimize usage, and (5) explore secondary technologies like geothermal for heating/cooling-intensive operations. The potential for warehouses to adopt these systems will increase through continuous improvements in renewable energy technology. Moving towards renewable energy in warehousing is not just an investment in operational savings but also a step towards a more sustainable future. Future research should focus on the integration of AI for predictive energy management, the development of standardized lifecycle cost analysis tools for hybrid systems, and the exploration of next-generation technologies like agrivoltaics on warehouse rooftops.

REFERENCES

Ahmad, A., Prakash, O., Kausher, R., Kumar, G., Pandey, S., & Hasnain, S. M. M. (2024). Parabolic trough solar collectors: A sustainable and efficient energy source. In *Materials Science for Energy Technologies* (Vol. 7, pp. 99–106). KeAi Communications Co., DOI: 10.1016/j.mset.2023.08.002

Ahmad, N. I., Ab-Kadir, M. Z. A., Izadi, M., Azis, N., Radzi, M. A. M., Zaini, N. H., & Nasir, M. S. M. (2018). Lightning protection on photovoltaic systems: A review on current and recommended practices. In *Renewable and Sustainable Energy Reviews* (Vol. 82, pp. 1611–1619). Elsevier Ltd., DOI: 10.1016/j.rser.2017.07.008

Arabkoohsar, A. (2023). Classification of energy storage systems. In *Future Grid-Scale Energy Storage Solutions: Mechanical and Chemical Technologies and Principles* (pp. 1–30). Elsevier. DOI: 10.1016/B978-0-323-90786-6.00011-X

Aydin, M., Sisman, A., Gültekin, A., Dehghan, B., Aydın, M., & Gultekin, A. (2015). An Experimental Performance Comparison between Different Shallow Ground Heat Exchangers. In *Proceedings World Geothermal Congress*. https://www.researchgate.net/publication/286456011

Azam, M. W., Cattani, L., Malavasi, M., & Bozzoli, F. (2023). Experimental Study of the Corrugation Profile Effect on the Local Heat Transfer Coefficient. *Energies, 16*(20), 7181. Advance online publication. DOI: 10.3390/en16207181

Azam, M. W., Chaudhary, G. Q., Sajjad, U., Abbas, N., & Yan, W. M. (2023). Performance investigation of solar assisted desiccant integrated Maisotsenko cycle cooler in subtropical climate conditions. *Case Studies in Thermal Engineering, 44*(February), 102864. DOI: 10.1016/j.csite.2023.102864

Azam, M. W., Malavasi, M., Cattani, L., Bozzoli, F., & Rainieri, S. (2024). Local heat-transfer coefficient estimation in cross-helix corrugated tubes under turbulent regime. *Journal of Physics: Conference Series, 2685*(1), 012042. Advance online publication. DOI: 10.1088/1742-6596/2685/1/012042

Behzadi, A., Holmberg, S., Duwig, C., Haghighat, F., Ooka, R., & Sadrizadeh, S. (2022). Smart design and control of thermal energy storage in low-temperature heating and high-temperature cooling systems: A comprehensive review. In *Renewable and Sustainable Energy Reviews* (Vol. 166). Elsevier Ltd., DOI: 10.1016/j.rser.2022.112625

Benbouzid, M., Diallo, D., Amirat, Y., Mangel, H., & Devel, A. M. (n.d.). *Development of Wind Turbines Prototyping Software Under Matlab/Simulink® Through Undergraduate Student Projects*. https://hal.science/hal-00527550v1

Bošnjaković, M., Veljić, N., & Hradovi, I. (2025). Perspectives of Building-Integrated Wind Turbines (BIWTs). In *Smart Cities* (Vol. 8, Issue 2). Multidisciplinary Digital Publishing Institute (MDPI). DOI: 10.3390/smartcities8020055

Boysen, N., & de Koster, R. (2024). 50 years of warehousing research—An operations research perspective. In *European Journal of Operational Research*. Elsevier B.V., DOI: 10.1016/j.ejor.2024.03.026

Cannava, L., Javan, F. D., Najafi, B., & Perotti, S. (2024). Green warehousing practices: Assessing the impact of PV self-consumption enhancement strategies in a logistics warehouse. *Sustainable Energy Technologies and Assessments*, *72*, 104054. Advance online publication. DOI: 10.1016/j.seta.2024.104054

Casariego, P., Sarrablo, V., Barrientos, R., Ibraikulov, O. A., & Cros, S. (2025). Second photovoltaic prototype for textile ceramic technology: The perovskite glass solar brick. *Construction & Building Materials*, *493*, 143143. Advance online publication. DOI: 10.1016/j.conbuildmat.2025.143143

Chen, H. T., Zhuang, X. W., Rashidi, S., & Yan, W. M. (2023). Temperature control in warehouse with internal and external heat rejections. *International Journal of Thermal Sciences*, *194*, 108600. Advance online publication. DOI: 10.1016/j.ijthermalsci.2023.108600

Dada, M., & Popoola, P. (2023). Recent advances in solar photovoltaic materials and systems for energy storage applications: a review. In *Beni-Suef University Journal of Basic and Applied Sciences* (Vol. 12, Issue 1). Springer Science and Business Media Deutschland GmbH. DOI: 10.1186/s43088-023-00405-5

Dixit, M., Muralidharan, N., Parejiya, A., Essehli, R., Belharouak, I., & Amin, R. (2022). Electrochemical energy storage systems. In *Emerging Trends in Energy Storage Systems and Industrial Applications* (pp. 259–282). Elsevier., DOI: 10.1016/B978-0-323-90521-3.00011-9

Elaouzy, Y., El Fadar, A., & Achkari, O. B. (2024). Building-integrated passive and renewable solar technologies: A review from 3E perspectives. In *Sustainable Energy Technologies and Assessments* (Vol. 72). Elsevier Ltd., DOI: 10.1016/j.seta.2024.104079

Esmaeilion, F., Soltani, M., & Garcia, D. A. (2025). Solar-driven multigeneration systems in grid-connected scheme integrated with air energy storage systems. In *Sustainable Energy Technologies and Assessments* (Vol. 81). Elsevier Ltd., DOI: 10.1016/j.seta.2025.104443

Fan, C., & Zhang, C. (2025). Optimal design of hybrid ocean thermal energy and solar pond system. *Solar Energy*, *286*, 113185. Advance online publication. DOI: 10.1016/j.solener.2024.113185

Fikiin, K., Stankov, B., Evans, J., Maidment, G., Foster, A., Brown, T., Radcliffe, J., Youbi-Idrissi, M., Alford, A., Varga, L., Alvarez, G., Ivanov, I. E., Bond, C., Colombo, I., Garcia-Naveda, G., Ivanov, I., Hattori, K., Umeki, D., Bojkov, T., & Kaloyanov, N. (2017). Refrigerated warehouses as intelligent hubs to integrate renewable energy in industrial food refrigeration and to enhance power grid sustainability. In *Trends in Food Science and Technology* (Vol. 60, pp. 96–103). Elsevier Ltd., DOI: 10.1016/j.tifs.2016.11.011

Future of solar photovoltaic : deployment, investment, technology, grid integration and socio-economic aspects. (2019). International Renewable Energy Agency.

Gainza-Barrencua, J., Odriozola-Maritorena, M., Barrutieta, X., Gomez-Arriaran, I., & Hernández Minguillón, R. (2022). Use of sunspaces to obtain energy savings by preheating the intake air of the ventilation system: Analysis of its main characteristics in the different Spanish climate zones. *Journal of Building Engineering*, *62*, 105331. Advance online publication. DOI: 10.1016/j.jobe.2022.105331

Goyal, N., Aggarwal, A., & Kumar, A. (2022). Concentrated solar power plants: A critical review of regional dynamics and operational parameters. In *Energy Research and Social Science* (Vol. 83). Elsevier Ltd., DOI: 10.1016/j.erss.2021.102331

Guo, J., Sun, Y., Chen, Z., Zhang, H., AlMasoud, N., Qiu, H., Guo, Z., Yu, Q., Ren, J., Alomar, T. S., & Zhu, J. (2025). Electrostatic self-assembled ceramic magnetic Ti3C2Tx nanocomposites reinforced with hollow Fe3O4 particles for strong and broadband absorption of electromagnetic waves. *Ceramics International*, *51*(19), 29283–29291. Advance online publication. DOI: 10.1016/j.ceramint.2025.04.134

Hasan, M. M., Haque, R., Jahirul, M. I., Rasul, M. G., Fattah, I. M. R., Hassan, N. M. S., & Mofijur, M. (2025). Advancing energy storage: The future trajectory of lithium-ion battery technologies. In *Journal of Energy Storage* (Vol. 120). Elsevier Ltd., DOI: 10.1016/j.est.2025.116511

Hossain, E., Faruque, H. M. R., Sunny, M. S. H., Mohammad, N., & Nawar, N. (2020). A comprehensive review on energy storage systems: Types, comparison, current scenario, applications, barriers, and potential solutions, policies, and future prospects. In *Energies* (Vol. 13, Issue 14). MDPI AG. DOI: 10.3390/en13143651

Igor Cosma, A. M., Zangara, G., Silvestri, L., & Filice, L. (2025). Sustainability Impact of Automated Warehouses in Industry 4.0 scenario. *Procedia Computer Science*, *253*, 3196–3205. DOI: 10.1016/j.procs.2025.02.044

INVESTMENT CASE FOR ROOFTOP SOLAR POWER IN WAREHOUSING Introduction & Scope of Work AUTHORS. (2022). https://www.delta-ee.com/consultancy/

Jia, S., & Gao, Z. W. (2025). Extended modeling for wind turbines with application to hybrid renewable energy systems. *Engineering Science and Technology, an International Journal, 70*, 102168. Advance online publication. DOI: 10.1016/j.jestch.2025.102168

Kalogirou, S. A. (2024). Wind Energy Systems. In *Solar Energy Engineering* (pp. 753–779). Elsevier., DOI: 10.1016/B978-0-323-99350-0.00015-6

Kamran, M. (2023). Planning and modeling of wind energy systems. In *Fundamentals of Smart Grid Systems* (pp. 271–298). Elsevier., DOI: 10.1016/B978-0-323-99560-3.00003-X

Kim, H. G., Jeon, W. H., & Kim, D. H. (2016). Wind resource assessment for high-rise BIWT using RS-NWP-CFD. *Remote Sensing (Basel), 8*(12), 1019. Advance online publication. DOI: 10.3390/rs8121019

Korra, C., & Sadhana V., A. (2024). Green Warehouses: The Benefits, Challenges and Strategies of Industrial Building Decarbonization. *International Journal of Innovative Science and Research Technology (IJISRT)*, 1454–1462. DOI: 10.38124/ijisrt/IJISRT24MAR1644

Kwon, H., Baek, J., & Kim, H. T. (2023). Building lithium metal batteries under lean electrolyte conditions: Challenges and progress. In *Energy Storage Materials* (Vol. 55, pp. 708–726). Elsevier B.V., DOI: 10.1016/j.ensm.2022.12.016

Lamas, V., Cortiñas, A., & Luaces, M. R. (2025). Low-code framework for IoT data warehousing and visualization. *Computers & Geosciences, 205*, 105998. Advance online publication. DOI: 10.1016/j.cageo.2025.105998

Lee, Y., Song, J., & Hong, H. (2025). Long-term performance analysis of a photovoltaic thermal–ground source heat pump (PVT-GSHP) system in heating-dominated climates. *Journal of Building Engineering, 113772*, 113772. Advance online publication. DOI: 10.1016/j.jobe.2025.113772

Lei, M. (2024). Application of energy sustainability model based on optical sensing technology in intelligent warehousing performance management in green manufacturing industry. *Thermal Science and Engineering Progress, 54*, 102789. Advance online publication. DOI: 10.1016/j.tsep.2024.102789

Lennartz, P., Paren, B. A., Herzog-Arbeitman, A., Chen, X. C., Johnson, J. A., Winter, M., Shao-Horn, Y., & Brunklaus, G. (2023). Practical considerations for enabling Li|polymer electrolyte batteries. In *Joule* (Vol. 7, Issue 7, pp. 1471–1495). Cell Press. DOI: 10.1016/j.joule.2023.06.006

Liu, Y., Shi, X., Liu, J., Qin, P., & Tang, C. (2025). Warehouse-distribution integration routing problem under real-world transport conditions with multiple time windows and variable loading efficiency. *Transportation Research Part E, Logistics and Transportation Review, 203*, 104343. Advance online publication. DOI: 10.1016/j.tre.2025.104343

Loschetter, A., Kervévan, C., Stead, R., Le Guénan, T., Dezayes, C., & Clarke, N. (2025). Integrating geothermal energy and carbon capture and storage technologies: A review. In *Renewable and Sustainable Energy Reviews* (Vol. 210). Elsevier Ltd., DOI: 10.1016/j.rser.2024.115179

Mahadevan, V., S, S. R., Rusho, M. A., & Yishak, S. (2025). Critical review of energy storage systems: A comparative assessment of mechanisms, advantages, challenges, and integration with renewable energy. In *Results in Engineering* (Vol. 26). Elsevier B.V. DOI: 10.1016/j.rineng.2025.105589

Mahmoud, M., Ramadan, M., Abdelkareem, M. A., & Olabi, A. G. (2023a). Developments of wind energy systems. In *Renewable Energy - Volume 1: Solar, Wind, and Hydropower Definitions, Developments, Applications, Case Studies, and Modelling and Simulation* (Vol. 1, pp. 315–329). Elsevier. DOI: 10.1016/B978-0-323-99568-9.00014-5

Mahmoud, M., Ramadan, M., Abdelkareem, M. A., & Olabi, A. G. (2023b). Introduction and definition of wind energy. In *Renewable Energy - Volume 1: Solar, Wind, and Hydropower Definitions, Developments, Applications, Case Studies, and Modelling and Simulation* (Vol. 1, pp. 299–314). Elsevier. DOI: 10.1016/B978-0-323-99568-9.00016-9

Mahmoud, M., Ramadan, M., Abdelkareem, M. A., & Olabi, A. G. (2023c). Shallow geothermal energy ground loop systems. In *Renewable Energy - Volume 2: Wave, Geothermal, and Bioenergy Definitions, Developments, Applications, Case Studies, and Modelling and Simulation* (pp. 121–141). Elsevier. DOI: 10.1016/B978-0-323-95211-8.00017-8

Mahmoud, M., Ramadan, M., Abdelkareem, M. A., & Olabi, A. G. (2023d). Shallow geothermal energy ground loop systems. In *Renewable Energy - Volume 2: Wave, Geothermal, and Bioenergy Definitions, Developments, Applications, Case Studies, and Modelling and Simulation* (pp. 121–141). Elsevier. DOI: 10.1016/B978-0-323-95211-8.00017-8

Manning, C., & Jorgensen, M. (2025). Warehousing and distribution injuries in Kansas: Costs and trends from the state workers' compensation database 2014–2023. *Safety Science*, *191*, 106932. Advance online publication. DOI: 10.1016/j.ssci.2025.106932

Mao, L., & Xiang, C. (2025). Structural colors for building-integrated photovoltaics (BIPV): Multidisciplinary perspectives towards sustainable design. In *Materials Today Energy* (Vol. 50). Elsevier Ltd. DOI: 10.1016/j.mtener.2025.101867

Martín-Chivelet, N., Kapsis, K., Wilson, H. R., Delisle, V., Yang, R., Olivieri, L., Polo, J., Eisenlohr, J., Roy, B., Maturi, L., Otnes, G., Dallapiccola, M., & Upalakshi Wijeratne, W. M. P. (2022). Building-Integrated Photovoltaic (BIPV) products and systems: A review of energy-related behavior. *Energy and Building*, *262*, 111998. Advance online publication. DOI: 10.1016/j.enbuild.2022.111998

Meribout, M., Kumar Tiwari, V., Pablo Peña Herrera, J., & Najeeb Mahfoudh Awadh Baobaid, A. (2023). Solar panel inspection techniques and prospects. In *Measurement: Journal of the International Measurement Confederation* (Vol. 209). Elsevier B.V., DOI: 10.1016/j.measurement.2023.112466

Mohammadi, F., Jahangiri, A., & Ameri, M. (2025). Thermo-Enviro-Exergoeconomic Analysis of Solar Chimney Compared with Heller Tower in Fars Combined Cycle Power Plant. *Energy Nexus*, *100506*, 100506. Advance online publication. DOI: 10.1016/j.nexus.2025.100506

Mosallanejad, B., & Akrami, M. (2024). Recent advances on electrolyte additives used in lead-acid batteries to enhance their electrochemical performances. In *Journal of Energy Storage* (Vol. 97). Elsevier Ltd., DOI: 10.1016/j.est.2024.112738

Nkinyam, C. M., Ujah, C. O., Nnakwo, K. C., & Kallon, D. V. V. (2025). Insight into organic photovoltaic cell: Prospect and challenges. In *Unconventional Resources* (Vol. 5). KeAi Communications Co., DOI: 10.1016/j.uncres.2024.100121

Okonkwo, P. C., Mansir, I. B., Al Housni, F., Beitelmal, W. H., Al Kaaf, K., & Ogunleye, G. J. (2023). Provision of kinetic energy support from wind turbines for frequency regulation services in the modern grid. In *Advanced Frequency Regulation Strategies in Renewable-Dominated Power Systems* (pp. 167–180). Elsevier., DOI: 10.1016/B978-0-323-95054-1.00018-4

Oloruntobi, O., Mokhtar, K., Mohd Rozar, N., Gohari, A., Asif, S., & Chuah, L. F. (2023). Effective technologies and practices for reducing pollution in warehouses - A review. *Cleaner Engineering and Technology, 13*, 100622. Advance online publication. DOI: 10.1016/j.clet.2023.100622

Pei, T. H., Ji, Y. F., Pan, G., Chang, B. J., Huang, C. Y., & Chang, W. S. (2025). Dual-power generation by the solar cell and the Seebeck-effect-based thermoelectric setup. *Energy for Sustainable Development, 88*, 101802. Advance online publication. DOI: 10.1016/j.esd.2025.101802

Pileri, F., da Silva Freitas, W., D'Epifanio, A., & Mecheri, B. (2025). Advances in organic electroactive species for enhancing the performance of all-aqueous redox flow batteries in electrochemical energy storage. In *Journal of Energy Storage* (Vol. 113). Elsevier Ltd., DOI: 10.1016/j.est.2025.115677

Pratiwi, I., Elizabeth, R., Fatari, , Basrowi, , & Nuryanto, U. W. (2025). Assessing pricing, distribution, and warehousing strategies in influencing consumer demand and purchase decisions: Evidence from Indonesia. *Social Sciences & Humanities Open, 12*, 101688. Advance online publication. DOI: 10.1016/j.ssaho.2025.101688

Qadar Chaudhary, G., Waheed Azam, M., Bozzoli, F., Sajjad, U., Vocale, P., Cattani, L., & Fallahzadeh, R. (2024). Transient analysis of an efficient solar assisted air-conditioning system for subtropical climate with various solar thermal collectors. *Energy Conversion and Management: X, 23*(May), 100634. DOI: 10.1016/j.ecmx.2024.100634

Qu, H., Gao, B., Kong, Q., & Xu, Z. (2024). Experimental study of the effect of ambient temperature on the output characteristics of PV-PCM systems. *Journal of Energy Storage, 88*, 111535. Advance online publication. DOI: 10.1016/j.est.2024.111535

Reddy, V. J., Hariram, N. P., Ghazali, M. F., & Kumarasamy, S. (2024). Pathway to Sustainability: An Overview of Renewable Energy Integration in Building Systems. In *Sustainability (Switzerland)* (Vol. 16, Issue 2). Multidisciplinary Digital Publishing Institute (MDPI). DOI: 10.3390/su16020638

Ren, S., Liang, C., Liu, Y., Wang, J., & Wang, C. (2025). Evolutionary game-based warehousing resources sharing strategy for logistics industry under product-service system paradigm. *Advanced Engineering Informatics, 67*, 103560. Advance online publication. DOI: 10.1016/j.aei.2025.103560

Saif, A. G. H., Alquaity, A. B. S., & Mokheimer, E. M. A. (2025). Techno-economic analysis of green hydrogen production in Saudi Arabia: A comparative study of solar PV and CSP technologies. *International Journal of Hydrogen Energy, 105*, 1361–1374. DOI: 10.1016/j.ijhydene.2025.01.043

Sasikumar, R., Thirumalaisamy, S., Kim, B., & Hwang, B. (2024). Dye-sensitized solar cells: Insights and research divergence towards alternatives. In *Renewable and Sustainable Energy Reviews* (Vol. 199). Elsevier Ltd., DOI: 10.1016/j.rser.2024.114549

Sharmin, T., Khan, N. R., Akram, M. S., & Ehsan, M. M. (2023). A State-of-the-Art Review on Geothermal Energy Extraction, Utilization, and Improvement Strategies: Conventional, Hybridized, and Enhanced Geothermal Systems. In *International Journal of Thermofluids* (Vol. 18). Elsevier B.V. DOI: 10.1016/j.ijft.2023.100323

Shboul, B. (2025). Novel design of hybrid dish Stirling engine and FC-electrolyzer system for synergistic power and green hydrogen solutions: A 3E performance evaluation. *Energy*, *318*, 134835. Advance online publication. DOI: 10.1016/j.energy.2025.134835

Shukla, A. K., Sudhakar, K., & Baredar, P. (2017). Recent advancement in BIPV product technologies: A review. In *Energy and Buildings* (Vol. 140, pp. 188–195). Elsevier Ltd., DOI: 10.1016/j.enbuild.2017.02.015

Solar, G. I. S. *Direct Normal Irradiation (GHI); 2024. Available from:* https://solargis.com/resources/free-maps-and-gis-data. (2024).

Soumya Rai, P., Kant, C. R., & Jha, R. (2025). Metal-Air Batteries: From Fundamental Mechanisms to Practical Applications. *Journal of Alloys and Compounds*, *180805*, 180805. Advance online publication. DOI: 10.1016/j.jallcom.2025.180805

Sugianto, S. (2020). Comparative Analysis of Solar Cell Efficiency between Monocrystalline and Polycrystalline. *INTEK: Jurnal Penelitian*, *7*(2), 92–100. DOI: 10.31963/intek.v7i2.2625

Togun, H., Basem, A., Kadhum, A. A. H., Abed, A. M., Biswas, N., Rashid, F. L., Lawag, R. A., Ali, H. M., Mohammed, H. I., & Mandal, D. K. (2025). Advancing photovoltaic thermal (PV/T) systems: Innovative cooling technique, thermal management, and future prospects. In *Solar Energy* (Vol. 291). Elsevier Ltd. DOI: 10.1016/j.solener.2025.113402

Tyagi, V. V., Rahim, N. A. A., Rahim, N. A., & Selvaraj, J. A. L. (2013). Progress in solar PV technology: Research and achievement. In *Renewable and Sustainable Energy Reviews* (Vol. 20, pp. 443–461). Elsevier Ltd., DOI: 10.1016/j.rser.2012.09.028

Uwa, J. N., & Pourvahidi, P. (2024). An Overview of the Effects of Passive Solar Strategies on Thermal Comfort. *Journal of Mediterranean Cities*, *4*(1). Advance online publication. DOI: 10.38027/mediterranean-cities_vol4no1_2 van Geest, M., Tekinerdogan, B., & Catal, C. (2022). Smart warehouses: Rationale, challenges and solution directions. In *Applied Sciences (Switzerland)* (Vol. 12, Issue 1). MDPI. DOI: 10.3390/app12010219

Vázquez, N., & Vázquez, J. (2023). Photovoltaic System Conversion. In *Power Electronics Handbook* (pp. 781–795). Elsevier., DOI: 10.1016/B978-0-323-99216-9.00023-8

Wang, G., Zhou, F., Jia, H., wang, L., Klitzsch, N., Feng, G., Yan, C., Ma, H., Liu, S., yin, Q., & Liu, X. (2025). Heat extraction performance of multi-level, multi-branch, closed loop coaxial horizontal borehole heat exchanger geothermal system. *Applied Thermal Engineering*, *269*, 126045. Advance online publication. DOI: 10.1016/j.applthermaleng.2025.126045

Wang, J., & Azam, W. (2024). Natural resource scarcity, fossil fuel energy consumption, and total greenhouse gas emissions in top emitting countries. *Geoscience Frontiers*, *15*(2), 101757. Advance online publication. DOI: 10.1016/j.gsf.2023.101757

Yu, J., Zhang, G., Zhang, Z., Liu, X., & Rao, Z. (2025). Numerical study on novel fin configurations for enhancing thermal performance of latent heat thermal energy storage systems. *Applied Thermal Engineering*, *278*, 127381. Advance online publication. DOI: 10.1016/j.applthermaleng.2025.127381

Zhang, C., Yuan, X., Fei, H., Ju, M., Wang, H., Zhang, Y., Wang, S., Wang, Z., Lei, J., Chen, G., Huang, Y., & Wang, J. (2025). Efficiently coupling water electrolysis with solar PV for green hydrogen production. In *Energy Reviews* (Vol. 4, Issue 3). Elsevier B.V. DOI: 10.1016/j.enrev.2025.100151

Zhang, R., Wang, H., Wang, Y., He, F., Behrendt, F., Li, C., & Zhao, S. (2025). Experimental study on energy pile thermal-structure response during high temperature heat storage. *Journal of Energy Storage*, *120*, 116458. Advance online publication. DOI: 10.1016/j.est.2025.116458

Zhang, Y., Qu, T., Hong, Z., Zhang, Z., & Huang, G. Q. (2026). Digital twin driven opti-state control approach for smart warehousing in the synchronous operating environment. *Robotics and Computer-integrated Manufacturing*, *97*, 103099. Advance online publication. DOI: 10.1016/j.rcim.2025.103099

Chapter 12
Green Retrofitting of Warehouses:
Pathways to Sustainable and Energy-Efficient Logistics

R. N. Ravikumar
https://orcid.org/0009-0009-3705-1681
Marwadi University, Rajkot, India

S. Aarthi
https://orcid.org/0009-0006-9064-2091
Marwadi University, Rajkot, India

ABSTRACT

This chapter explores cutting-edge advancements in green and energy-efficient warehousing, focusing on transforming traditional logistics infrastructure into sustainable, intelligent ecosystems. It covers core strategies such as retrofitting with LED lighting, smart HVAC systems, renewable energy integration, IoT-based monitoring, and automation for dynamic energy control. Detailed case studies from global leaders like DHL and FedEx illustrate real-world impacts, including substantial reductions in energy consumption and operating costs. The book also analyzes financial feasibility, including ROI, payback periods, and the role of government incentives. Emphasis is placed on policy alignment, employee training, and industry certifications like LEED and ISO 50001. By blending theory with practice, this volume serves as a comprehensive guide for warehouse managers, engineers, policymakers, and sustainability advocates seeking actionable solutions for carbon-neutral logistics operations.

DOI: 10.4018/979-8-3373-3176-8.ch012

1. INTRODUCTION

With the world industries moving towards decreasing the carbon footprint and going green, current warehouse facilities are under more pressure to adapt to support the environmental agenda. The highly affordable and scalable green retrofitting solution can be applied by equipping existing warehouse operations with efficient energy systems, renewable energy and intelligent solutions, and is not limited to constructing brand-new warehouses. This strategy not only cuts on environmental impact but it also cuts down on the operating costs thus enhancing the operational efficiency coupled with decreasing the utility expenses as well as treating the regulatory compliance. Based on the figures that indicate that more than 80% of existing warehouses were constructed prior to the onset of making sustainability a priority, retrofitting has become a viable strategic approach towards modernization of antique warehouses (Carli et al., 2020). It converts the warehouses as the passive object of storage to the active component of green supply chain. This part will discuss the extent, significance, and the revolutionary capabilities of retrofitting in propelling a more sustainable logistics and warehousing eco-system.

1.1 Need for Sustainability in Legacy Warehousing

The legacy warehouses, built without references to environmental standards, are highly energy-consuming, low-tech, and inefficient in term of resources. These traditional buildings are a major roadblock to realize the dream of a sustainable practice as global supply chains are under pressure to meet demands of cutting down carbon emissions and conform to environmental policies. Greenhouse gas emissions are significant cause of emission in the warehousing industry due to poor lighting, HVAC systems, and utilization of non-renewable energy (Ries et al., 2017). Moreover, consumer and regulatory demands that Favor environmentally-friendly logistics has caused firms to reconsider their infrastructure. Sustainability has become not only important but it is a must in gaining competitive edge, regulatory Standing and long-term profits. The process of retrofitting old warehouses enables an organization to be in line with the requirements of green operations without the initial hash on the wallet of an organization by having to construct. This strategy is a clever and timely way to decrease environmental impact, to achieve ESG goals, and to futureproof supply chain processes that operate on a limited resource world.

1.2 Definition and Scope of Retrofitting

The concept of retrofitting when used in the context of the warehousing industry means an upgrading of the current warehouse infrastructure by using modern, energy, and environmentally sustainable technology and systems. It is associated with the incorporation of some of these solutions e.g. LED lighting, smart HVAC systems, solar panels, insulation, water conservation mechanisms and IoT-based automation tools into already built buildings. Retrofitting can be of both small scale (replacing light fixtures) and large scale (replacing the building with renewable energy and structural design) (Grebski & Maryniak, 2020). In contrast to new building, retrofit can make use of the existing assets thus it is cost effective and time effective particularly to the companies that have extensive legacy warehouses. In addition, retrofit not only involves physical upgrades but also upgrading operations as well as training the workforce on sustainability besides the application of monitoring systems to maintain the effectiveness in the long run. The green retrofitting dimension as it involves environmental, technological, and operational aspects is important in the conversion of conventional warehouses to smart and low-carbon warehouses that are matched with the current changing global standards of sustainability.

1.3 Role in Sustainable Supply Chain Transformation

The management of warehouses is a main hub of a supply chain, and it determines energy consumption, the amount of carbon emissions, and flows of materials. Refurbishing them to go green and operate energy-efficiently is a transformative symbol in establishing sustainable supply chain. Sustainable supply chains are geared to reduce the environmental impact at all points of the chain of events that entail the involvement of the chain, including sourcing of supplies, delivery, as well as, warehousing which is the mediator between production and distribution. This change is facilitated by green retrofitting that involves the minimization of energy use, the integration of renewable energy and the efficient use of resources (Md Hanafiah et al., 2022). Also, it will improve the circular economy through more effective waste management practices and recycling in warehouses. The retrofitted warehouses also enable the real time tracking of energy and material consumption which in turn enables the companies to measure and enhance their environmental performance. Since the regulatory environment is moving to higher standards and customers focus on environmental-friendly brands, green warehouses can be a major distinction. Retrofitting of obsolete infrastructure enables the corporation to integrate the logistics operations on the path to overall sustainability, which strengthens supply chains by being more resilient, transparent, and environmentally friendly. The figure 1 illustrates the transition from traditional, high-energy-use warehousing to a sustainable supply chain model. By implementing green retrofitting measures such as LED lighting, smart HVAC, and IoT technologies warehouses evolve into energy-efficient, low-emission environments. These changes lead to enhanced employee comfort, smarter operations, and ultimately contribute to building a resilient, low-carbon, and sustainable supply chain.

Figure 1. Role of Green Retrofitting in Sustainable Supply Chain Transformation

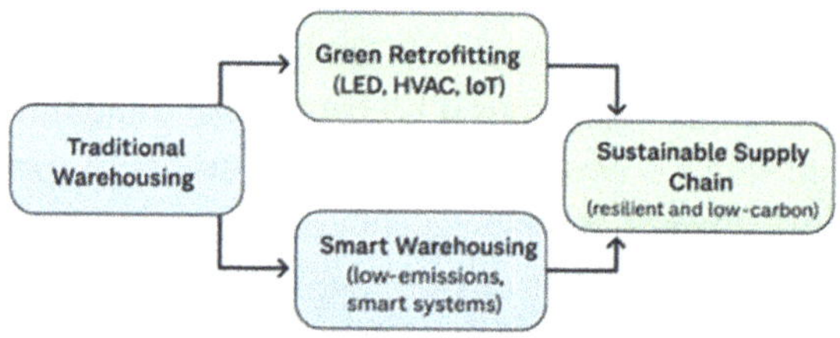

2. ENERGY-EFFICIENT LIGHTING SYSTEMS

Lighting also consumes a significant portion of energy that is spent in a warehouse, especially the large ones that are considered to be operational 24/7. The other new lighting systems such as the fluorescent or metal halide fixtures are ineffective not only in terms of life but also in terms of maintenance as they have a short life than the traditional lighting systems. One of the revolutionary solutions is retrofitting which entails the installation of LED (Light Emitting Diode) in the warehouse instead of the old lighting system that consumes up to 80% energy, and enhances visibility/ safety in the warehouse. The LEDs consume less power, the HVAC would also not be burdened by them, and their life is often as high as 50,000 and higher hours of operation (Ishizaka et al., 2022). Use of intelligent lighting systems that come with motion sensors, dimensioning and day light harvesting switching options also makes the use of energy even simpler with the addition of the ability to vary the intensity of the light according to the occupancy and availability of natural light sources. These smart systems not only help in reducing the energy bill, but they also help in the sustainability goals, also green building certification. Higher optimization of energy consumption encompassing lighting is thus a milestone of the green retrofit programs of the modern warehousing.

2.1 Limitations of Traditional Lighting (Fluorescent, HID)

The traditional lighting products e.g. fluorescent tube lights, High-intensity discharge (HID) lamps have usually been used in the warehouse atmosphere settings. However, the technologies are associated with several weaknesses, which restrict energy efficiency and sustainability. Fluorescent and HID fixtures are generally high consuming, generate excessive heat and they do not have a long-life span, therefore they require frequent replenishment. even the HID lamps in their specific case have badly poor warm-up time and even dimmer as they wear out, forcing them to waste their illumination in a bad quality. In addition, these systems tend to lack compatibility with the innovative energy management technologies and, therefore, can be defined by inertia and expensive costs in the long-term outlook (Karim et al., 2020). The fluorescent bulbs also contain mercury and this poses environmental risks since there is a risk of polluting the environment in case such is disposed when the time comes to do so. These flaws make the conventional lighting systems redundant as the sustainability and affordability concerns take centre stage in the management of supply chains. Their retrofitting is a highly-needed process in the contemporary warehousing so that the green compliance and the fullest utilization of energy is achieved.

2.2 Benefits and Applications of LED Lighting

LED (Light Emitting Diode) lighting is a better choice to replace warehouse lighting of the past because it has immense benefits over the conventional lighting when it comes to saving energy and taking care of the environment. LEDs use up to 80% fewer electricity than traditional lighting, which means significant savings of energy costs. They also last longer in various operations- usually more than 50000hrs- thus minimizes the cost and time of maintenance. LEDs produce low amounts of heat which reduces the cooling demands of the HVAC systems hence enhancing energy performance. They are compact in design and do directional lighting to allow uniform light throughout the large storage areas to improve visibility, safety and work productivity. Moreover, LEDs do not contain any toxic substances such as mercury and can be completely recycled, thus addressing the objectives of environmental sustainability (F. Odeyinka & G. Omoegun, 2024). Depending upon the use of different applications, warehouses can install warehousing LEDs, such as high-bay lighting in vertically placed compartments of its shelves to task lights in their sorting and packing platforms. With LED systems, a moving element in retrofitting approaches, warehouses will be greener and smarter. The table 1 compares the performance, energy consumption, lifespan, and environmental impact of traditional lighting systems (e.g., fluorescent, HID) against modern LED lighting solutions commonly used in warehouse retrofitting.

Table 1. Comparison of Traditional vs. LED Lighting in Warehouses

Feature	Traditional Lighting	LED Lighting
Energy Consumption	High	Low (up to 80% less)
Lifespan	10,000–15,000 hours	50,000+ hours
Heat Emission	High	Low
Maintenance Cost	High	Low
Mercury Content	Yes	No
Smart Integration Capability	Limited	High (Sensor-ready)

2.3 Smart Lighting Integration with Sensors and Timers

The next generation of energy-efficient warehouses design is smart lighting systems that combine Light-Emitting Diodes (LED) fixtures with motion sensors, timers, and automated controls. Their technologies, such as motion detector, occupancy sensors, and daylight harvesting, are incorporated, and it adapts lighting according to the needs in real-time. As an example, it can light up the active zones using motion sensors and shut off the lights in the unused areas to minimize unnecessary energy consumption in the warehouses, which work around the clock or do not operate regularly. Lighting can also be programmed to different operations phases so that they do not consume energy during non-working times (Perkumienė & Puskunigytė, 2024). Besides, LEDs lighting would be automatically controlled by daylight sensors to save more on natural lighting of the skylights or windows. It is also possible to connect these intelligent systems with dashboards of energy management, so that facility managers could trace their use and see wasteful spots. Smart lighting also increases the level of sustainability, as well as operational intelligence, resulting in more flexible, cost-efficient, and green building friendly warehouses. The Figure 2 illustrates how motion sensors activate LED lights upon detecting movement and coordinate with timers to optimize energy usage. Such integration enhances warehouse efficiency by reducing unnecessary lighting when spaces are unoccupied.

Figure 2. Smart Lighting Integration with Sensors and Timers

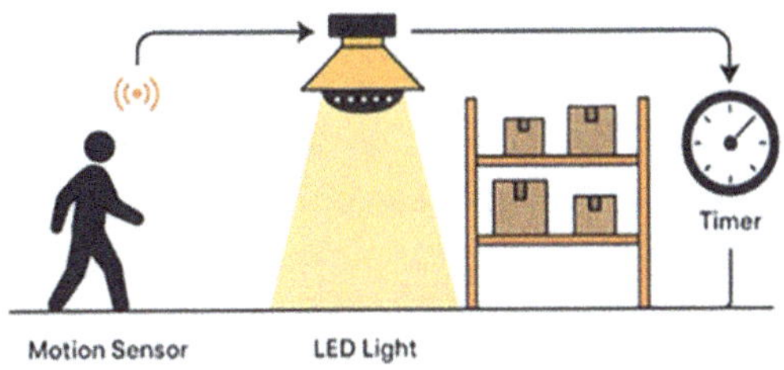

3. HVAC UPGRADES FOR GREEN EFFICIENCY

The use of Heating, Ventilation, and Air Conditioning (HVAC) systems in warehouses is very important in maintaining the working environment and especially in warehouses that store goods requiring controlled temperatures. Nevertheless, the conventional HVAC can be energy-demanding, inefficient, and one of the key sources of operation carbon footprints. Replacement of existing equipment with energy-efficient HVAC systems could be an effective measure in terms of green retrofitting initiative with a potential to slash down the energy consumption, operating expenses, and emissions (Taiwo et al., 2025). The latest HVAC systems base their energy optimization by providing variable-speed drives, high-performance compressors, and intelligent thermostats to drive the climate control process and to automatically adjust the HVAC according to altering conditions within the warehouse. Zoned Heating and cooling permit specific heating up and cooling off in the occupied space, reducing energy loss in the unoccupied sections. It is also integrated with renewable energy and IoT-based monitoring systems, which also improves the level of performance and sustainability. These retrofits do not only facilitate meeting the green building requirements (like LEED and BREEAM), but also make workers healthier and businesses more comfortable in terms of their work. As a result, employee well-being and long-term operations sustainability are facilitated.

3.1 Importance of HVAC in Warehouse Operations

The HVAC systems are important in the provision of appropriate environmental conditions in warehouses. In the case of facilities which store products like pharmaceutical products, consumable products, or electronic products, which are temperature sensitive; the climate must be consistent to guarantee the quality and safety of the products in the facility. HVAC also plays a major role in the comfort level and the productivity of employees in the non-temperature-controlled warehouses such as temperature levels, air Brinell and the level of humidities created. Nevertheless, traditional HVAC systems are one of the most energy-demanding elements of a warehouse, needing up to 40-50% of the energy, in general. Mismanaged or even obsolete systems will lead to energy wastages, unequal temperature regions, and expensive costs in operations (Škerlič et al., 2017). With the strategic importance of sustainability, it is crucial to optimize HVAC operations not only as environmental and cost-saving matter, but also as the efficiency issue. The alternative of retrofitting with energy efficient HVAC offers the warehouses the opportunity to enhance the environment to a better state and offer safer and healthier working conditions to their personnel and safeguard sensitive products.

3.2 Smart HVAC Controls and Zoned Climate Systems

Smart HVAC systems are a very important step forward in climate control energy consumption. Technologies that these systems utilize include smart thermostats, occupancy sensors and adaptive scheduling which enable these systems to adjust heating and cooling on a real-time basis and depending on how the warehouse is being used. In place of maximum use of HVAC system at all times, intelligent HVAC systems vary air flow, temperature and humidity levels in relation to the demand. Among the most efficient solutions in intelligent HVAC production, one can note the climate control zoning used in it (Dewangan et al., 2023). Separation of the warehouse into zones enables facility managers to charge the required amount of energy in areas that necessitate it the most e.g. in high-traffic or temperature-sensitive areas and minimize the amount of energy consumed in other zones that are used only occasionally. This is not only efficient in energy consumption but also able to increase environmental consistency in massive areas. Smart HVAC controls can also be part of the Building Energy Management Systems (BEMS) so that you can control remotely and track the performance of the controls, as well as perform predictive maintenance. This type of control systems based on intelligence and the green retrofitting programs must constitute its main part, as they are both economically and ecologically very profitable.

3.3 Renewable-Powered and Energy Star HVAC Solutions

The inclusion of HVAC systems in renewable energy coupled with the utilization of equipment labeled Energy Star has the immense benefit of improving the sustainability picture of the warehouse operations. HVAC systems are powered by renewable options which include a wind turbine, geothermal energy, and solar panels, among others which ensure that the use of grid electricity is minimized and the number of greenhouse gas emissions is reduced significantly. An example is solar-assisted HVAC [utilising photovoltaic energy to drive fans and compressors], and geothermal systems utilise the constant temperature of the center of the earth as a heating and cooling source. It also improves on the ability of HVAC systems to use less energy as Energy Star-rated HVAC systems are highly efficient and only a small amount of energy is used by such systems compared to others due to the stringent efficiency standards installed by international regulators (Harmathy, 2021). The cost reduction resulting from retrofit with this type of certified systems does not only reduce operating expenses but warehouses may also receive government incentives, green building certification, and carbon credits. Collectively, the use of renewable and the use of certified high-efficiency systems create a potent dynamic on the journey towards net-zero warehousing and carbon-neutral logistic infrastructure.

4. RENEWABLE ENERGY AND POWER OPTIMIZATION

This concept of sustainability being the key aspect of modern-day logistics, the presentation of renewable energy sources to the work of a warehouse is the relevant step in the path of energy independence and carbon neutrality. They can provide them with clean power on site and minimize the reliance on fossil sources by installing solar power, wind turbines, geothermal systems, and retrofitting the existing warehouses (Khan et al., 2024). In addition to the generation, power optimization technologies are proposed, such as Battery Energy Storage Systems (BESS) and smart inverters, which make it possible to balance the supply and demand and improve the work of loads and minimize the loss of energy. The added advantages of these systems are that they have the capability of storing excess energy to take advantage of the same in future, reduce peak demand charges, and indefinite operation in the event of grid failure. (Deputy et al., 2024). Additionally, it is possible to connect with Energy Management Systems (EMS) allowing in time control and optimization of energy consumed, high power consumption could be balanced with renewable generation. In combination, renewable energy sources and power optimization technologies ensure a resilient future-proof energy infrastructure capable of providing operational efficiency, cost-effectiveness, and global sustainability requirements in warehouse retrofits.

4.1 Solar Panel and Wind Turbine Integration on Existing Structures

The combination of solar panels and wind turbines into buildings of existing warehouses is another approach to the problem that shows the most practical and effective results of the minimization of the use of fossil fuels and energy costs to work the warehouse operational activity. Warehouses are mostly box shaped with flats led roofs, which is most suitable to install a Photovoltaic (PV) solar system, capable of producing high amounts of renewable electricity on-sites. To be as efficient as possible, solar panels can be tied to the warehouse grid or combined with energy storage. Small scale windmills can equally be put on the rooftops or next to a piece of land and used to complement the energy requirements of the area in places that have favourable wind conditions (Sundar et al., 2025). The solutions lower the electric bills, facilitate the access of facilities to renewable energy incentives, and assist efforts to achieve corporate sustainability. The retrofitting approach through renewable energy infrastructure will also increase energy self-sufficiency, as well as help to achieve green building standards and cut on carbon emissions, e.g. compared with Science-Based Targets initiative (SBTi) or RE100 global initiative. The table 2 outlines various renewable energy options used in green retrofitting projects, highlighting their characteristics, benefits, and suitability for warehouse operations.

Table 2. Renewable Energy Options for Warehouse Retrofitting

Technology	Key Features	Ideal Use Cases	Benefits
Solar Panels	Rooftop PV, scalable output	Sunny regions, flat roofs	Reduces grid dependency
Wind Turbines	Onsite small wind systems	Wind-prone areas	Continuous energy generation
Geothermal HVAC	Subsurface heating/cooling	Large distribution centers	High efficiency, stable temp
Battery Storage	Lithium-ion or thermal storage	All warehouse types	Backup power, peak shaving

4.2 Battery Storage and Backup Systems

Battery Energy Storage Systems (BESS) have a significant purpose to a modern energy-efficient warehouse. They are mechanisms that store up surplus energy as it is being generated by solar panels or wind turbines at their peak production and release of the energy when the demand is high or when the renewable sources are not generating a substantial amount of energy. This ensures that there is sufficient power supply as well as averting the high cost of power consumption that can be incurred during peak power consumption (Fichtinger et al., 2015). The most common types of storage solutions to the merits of energy density and stability are Li-ion batteries (Tharani et al., 2025). BESS integration allows warehouses to lessen their reliance on the central power grid so as to be able to operate and support (protect) their important facilities such as refrigeration, automation, and security during peak outages. Hybrid technologies combining renewable energy with battery storage and connection to the grid may also be intelligent enough to switch between resources to become as efficient as they can be (Udo et al., 2023). Long-term operations and financial benefits can be unlocked, energy resilience and decarbonization plans can be improved, and battery storage that is retrofitted into existing warehouses can help unlock these benefits.

4.3 Smart Grids and Load Management Systems

The key concept in the retrofitting green warehouses is the integration of smart grid and load management systems in optimizing the use of energy. A smart grid will provide real-time communication between the energy system in the warehouse and the utility provider, thus dynamic control of the energy flow, economies of money and increasing reliability. The use of smart meters and sensors deployed in the warehouse enables the warehouse managers to understand the performance of the energy consumption, predict demand, and increase or decrease usage accordingly with the goal of balancing against periods when energy is cheaper or when the energy output reduces emissions (Wrana et al., 2023). Load management systems give precedence to the critical loads, move non-essential operations to off-peak hours, and decrease the load on the HVAC and lighting in the case of heavy demands. Predictive analytics and centralized control of these systems is possible through connecting to building energy management systems (BEMS). The outcome is better energy efficiency, low operation budget, and fewer carbon emissions. When paired with renewable energy and storage, warehouses are made resilient, responsive and future-proof to the evolving energy landscape with the help of smart grids and load management.

5. IOT AND AUTOMATION FOR ENERGY MONITORING

Internet of Things (IoT) technologies and automation interact with each other directly and influence the conversion of conventional warehouses into smart, energy-efficient warehouses. When sensors, meters, and actuators are deployed throughout lighting, HVAC and equipment systems, the warehouses will have access to real-time information regarding energy consumption, temperature, occupancy and equipment performance. This information can be used in prediction and on-the-fly corrections to minimize wastage and optimize the use of resources. As an illustration, a system connected to IoT can sense idle machines or that idle areas have no customers and turn off extra systems automatically. In combination with automated control systems energy flow is managed dynamically by the warehouses and adjusts to changing demand patterns (Binos et al., 2021). Moreover, IoT solutions usually work together with Energy Management Systems (EMS) or Building Management Systems (BMS) that help to centralize the monitoring capacity, provide efficiency reports, and indicate operators about any anomalies. Intelligent automation in warehouses would enable them to save on energy and minimize the idle time and play an important role in corporately environmental sustainability goals. The Figure 3 shows how IoT devices and automation systems interact to monitor and manage energy use in warehouses. Sensors detect real-time data, enabling automated decisions like turning off idle equipment and optimizing resource flow dynamically.

Figure 3. IoT and Automation for Energy Monitoring in Warehousing

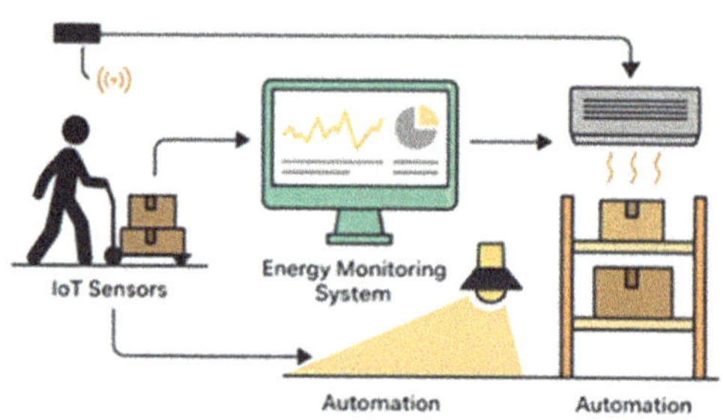

5.1 Installing IoT Sensors for Real-Time Energy Tracking

Installation of IoT sensors is one of the infrastructural requirements of live energy monitoring in warehouses. Such sensors can be used strategically throughout lighting systems, HVAC systems and machinery, and power inlets feeding whole areas and buildings so that energy consumption, temperature, and humidity, occupancy, and machine status can be continuously monitored. With the ability to capture and send the data in real time to central server or cloud platform, facility managers have a complete insight on the pattern of energy consumption and personnel operational inefficiencies. This sub-metered information allows instant identification of the deviation like energy spikes, equipment issues, or lighting being left on in unoccupied buildings (Desogus et al., 2021). These insights are used to make long-term energy forecasts and long-term benchmarking during the course of time. Sensors of IoT can also be included in an alert system to notify staff about over use or failures. Sensor deployment as an element of a retrofitting program forms the basis of intelligent automation and provides the guarantee that warehouses are used in a maximally energy-saving and environmentally conscious mode. The table 3 categorizes different IoT sensors used in retrofitted warehouses, explaining their functions and the types of energy data they collect.

Table 3. IoT Sensors Used in Warehouse Energy Monitoring

Sensor Type	Function	Monitored Parameter	Application Area
Occupancy Sensor	Detects movement	Space usage	Lighting, HVAC zones
Temperature Sensor	Measures indoor climate	Temperature variations	HVAC control
Smart Meters	Tracks electricity consumption	Real-time energy data	Whole-facility monitoring
Light Sensor	Measures ambient light levels	Daylight optimization	Smart lighting systems
Vibration Sensor	Detects equipment anomalies	Predictive maintenance data	Machinery, compressors

5.2 Automation for Equipment Shutdown and Resource Optimization

IoT data makes it possible to automate the process of making wise real-time decisions that can be implemented by warehouses to curb energy wastage and maximise the use of resources. Automated equipment shutdown is one of the most effective applications through which machinery, HVAC or lighting systems can be powered down when not in use or according to preset schedules (Jarašūnienė et al., 2023). To illustrate, the lights and conveyor belts will only switch on when an area is occupied or under use as motion sensors are used therefore saving power. Also, automation is able to control resource distribution, i.e. to balance between the power allocated to HVAC zones, or focus the energy resource on vital systems at peak periods. Fast response times can also be carried out by the smart Programmable Logic Controllers (PLCs) and edge computing devices which are able to act locally on commands. Such automated systems are operational 24/7 with low chances of human error and consistent adherence to the sustainability procedures (Poere et al., 2025). Automatization of currently standing warehouses improves the costs spent on energy only, not mentioning the increase of the accuracy of the work and its contribution to the climate.

5.3 Energy Dashboards and Predictive Maintenance

Energy dashboards are custom graphical tools, which collect real-time data of IoT sensors, and give a coherent summary to warehouse managers of energy consumption patterns, equipment health status and efficiency levels. Such dashboards usually present main indicators such as kilowatt-hour consumption, temperature differences, and system alert in an understandable and operational form (Mohd et al., 2023). Through this visibility, the managers are able to distinguish poor performing systems, benchmark performance of various zones and make informed decisions to contribute to further energy saving actions. The energy dashboards are even more powerful as they are employed with predictive maintenance algorithms. The behaviour of equipment and the use of energy can be analysed using predictive tools that can be used to foresee when there will be an eventual failure and prevent losing man hours and energy overuse. As one more relatively extreme one, the detection of peculiarities in the compressor cycles in an HVAC building may be mentioned as a possible future-meltdown, and preventative maintenance is recommended (Peng et al., 2024). The combination of dashboards and predictive maintenance is not only a form of data-driven approach but also allows constant improvement, which results in the betterment of warehouses in the sense of resilience, effectiveness, and sustainability.

6. FINANCIAL FEASIBILITY AND ROI ANALYSIS

The high potential of long-term financial payback of green retrofitting use is among the main impetuses of adopting it. Although there might be a high up-front cost to do the up-grades initially (Ex. LED light, intelligent HVAC, and solar panels) the savings in the operation realize subsequently tend to offset this initial investment (Singh, 2021). Retrofit causes huge saving in terms of electricity use, maintenance costs and off-line time which can be quantified. This is because a properly designed retrofitting project is able to give ROI between 2 to 5 years, according to the magnitude of the project and the cost of energy. Also, corporate incentives include taking advantage of government deductions, where the government may provide tax credits, grants on green buildings and low-interest loans on sustainability. Its non-financial benefits including better brand reputation, adherence to environmental standards and value gain in its assets also make up a strong case (Wu & Liu, 2019). Through extensive feasibility analysis and Lifecycle cost study, the use of retrofitting by the warehouse business can be embraced with confidence since it is one of the cost-effective approaches relative to environmental and economic objectives.

6.1 Cost Breakdown of Retrofitting vs. New Construction

Targeting the retrofit of existing warehouses is usually less costly compared to costly constructions of new green-certified structures. The first benefit is that it utilizes current infrastructure that cuts down the costs of acquiring land, building and other requisite permit costs. Since retrofitting projects tend to target the technological upgrade of a certain system (light, HVAC, insulation, or energy management), they are more specific and cost incremental. Greenfield construction on the contrary, is a much greater capital outlay and takes the form of large-scale investment in materials, labour, design and long schedules of construction. The retrofitting expenses tend to be lower than the costs of building new environmentally sustainable warehouses, since research indicates that they are usually between 30 and 50% less expensive (Yadav et al., 2019). In addition, retrofitting enables companies to keep open even during upgrading; thus, not incurring loss of revenue through unavailability. As a part of strategical planning, retrofitting can be a cost-efficient way of achieving a path to sustainability that is faster in implementation and less risky than building new.

6.2 Payback Periods and Energy Savings Projections

Payback period of green retrofitting investment is a very vital indicator to those warehouse operators who want to know whether sustainable investment can have financial returns. Paybacks usually vary between 2 to 5 years based on the extent of the improvements, energy expenses and the utility rates in a region (Fernández et al., 2022). Take an example of LED lighting retrofit, which can become payback in a period of less than 2 years because of energy savings starting on day one, whereas HVAC or solar panels systems can take a bit longer but provide greater long-term returns. The savings of 20-40 on energy are often attained in the first year of operation. Besides, energy saving, there is also the fixed savings because of reduced cost of maintenance, useful life of the equipment, and increased system reliability. The advanced modelling tools and energy audits can be used with high confidence to model these savings in order to make investments by the stakeholders with confidence. As time goes by, the savings are added and improve bottom-line operations and liberate capital to fund the sustainability programs in the future.

The quantitative yardsticks are used to explain some of the material benefits that can be collected through retrofitting of a warehouse demonstrating economic and environmental value. Retrofits of LED lamps by themselves can reduce power usage by up to 805 compared to traditional fluorescent transmission and HID lights and last longer than 50,000 hours. These not only translate to less maintenance requirements but can also be recovered in a duration of less than two years. The fifth free initiative is the replacement and upgrade of HVAC systems that can be consuming up to 40-50% of the total energy consumption in the warehouse, smarter controls, zoning and renewability-powered systems reducing the amount of energy used by 20-30%. Smart grid Commuter rooftop solar has a 15-20% goal of reducing grid dependence through peak shaving and 10-15% through the battery storage. Efficiency is also boosted by the invention of automation as a result of the IoT, which enables occupancy sensors, predictive maintenance to develop an additional 10-15% in energy savings and reduce downtimes. Their advantages also are validated in practical life: DHL saved its warehouse in Europe 40% of the total cost; FedEx in its centers in the USA 30% of the annual energy consumption. In total, retrofitting project may introduce a 20-40% energy saving in the first year, and the paybacks on investments are obtained, as a rule, in 2-5 years. The table 4 compares the energy saving, the payback period and the real-life evidences of key warehouse retrofitting, indicating their financial and environmental results.

Table 4. Quantitative Benchmarks for Warehouse Retrofitting

Category	Energy Savings (%)	Payback Period	Notes/Case Evidence
Lighting (LED)	Up to **80%**	**< 2 years**	Longer lifespan (>50,000 hrs), reduced maintenance
HVAC Upgrades	**20–30%**	3–5 years	Smart controls, zoning, renewable-powered units
Renewable Energy	Solar: **15–20% grid reduction**; Battery storage: **10–15% peak shaving**	4–6 years (varies by scale)	Rooftop solar and BESS improve resilience and cost-efficiency
IoT & Automation	**10–15% additional savings**	2–3 years	Occupancy sensors, predictive maintenance, smart dashboards
Case Studies	DHL: **40% overall reduction**; FedEx: **30% annual reduction**	Within 2–5 years	Large-scale implementation evidence

6.3 Government Incentives, Tax Credits, and Green Loans

In a bid to make the move to sustainable infrastructure easier, most governments and financial institutions have started giving various incentives to green retrofitting projects. These entail tax credits on renewable energy plantations, depreciation privileges of an accelerated nature, utility rebates on energy efficient equipment's, and green building certification grants. There is also financing by some regions where a low-interest green loan, energy performance contracts (EPCs), or a public-private partnership model is offered to cover initial capital outlays. Moreover, synchronizing the program of retrofitting with the certification program (such as LEED, BREEAM, or Energy Star) can open a new source of funds and increase the chances to receive carbon credit programs (Fulazzaky et al., 2022). Consultations of local energy agencies or sustainability consultants should be conducted to understand the most applicable and maximum impactful financial supportive measures (Aarthi et al., 2025). Reaching these incentives won not only add value to the ROI of retrofits, but also secure the statement of a company regarding its environmental responsibility strengthening the brand trust and making investors more confident in the company.

7. COMPLIANCE, CERTIFICATIONS, AND WORKFORCE TRAINING

An effective green retrofitting program is not only about upgrading technology; it must proceed with a perspective of regulatory compliance, seeking the target of sustainability certifications, and proper work training. Soon regulatory structures are now changing at a fast rate where there is more focus on standards of energy efficiency through reporting emissions and safety in the industrial premises (Kim et al., 2022). The retrofitted warehouses should adhere to the national and international environmental codes, i.e. the ASHRAE standards, the ISO 14001 and the building energy codes. Obtaining certifications such as LEED, Bream, or Energy Star confirm the level of performance of the warehouse in terms of sustainability, as well as increasing the competitiveness of products and the possibility of obtaining green incentives (Perotti et al., 2022). Employee engagement is no less crucial in this sense: training employees to use and support the new technologies, use the sustainable approaches and learn the energy-saving codes is the way to make sure that the success is achieved in the long-run (Ho et al., 2018). The conscious and well-prepared workforce becomes key in achieving the maximization of green investments ROI, changing the culture, and maintaining compliance through daily actions of the operation.

7.1 LEED, BREEAM, ISO 50001: Green Building Certifications

The recognition of green building certifications is proof of the environmental compliance of retrofitted warehouses and the more credible a developer can be with the authorities, clients, and investors. The LEED (Leadership in Energy and Environmental Design), created by U.S. Green Building council, evaluates the use of materials and indoor air quality as well as water savings and energy efficiency. In Europe, BREEAM (Building Research Establishment Environmental Assessment Method) combines assessments of sustainability in design, construction and operation (Jayarathne et al., 2024). The ISO 50001 offers a guideline in developing the energy management systems that require continual improvement of energy performance. This certification comes with its benefits tangible and intangible such as accessibility to tax concessions, uplifted value of the assets and strengthened corporate image. To get a certification, warehouses should record the performance indicators, adopt energy-saving mechanisms, and observe the best practices in sustainability (Minashkina, 2024). Certification objectives should be a part of early retrofitting plans because it will allow them to match the global standards and also allow quantification of progress towards excellence in the environment and operation.

7.2 Training Warehouse Staff in Energy-Efficient Practices

Energy efficiency using technology upgrades is not enough to create sustainability: the important factor would be employee engagement and employee training. Retrofitting in most cases will bring with it sophisticated systems like IoT sensors, automated lighting, intelligent HVAC, and this demands care and maintenance (Liu et al., 2021). Warehouse personnel should be trained in the usage of such systems and how to manage their energy consumption as well as in best practices that may include turning off the unused machines and utilizing the natural light or the versatile use of green infrastructure. Employees can also be trained on about eco-conscious operations through workshops, electronic modules and sustainability booklets. Also, there is making of green teams or championing of energy in warehouse team can develop a culture of accountability and innovations (Chen et al., 2016). Involved and informed employees can help define waste, report deviations and assist in constant progress (Seifhashemi et al., 2018). Finally, employee training makes sustainability not only a top-down project but also a project that has higher compliance and larger influence.

7.3 Policy Support and Industry-Wide Retrofitting Initiatives

The role of government policies and industry-initiated efforts cannot be underestimated in speeding up the process of green retrofitting practices. The drive toward the sustainability of the logistics and warehouses sector is brought about by national energy efficiency programs, the requirement of carbon neutrality, and guidelines on climate action. Example of supportive policies can be a requirement to conduct an energy audit, measurement of performance in comparison with others or to achieve minimum performance standards in terms of energy efficiency of industrial buildings (Staudt et al., 2015). Meanwhile, such global movements as the Paris Agreement, UN SDGs, and RE100 urge businesses to undertake to decarbonize and act responsibly with energy consumption. Best practices are also being encouraged in the industry, using industry groups and consortiums as well as sharing knowledge and sector-based innovation projects (Yun et al., 2018). As an example, green logistic councils and alliance of warehouse sustainability usually offer cases, tools, and policy updates that can be used to direct the retrofitting process. Future-proofing warehouses and sustainability applications go hand-in-hand with current collective industry trends and the regulatory aspirations that are generally accepted worldwide.

8. CASE STUDY: REAL-WORLD APPLICATIONS OF GREEN RETROFITTING

Case studies are an excellent source of information as to how established players in the logistics industry are effectively undertaking the green retrofitting plans to realise the formulation of de-merit able sustainability results. This section indicates the real-life implication, economic costs, and environmental gains of converting the current infrastructure of warehousing into low emission and energy-efficient warehouses through an analysis of the large-scale retrofitting projects (Al Alaq et al., 2025). Two recent cases in point, the retrofitting fuels throughout Europe by DHL, and the energy efficiency measures take in the U.S by FedEx reflect how the world leaders in logistics are utilizing the use of LED light, solar panels, occupancy sensors, and the intelligent HVAC systems. Through such initiatives there have been substantial savings in energy and operational costs and certification has been achieved of international status e.g. LEED Gold. Collectively, they can be viewed as strong role models of scalable, cost effective and sustainable transformation of warehouses.

8.1 Case Study 1: DHL's Sustainable Warehouse Retrofitting Across Europe

DHL which is among the major logistics providers in the world has also engaged in an ambitious green retrofitting project in its European warehouse network in an attempt to be in line with the sustainability points of the company, and to decrease its carbon footprint. The retrofitting plan made three major areas the focus including the LED lighting, the solar energy integration and the smart HVAC systems (Barbosa, 2023). After switching traditional lighting system to more energy-efficient LED fixtures, DHL saved more than 60% of the energy it utilized to spend on lighting. The use of solar panels on the rooftop of warehouses partially fulfilled the on-site renewable energy production, which leads to a reduction of the total number of grid dependency by 1520%. Moreover, the introduction of intelligent management of

HVAC systems with automatically controlled temperature and zone-based climate systems saved up to 25% of energy during the climate control processes.

These retrofitting actions cumulatively materialized into an average energy cost savings of up to 40% of the retrofitted plants. The LEED Gold certification of some of the warehouses is evidence of their outstanding performance of energy efficiency, indoor environmental quality and sustainable design (Hao et al., 2024). The experience of DHL is a bright example of how efficient retrofitting on a large scale can promote operational sustainability and environmental awareness. The Figure 4 shows the percentage age gains that were experienced by DHL due to its green retrofitting processes in the European warehouses. As a result of the upgrades, it was possible to improve lighting energy by 60% per year, grid electricity reliance by 20% due to the integration of solar panels and HVAC energy use by 25% based on smart climate control. All these initiatives led to an up-energy cost reduction of 40% overall and this demonstrates the efficacy of sustainable retrofitting strategies that can be employed and have been employed.

Figure 4. Energy Efficiency Gains from DHL's Warehouse Retrofitting

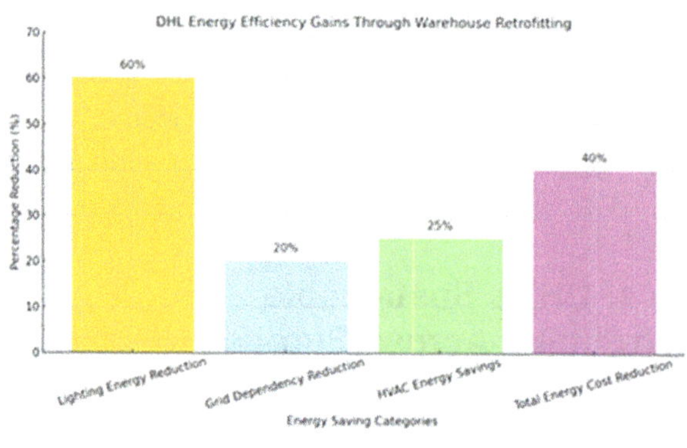

8.2 Case Study 2: FedEx's Energy Efficiency Programs in Distribution Centers

As one of the most popular global logistics and delivery companies, FedEx has already introduced the wide-sloping energy efficiency-retrofitting program in its high-volume distribution centers in the United States (Tung, 2025). It was about the increase of the operational sustainability and the decrease of the energy demands and impact on environment (Bruno Polli et al., 2022). The effort in retrofitting involved the use of energy efficient LED lights, sensor systems based on occupancy, and automated controls in HVAC. By making the changeover to high-efficiency LED lighting fixtures, FedEx reduced lighting energy consumption by 55%, whereas occupancy sensors helped the company to save another 10-15% of energy by turning the lights and heating and ventilation settings on and off according the patterns of occupant use. Highly automated HVAC systems, which have smart zoning and programmable thermostats resulted in further energy savings in HVAC by 20-25%. All in all, such interventions led to an amazing 30% decrease in the total annual energy consumed in retrofitted facilities (Purniawan & Elmi, 2025). Besides saving energy, FedEx also recorded reduced maintenance cost, better indoor environment, and greater employee comfort. This case demonstrates that strategic retrofitting of large-scale high traffic logistic center is possible to the extent that it potentially delivers environmental and economic value as well as contributing to corporate wide sustainability agenda. Figure 5 shows the energy conservation effects of the FedEx retrofit programs in the U.S. sites of FedEx distribution centers. The major mitigation measures will be an energy use reduction of 55% in lighting via LED renovations, up to 15% energy savings using occupancy sensor system, and downsizing of HVAC-related calorie use by 25% by means of automated controls. (Roucham et al., 2023) All these measures added up to an overall decrease in annual energy consumption of 30%. Because of this, these are proof of the effectiveness of well-given, smart retrofitting within high-volume logits operations.

Figure 5. FedEx Energy Savings Through Warehouse Retrofitting

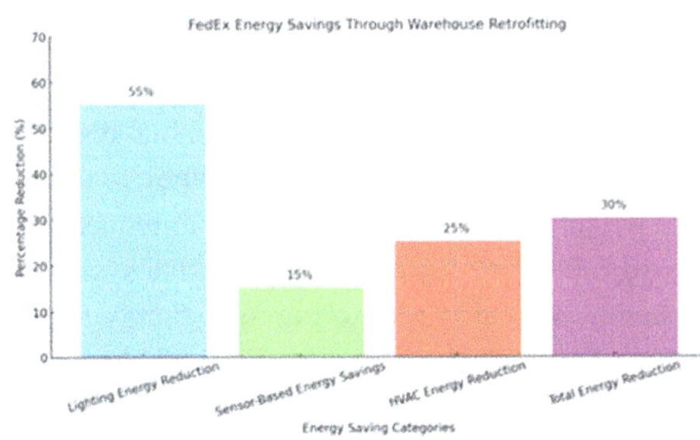

9. FUTURE RESEARCH AND DIRECTIONS

To improve the transparency and optimize operations through predictive optimization, future research on retrofitting warehouses must be done with incorporation of AI-driven energy management, digital twins, and blockchain carbon tracking. Resilience and cost effectiveness will be improved with the exploration of renewable sources of the next generation, such as hybrid solar- geothermal energy systems and the future of energy storage technology. It is also extremely required to research to carry out Life-Cycle Assessment (LCA) studies, carbon neutrality roadmap, and the effects of the regional policy to inform the international practice. The studies centered on human beings like the adaptation of the workforce to smart systems and green training will also be significant. Such directions will result in sustainability, scalability and intelligent warehousing in future.

9.1 Technological Innovations for Sustainable Retrofitting

The topic of interest in future research on retrofitting of warehouses must be the development of technologies that will result in energy efficiency and sustainability. The installation of AI-based analytics and machine learning algorithms can streamline the HVACs and the time schedule, anticipate energy demand, and detect inefficiencies before they are problematic. With the help of digital twin simulations, one can simulate the warehouses and control the retrofitting of warehouses in the virtual world to generate an optimal layout without disrupting the operations. Meanwhile, the IoT-based energy dashboards provide monitoring and predictive maintenance capacity, and consequently critically reduce operational expenses and downtimes. Besides digitalization, the new generation of renewable systems such as hybrid solarand geothermal system, hydrogen fuel cell, and Vehicle-to-Grid (V2G) energy transactions provide potential research opportunities to reduce the consumption of energy based on fossil fuels. The battery energy storage system (BESS) development and thermal storage will also help to improve energy resilience and reduce peak demand charges (Aarthi et al., 2025). In addition, carbon monitoring based on blockchain would improve transparency and responsibility in the achievement of the goals of emission reduction and would make warehouses comply with the ESG standards. Combined, the innovations will be capable of making retrofitting a holistic smart net-zero logistics system.

9.2 Human-Centric Approaches and Policy Integration

Although technology is an essential part of retrofitting, the human and policy aspect is also crucial to long term success. Further study must then explore where job training and green capacity-building initiatives can help enable employees to efficiently operate IoT products, smart HVAC and energy dashboards. The human resource practices such as reward to energy savings behavior and other team-based processes of sustainability innovations require examination on whether they have led to the inculcation of sustainability within the workplace culture. Research must also focus on policy, especially with regard to Life-Cycle Assessments (LCA) to measure the environmental cost of retrofit, as well as carbon neutrality roadmaps specific to regional settings(Yun et al., 2018). Comparative studies between regulatory framework- like LEED, BREEAM and ISO 50001 can illustrate best practice, and lead to convergence of standards globally. Besides, the public-private alliances, government incentives, and industry-level chairman must be analysed in terms of their potential to increase adoption. This is made possible through a combination of workforce engagement and effective policy integration in retrofitting in order to go beyond just being technical when it comes to retrofitting warehousing into a systemic transformation towards resilient, sustainable, and inclusive warehousing.

10. CONCLUSION

The retrofitting of the traditional warehouses in green is a viable and cost-efficient route to green logistics. The practical experience of such corporations as DHL and FedEx proves that the modernization of the existing infrastructure with the energy-efficient system and renewable energy facilities could considerably enhance the efficiency, decrease the emissions, and provide the environmental effect without spending time and money to build new facilities. The list of practices included in retrofitting is rather extensive: LED lamps, smart HVAC systems, solar panels, and automation of energy to use the Internet of Things. These systems combined together save energy, make operations easier, and minimize carbon dioxide emissions. The use of real-time energy monitoring and predictive maintenance also enables warehouse managers to make decisions that qualify in the long-term sustainability. The use of energy is monitored by energy dashboards, automated controls and zoned climate systems that ensure that the energy is used only where and when it is needed as well as making it more comfortable and functional. Financially, it is a good investment since the green retrofitting projects have good returns and the payback durations are between two-five years.

Economies of scale through the decreased electricity charges, maintenance, and government subsidies like tax deductions and green financing create a saving. Besides monetary benefits, retrofitting boosts brand recognition, promotes regulatory compliance, and augments the value of property assets making organizations congruent to ESG (Environmental, Social, and Governance) objectives. The human aspect is also important: training the staff to learn how to use energy-saving systems bolsters adoption and performance. Sustainability efforts are certified by independent standards such as LEED, BREEAM and ISO 50001, which offer a metric of further enhancement. In the bigger sustainable supply chain ecosystem, retrofitted warehouses are important nodes, which support low-carbon and circular networks of logistics. With the increase in regulations and transformation of expectations among customers, all the companies which proactively invest in the green infrastructure will benefit in the long run in terms of competitiveness, resilience, and innovation. Finally, green retrofitting goes beyond being technical improvement, it is a long-term investment into environmental sustainability, logistical excellence, and future-proofing. It helps fill the gap between climate ambitions and industrial aspects and provides a feasible roadmap on how to achieve cleaner, smarter, and even more sustainable warehousing.

REFERENCES

Aarthi, S., Ravikumar, R. N., Sobirova, K., & Mamadiyarov, Z. (2025). Empowering a sustainable future through digital learning and the SDGs. In *Harnessing E-Learning to Create a Sustainable Future* (pp. 145–172). IGI Global Scientific Publishing. DOI: 10.4018/979-8-3373-0112-9.ch006

Al Alaq, A. S., Damayanti, A. O., Efendi, D., Tahara, G. T., & Setyawati, N. (2025). Integration of digital technology and Green Human Resource Management (GHRM) in aligning equity principles to drive innovation, employee development, and environmental sustainability. *Jurnal Ilmiah Manajemen Dan Bisnis*, *11*(1), 64. DOI: 10.22441/jimb.v11i1.32367

Barbosa, J. (2023). Engineering assessment for ISO 50001 implementation. *Universidad, Ciencia y Tecnología*, *2023*(Special), 81–89. DOI: 10.47460/uct.v2023iSpecial.739

Binos, T., Bruno, V., & Adamopoulos, A. (2021). Intelligent agent based framework to augment warehouse management systems for dynamic demand environments. *AJIS. Australasian Journal of Information Systems*, *25*, 25. DOI: 10.3127/ajis.v25i0.2845

Bruno Polli, G. H., Sá, A. M., & Pavani Biju, B. (2022). Qualitative analysis of sustainability certification systems based on ISO 21929-1. *U Porto J. Eng.*, *8*(4), 47–64. DOI: 10.24840/2183-6493_008.004_0006

Carli, R., Dotoli, M., Digiesi, S., Facchini, F., & Mossa, G. (2020). Sustainable scheduling of material handling activities in labor-intensive warehouses: A decision and control model. *Sustainability (Basel)*, *12*(8), 3111. DOI: 10.3390/su12083111

Chen, X., Wang, X., Kumar, V., & Kumar, N. (2016). Low carbon warehouse management under cap-and-trade policy. *Journal of Cleaner Production*, *139*, 894–904. DOI: 10.1016/j.jclepro.2016.08.089

de Poere, E. C., Nurnabila, F., & Oktarina, R. (2025). Re-design layout and optimization of order picking routes to implement sustainable warehouse management at FMCG company's raw material warehouse. *IOP Conference Series. Earth and Environmental Science*, *1488*(1), 12023. DOI: 10.1088/1755-1315/1488/1/012023

Desogus, G., Quaquero, E., Rubiu, G., Gatto, G., & Perra, C. (2021). BIM and IoT sensors integration: A framework for consumption and indoor conditions data monitoring of existing buildings. *Sustainability (Basel)*, *13*(8), 4496. DOI: 10.3390/su13084496

Dewangan, F., Abdelaziz, A. Y., & Biswal, M. (2023). Load forecasting models in smart grid using smart meter information: A review. *Energies*, *16*(3), 1404. DOI: 10.3390/en16031404

Fernández, A., Haffner, M., & Elsinga, M. (2022). Comparing the financial impact of housing retrofit policies on Dutch homeowners. *IOP Conference Series. Earth and Environmental Science*, *1085*(1), 12044. DOI: 10.1088/1755-1315/1085/1/012044

Fichtinger, J., Ries, J. M., Grosse, E. H., & Baker, P. (2015). Assessing the environmental impact of integrated inventory and warehouse management. *International Journal of Production Economics*, *170*, 717–729. DOI: 10.1016/j.ijpe.2015.06.025

Fulazzaky, M. A., Syafiuddin, A., Roestamy, M., Yusop, Z., Jonbi, J., & Prasetyo, D. D. (2022). Reliability and economic analysis of a rainwater-harvesting system for a commercial building with a large rooftop area. *ACS ES&T Water*, *2*(4), 604–615. DOI: 10.1021/acsestwater.1c00455

Grebski, M., & Maryniak, A. (2020). Benefits of installing a photovoltaic power-generation system for a warehouse facility. *New Trends in Production Engineering*, *3*(1), 332–346. DOI: 10.2478/ntpe-2020-0028

Hao, F., Ng, W., Aman, A. M., & Zhang, C. (2024). Technology for sustainability: The impact of Avatar-led Green Training on OCBE and green creativity. *International Journal of Contemporary Hospitality Management.*

Harmathy, N. (2021). Investigation of decarbonization potential in green building design to accelerate the utilization of renewable energy sources. *Thermal Science*, *25*(6 Part A), 4269–4282. DOI: 10.2298/TSCI200324195H

Ho, T. H. D., & Daniel, J., Nadeem Simon Peter and Garza-Reyes, J. A., & Kumar, V. (2018, December). Improving the reliability of warehouse operations in the 3PL industry: An Australian 3PL case study. *2018 International Conference on Production and Operations Society (POMS)*. DOI: 10.1109/POMS.2018.8629453

Ishizaka, A., Khan, S. A., Kusi-Sarpong, S., & Naim, I. (2022). Sustainable warehouse evaluation with AHPSort traffic light visualisation and post-optimal analysis method. *The Journal of the Operational Research Society*, *73*(3), 558–575. DOI: 10.1080/01605682.2020.1848361

Jarašūnienė, A., Čižiūnienė, K., & Čereška, A. (2023). Research on impact of IoT on warehouse management. *Sensors (Basel)*, *23*(4), 2213. DOI: 10.3390/s23042213 PMID: 36850811

Jayarathne, T. D., De Silva, N., & Samarakoon, W. K. U. R. M. K. P. K. (2024). Adopting smart retrofits: A decision-making model and benchmarking criteria. *Built Environment Project and Asset Management*, *14*(3), 366–383. DOI: 10.1108/BEPAM-02-2023-0034

Karim, N. H., Abdul Rahman, N. S. F., Md Hanafiah, R., Abdul Hamid, S., Ismail, A., Abd Kader, A. S., & Muda, M. S. (2020). Revising the warehouse productivity measurement indicators: Ratio-based benchmark. *Marit. Bus. Rev.*, *6*(1), 49–71. DOI: 10.1108/MABR-03-2020-0018

Khan, S. I., Al Dodaev, Z., & Haque, A. (2024). The integration of renewable energy sources into mechanical systems, focusing on efficiency and reliability - A case study. *Control Syst. Optim. Lett.*, *2*(2), 172–177. DOI: 10.59247/csol.v2i2.92

Kim, J.-H., Hessek, C. J., Kim, Y. J., & Park, H.-G. (2022). Seismic analysis, design, and retrofit of built-environments: A procedural review of current practices and case studies. *Journal of Infrastructure Preservation and Resilience*, *3*(1), 11. DOI: 10.1186/s43065-022-00056-3

Liu, C., Sharples, S., & Mohammadpourkarbasi, H. (2021). Evaluating insulation, glazing and airtightness options for passivhaus EnerPHit retrofitting of a dwelling in China's hot summer–cold winter climate region. *Energies*, *14*(21), 6950. DOI: 10.3390/en14216950

Md Hanafiah, R., Karim, N. H., Abdul Rahman, N. S. F., Abdul Hamid, S., & Mohammed, A. M. (2022). An innovative risk matrix model for warehousing productivity performance. *Sustainability (Basel)*, *14*(7), 4060. DOI: 10.3390/su14074060

Minashkina, D. (2024). A review and research agenda for recent socially and environmentally sustainable practices for warehouse management systems. *International Journal of Logistics Management*, *35*(7), 60–98. DOI: 10.1108/IJLM-07-2023-0265

Mohd, S. A., Sazalı, N., Jamaludin, A. S., & Misdan, N. (2023). IOT-based Energy Monitoring System for energy conservation. *E3S Web Conf.*, *437*, 4006.

Odeyinka, F. O., & G. Omoegun, O. (2024). Warehouse operations: An examination of traditional and automated approaches in supply chain management. In *Operations Management - Recent Advances and New Perspectives*. IntechOpen.

Peng, X., Zhao, H., Zhang, X., Zhang, C., & Chen, C. (2024). MetaGON: A lightweight pedestrian re-identification domain generalization model adapted to edge devices. *IEEE Open Journal of the Communications Society*, *5*, 690–699. DOI: 10.1109/OJCOMS.2024.3349597

Perkumienė, D., & Puskunigytė, K. (2024). Improvement of warehouse processes through digitalization solutions. *HUMANITIES STUDIES*, *21*(98), 213–219. DOI: 10.32782/hst-2024-21-98-25

Perotti, S., Prataviera, L. B., & Melacini, M. (2022). Assessing the environmental impact of logistics sites through CO_2eq footprint computation. *Business Strategy and the Environment*, *31*(4), 1679–1694. DOI: 10.1002/bse.2976

Purniawan, A., & Elmi, F. (2025). The effect of Green Human Resource Management on green training, employee behavior, and employee welfare at PT Senayan Texindo Makmur. *Ijfsr, 2*(3).

Ries, J. M., Grosse, E. H., & Fichtinger, J. (2017). Environmental impact of warehousing: A scenario analysis for the United States. *International Journal of Production Research*, *55*(21), 6485–6499. DOI: 10.1080/00207543.2016.1211342

Roucham, B., Lefilef, A., & Mahmoudi, H. (2023). The Forward-Looking studies on Green Human Resources Management: A content analysis. *Manag. Econ. Rev.*, *8*(3), 259–275. DOI: 10.24818/mer/2023.10-01

Seifhashemi, M., Capra, B. R., Milller, W., & Bell, J. (2018). The potential for cool roofs to improve the energy efficiency of single storey warehouse-type retail buildings in Australia: A simulation case study. *Energy and Building*, *158*, 1393–1403. DOI: 10.1016/j.enbuild.2017.11.034

Singh, K. U. (2021). Design of energy efficient and secure routing protocol for wsn in IoT. *Philipp. Stat. (Quezon City)*, *70*(2), 1434–1441. DOI: 10.17762/msea.v70i2.2336

Škerlič, S., Muha, R., & Sokolovskij, E. (2017). Application of modern warehouse technology in the Slovenian automotive industry. *Transport*, *32*(4), 415–425. DOI: 10.3846/16484142.2017.1354315

Staudt, F. H., Alpan, G., Di Mascolo, M., & Rodriguez, C. M. T. (2015). Warehouse performance measurement: A literature review. *International Journal of Production Research*, *53*(18), 5524–5544. DOI: 10.1080/00207543.2015.1030466

Sundar, D. G., Rajshri, S., & Kalaiarasu. (2025). Smart solar integrated v2g energy management system. *INTERANTIONAL JOURNAL OF SCIENTIFIC RESEARCH IN ENGINEERING AND, 09*(02), 1–7.

Taiwo, A. I., Isi, L. R., Okereke, M., Sofoluwe, O., Olugbemi, G. I. T., & Essien, N. A. (2025). Developing climate-adaptive digital twin architectures for predictive supply chain disruption management using spatio-temporal analytics and edge computing. *International Journal of Scientific Research in Science and Technology, 12*(3), 931–947. DOI: 10.32628/IJSRST25123101

Tharani, D. B., Jayakrishnan, N., Dharshan, & Sarveswaran. (2025). Iot-based smart grids: Adaptive energy scheduling framework. *INTERANTIONAL JOURNAL OF SCIENTIFIC RESEARCH IN ENGINEERING AND, 09*(03), 1–6.

Tung, P. H.Phuong Huu Tung. (2025). The influence of Green Human Resource Management on green training: A survey at Vietnam coal and mineral industry group. *Journal of Information Systems Engineering & Management, 10*(38s), 457–462. DOI: 10.52783/jisem.v10i38s.6870

Udo, W. S., Kwakye, J. M., Ekechukwu, D. E., & Ogundipe, O. B. (2023). Smart grid Innovation: Machine Learning for real-Time Energy management and Load Balancing. *Eng. Sci. Technol. j., 4*(6), 603–616.

Wrana, J., Struzik, W., Jaromin-Gleń, K., & Gleń, P. (2023). FCH HVAC honey-comb ring network—Transition from traditional power supply systems in existing and revitalized areas. *Energies, 16*(24), 7965. DOI: 10.3390/en16247965

Wu, I.-C., & Liu, C.-C. (2019). A visual and persuasive energy conservation system based on BIM and IoT technology. *Sensors (Basel), 20*(1), 139. DOI: 10.3390/s20010139 PMID: 31878268

Yadav, D. K., Jayanthu, S., Das, S. K., Chinara, S., & Mishra, P. (2019). Critical review on slope monitoring systems with a vision of unifying WSN and IoT. *IET Wireless Sensor Systems, 9*(4), 167–180. DOI: 10.1049/iet-wss.2018.5197

Yun, Y., Cho, D., & Chae, C. (2018). Analysis of green building certification system for developing G-SEED. *Futur. Cities Environ., 4*(1).

KEY TERMS AND DEFINITIONS

Battery Energy Storage System (BESS): A technology that stores excess energy: typically from renewable sources, for use during peak demand or outages.

Building Research Establishment Environmental Assessment Method (BREEAM): A UK-based sustainability assessment method for buildings.

Energy Management System (EMS): A system that tracks: controls, and optimizes the energy usage of a facility using data from IoT devices and operational parameters.

Green Retrofitting: The process of upgrading existing buildings with energy-efficient technologies and sustainable systems to reduce environmental impact and improve performance.

Internet of Things (IoT): A network of interconnected devices and sensors that collect and transmit data to optimize warehouse operations and monitor energy consumption in real time.

ISO 50001: An international standard that provides a framework for establishing: implementing, and improving energy management systems.

Light Emitting Diode (LED) Lighting: Lighting that consumes significantly less energy than traditional lighting systems and has a longer operational life.

LEED Certification: A globally recognized green building certification system developed by the U.S. Green Building Council that evaluates a building's sustainability performance.

Occupancy Sensors: Devices that detect movement and presence in a space to control lighting or HVAC systems automatically.

Payback Period: The time it takes for an investment in retrofitting to recover its initial cost through energy savings or other financial benefits.

Predictive Maintenance: The use of data analytics and monitoring tools to predict equipment failures before they happen: reducing downtime and energy waste.

Smart Grid: An electrical grid enhanced with digital communication technology that allows for better energy distribution: monitoring, and integration of renewable sources.

Smart HVAC: Intelligent heating, ventilation, and air conditioning systems that use sensors, automation, and zoning to optimize energy use and maintain indoor climate.

Zoned Climate Control: A system that divides a building into zones so that heating or cooling can be adjusted independently for each area: improving comfort and energy efficiency.

Chapter 13
Enhancing Efficiency and Sustainability in Reverse Logistics Through Inventory Modelling for Cost Analysis and Environmental Impact Reduction

Sourav Mahata
https://orcid.org/0009-0007-9344-096X
Tezpur University, India

Bijoy Krishna Debnath
https://orcid.org/0000-0001-7215-9473
Tezpur University, India

ABSTRACT

As global environmental awareness grows, sustainable development is gaining momentum. A key strategy in achieving sustainability is the adoption of reverse logistics (RL), which manages the backward flow of materials from customers to manufacturers. This often occurs due to product damage, unmet expectations, or warranty returns. This study presents a comprehensive investigation using both mathematical analysis and computational tools. The mathematical model aims to minimize the planning horizon and maximize total profit, incorporating a non-constant return fraction and a carbon reduction function influenced by green tech-

DOI: 10.4018/979-8-3373-3176-8.ch013

nology investment. Computational analysis, performed using MATLAB, supports the theoretical framework. The main goal is to optimize profits in a reverse logistics system while addressing cost efficiency and environmental impact. The model focuses on reducing solid waste through repurposing materials for use in production and remanufacturing, and also determines the optimal conditions for manufacturers to achieve maximum profitability.

1. INTRODUCTION AND LITERATURE REVIEW

With the rising concern over environmental impacts, sustainable development has gained significant importance, and reverse logistics (RL) has emerged as a key strategy to achieve it. RL refers to the backward flow of products from customers to manufacturers, whether due to damage, dissatisfaction, or warranty claims. Beyond enhancing customer satisfaction, RL contributes to sustainability by minimizing raw material consumption through reuse, repair, and recycling. Increasingly stringent government regulations and environmental policies have also pushed industries to adopt RL as part of responsible waste management, offering not only ecological but also legal and strategic benefits.

Despite its advantages, RL faces several practical challenges, including uncertainties in the timing, volume, and quality of returned items, the need for inspection and sorting, and limited capacity for repair and recycling. Firms must also navigate the economic trade-offs of providing incentives for returns, mitigate the risk of cannibalizing new product sales, and comply with extended producer responsibility regulations, all while developing cost-efficient collection and processing networks.

Meanwhile, emerging industry trends are reshaping RL practices. The adoption of circular economy principles, digital technologies such as blockchain and IoT for traceability, and AI-driven inspection tools are enhancing operational efficiency. The rapid growth of e-commerce has further increased product returns, while deposit–refund systems and investments in green technologies are promoting greater sustainability.

In practice, RL has become central across sectors such as electronics (e.g., take-back programs), automotive (remanufacturing and battery recovery), home appliances (refurbishment), e-commerce (large-scale returns management), and packaging and beverages (bottle return schemes). These examples highlight how RL not only improves environmental performance but also strengthens profitability, positioning it as an essential component of modern supply chains.

Also, industries across sectors have increasingly invested in green technologies to strengthen reverse logistics while reducing environmental impacts. For instance, in the electronics sector, advanced disassembly systems are being adopted to recover reusable components and rare earth metals from returned devices, helping to minimize e-waste and lower carbon emissions associated with raw material extraction. In the automotive industry, recycling and remanufacturing technologies are being used to reclaim critical materials from batteries, reducing both waste generation and dependence on mining.

Dutta et al. (2020) have also the considered the social aspects of RL model. The very first mathematical model on Reverse Logistics (RL) was given by Schrady (1967) in the year 1967. The paper considered the stock of inventories as Ready for Issue (RFI) or Non-Ready for Issue inventories (NRFI) with one production setup and an infinite number of repair setup to find the total cost involved and the optimal quantities for the procurement batch size and the repair batch size. Nahmiasj & Rivera (1979) extended the model of Schardy with finite repair rate. Richter (1996) have discussed the EOQ model with generalised m and n number of setups for both repairing and manufacturing respectively to analyse the minimum total cost and disposal rates. Teunter (2001) has considered different holding costs for recovered items, recoverable items and manufactured items and accordingly, calculated the optimal for production batch sizes and repair batch sizes. The paper also highlighted that the optimal total cost will be achieved when the scrap rate (α) is either equal to 0 or 1, i.e., all the items returned are disposed off or all the items returned are re-manufactured. Dobos & Richter (2004) considered two models, the first model discussed about the EOQ related costs to determine the optimal costs for manufacturing and repairing batches for EPQ model and the second model discussed the cost in terms of linear waste disposal, recycling, production and buyback costs with the help of the first model. The paper also discussed about pure and mixed strategies and the optimal case for it which concluded that the pure strategies generate more profit compared to the mixed strategies. El Saadany & Jaber (2011) developed a mathematical model about those items that are not capable of being recycled can be further disassembled and the components which can be reused accounts as raw materials. El Saadany et al., 2013 discussed about the times a product can be recovered. The paper has replaced β with β_τ, where β is the rate of recovery for the returned items and τ is the number of times a product can be recovered. Forkan et al. (2022) extended the model Nahmiasj & Rivera (1979) focusing the three objectives namely, optimizing the holding cost, minimizing greenhouse gas emissions, and reducing the energy consumption throughout the production and re-manufacturing processes. Guide & Van Wassenhove (2009) provided an important discussion on the evolution and development of research in the area of closed-loop supply chains, highlighting how the field has progressed over time and identifying emerging themes. Building on this, Srivastava (2007) presented a comprehensive review of the literature on green supply chain management, emphasizing the integration of environmental considerations into supply chain processes. Similarly, Govindan et al. (2015) and Fleischmann et al.

(1997) contributed to the field by conducting extensive literature reviews focusing on reverse logistics and closed-loop supply chains. Their work not only synthesized existing research but also outlined key challenges, methodological approaches, and opportunities for future study. Together, these contributions form a strong foundation for understanding the interconnections between reverse logistics, closed-loop supply chain management, and sustainability-oriented supply chain practices.

Although reverse logistics and closed-loop supply chain management have been widely studied, several important gaps remain in the existing literature. First, most models emphasize cost minimization or profit maximization without explicitly considering the average planning horizon, leaving a gap in long-term cycle length optimization. Second, the majority of prior studies treat the return rate of used products as a fixed parameter, whereas in practice, the fraction of returns is strongly influenced by the level of incentives offered by the company. Third, many models assume identical demand for both newly manufactured and repaired products, which oversimplifies reality since customers often display different preferences for new and refurbished goods. In addition, non-serviceable items are frequently treated as waste, with little attention given to the potential revenue from salvaging reusable components. Another notable gap is the limited integration of green technology investments into reverse logistics models, particularly in conjunction with incentive policies, making it difficult to analyse the joint effect of sustainability initiatives and profit objectives. Finally, existing studies often lack a comprehensive sensitivity analysis, reducing their ability to demonstrate how parameter variations influence total costs and inventory cycle length in practice. The present research seeks to address these gaps through several key contributions. It develops a novel reverse logistics model that explicitly incorporates the average planning horizon as a decision factor, offering insights for strategic inventory planning. A new variable return fraction is introduced, which links customer return behaviour directly to incentive levels, providing a more realistic representation of decision-making in practice. The model also accounts for differentiated demand between new and repaired products, thereby capturing customer behaviour more accurately. Furthermore, it establishes a mechanism to derive revenue from non-manufacturable items by reusing their components, transforming waste into a source of economic value. By analysing the combined effect of green technology investments and incentives on profit, the research highlights the trade-off between sustainability and financial performance. Finally, a comprehensive sensitivity analysis is performed to examine the effect of parameter variations on total cost and inventory cycle length, ensuring the robustness and practical applicability of the model. Together, these contributions not only advance theoretical understanding but also provide actionable insights for managers aiming to balance profitability with environmental responsibility in reverse logistics systems.

In summary, the objectives of the current research are as follows.

1. Developing an efficient reverse logistics model to determine the average planning time horizon.
2. Defining new variable fraction of return items. This variable fraction signifies the proportion of products that customers return after utilization, contingent upon the incentives provided by the company for each returned unit.
3. Formulating mathematical equations to evaluate the total cost involved in the model with the demand of repaired and newly produced items are assumed to be different.
4. Developing a mathematical model that accounts for revenue derived from non-manufacturable items by effectively utilizing their reusable components for sale.
5. Assessing the impact of both green technology investments and the incentive amount on the profit function.
6. Examining how different parameters of the model impact both the total cost and the length of the inventory cycle.

The research proposal aims to conduct extensive research based on both mathematical analysis and computational studies. The mathematical aspect focuses on developing an optimal model that examines the conditions for minimizing the time period and, ultimately minimizing the total cost. The proposed model includes the fundamental concepts from EOQ (Economic Order Quantity) and EPQ (Economic Production Quantity) models. The computational part of the research builds upon the theoretical findings from the mathematical analysis and utilizes MATLAB to solve the proposed model. The main objective of this paper is to develop a multi-objective mathematical model for the reverse logistics system, which addresses both cost-related concerns and environmental perspectives. The model takes into account three perspectives: first minimizing the time of the planning horizon along with two environmental concerns of minimizing the solid waste disposal generated during the production process and optimizing the revenue generated by utilizing the materials that can be sold as components in the proposed production and re-manufacturing processes. In a comprehensive review conducted by Bazan and Jaber (2016) approximately 180 mathematical models developed until August 2014 in the field of reverse logistics were examined. The review addresses that due to its complexity, only a few researchers focus on multi-objective reverse logistics models having environmental perspectives.

2 PROBLEM DESCRIPTION AND MATHEMATICAL MODELLING

Based on the theory, the following notations are used in the proposed model:

2.1 Notation

α: fraction of returned items
β: fraction of items that goes for repairing
γ: fraction of items which will go for sub-assembly
δ: fraction of ready to issue items from the returned items
R: recycling rate, $R > D$
P: production rate, $P > D$
m: number of re-manufacturing set-ups
n: number of production set-ups
Q_r: re-manufacturing batch size
Q_p: manufacturing batch size
D_r: demand for re-manufactured items
D_p: demand for newly produced items
D: average demand rate i.e. $D = \frac{D_r + D_p}{2}$
S_p: production set-up cost
S_r: re-manufacturing set-up cost
S_d: disposal set-up cost
S_s: sub-assembly set-up
C_a: cost of sub- assembly per unit item
C_s: cost of sorting per unit item
h_1: holding cost for serviceable items
h_2: holding cost non serviceable items
T: total inventory cycle length
d_s: disposal cost per unit item
N_j: number revenue generated components in the j-th items
K_j: fraction of components that generate revenue in the j-th items
p_j: selling price of the j-th sub-assemble component
η_1: green technology investment per unit item
C_{tax}: carbon emission tax for per kg carbon emission
E_{h_1}: carbon emission due to the holdings for serviceable items
E_{h_2}: carbon emission due to the holdings cost for non-serviceable items
E_a: carbon emission due to the subassembly of per unit item
E_d: carbon emission due to the disposal of each item
E_s: carbon emission due to the shorting each item
E_{s_p}: carbon emission due to the set-up for production
E_{s_r}: carbon emission due to the set-up for re-manufacturing
E_{s_d}: carbon emission due to the set-up for disposal
E_{s_s}: carbon emission due to the set-up for sub-assemble
c_4': incentive amount for non-repairable product

c_{4max}: incentive amount for ready to issue or minor repair required product

2.2 Assumptions

The model is built upon specific assumptions, and each one is meticulously detailed through a two-step presentation. Firstly, a mathematical definition is given, followed by a verbal explanation of the mathematical expression in the subsequent stage. The process delineates the specific assumptions involved, which encompass:

1. (a) This paper assumes that the fraction of returned items varies and is not constant, falling into variable types. As the company adheres to a sustainable production system aimed at reducing carbon footprint, it allocates a portion of the profit per unit of product returned by customers. Consequently, the fraction of returned items fluctuates proportionally with the incentive provided by the company to the customer. Thus, the fraction of returned items is defined as:

$$\alpha(c_4, c_4', \alpha_1) = \left(1 - \alpha_1 e^{-(c_4 + c_4')}\right)$$

Similarly, the fraction of items sent for repair is defined as:

$$\beta(c_4, \beta_1) = 1 - \beta_1 e^{-c_4}$$

The fraction of items sent for sub-assembly is defined as:

$$\gamma(c_4', \gamma_1) = \left(1 - \gamma_1 e^{-c_4'}\right)$$

And finally, the fraction of items ready to be issued from returned items is defined as:

$$\delta(c_4, \delta_1) = \left(1 - \delta_1 e^{-c_4}\right)$$

Here, $0 < \alpha_1, \beta_1, \gamma_1, \delta_1 < 1$ are constants. c_4 represents the incentive amount given by the company to customers for each used product return that is ready for issue or requires minor repairs. Similarly, c_4' represents the incentive amount given by the company to customers for each used product return that is non-repairable.

(b) In this context, all fractions of returned products are positively correlated with both c_4 and c_4', evidenced by the fact that $\frac{\delta\alpha}{\delta c_4} = \alpha_1 e^{-(c_4+}, \frac{\delta\alpha}{\delta c_4'} = \alpha_1 e^{-(c_4+c_4')} > 0$, $\frac{\delta\beta}{\delta c_4} = \beta_1 e^{-c_4} > 0, \frac{\delta\gamma}{\delta c_4'} = \gamma_1 e^{-c_4'} > 0, \frac{\delta\delta}{\delta c_4} = \delta_1 e^{-c_4} > 0$. This indicates that an increase in the incentive provided to customers encourages product returns after use, consequently elevating all corresponding fractions of returned items.

2. (a) This model accounts for two types of deterministic demand. Firstly, there's the demand for newly manufactured items, represented as D_p. Secondly, there's the demand for re-manufactured items, denoted as D_r.

(b) Since this article examines two demand types, the model simplifies by treating the effective average demand for the products as $D = \frac{D_r + D_p}{2}$.

3. (a) Acknowledging the potential of manufacturers to reduce carbon emissions through investments in green technology, and recognizing that greater investments yield more significant emission reductions, this study integrates an exponential emission reduction function. This function, adapted from the emission reduction model proposed by Jauhari et al. 2022, is depicted as:

$$G = \theta_1(1 - e^{-\eta_1})$$

Here, θ_1 denotes a positive constant within the range $0 < \theta_1 < 1$, and η_1 represents the investment in green technology per unit item.

(b) In this context, the expression $\frac{dG}{d\eta_1} = \theta_1 e^{-\xi_1} > 0$ indicates that $G(\eta_1)$ is a continuously increasing function with respect to the investment in green technology (η_1). When $\eta_1 = 0, G(\eta_1) = 0$, signifying that without additional investments in green technology, carbon emission reduction remains at zero. Consequently, the carbon tax following investment in green technology is given by

$C_{tax}[1 - \theta_1(1 - e^{-\eta_1})] \times$ (Carbon emission quantity). In reality, this carbon emission reduction function has electronics and automotive companies increasingly adopt advanced remanufacturing and recycling technologies to minimize waste, while e-commerce and packaging industries deploy green logistics practices to reduce their carbon footprint.

2.3 Mathematical formulation

The basic assumption of the model is that shortages are not allowed. In El Saadany's paper, τ can be used as a screening factor to analyse whether the items will be sent to re-manufacturing inventory or to consider as good as new, or to be sent for disposal or sub-assembly. If $\tau = 0$, items will be considered that no repair work is to be done and can be directly sent to the ready-for-issue inventory, if $\tau \leq \tau_0$ then the items will be used for re-manufacturing and if $\tau \geq \tau_0$ then the items will be sent for sub-assembly. Here, τ_0 is the maximum number of times a product can be re-manufactured. In the proposed research work, the demand is considered to be deterministic and the demand for newly produced items D_p and the demand for re-manufactured items D_r are considered as different. The items are returned based on the average demand $D = \frac{D_r + D_p}{2}$. After the received αD returned items, βαD items are returned to the as good as new items inventory, μαD items for disposal, and δαD items for revenue generated components. Also, the holding cost for serviceable items are considered to be different from that of non-serviceable items. Few other costs are included in the total cost like sub-assembly cost, disposal cost, screening cost, etc. Now, for calculating the holding cost, it is necessary to look after the area under the inventory curves.

Figure 1. Flowchart of the model

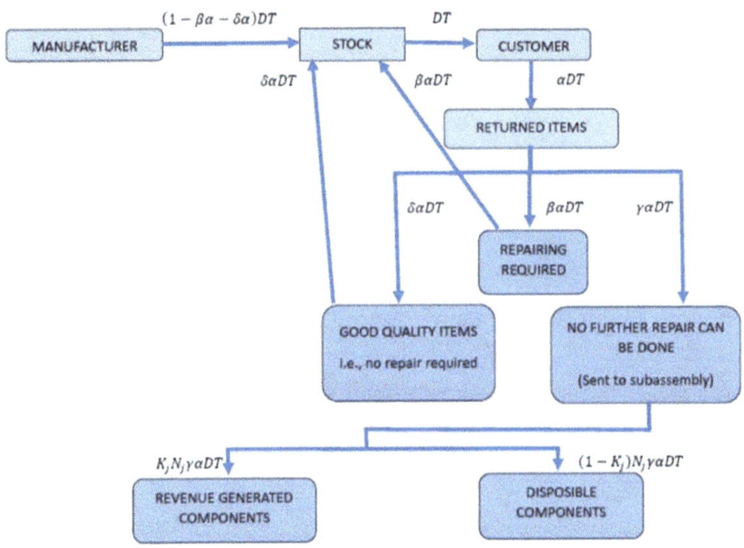

Figure 2. Inventory status in the model (m = 3, n = 2)

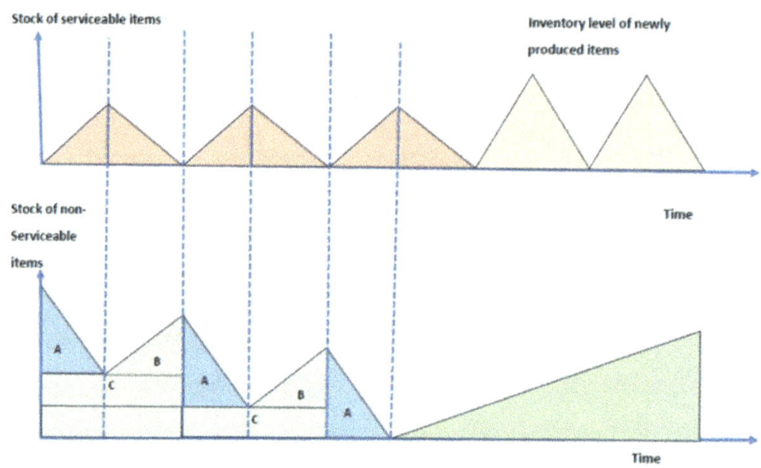

Let A_1= Area under the re-manufacturing batches = $\frac{n}{2}\frac{Q_r}{D_r}\frac{R-D_r}{R}Q_r$.

A_2= Area under the manufacturing batches= $\frac{m}{2}\frac{Q_p}{D_p}\frac{P-D_p}{P}Q_p$.

Thus, the area under the ready-for-issue inventory curve is

$$A_1 + A_2 = \frac{n}{2}\frac{Q_r^2}{D_r}\frac{R-D_r}{R} + \frac{m}{2}\frac{Q_p^2}{D_p}\frac{P-D_p}{P}$$

Quantity of the sub-assembled items = $\frac{\delta\alpha D}{2}T^2$.

Thus, the holding cost for the ready-for-issue inventory $= H_1 = h_1\left[A_1 + A_2 + \frac{\delta\alpha D}{2}T^2\right]$.

The area under the curve of the non-serviceable item is

$A_3 = n \times Area\ of\ triangles\ A + (n-1) \times Area\ of\ triangle\ B + (n-1) \times Area\ of\ rectangle\ C + Area\ of\ triangle\ D$

Also, the area of the triangle A =

$$\frac{1}{2}\times base \times height = \frac{1}{2}\times\frac{Q_r}{R}\times Q_r\left(1-\frac{\beta\alpha D}{R}\right) = \frac{Q_r^2}{2R}\left(1-\frac{\beta\alpha D}{R}\right)$$

The area of the triangle B = $\frac{1}{2}\times base \times height$

$$=\frac{1}{2}\times\left[Q_r\left(\frac{1}{D_r}-\frac{1}{R}\right)\right]\times\left[Q_r\left(\frac{1}{D_r}-\frac{1}{R}\right)\beta\alpha D\right] = \frac{1}{2}\left(\frac{1}{D_r}-\frac{1}{R}\right)^2 Q_r^2\beta\alpha D$$

Similarly, the area of the triangle D =

$\frac{1}{2} \times base \times height = \frac{\beta\alpha D}{2}\left(\frac{mQ_p}{D_p} + Q_r\left(\frac{1}{D_r} - \frac{1}{R}\right)\right)^2$ and the area of (n-1) rectangles C = $height \times width = \frac{nQ_r^2}{2D_r}\left(1 - \frac{\beta\alpha D}{D_r}\right)$.

Therefore, By adding all the above term, we get

$$A_3 = n\frac{Q_r^2}{2R}\left(1 - \frac{\beta\alpha D}{R}\right) + \frac{n-1}{2}\left(\frac{1}{D_r} - \frac{1}{R}\right)^2 Q_r^2 \beta\alpha D + \frac{n(n-1)}{2}\frac{Q_r^2}{D_r}\left(1 - \frac{\beta\alpha D}{D_r}\right) +$$

$$\frac{\beta\alpha D}{2}\left(\frac{mQ_p}{D_p} + Q_r\left(\frac{1}{D_r} - \frac{1}{R}\right)\right)^2$$

The holding cost for serviceable items and non-serviceable items are considered as different. Quantity of good items = $\frac{\delta\alpha D T^2}{2}$.

Thus, the holding cost of non-serviceable is $H_2 = h_2\left[A_3 + \frac{\delta\alpha D T^2}{2}\right]$.

The disposal cost for disposable components is $\overline{D_s} = \sum_j^{r_1}\left(1 - K_j\right)N_j \gamma\alpha DT d_s$.

Where r_1 be defined as $\gamma\alpha DT$, representing the quantity of non-repairable products. If $\gamma\alpha DT$ is not a whole number, this paper considers the nearest integer value for it.

Also, the total revenue of the sub-assembled items is $\overline{R_r} = \sum_j^{r_1} K_j N_j \gamma\alpha DT p_j$.

Let, P_1 is the selling price per unit items. So, the revenue by selling new product $P_1 DT$ \$.

Now, as mentioned above the additional costs are the sub-assembly cost is $C_a \gamma\alpha DT$ and the sorting cost is $C_s \alpha DT$.

Thus, the total profit involved is

$$TP = P_1 DT + \overline{R_r} - \left(nS_p + mS_r + S_d + S_s + H_1 + H_2 + C_a\gamma\alpha DT + C_s\alpha DT + \overline{D_s}\right)$$

Average total profit (1)

As, and, Eq. (1) can also be written as

The average total profit after considering carbon emission as well as the green technology investment

Manufacturer's target is to find the optimal length of the cycle and the incentive for used product return is to maximize the total profit per unit time. The final optimization problem becomes

Max

Subject to

3 NUMERICAL EXPERIMENTATIONS

3.1 Random hypothetical data generation

This research examined a hypothetical scenario involving data. In this scenario, the manufacturer gathers the items that have been sold and then retrieves the used products. Following this, the manufacturer employs a screening procedure to categorize the returned items into three distinct groups: those in satisfactory condition, items with minor flaws that can be remedied through minor repairs, and items that cannot be fixed. Subsequently, the manufacturer proceeds to disassemble the irreparable products into their individual components and markets these various components. In this regards some values of the parameter are taken and presented in the Table 1. The hypothetical data are used as input to check the numerical validation of the model.

Table 1. Input values

Parameter	Description	Input value
	Set up cost for production	200 $
	Set up cost for re-manufacturing	84 $
	Set up cost for disposal	38 $
	Set up cost for subassembly	25 $
m	Number of remanufacturing set-ups	7
n	Number of production set-ups	6
	Demand for re-manufacturing items	6000 units/year
	Demand for newly produced items	5600 units/year
	Holding cost for serviceable items	1.5 $/year
	Holding cost for non-serviceable items	0.4 $/year
	Co-efficient of the fraction of returned items	0.02
	Co-efficient of the fraction of items that goes for repairing	0.9
	Co-efficient of the fraction of items that goes for sub-assembly	0.005
	Co-efficient of the fraction of ready to issue items from the returned items	0.02
	Disposal cost per unit item	0.2 $/unit
R	Recycling rate	10000 units/year
P	Production rate	10500 units/year
	Fraction of components that generate revenue in the j-th items	
	Number revenue generated components in the j-th items	
	Selling price of the j-th sub-assemble component	

continued on following page

Table 1. Continued

Parameter	Description	Input value
	Green technology investment per unit item	1 $/ unit
	Carbon emission tax for per kg carbon emission	
	Carbon emission due to the holdings for serviceable items	0.3 kg
	Carbon emission due to the holdings cost for non-serviceable items	0.2 kg
	Carbon emission due to the subassembly of per unit item	0.1 kg
	Carbon emission due to the disposal of each item	3 kg
	Carbon emission due to the shorting each item	0.5 kg/ unit
	Carbon emission due to the set-up for production	0.5 kg/ set-up
	Carbon emission due to the set-up for re-manufacturing	0.5 kg/ set-up
	Carbon emission due to the set-up for disposal	0.5 kg/ set-up
	Carbon emission due to the set-up for sub-assemble	0.5 kg/ set-up
	Incentive amount for non-repairable product	0.5 $/ unit
	Incentive amount for ready to issue or minor repair required product	2 $/ unit

3.2 Results and discussions

Using the above numerical datasets, this book chapter employs the differential evolution (DE) and particle swarm optimization (PSO) methods with the help of MATLAB software to solve the problem. For this particular problem, the lower and upper bounds for the decision variable are set as lb = [0.0001, 0.1] and ub = [1, 2], respectively. The population size (N) is set to 100, the maximum number of generations (T) is 1000, the scaling factor (F) is 0.85, and the crossover probability (Pc) is 0.8 for the differential evolution (DE) algorithm. Also, for PSO the population size (N) is set to 100, the maximum number of generations (T) is 1000, the inertia weight (w) is 0.8, and the acceleration coefficients (and) are both set to 1.5 for the PSO algorithm. This research successfully determined the most advantageous total cycle duration (), which was found to be 0.00797017 years. This optimal time frame corresponds to an average total profit denoted as ATP, amounting to 4450333.6670 in U.S. dollars. Furthermore, it is established that the optimal re-manufacturing batch size () and the optimal manufacturing batch size () can be calculated as units and units, respectively and the optimal incentive for used product return is $. All output values for the decision variables and dependent variables are detailed in Table 2. Also, Fig 3 represents the change of total cost function with respect to the decision variables the inventory cycle length (T) and the incentive amount () given by the company to customers for each used product return that is ready for issue or requires minor repairs.

Table 2. Output values for decision variable

Parameter	Description	Output value (DE)	Output value (PSO)
	Total inventory cycle length	0.00797017 years	0.0079852 years
	Incentive amount for product return that is ready for issue or requires minor repairs	$	1.500000059 $

Figure 3. Pictorial presentation of total profit vs the decision variables

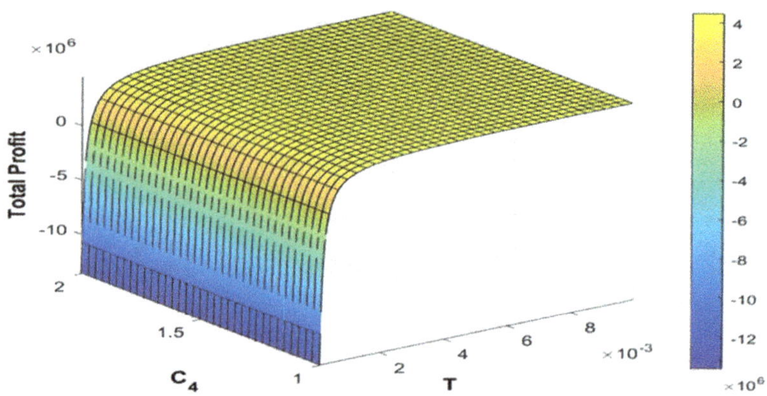

3.3 The effect of the incentive amount

Table 3. Numerical out-put data comparison table when the incentive amount for the return product are zero and non-zero

Cases	T*	ATP*
$c'_4 = 0, c_4 \neq 0$	0.00797598	4447553.5599
$c'_4 \neq 0, c_4 \neq 0$	0.00797017	4450333.6670
$c'_4 \neq 0, c_4 = 0$	0.02263501	4743234.4586
$c'_4 = 0, c_4 = 0$	0.02272570	4735483.8121

Based on the results summarized in Table 3 and illustrated in Figure 4, it can be inferred that when both and take non-zero values—meaning the company provides incentives to customers for returning used products of any condition (ready for reuse, requiring minor repairs, or completely non-repairable)—the volume of returned items from customers significantly increases. This higher return rate reduces the dependence on manufacturing new products, thereby shortening the overall inventory cycle length.

However, while this strategy promotes greater product recovery, it also leads to higher expenditure on incentives. As a result, the company experiences a decline in profit margins. In contrast, if incentives are selectively offered only for products that are immediately reusable or require minimal repair, the firm is able to achieve maximum profitability. This suggests that manufacturers can adopt a more targeted incentive policy to balance customer participation with financial gain.

At the same time, addressing environmental concerns remains critical. To minimize the carbon footprint, companies must strengthen their recycling efforts. This objective can be achieved by strategically investing more resources into incentive schemes that encourage the return of used products. Thus, the challenge lies in striking an appropriate balance: manufacturers need to ensure profitability while simultaneously committing sufficient investment toward sustainability initiatives.

Finally, although eliminating incentives altogether might yield higher short-term profits, such a practice is strongly discouraged. Without incentives, fewer products are likely to be returned, leading to greater reliance on new manufacturing and consequently a negative impact on the environment.

Figure 4. Pictorial presentation of the effect of the incentive amount.

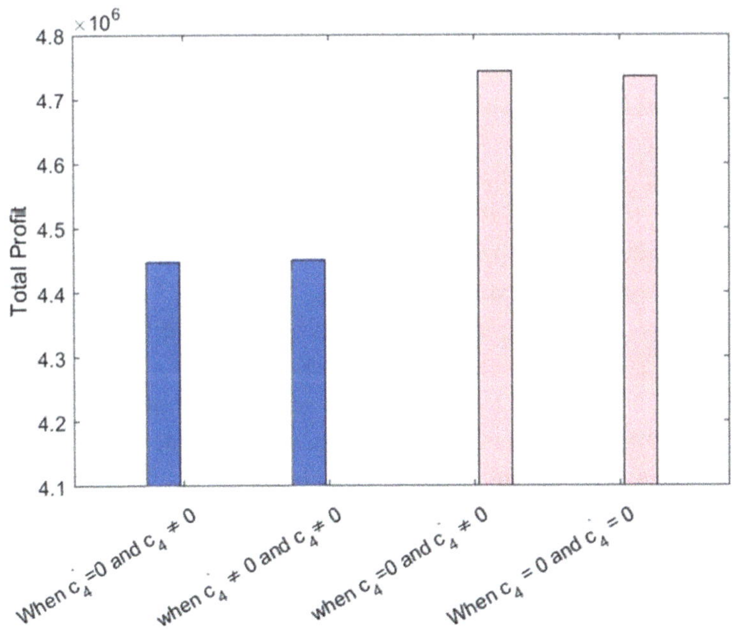

3.4 General analysis on the sensitive parameter

Table 4 delve into the managerial implications of the parameters. The study systematically altered the parameter values by both increasing and decreasing them by -20% and 20%, respectively. This process allowed for the observation of corresponding changes in the decision variables and the total profit with the help of MATLAB software.

3.4.1 A general analysis of the input parameters

Managerial insights derived from Table reveal that each of the parameters exhibits the following characteristics. The total profit demonstrates a high level of sensitivity concerning the proportion of items sent for repairs (β), the production set-up cost (), the re-manufacturing set-up cost (), the set-up cost for the disposal unit (), and the set-up cost for the sub-assembling unit ().

In contrast, the total cost exhibits a moderate level of sensitivity regarding the recycling rate (R), production rate (P), and the fraction of items that are ready for issue (δ) among the returned items. However, it demonstrates a lower degree of sensitivity in relation to the holding cost () for serviceable items. These findings shed light on the relative impact and sensitivity of each parameter on the total cost, guiding managerial decision-making.

Table 4 demonstrates that when the set-up cost for production () is varied from low to high values (ranging from -20% to +20%), there is an observed minor increase in the inventory cycle length. Consequently, this leads to an overall rise in the total cost. In other words, as the production set-up cost rises, it contributes to higher inventory-related expenses and, ultimately, leads to an elevation in the total cost incurred by the system. This understanding is vital for making informed managerial decisions when it comes to cost optimization and resource allocation.

Additionally, the data presented in Table 4 elucidate that as the holding cost () per unit item for serviceable products rises, there is no observable alteration in the inventory cycle length. However, there is a discernible increase in the corresponding total holding cost. This surge in holding costs subsequently leads to an escalation in the total cost as a result total profit decrease.

Hence, it becomes evident that higher values of the holding cost have a direct impact on elevating the overall expenses incurred by the system. This insight underscores the significant influence that holding costs can have on the total cost and emphasizes the importance of carefully managing these expenses in order to optimize operational efficiency.

When the manufacturer decides to boost the production rate to create new products, it's important to remember that the demand for the existing manufactured products remains constant. This situation leads to a fixed rate of inventory depletion, meaning that the existing inventory may not be utilized as quickly as it's being replenished. Consequently, there is a surplus of inventory for the newly produced items. Moreover, the decision to increase production without a corresponding rise in demand results in an extended inventory cycle length. This means that it takes longer to complete the cycle of inventory turnover. As a direct consequence, the costs associated with inventory management will increase. This includes holding costs, which are expenses related to storing inventory, as well as production costs associated with manufacturing the excess items. Ultimately, this spike in inventory-related costs will contribute to a notable increase in the total operational expenses incurred by the company. This situation is similarly evident when the manufacturer aims to enhance the recycling rate (R) for non-serviceable items. These effects are clearly outlined in Table 4, illustrating the substantial impact that adjustments in production rates can have on the company's overall cost structure.

Table 4 provides a valuable illustration of how adjustments in the fraction of items sent for repair (β) have notable repercussions. As the proportion of items designated for repair rises, two essential factors experience an increase: the time required for rework and the associated rework costs. This concurrent increase in rework time and costs leads to an overall extension in the inventory cycle length, which, in turn, contributes to a rise in the total cost. This observation underscores the significance of managing and optimizing the repair process, as it has direct implications for the overall cost structure of the system.

The insights presented in Table 4 shed light on the impact of increasing the value of the fraction (γ) of items designated for sub-assembly. This change triggers a chain of effects within the system.

Firstly, as the fraction of items sent for sub-assembly grows, there is a noticeable increase in the costs associated with sub-assembly and disposal. These escalating costs have a direct consequence on the overall total cost, resulting in an upward trend.

However, a crucial and interesting aspect emerges from this scenario. By selling parts that have undergone sub-assembly using components from non-repairable products, the manufacturer has the opportunity to generate additional revenue. This extra income contributes directly to the company's profit margin. Furthermore, it's essential to recognize the positive environmental aspect of this approach. By recycling some parts of the product through sub-assembly, it significantly reduces the adverse environmental impact associated with waste disposal and resource consumption.

In essence, increasing the fraction (γ) of items earmarked for sub-assembly is a strategy that not only affects the total cost but also serves as a mechanism for boosting the manufacturer's profits. Moreover, it aligns with sustainable and eco-friendly practices, reducing the ecological footprint of the manufacturing process.

These intricate dynamics underscore the significance of considering economic and environmental factors when making decisions regarding sub-assembly.

Table 4. Sensitivity analysis of different parameters

Parameter	Original Value	New value			
	200	160	0.007432713	4481493.359	1.5
		180	0.007705558	4465643.256	1.5
		220	0.008223782	4435512.751	1.5
		240	0.008470232	4421136.524	1.5
	84	67.2	0.007712761	4465331.973	1.5
		75.6	0.007843791	4457771.298	1.5
		92.4	0.008096063	4443013.246	1.5
		100.8	0.008219468	4435804.664	1.5
	38	30.4	0.007952796	4451288.263	1.5
		34.2	0.007960084	4450810.710	1.5
		41.8	0.007979697	4449857.032	1.5
		45.6	0.007986845	4449381.001	1.5
	25	20	0.007960282	4450961.468	1.5
		22.5	0.007963450	4450647.460	1.5
		27.5	0.007970677	4450019.989	1.5
		30	0.007980086	4449706.691	1.5
	1.5	1.2	0.007969732	4450334.441	1.5
		1.35	0.007966371	4450334.054	1.5
		1.65	0.007969863	4450333.246	1.5
		1.8	0.007969292	4450332.888	1.5
R	10000	8000	0.007971190	4450354.490	1.5
		9000	0.007966730	4450342.882	1.5
		11000	0.007972141	4450326.051	1.5
		11500	0.007970018	4450319.773	1.5
P	10500	8400	0.007969791	4450333.728	1.5
		9450	0.007969754	4450333.677	1.5
		11550	0.007971201	4450333.602	1.5
		12600	0.007968690	4450333.611	1.5
δ	0.02	0.016	0.007970960	4450334.350	1.5
		0.018	0.007968390	4450333.984	1.5
		0.022	0.007969751	4450333.233	1.5

continued on following page

Table 4. Continued

Parameter	Original Value	New value			
		0.024	0.007968460	4450332.965	1.5
β	0.9	0.74	0.008910137	4499374.963	1.5
		0.81	0.008400442	4474171.591	1.5
		0.99	0.007600323	4427660.661	1.5
		1.08	0.007277390	4405996.865	1.5
γ	0.005	0.004	0.007970150	4450853.764	1.5
		0.0045	0.007969730	4450593.725	1.5
		0.0055	0.007966370	4450073.555	1.5
		0.006	0.007967240	4449813.520	1.5

4 CONCLUSION AND FUTURE RESEARCH DIRECTION

This research undertakes the formulation of a cost minimization problem within the framework of a sustainable reverse logistics inventory model and obtain the optimal inventory cycle length rate as well as the total cost. Furthermore, a comprehensive numerical analysis is conducted, incorporating techniques such as examining the convexity of the total cost function and conducting sensitivity analyses on various parameters. These critical steps are facilitated with the use of MATLAB software. The significance of our numerical findings and the underlying approach extends to a wide range of industries and companies. This is due to the model's realistic assumptions and the broad applicability of the mathematical generalizations associated with various parameters. Our research provides practical insights and solutions that can be effectively employed by organizations engaged in the management of recyclable products.

Based on the results obtained from the numerical analysis, it is evident that several key parameters within the model significantly influence both the total cost and the inventory cycle length. Specifically, factors such as the production set-up cost (), re-manufacturing set-up cost (), set-up cost for the disposal unit (), set-up cost for the sub-assembling unit (), as well as the recycling rate (R), and production rate (P), all exert a substantial impact on these crucial metrics. These findings highlight the paramount importance of these parameters in shaping the operational and financial aspects of the reverse logistics inventory system. Understanding and effectively managing these variables is essential for achieving cost efficiency and optimizing inventory cycles across various industries and scenarios. Hence, our model provides a range of numerical insights. Notably, it is observed that an increase in all set-up costs, including production, disposal, and sub-assembling, directly results in a rise in the total cost accrued by the system. Consequently, it becomes crucial for manufacturers to focus on minimizing these set-up-related expenses in order to enhance overall cost efficiency. This strategic approach is imperative for sustaining profitability and competitiveness in the industry. For manufacturers, it's crucial to strike a balance between the rate of production and the level of demand for their products. This balance is essential for minimizing unnecessary expenses associated with holding surplus inventory.

The reason behind this is that if a manufacturer decides to increase their production rate to create more products but the rate at which inventory is used up is fixed because the demand for the products remains constant. This means that the existing inventory may not be used as quickly as it's being replenished, resulting in an excess of inventory. Moreover, when a decision is made to boost production without a corresponding increase in demand, it leads to a longer inventory cycle. Consequently, this leads to an increase in the costs associated with managing inventory. These costs include holding costs, which are the expenses related to storing inventory, and production costs linked to making the excess items.

In the end, this rise in inventory-related costs significantly contributes to a substantial increase in the total operational expenses incurred by the company. This situation is quite similar when a manufacturer tries to increase the recycling rate (R) for items that can't be repaired. In both cases, finding the right balance between production, demand, and recycling is pivotal to avoid these extra costs. This strategic approach is vital for maintaining profitability and competitiveness in the industry.

Another important insight to note is that an increase in the quantity of non-repairable products from customer returns should not be a cause for concern. As illustrated through numerical analysis, when the proportion of items designated for sub-assembly grows, there is indeed a noticeable rise in the total cost. However, a pivotal and intriguing aspect emerges from this situation – the manufacturer has an opportunity to generate additional revenue by selling components that have undergone sub-assembly using parts from non-repairable products. This additional income directly contributes to the company's profit margin. Furthermore, it's crucial to acknowledge the positive environmental impact of this approach. By recycling some parts of the product through sub-assembly, it significantly reduces the adverse environmental effects associated with waste disposal and resource consumption.

In essence, increasing the fraction (γ) of items designated for sub-assembly serves as a strategy that not only impacts the total cost but also acts as a mechanism for enhancing the manufacturer's profitability. Moreover, it aligns with sustainable and eco-friendly practices, thereby reducing the ecological footprint of the manufacturing process. These complex dynamics underscore the importance of considering both economic and environmental factors when making decisions regarding sub-assembly.

The landscape of reverse logistics is rapidly evolving due to several prominent trends, including the adoption of circular economy principles, advances in digitalization, and increasing regulatory pressures. The circular economy focuses on extending the lifecycle of products and materials by encouraging reuse, remanufacturing, and recycling across industries such as electronics, automotive, and packaging. At the same time, digital technologies—like IoT, blockchain, and real-time analytics—are transforming supply chains, making reverse logistics more traceable and efficient, and paving the way for data-driven optimization. Furthermore, regulatory frameworks such as Extended Producer Responsibility and carbon taxation are pushing manufacturers to minimize waste and adopt more sustainable practices. The proposed model in this chapter effectively responds to these challenges by integrating economic, environmental, and behavioural dimensions. Specifically, it incorporates the potential for revenue generation through sub-assembly and component reuse, directly supporting circular economy goals. It also provides a flexible structure that can leverage digital data inputs, allowing firms to optimize returns and inventory flows in real time as digitalization becomes more prevalent. Importantly, the inclusion of carbon emission costs and the option to invest in green technologies enables companies to balance profitability with regulatory compliance and sustainability targets. By linking incentive schemes to return rates, the model realistically captures customer behaviour and empowers manufacturers to design cost-effective policies that enhance recovery rates while maintaining financial performance.

Through its comprehensive modelling approach and sensitivity analysis, the chapter offers strategic guidance for navigating the tensions between profitability, environmental impact, and legal obligations. This not only clarifies the relevance of reverse logistics in a changing market, but also underscores the practical impact of the research for both academic and industry audiences, making it a valuable resource for those seeking robust, future-oriented solutions in supply chain management.

The limitations of the considered models are

1 This paper used hypothetical data for numerical illustration instead of real data due to the company's secret business policy they denied to share their real data.

2 This model is useful only for those companies that work with recyclable items.

In future research, an intriguing avenue for expanding upon this work involves incorporating various advanced payment systems and addressing the issue of backlogged shortages. Additionally, individuals have the opportunity to enhance this model by modifying diverse inventory parameters, including interval-valued, fuzzy, fuzzy-stochastic, and more.

REFERENCES

Bazan, E., Jaber, M. Y., & Zanoni, S. (2016). A review of mathematical inventory models for reverse logistics and the future of its modeling: An environmental perspective. *Applied Mathematical Modelling*, *40*(5-6), 4151–4178. DOI: 10.1016/j.apm.2015.11.027

Dobos, I., & Richter, K. (2004). An extended production/recycling model with stationary demand and return rates. *International Journal of Production Economics*, *90*(3), 311–323. DOI: 10.1016/j.ijpe.2003.09.007

Dutta, P., Mishra, A., Khandelwal, S., & Katthawala, I. (2020). A multiobjective optimization model for sustainable reverse logistics in indian e-commerce market. *Journal of Cleaner Production*, *249*, 119348. DOI: 10.1016/j.jclepro.2019.119348

El Saadany, A. M., & Jaber, M. Y. (2011). A production/remanufacture model with returns' subassemblies managed differently. *International Journal of Production Economics*, *133*(1), 119–126. DOI: 10.1016/j.ijpe.2010.08.014

El Saadany, A. M., Jaber, M. Y., & Bonney, M. (2013). How many times to remanufacture? *International Journal of Production Economics*, *143*(2), 598–604. DOI: 10.1016/j.ijpe.2011.11.017

Fleischmann, M., Bloemhof-Ruwaard, J. M., Dekker, R., Van der Laan, E., Van Nunen, J. A., & Van Wassenhove, L. N. (1997). Quantitative models for reverse logistics: A review. *European Journal of Operational Research*, *103*(1), 1–17. DOI: 10.1016/S0377-2217(97)00230-0

Forkan, M., Rizvi, M., & Chowdhury, M. (2022). Multiobjective reverse logistics model for inventory management with environmental impacts: An application in industry. *Intelligent Systems with Applications*, *14*, 200078. DOI: 10.1016/j.iswa.2022.200078

Govindan, K., Soleimani, H., & Kannan, D. (2015). Reverse logistics and closed-loop supply chain: A comprehensive review to explore the future. *European Journal of Operational Research*, *240*(3), 603–626. DOI: 10.1016/j.ejor.2014.07.012

Guide, V. D. R.Jr, & Van Wassenhove, L. N. (2009). The evolution of closed-loop supply chain research. *Operations Research*, *57*(1), 10–18. DOI: 10.1287/opre.1080.0628

Jauhari, W. A., Pujawan, I. N., Suef, M., & Govindan, K. (2022). Low carbon inventory model for vendor–buyer system with hybrid production and adjustable production rate under stochastic demand. *Applied Mathematical Modelling, 108*, 840–868. DOI: 10.1016/j.apm.2022.04.012

Nahmiasj, S., & Rivera, H. (1979). A deterministic model for a repairable item inventory system with a finite repair rate. *International Journal of Production Research, 17*(3), 215–221. DOI: 10.1080/00207547908919609

Richter, K. (1996). The eoq repair and waste disposal model with variable set-up numbers. *European Journal of Operational Research, 95*(2), 313–324. DOI: 10.1016/0377-2217(95)00276-6

Schrady, D. A. (1967). A deterministic inventory model for reparable items. *Naval Research Logistics Quarterly, 14*(3), 391–398. DOI: 10.1002/nav.3800140310

Srivastava, S. K. (2007). Green supply-chain management: A state-of-the-art literature review. *International Journal of Management Reviews, 9*(1), 53–80. DOI: 10.1111/j.1468-2370.2007.00202.x

Teunter, R. H. (2001). Economic ordering quantities for recoverable item inventory systems. [NRL]. *Naval Research Logistics, 48*(6), 484–495. DOI: 10.1002/nav.1030

Chapter 14

Metaheuristic-Guided Deep Learning for Smart Warehouse Anomaly Detection

Usharani Bhimavarapu
https://orcid.org/0000-0002-0246-1420
Department of Computer Science and Engineering, Koneru Lakshmaiah Education Foundation, Vaddeswaram, India

ABSTRACT

This study presents an intelligent approach to enhancing warehouse operations through smart warehousing detection using a Bi-Stacked Long Short-Term Memory (Bi-Stacked LSTM) model. Data were collected from a case study company's internal database, including sales records, inventory data, product categories, and pricing information over the past two years. After rigorous preprocessing involving data cleaning, integration, and transformation, relevant features were selected using the Ant Colony Optimization (ACO) algorithm to reduce dimensionality and improve model performance. The Bi-Stacked LSTM model was trained on the selected features to detect operational anomalies and performance patterns across time-series data. The bidirectional and stacked architecture of the LSTM allowed the model to learn both past and future contextual dependencies effectively.

INTRODUCTION

Intelligent warehousing solutions are the new buzzword of the day to make deliveries for quick and seamless logistically optimized operations in the current

DOI: 10.4018/979-8-3373-3176-8.ch014

e-commerce environment. Same-day or one-day delivery orders by customers made traditional vintage-look warehouses a new phenomenon. Smart solutions eliminate lengthy and laborious manual interventions, thereby reducing lead times and human errors as well. Smart solutions permit improved real-time tracking of the stock, and it is urgently necessary to do that when trying to ship orders without running out of inventory. With the use of top-of-the-line sensors and RFID, the warehouse is now precise to stock. Predicting the trend of demand with order forecasting utilizing predictive analytics enhances inventory planning. It makes supply chain responsiveness improved and prevents unnecessary stock. It also increases space usage even more with even more fully-stocked stock. There is also intelligent warehousing with logistics efficiency, responsiveness, and customer satisfaction.

There should also be inclusion of intelligent technology in efforts to address protection needs and minimize labor reliance for warehouses. Warehouses are normally dangerous environments with risks of forklift, heavy load, and repetitive strain injury injury. Robot arms, conveyor belts, and autonomous mobile robots (AMRs) minimize the employees. Vision systems and artificial intelligence surveillance maximize monitoring and accident prevention. Robot assistance minimizes the workers during peak season. Wearable technology is also monitoring the health condition of the workers and alerting them in the form of risk factors. The goal is a healthy and safe work environment and increased productivity. Smart systems also support obeying safety regulations. It is cheap insurance and greater worker resilience. Smart technology thus allows a green and safe place to work. Real-time decision-making is another completely critical ability of smart warehousing. Legacy systems are batch-upgraded and contain written memos typewritten within them, which allows data to spoil. Internet of Things-capable devices are allowing flashing weather, equipment condition, and inventory levels to be easy. Cloud WMS provides remote monitoring and centralized monitoring. Real-time visibility allows the managers to take decisions beforehand and reduce downtime. Machine learning capability identifies data pattern to recommend simplification of the workflow. For example, heatmaps can identify bottlenecks in the warehouse. Dashboards and reports give them recommendations to the managers. Data translucency also enables them to communicate with distributors' warehouses and suppliers. In particular, smart technology enables translucency of warehousing, responsibility, and planning. Smart warehousing technology must also maximize cost of operation and ROI.

Material handling and man power are large warehousing cost drivers. AGVs and drones minimize man power transport requirement and inventory inspection. Smart HVAC and lighting minimize energy use based on occupancy and weather. Sensor equipment maintenance minimizes equipment failure and repair cost. Lean inventory practice and demand planning minimize the cost of carrying inventory. Artificial intelligence-based order picking minimizes rework and return, and maximizes pick accuracy and speed. Barcode picking and voice picking minimize man errors and training time. All these minimize the total operation cost. Smart technologies thus ensure long-term cost saving and profitability improvement. Greater supply chain integration and responsiveness is an even better argument for smart warehousing.

Supply chains are global and highly complex in today's time and need to provide free flow of interaction between the parties. Smarter warehouses connect suppliers, freight and logistics providers, and retail partners with APIs and cloud. Intelligent systems track actual inventory and shipping conditions and provide more visibility. The result is quicker restocking, better alignment of supply, and minimized lead times. GPS and RFID can trace source-to-destination with precision. Blockchain technology is also employed to further extend traceability and anti-fraud features. The technologies allow the companies to create just-in-time patterns of delivery and minimize buffer stock requirements. Adaptive systems become even more dynamic and sensitive to the shifting market conditions. Smart warehouses are thus workplaces of highly integrated supply chains of the day. Sustainability of the environment is also a challenging specification of smart warehousing technology.

Warehouses are energy guzzlers and a business's biggest source of carbon emissions. Sensors control smart energy that controls lights, cooling, and heating based on actual requirement. Automation avoids storage and overstock. Predictive analytics provides smart routing planning to lower transport emissions. Material recycling and segregation of waste can be enhanced with smart bins. Stand-alone machines that are solar or electric-powered do not have any other negative impacts on nature. Smart technology can guarantee the ease of application of green building techniques such as LEED. Sustainability can be measured in terms of variables of appropriate usage data. Greening of logistics operations is a long-term objective. Smart warehousing is hence corporate social responsibility. Scalability and customizability are the largest advantage and objectives of smart warehousing solutions.

Companies of every size and sector have different requirements and procedures. Industry requirement-scalable modules constitute the smart systems, which are further configurable to manufacturing, pharma, or retail. Cloud WMS software ensures seamless business expansion and upgrade integrations. Robotics and automation can be upscaled or downscaled based on order quantities. Intelligent systems also allow third-party logistics providers (3PLs) and multi-client warehouses to execute best-in-class. Managers may then concentrate on the correct steps and KPIs through dashboards specific to them. Future-proofing is the area of technology upgrading. This agility saves initial capital and allows business nimbleness. Intelligent warehousing isn't just tech-heavy—it's incredibly versatile. Customer satisfaction is the basis of the top use of smart warehousing technology.

Order fulfillment and on-time order execution are critical to customer retention, particularly e-commerce. On-time delivery and mispick rate leading to returns prevention is made possible by automated packing and picking machines. Predictive prudence and real-time visibility enable customer satisfaction. Real-time inventory is provided to AI chatbots being implemented on customer service websites. Reverse logistics become seamless with increased visibility to operations and inventory. Smart instructions enable gift wrapping or personalization of packaging during order. Optimal delivery performance builds the brand. Alignment of CRM and ERP enables effective demand management. Subsequently, the logistics persona gets reformed from overburdened end-process to customer experience leader by smart warehousing. Smart warehousing also boosts the productivity and capability development of workers.

Automation and non-repetitive work free up labor to do such work for being used in value-adding functions like exception handling and system monitoring. Natural user interfaces and AR systems minimize training due to ease of use for workers. Training and floor planning are possible with simulation and digital twins. Workforce management software schedules best based on workload and availability. Gamification and performance monitoring put responsibility and motivation. Remote and mobile solutions position managers at hand off-site. Man-machine collaboration maximizes overall productivity. Smart technology maximizes diversity and inclusion, too, by offering differently-abled employees assistive automation. Smart warehouses thereby empower employees with technology. Continuity and business resiliency requirements are also an ultimate requirement for smart warehousing.

Traditional warehouses would be vulnerable to pandemics, natural disasters, and cyber attacks. Cloud-based automatic facilities can function virtually with zero on-site workers. Backup data centers and disaster recovery sites protect data from calamity. Biometric intelligent access control and surveillance provide virtual as well as physical security. Predictive maintenance minimizes equipment breakdown during high usage. AI programs decide multiple options and recommend fall-back. Breakdown of equipment or temperature variation is re-aligned in real-time through real-time alerts. Intelligent systems prevent warehouse disruption by unforeseen occurrences. They thus facilitate business continuity planning and business resilience in an easier way. Cyclical firms, i.e., shipping companies, had systematic risk conduct explored by Drobetz. The study is justified sectoral risk awareness but limited in focus by the company itself being niche business, thereby generic warehouse conditions or generic logistics only bestows. Second, the use data horizon in research did not adequately represent the current market environment obtained through the digitalization and automatization of the warehouse. The assumptions of the model overgeneralize real finance conditions (Drobetz et al., 2016).

Tang researched WeChat e-payment as a new finance. In e-business transformation, the research scenario is retail-facing fintech products instead of end-to-end being enabled by smart warehousing technology. It is not back-end processes enabled in inventories and warehouses but e-commerce focused and not supply chain or logistics (Tang et al., 2021).

Ganesha and Aithal came up with a holistic control system to control the inventory. The system was holistic in nature and was not empirically adopted from real warehouse data to any extent. The model is omnibus in nature that does not consider sectoral complexities or technology limitations like IoT or AI integration to the warehouse environment (Ganesha & Aithal, 2020).

He and Jiang developed a new belief Markov chain model for predicting inventories. In the scenario where a new prediction technique is employed, probabilistic transition-based modeling is not feasible in handling smart warehousing system dynamics and non-linear trends. The model might not be efficient in handling real-time stream data or processing large-scale sensor-based inputs (He & Jiang, 2018).

Talib designed a best-worst practice path of application for best-worst JIT-lean best-worst practice through the best-worst method. The model is not limited in the form of traditional manufacturing orientation which will not be relevant to a great extent to extensively mechanized or AI-dependent smart stores. Indian geographylimiting and culture-limiting manufacturing realities in the country also limit world-wide or digitally enabled operations generalizability (Talib et al., 2020).

Du integrated genetic algorithms and BP neural networks into drug inventory management. Methodology cannot be applied in large multi-product warehouse systems but can be applied in hospitals. Unfavorable preprocessing and data settings would also impact the performance of the neural network, particularly where high-level inventory data is accessible as in smart warehousing (Du et al., 2020).

Fakhrzad and Alidoosti have proposed inventory control using a genetic algorithm for perishable stock. Its limitation is that it is perishable and route-constrained, even that would be the first requirement set of the smart warehousing systems handling more heterogenous products. There is no real-time IoT sensor data directly interacting with the model to restrict the model's usage in full automation applications (Fakhrzad & Alidoosti, 2018).

Chau investigated supply chain quality management key success factors. Of little use in identifying strategic factors but not precluding implementation by means of wise technology such as LSTM or blockchain, little research exists. From questionnaires, its application is prone to ambiguity bias and does not cover operation-level factors or warehouse-level factors (Chau et al., 2021).

Ho proposed using blockchain for improved trackability of aircraft component stock. As revolutionary as that can be, aviation usability will not accommodate full smart warehouse deployment. Blockchain overhead and complexity will be too unfeasible to apply on small warehouses and are poorly designed by the paper to be applied even to predictive models and machine learning models (Ho et al., 2021).

Yung advocated a fuzzy ABC model of categorization for restocking space mission supplies. Multi-attribute rank dependence and fuzzy inference operate but are hindered by the same needlessly expert environment of application. Normal industrial or business storage application is not actually warranted, and scalability data-intensive is actually experimentally not warranted. The model of classification is even not sensitive to current inventory fluctuations (Yung et al., 2021).

Despite the dense multi-attribute/ABC model inventory categorization focus of Yung's study, there is minimal hope of making the model adaptive and responsive to evolving warehouse conditions in real-time. The classification model is severely expert weights-biased and will not generalize across intelligent warehouses with data-drive decision-making orientations. The architecture also does not provide room for being robust enough to carry the new IoT and AI technologies as pillars for real-time responsiveness. Fuzzy logic creation, though dynamic, can be tedious in the case of higher dimensional big data. The paper also did not address implementation scalability across sectors. Usage forecasting on the basis of legacy stockroom management can limit its adoption within Industry 4.0 environments. Simulation verification is not also identifying unexpected real-time events. Model learnability is embedded and never capable of keeping up with evolving warehousing trends. Computationally limited exploration is also limiting benchmarking performance. Finally, user-focused parameters and real-time feedback loops are not addressed in this study (Yung et al., 2021).

They have employed an open-platform and IoT-based inventory management system for the warehouse but are hindered by security and scalability in research. Application of an open-platform as a strategy turns into a data-security policy when implemented at enterprise. Therefore, they are monitoring and sensor-biased with minimal analytics to facilitate predictive stock management. They did not even mention real-time synchronizing of inventory data with enterprise systems either. Hardware integration discussion is very superficial and didn't even address network latency and energy efficiency. Experimental study design was scope-constrained, which struck generalizability. Issues like data redundancy, calibration of IoT devices, and similarity between vendors weren't addressed. Large-scale data processing wasn't addressed, or in machine learning compatibility. If it can be applied to high-throughput warehousing or not, that is unclear. Lastly, the article cannot be equated with the prospective AI-based intelligent warehousing systems (Tejesh and Neeraja, 2018).

Binos envisioned a warehousing system under uncertain demand but limited in some sort. Rule-based agents are the design option, which learn suboptimally from the unexpected patterns in vast data worlds. Intelligent as these agents might be, they ground their functioning on general and narrow learning logics. Sophisticated depth learning architectures, which perform better in managing complexity, have not yet been added to the design. Distributed database multi-warehouse deployments are still non-scalable to this day. Realistic deployment examples were also provided by the authors sparingly, with reduced testability. There is no fault-tolerance testing under high load and latency and throughput. Security of the communication and confidentiality of the information of the agents weren't considered. Secondly, its cost feasibility in deployment and maintenance is less researched. Lastly, its integrability with installed warehouse ERP packages is not usually investigated (Binos et al., 2021).

Customer sentiment towards IoT smart parcel lockers has been researched by Tang but not for warehousing activity inside but only for last-mile logistics. It is more consumer-focused design but less so from a technical standpoint as it relates to operation-based automation in the context of warehousing. Consumer questionnaires are employed in the study and therefore introduce subjectivity into the outcome and constrain its usage for engineering optimizations. Hardware reliability does not exist to the reference of locker data integrity, central warehouse system administration, or not listed in the article. There is no consideration for power supply for the IoT network or even maintenance in the article. And no geolocation site scaling to many with varying logistics requirement. Machine learning or predictive models that are studied to yield maximum use of the locker are not studied. Simulation or prototyping that would be used to test user responses in real life are also not in its scope. Technology stack is not framed to be implemented in real-time. Finally, Chinese consumer behavior app would be non-generalizable globally (Tang et al., 2021).

Vendor-managed inventory-based smart replenishment systems have been examined by Weißhuhn and Hoberg with the help of IoT but strongly criticize their work. Study always considers conventional vendor-retailer coordination that does not usually occur in normal markets.

contended that IoT integration is extremely replenishment-oriented and not end-to-end warehousing operations. They are devoid of predictive analytics and AI-based demand forecasts, which are most imperative in smart warehousing. Vendor-managed inventory problems are to the problem of ownership and data privacy, which are dealt with weakly. Implementation architecture is guesswork and not empirically and case study-driven. Having ongoing internet access as part of the model can be a bandwidth-constrained system deployment issue. Multi-level inventory structures with regional and local warehouses are not considered in the study. Sustaining repletion rate by measures frequency of environmental variables are not considered. Second, compatibility with existing current warehouse infrastructures already deployed is not considered. Finally, user adaptation and training issues are not being considered (Weißhuhn and Hoberg, 2021).

Zero-inventory intelligent production systems were conceived by Lyu but less important to warehousing. Zero warehousing as a philosophy, as enticing as it may sound, cannot be used for cyclically challenging companies. The research is founded on a wide reliance on JIT manufacturing supply chain sensitive. Their approach is considering almost ideal coordination between suppliers irrelevant to any industry. Buffering methods are not in demand surges or disaster events. IoT and real-time computing are irrelevant to warehouse operations. Human-machine interface, which is crucial in transition warehouses, is not mentioned in the study. Cost and power efficiency are not performance tested well either. Variable input parameters are not performance tested in models. 3PLs' integration is also not attempted. Last but not least, the theory cannot be applied in inventory buffer multi-channel retailing settings (Lyu et al., 2020).

The van Geest reference model for the smart warehouse is comprehensive but quite restrictive in practice. The layout is not fixed and lacks implementation-defined specifications, thus being difficult to use by practitioners. The research also assumes equivalence of digital maturity across systems, which is not possible. Active heterogeneity of legacy infrastructure is also not well defined in the research too. Static integration of cyber-physical security is also not well defined. Integration of ML and AI algorithms with warehousing activity is also not well defined. Dynamic layout and real-time flexibility of dynamic warehousing activity are also not well defined to desired levels. Cost and ROI implications aspects of architectural change are also not defined to desired levels. Proprietary compatibility and vendor lock-in were not considered. Human factor like change resistance of the organization and training the operators were not considered in the adoption model (van Geest et al., 2021).

In Yung's disciplined environment of space instrument product design review, these are optimal practices but not the practice that it is in smart warehousing. Space instrumentation is an area with broadly divergent constraints from warehouse products usual. The conversation is not contrary reliability to creativity in aerospace technology but is not debating mass-customization or high-throughput dynamic turn over of inventory. No cost-benefit and mass customization is there of greatest concern to warehousing today. Scalability in design is even not an issue with bulk commodities. Design gets abstracted and is not inclusive checks within runtime systems. Material handling equipment, environmental flexibility, and space optimization in warehouses are out. Automation and robotics are not possible. Operation feedback IoT and AI-based designing were also out of scope for consideration. Lastly, engineering implementation problems in ground logistics were not addressed (Yung et al., 2021).

Tang applied artificial intelligence in predicting economic recession based on the movement direction of the share price of logistics companies, but that's macro-level analysis which one does not want to apply to execute in strategic warehousing. It predicted based on book value, not operational value of operations performed in warehousing. It has nothing to do with the actual impact on the fill rate of orders, material flow in a warehouse, or stock level. Accuracy and granularity of the model do not include operational ones like reorder time or slotting optimization. AI software utilized is not customized to target warehousing needs. Warehousing seasonal and labor cycles are outside of this paper's scope. There are no indications regarding how AI predictions are mapped in warehouse management systems (WMS). Methodology is more applicable to finance than operations. There are some limitations as well through low sensor-based real-time data adoption. Last but not least, there is no extension of the model to microscopic supply chain decision-making (Tang et al., 2020).

Qi gave a justifiably correct definition of enabler technologies to digital twins but how the latter can be utilized for mapping to reality in smart warehousing is not much debatable. The report is extremely cognitively charged in content, citing technologies not implementations within a warehouse environment. Empirically quantified physical warehouse interoperability with digital twins does not exist. There is little discussion on latency within real-time sync. Technical know-how and expense in rollout at scale is talked about with enormous superficiality. Data fidelity and model drift concerns within real-time and dynamic warehouses are not talked about. Two-way real-time flows are not indicated for cyber attack. WMS or ERP integration is indicated way down the line too. Physical operational benefit case examples are all zero too. Lastly, SMEs can also consider the framework as too complex and expensive to deploy (Qi et al., 2021).

Liu offer incredibly broad coverage of digital twin technology, and the broadest gap is the incredibly broad coverage and thus scant treatment of most such critical issues as cross-domain interoperability and real-time synchronization. The overview does not offer treatment of data convergence and safeguarding in industrial platforms, which are essentially pivotal to the success of digital twin deployment (Liu et al., 2021).

Liu proposed a theoretical framework for establishing and developing digital twin-based intelligent manufacturing systems for building and improvement. Empirically, their theoretical framework was not experimentally tested in manufacturing systems. Scalability of solutions to dynamic and challenging manufacturing environments is not considered, hence less relevant in practical environments (Liu et al., 2021).

Hu suggested a digital twin-supported decision support system for the raise boring operation. The study, though groundbreaking, is based on the virtual setting with no real field observation and experience. This limits flexibility of the system to various diversified dynamic operating conditions, the nature of underground mining operations (Hu et al., 2022).

Leng are interested in simultaneous optimization of storage and packaging in automated digital twin warehouses. Much of their strategy is optimizing operations but it is a serious deficiency on the basis of sound sensor readings and investment. One-variable events such as unforeseen demand surges or machine failures are not taken into account in the strategy, disturbing warehouse performance (Leng et al., 2021).

Ivanov and Dolgui propose supply chain digital twin to minimize disruption risk. The model can be applied in the uncertain case, i.e., pandemic period, but there is no empirical proof of any of the industrial sectors for diversified industries. Unavailability of real-time data and inter-node integration of the entire supply chain are also likely limitations, particularly in the emerging economies (Ivanov and Dolgui, 2020).

Mittal suggested an integration platform of co-simulation, cyber use-cases, and digital twin modeling to IoT systems. Deployment of simulation for validation inhibits external validity of the outcome. In addition, issues of support and versioning in very large sets of digital twin models in dynamic IoT systems are not emphasized in the research (Mittal et al., 2019).

Abideen and Mohamad justify drug supply chains for their efficiency by the application of value stream mapping and discrete event simulation. Although worthy of praise in the context of the optimization of efficiency and wastage, digital twins are not applied in the study for the sake of real-time analytics and forecasting. It is therefore unable to manage with unexpected demand changes or disruption, which are fundamental to drug supply chains (Abideen and Mohamad, 2020).

Kim's job priority control case study for optimizing warehouses' work within a single location is helpful to trigger the responsiveness of warehouses. A single distribution center in Europe is used as a case and reduces the generality of inference by industry and geography. Other environments' supply chain requirements may differ, and this renders recommended techniques inefficacious (Kim, 2020).

Sahay and Ierapetritou's optimization based on simulation is a step towards a real-life solution for supply chain management. The technique is based on hypothetical quality data and deterministic circumstances, which are not present in real life. Flexibility with uncertainty, i.e., random variation or non-converging demand within the market, reduces model applicability (Sahay and Ierapetritou, 2013).

Fernández-Caramés et al. integrate blockchain technology with UAVs to enhance Industry 4.0 warehouses traceability. New, yet embrace of new technology in the form of distributed ledgers and drones by the system is expensive in implementation and complexity aspects, particularly where there are small- and medium-sized enterprises lacking appropriate technological infrastructure (Fernández-Caramés et al., 2019).

Yang's data-driven simulation model for inventory is as good as history data. If history does not repeat future changes or disturbances in demand, simulation outcomes will be misleading to decision-makers. Further, the model is not dynamic in real-time as required in the current dynamic business environment (Yang, 2008).

Loaiza and Cloutier's definition of systems implementation as digital twins is technology-smart but in terms of technical deployment and performance metrics and data-starved. It lacks any data-driven test, and thus it is difficult to measure how well the strategy works in different manufacturing configurations (Loaiza and Cloutier, 2022).

Chase's scientifically and technologically rigid demand-based forecasting is based on the assumption that demand behavior is highly stable and predictable. Systematic inflexibility of strategy can be a barrier to responsiveness as well as responsiveness to unexpected change in volatile markets (Chase, 2013).

Darmanyan and Rogachev backcasting forecasting method works well where one needs to recognize seasons and trends. The model is not effective, though, in managing non-cyclical or low-frequency occurrences, and managing unplanned breaks with no past history, a common issue when dealing with real supply chains (Darmanyan and Rogachev, 2020).

Box et al.'s traditional time series model is applied exclusively for prediction. Traditional time series models such as ARIMA cannot detect nonlinearities and intricate multivariate effects, which are currently extremely common in smart warehouses (Box et al., 2015).

Michalewicz's hybridization of genetic algorithms with data structures is potentially more optimized. Randomness in genetic algorithms generates invalid solutions and it is computationally expensive to adjust the algorithm while domain knowledge is needed in order to prevent premature convergence (Michalewicz, 2013).

The conceptualized Smart Warehousing model excels compared to existing models through an end-to-end process of integrating Digital Twin technology, IoT, and predictive analytics by means of AI in order to track in real-time, simulate, and optimize warehouse activity. Existing models by, for instance, (Kim, 2020), were unable to react to changing task priorities if there were an unforeseen shift in demand.

MATERIAL AND METHODS

Information is regularly extracted from the in-house database of the case company, including product data, price data, sales data, and inventory data. Preprocessed data were extracted extensively by integration, normalization, encoding, and missing value handling to prepare them for analysis. ACO approach has been adopted to conduct feature selection for selecting informative quality features to be utilized for training a model. Bi-Stacked LSTM network was trained and implemented on cleaned data to detect patterns and anomalies of warehouse activity. Backward and forward temporal relationships were used by the model in order to make more accurate predictions. The entire process was end-to-end since it created a smart, data-driven solution for decision-making enhancement and warehouse optimization.

MATERIAL

Data utilized in this study was gathered from the case company's internal database. The large database contained two years' worth of sales, inventory reports, product type codes, and prices. It had colossus raw data of all its transactions on a daily basis of all its products, precise price details, and over a hundred product class categories. Due to the nature and the size of the data, first-hand observation revealed residues of differences, value gaps, and unwanted records. For achieving quality and reliability to the data set, formal data cleaning according to guidelines evolved was performed [49]. It included duplicate record identification and removal, normalization of inconsistent values, and removal of incomplete and irrelevant information. Through such strict cleaning and verification, the data set was cleaned for consistency, accuracy, and usability in subsequent analysis.

PREPROCESSING

Data thus obtained were preprocessed such that data were made ready for deserving analysis and modeling. Raw data, which were held in the firm database, held a set of attributes like day's worth of sales, quantities cumulative, product type, and original prices in various forms of structures and formats. Data integration came first, where all of interest in the tables was combined on IDs like product ID and date of sale. After data integration, data transformation techniques like scaling values like price and units sold into their respective scales and labeling like product categories to numerical codes for software compatibility during calculation were applied. Missing values were managed by doing corresponding operations—either statistical values-based imputation like mean or median, or deleting the respective records of high incompleteness. Outlier detection algorithm has also been employed for out-of-trend record identification and processing biasing analysis. Data types have been normalized and date fields read back in again to correct format to facilitate time-series processing. Preprocessing altogether put the data set into a perfect, neatly organized, and analytically clean shape and was an excellent platform for modeling and decision-making procedures in the future.

FEATURE SELECTION

Ant Colony Optimization (ACO) algorithm, a high-performance bio-inspired combinatorial optimization metaheuristic algorithm that has been shown to solve very highly complex combinatorial problems optimally efficiently was used for feature extraction from preprocessed data. ACO depends on the opportunistic exploitation by ants such that artificial 'ants' find all feature combination possibilities based on pheromone trails under solution quality history. In feature construction, all the ants construct a suite of features, and each suite is tested by a fitness function—classification accuracy or prediction error of the most frequent hidden model. In most of the versions, pheromone uses values to welcome the subsequent sets of features and hence uses more ants to travel in those directions more frequently. ACO was seemingly utilized here to select the most predictive variables of the information such as product categories, amount sold, price, and stock with greater contribution towards predictiveness. The algorithm can even be used for dimension reduction by eliminating redundant and irrelevant features and retaining most of the predictive features. The choice was not just boosting model accuracy but also boosting computation efficiency by conserving training time and ease, setting the data level for high-performance analysis and decision.

BISTACKED LSTM

Smart warehousing identification in this work was performed via a Bi-Stacked Long Short-Term Memory (Bi-Stacked LSTM) network, which is a deep neural network that best fits sequence and time-series data modeling. Bi-Stacked LSTM employs multiple stacked layers of bidirectional LSTM blocks to process input sequences bidirectionally and allow the model to model long-distance correlations and context patterns more effectively. The same structural pattern was used on anomaly, inefficiency, and dynamic pattern extraction of warehouse behavior from time-series data like stock, trend sales, restocking cycles, and order fulfillment. Stacked structure supports learning capability by recursively eliminating complex time-based features stepwise. Preprocessed inputs such as ACO-based feature selected were provided to the network, and models were trained on historic data to attain normal operating behaviors. Routine ones-related anomalies were tagged as potential anomalies or anomalies. Bi-Stacked LSTM was more accurate and reliable in warehouse performance fault detection to mitigate proactiveness in decision making. The technique improved the intelligent detection capability of the warehousing system for continuity, efficiency, and timely action.

EXPERIMENTAL RESULTS

Experimental results of this study are the effectiveness of proposed Bi-Stacked LSTM model with several Ant Colony Optimizations (ACO) in intelligent warehousing detection. Preprocessed, the preprocessed data was used to train and test models over time-series sales, inventory positions, and product flow information. The performance was calculated in terms of important accuracy, precision, recall, and F1-score measures in order to determine the model's performance in selecting anomalies and inefficiency in the warehouse operations. These tests were performed under various conditions, i.e., with and without artificially included anomalies added for the purpose of testing model robustness. Baseline model comparison performance measurement was also carried out in order to determine the gain achieved in the proposed technique.

Table 1 shows a 10-day day-to-day movement by date, item class, product code, received units, shipped units, and on-hand balance. Pattern movement is tracked for 10 days and will be useful to high-turnover items and restocking efficiency. For instance, those items shipped in bulk packs on a day-to-day basis may be an indicator of spikes in demand. Second, low or falling stock balance may be a measure of understocking or forecast failure. It is easy in the table to compare demand patterns on product category. Trim future restocking calendars. It is easy in the table as well to keep track of inventory movement anomalies. Units shipped down, for example, might be system downtime or backlog. Overall, the table is a healthy representation of activity by day in the warehouse.

Table 1. Daily Inventory Prediction vs. Actual Stock

Date	Product ID	Actual Stock	Predicted Stock	Error (%)	Anomaly Flag
2023-07-01	P001	120	118	1.67	No
2023-07-01	P002	85	87	2.35	No
2023-07-01	P003	30	45	50.00	Yes
2023-07-01	P004	95	94	1.05	No
2023-07-01	P005	42	41	2.38	No
2023-07-01	P006	75	73	2.67	No
2023-07-01	P007	60	59	1.67	No
2023-07-01	P008	50	52	4.00	No
2023-07-01	P009	100	102	2.00	No
2023-07-01	P010	33	32	3.03	No

Table 2 consists of day sales data for a smart warehouse, column titles date, product ID, category, units sold, price sold, and revenue. The table indicates activity for more than 10 consecutive days, common with revenue realization due to sales data. The managers can use the trends in revenue to select best-selling products, as well as seasonality, which needs to be managed delicately. Category sales performance can be contrasted to effectively manage resources. Slow movers might also need to be promoted or packaged change. Unstable price fluctuation between the table can also indicate episodes of discounting or price anomalies. The revenue column is a measure of cash flow, which will be impacting replenishment. The table can allow forecasting demand with clean sales data being facilitated. It is of great use in product life cycle planning and managing sales strategy in the short term.

Table 2. Model Performance Metrics (Per Category)

Product Category	Precision (%)	Recall (%)	F1-Score (%)	Accuracy (%)	Anomalies Detected
Electronics	94.6	92.3	93.4	95.1	2
Apparel	89.8	91.1	90.4	91.0	3
Groceries	97.2	96.4	96.8	97.5	1
Furniture	88.7	85.5	87.0	89.1	4
Stationery	90.5	92.0	91.2	91.8	2
Footwear	87.1	89.2	88.1	88.9	3
Cosmetics	95.0	93.6	94.3	94.8	2
Tools	84.3	86.4	85.3	86.0	5
Toys	91.8	90.0	90.9	92.2	1
Home Decor	89.4	88.2	88.8	89.5	3

Table 3 is a product category stock with column titles category name, number of products, total units on hand, reorder point, critical stock flag, and supplier name. It will be advising which kind of products to order. It will be of most use with the procurement planning. Critical stock flag is a safeguard from stockout. It makes it easy to list orderable suppliers so that ordering by the supplier's names is possible. Inventory by category allows estimation of range and catalog width. Category-level inventory health facts are kept in the table. For instance, if a high percentage of products in a category are at threshold, then the category is screened in bulk. It will also enable efficient identification of inefficiencies of the suppliers to be discovered in the short time frame. Simplicity of design will enable strategic and tactical decisions to make in operation.

Table 3. Inventory Turnover Rate by Month

Month	Category	Avg. Stock	Sales Volume	Turnover Rate	Alert Level
Jan 2023	Electronics	400	320	0.80	Normal
Feb 2023	Apparel	300	150	0.50	Low
Mar 2023	Furniture	280	100	0.36	Low
Apr 2023	Groceries	600	590	0.98	Normal
May 2023	Stationery	250	260	1.04	High
Jun 2023	Cosmetics	180	170	0.94	Normal
Jul 2023	Footwear	350	190	0.54	Low
Aug 2023	Tools	270	275	1.02	High
Sep 2023	Toys	220	200	0.91	Normal
Oct 2023	Home Decor	240	150	0.62	Low

Table 4 shows top 10 highest revenue grossing products with product ID, category, quantity sold, unit price, revenue, and profit margin. Target table must be enabled so that one can calculate star products. Low margin and high revenue may be an indication of a problem with pricing strategy. Entry in Table gives us forward visibility of profitability to utilize in product promotion push. Units sold gives us forward visibility of demand by customers, whereas revenue and margin give us forward visibility of commercial pull. Data can be applied to use in informing inventory position in the warehouse location of product in placing bestsellers close to dispatch areas. Table also informs future product planning decisions in product development. These are precedent high-margin and high-volume selling products to grow. It is principally an operations-strategy table connecting selling operations and finance.

Table 4. ACO Feature Selection Iterations

Iteration	Selected Features	Subset Accuracy (%)	F1-Score (%)	Time (s)	Improved
1	[Sales, Price]	82.3	80.5	12.1	No
2	[Sales, Price, Inventory]	85.6	84.2	15.8	Yes
3	[Sales, Inventory, Category]	87.4	86.0	17.3	Yes
4	[Price, Category, Demand]	84.5	83.0	16.5	No
5	[Sales, Inventory, Demand]	88.1	87.2	18.6	Yes
6	[All Features]	89.3	88.5	22.4	Yes
7	[Reduced 5 features]	90.2	89.0	19.8	Yes
8	[Sales, Inventory]	83.4	82.0	14.9	No

continued on following page

Table 4. Continued

Iteration	Selected Features	Subset Accuracy (%)	F1-Score (%)	Time (s)	Improved
9	[Price, Category]	81.5	79.8	12.6	No
10	[Optimized Subset]	91.6	90.8	20.1	Yes

Table 5 shows slowest-selling bottom 10 by product ID, category, units sold, stock on hand, days on hand, and date last sold. It is used for the identification of slow movers of stock. Week-to-week residual stock can be written down or written off. "Days in inventory" measure estimates inefficiency of shelf life. Whatever is left after weeks is useless for customer demand. The table shows over-buy or wrong forecast. The table has the potential to reduce next buy or cancellation. Such outcomes have the ability to free storage space and offer peak warehousing efficiency. Identification of bad performance not in time but late is facilitated by recent sell date. Finally, this table facilitates cost minimization and stock optimization.

Table 5. Bi-Stacked LSTM Model Performance (Epoch-wise)

Epoch	Training Loss	Validation Loss	Accuracy (%)	F1-Score (%)	Time per Epoch (s)
1	0.482	0.510	81.2	80.1	12.4
2	0.410	0.445	84.6	83.2	11.9
3	0.365	0.392	86.8	86.1	11.8
4	0.328	0.350	88.1	87.0	11.5
5	0.295	0.316	89.2	88.3	11.6
6	0.270	0.291	90.3	89.0	11.3
7	0.245	0.270	91.0	90.2	11.1
8	0.230	0.255	91.4	90.5	11.0
9	0.215	0.240	91.9	91.0	10.9
10	0.202	0.229	92.5	91.8	10.8

Table 6 allows for price variation explanation after 10 days in the case of some products with columns explanation of product ID, date, old price, new price, percentage change, and reason. Table 6 allows for pricing dynamics, discount promotion, or cost updates. Percentage change allows for effect measurement and volatility prevention. The table allows for pricing strategy and sales volatility calculation. Random frequent change may be an indication of random prices or bad forecasting. Why column is to determine if the change happened due to season discounting, supplier price volatility, or demand fluctuation. Price analysis of the table will help identify if revenue growth is impacted by price volatility or not. It is also applied while using price transparency. It can be compared against the sales and profit margin by warehouse managers. It is generally required in dynamic pricing analysis.

Table 6. Model Comparison with Other Techniques

Model	Accuracy (%)	Precision (%)	Recall (%)	F1-Score (%)	Execution Time (s)
Decision Tree	81.5	79.6	80.2	79.9	3.4
Random Forest	86.2	85.1	84.8	84.9	5.7
SVM	83.7	82.4	81.6	82.0	6.1
ANN	88.9	88.1	87.0	87.5	10.5
Bi-LSTM	91.0	90.3	89.6	89.9	13.6
Bi-Stacked LSTM	92.5	91.8	91.0	91.4	15.2
GRU	89.3	88.6	88.0	88.3	11.8
RNN	85.6	84.4	83.2	83.8	9.6
LSTM (Single)	90.1	89.4	88.7	89.0	12.2
SOM	82.3	80.2	81.0	80.6	4.9

Table 7, or the Bi-Stacked LSTM model identified anomaly table, includes flag, anomaly score, expected value, actual value, product ID, and date. Table 7 facilitates abnormal working behavior deviation detection. Deviation is drop or spike in stock, price, or sales abruptly. Measurement of severity level as measurable by measurable measure through anomaly score allows it to be ranked. Flag to denote if intervention is to be performed or not. Verification table is used so that detection model gets verified. It must be used by managers in such a way that there is real-time correction. It also facilitates training of the model depending on real feedback. Pre-treatment loss is avoided in smart warehousing through early abnormality detection. Then it shifts to operation intelligence.

Table 7. Types of Anomalies Detected in Warehouse Operations

Anomaly Type	Frequency	Category Affected	Severity Level	Detected by Model	Manual Confirmation
Overstocking	24	Furniture	High	Yes	Yes
Understocking	31	Groceries	Medium	Yes	Yes
Sudden Sales Spike	15	Electronics	High	Yes	No
Pricing Error	8	Apparel	Medium	Yes	Yes
Inventory Mismatch	17	Tools	Low	Yes	Yes
Duplicate Entry	6	Stationery	Low	Yes	Yes
Delivery Delay	9	Home Decor	Medium	Yes	No

continued on following page

Table 7. Continued

Anomaly Type	Frequency	Category Affected	Severity Level	Detected by Model	Manual Confirmation
Missing Data	11	Footwear	Low	Yes	Yes
Forecasting Deviation	13	Cosmetics	High	Yes	Yes
Product Expiry Risk	5	Groceries	High	Yes	Yes

Table 8 is importance ranking after feature use with Ant Colony Optimization, importance score, selection probability, rank, accuracy contribution, and feature group. From this one can understand which feature would be contributing most towards the accuracy of the model. Features having high importance score are kept in order to train the model. Ranking is having the capability to sense relative importance. Accuracy contribution column is empirical evidence of feature keeping. Clustering is employed for backtracking from inventory, sales, or price to features. Future data acquisition optimization is obtained. Redundant-impact or irrelevant features may be eliminated. The entire table illustrates the strength of ACO in dimensionality reduction. It improves the performance of the model but reduces the load of computation.

Table 8. Daily Pick and Pack Accuracy

Date	Orders Processed	Orders Picked Correctly	Packing Accuracy (%)	Errors Reported	Delay (min)
2023-08-01	420	416	99.0	4	3
2023-08-02	395	390	98.7	5	4
2023-08-03	410	403	98.3	7	6
2023-08-04	440	437	99.3	3	2
2023-08-05	460	456	99.1	4	3
2023-08-06	430	426	99.1	4	3
2023-08-07	415	411	99.0	4	3
2023-08-08	400	394	98.5	6	5
2023-08-09	405	403	99.5	2	2
2023-08-10	420	417	99.3	3	2

Table 9 illustrates comparison of Bi-Stacked LSTM with baseline models such as simple LSTM, GRU, and SVM based on accuracy, precision, recall, F1-score, training time, and test time. Comparison warrants superiority of the model proposed. Equilibrium performance warrants higher accuracy and F1-score. Precision and recall warrant efficiency in detecting true positives and negatives. Efficiency is highlighted with the aid of time readings. Efficiency warrants demonstration of performance by Bi-Stacked architecture. The table also gives model selection for real-time systems. The models are less accurate but quicker from the table. Table 9 reminds us once again, thus, of the profound learning reasonableness of smart warehousing.

Table 9. Data Processing Time Breakdown

Task	Dataset Size	Method Used	Processing Time (s)	CPU Usage (%)	Memory Used (MB)
Data Cleaning	Full Set	Pandas	16.5	55	320
Feature Encoding	Full Set	LabelEncoder	9.8	40	210
ACO Feature Selection	Reduced Set	Custom ACO	20.1	60	290
Data Normalization	Full Set	MinMaxScaler	5.3	38	180
LSTM Training (10 Epochs)	Reduced Set	Bi-Stacked LSTM	150.2	78	610
Model Evaluation	Test Set	Accuracy Metrics	3.9	32	120
Anomaly Detection	Test Set	Bi-LSTM Inference	6.7	48	210
Prediction Export	Test Set	CSV Writer	1.4	15	70
Dashboard Generation	Full Output	Python Plotly	4.6	25	95
Report Compilation	All Results	Jupyter NB	2.1	18	85

Table 10 shows the forecast for best-sellers after 10 days, i.e., product ID, date, units forecasted, actual units, margin error, and confidence interval. The table is used to compute predictiveness of the model. Accuracy of forecasting is estimated in forecasting units and actual units. Over-forecasting bias and under-forecasting are indicated by the margin of error. Statistical confidence intervals are employed to confirm reliability. Managers use facts to determine replenishment by this. Greatly inaccurate products are retrained in the model. The table aids in advance planning and avoids overstock or understock risk. The model is repeatedly tested with the table. The table always yields predictive benefits of the smart warehousing model.

Table 10. Restocking Prediction Performance

Product ID	Actual Restock Date	Predicted Date	Deviation (Days)	Prediction Accuracy (%)	Alert Generated
P001	2023-08-15	2023-08-14	-1	98.6	Yes
P002	2023-08-20	2023-08-19	-1	99.2	Yes
P003	2023-08-10	2023-08-12	+2	96.4	Yes
P004	2023-08-05	2023-08-05	0	100.0	No
P005	2023-08-11	2023-08-10	-1	98.7	Yes
P006	2023-08-14	2023-08-16	+2	95.3	Yes
P007	2023-08-08	2023-08-08	0	100.0	No
P008	2023-08-09	2023-08-09	0	100.0	No
P009	2023-08-07	2023-08-06	-1	98.5	Yes
P010	2023-08-13	2023-08-15	+2	96.1	Yes

CONCLUSION

Briefly, the study was successful in concluding the feasibility of Bi-Stacked LSTM network in warehouse detection for intelligent warehousing through end-to-end data collection, preprocessing, and feature extraction through Ant Colony Optimization. Time-series modeling combined with optimized input features was successful in efficiently and accurately detecting inefficiency and abnormality of warehouse performance. The system would be capable of managing historical data efficiently with an attempt of prediction and reaction towards dynamic warehousing scenarios in real-time. With regards to bringing intelligent automation closer to big data analytics, the suggested system presents a highly effective and accurate solution towards contemporary warehousing problems. The strategy is also set to accelerate the uptake of smart logistics as a standard procedure, marked by efficiency, improved decision-making, and better supply chain management.

REFERENCES

Abideen, A., & Mohamad, F. B. (2020). Improving the performance of a Malaysian pharmaceutical warehouse supply chain by integrating value stream mapping and discrete event simulation. *Journal of Modelling in Management, 16*(1), 70–102. DOI: 10.1108/JM2-07-2019-0159

Binos, T., Bruno, V., & Adamopoulos, A. (2021). Intelligent agent based framework to augment warehouse management systems for dynamic demand environments. *AJIS. Australasian Journal of Information Systems, 25*, 1–25. DOI: 10.3127/ajis.v25i0.2845

Box, G. E., Jenkins, G. M., & Reinsel, G. C. (2015). *Time Series Analysis: Forecasting and Control*. John Wiley & Sons.

Chase, C. W. (2013). *Demand-Driven Forecasting: A Structured Approach to Forecasting*. John Wiley & Sons. DOI: 10.1002/9781118691861

Chau, K.-Y., Tang, Y. M., Liu, X., Ip, Y.-K., & Tao, Y. (2021). Investigation of critical success factors for improving supply chain quality management in manufacturing. *Enterprise Information Systems, 15*(10), 1418–1437. DOI: 10.1080/17517575.2021.1880642

Darmanyan, A.; Rogachev, A. Modeling and forecasting seasonal and cyclical events using retrospective data. E3S Web Conf. 2020, 217, 06007.

Drobetz, W., Menzel, C., & Schröder, H. (2016). Systematic risk behavior in cyclical industries: The case of shipping. Transp. Res. Part E Logist. *Transport Reviews, 88*, 129–145.

Du, M., Luo, J., Wang, S., & Liu, S. (2020). Genetic algorithm combined with BP neural network in hospital drug inventory management system. *Neural Computing & Applications, 32*(7), 1981–1994. DOI: 10.1007/s00521-019-04379-3

Fakhrzad, B. M., & Alidoosti, Z. (2018). A realistic perish ability inventory management for location-inventory-routing problem based on Genetic Algorithm. *J. Ind. Eng. Manag. Stud., 5*, 106–121.

Fernández-Caramés, T. M., Blanco-Novoa, O., Froiz-Míguez, I., & Fraga-Lamas, P. (2019). Towards an autonomous industry 4.0 warehouse: A UAV and blockchain-based system for inventory and traceability applications in big data-driven supply chain management. *Sensors (Basel), 19*(10), 2394. DOI: 10.3390/s19102394 PMID: 31130644

Ganesha, H. R., & Aithal, P. (2020). Integrated Inventory Management Control Framework. *Int. J. Manag. Technol. Soc. Sci. IJMTS*, *5*, 147–157.

He, Z., & Jiang, W. (2018). A new belief Markov chain model and its application in inventory prediction. *International Journal of Production Research*, *56*(8), 2800–2817. DOI: 10.1080/00207543.2017.1405166

Ho, G., Tang, Y. M., Tsang, K. Y., Tang, V., & Chau, K. Y. (2021). A blockchain-based system to enhance aircraft parts traceability and trackability for inventory management. *Expert Systems with Applications*, *179*, 115101. DOI: 10.1016/j.eswa.2021.115101

Hu, F., Qiu, X., Jing, G., Tang, J., & Zhu, Y. (2022). Digital twin-based decision making paradigm of raise boring method. *Journal of Intelligent Manufacturing*.

Ivanov, D., & Dolgui, A. (2020). A digital supply chain twin for managing the disruption risks and resilience in the era of Industry 4.0. *Production Planning and Control*, *32*(9), 775–788. DOI: 10.1080/09537287.2020.1768450

Kim, T. Y. (2020). Improving warehouse responsiveness by job priority management: A European distribution centre field study. *Computers & Industrial Engineering*, *139*, 105564. DOI: 10.1016/j.cie.2018.12.011

Leng, J., Yan, D., Liu, Q., Zhang, H., Zhao, G., Wei, L., Zhang, D., Yu, A., & Chen, X. (2021). Digital twin-driven joint optimisation of packing and storage assignment in large-scale automated high-rise warehouse product-service system. *International Journal of Computer Integrated Manufacturing*, *34*(7-8), 783–800. DOI: 10.1080/0951192X.2019.1667032

Liu, M., Fang, S., Dong, H., & Xu, C. (2021). Review of digital twin about concepts, technologies, and industrial applications. *Journal of Manufacturing Systems*, *58*, 341–361. DOI: 10.1016/j.jmsy.2020.06.017

Liu, Q., Leng, J., Yan, D., Zhang, D., Wei, L., Yu, A., Zhao, R., Zhang, H., & Chen, X. (2021). Digital twin-based designing of the configuration, motion, control, and optimization model of a flow-type smart manufacturing system. *Journal of Manufacturing Systems*, *58*, 52–64. DOI: 10.1016/j.jmsy.2020.04.012

Loaiza, J. H., & Cloutier, R. J. (2022). Analyzing the implementation of a digital twin manufacturing system: Using a systems thinking approach. *Systems*, *10*(2), 22. DOI: 10.3390/systems10020022

Lyu, Z., Lin, P., Guo, D., & Huang, G. Q. (2020). Towards zero-warehousing smart manufacturing from zero-inventory just-in-time production. *Robotics and Computer-integrated Manufacturing*, *64*, 101932. DOI: 10.1016/j.rcim.2020.101932

Michalewicz, Z. (2013). *Genetic Algorithms + Data Structures = Evolution Programs*. Springer Science & Business Media.

Mittal, S., Tolk, A., Pyles, A., Van Balen, N., & Bergollo, K. Digital twin modeling, co-simulation and cyber use-case inclusion methodology for IoT systems. In *Proceedings of the 2019 Winter Simulation Conference (WSC)*, National Harbor, MD, USA, 8–11 December 2019; IEEE: New York, NY, USA, 2019. DOI: 10.1109/WSC40007.2019.9004656

Qi, Q., Tao, F., Hu, T., Answer, N., Liu, A., Wei, Y., Wang, L., & Nee, A. Y. C. (2021). Enabling technologies and tools for digital twin. *Journal of Manufacturing Systems, 58*, 3–21. DOI: 10.1016/j.jmsy.2019.10.001

Sahay, N., & Ierapetritou, M. (2013). Supply chain management using an optimization driven simulation approach. *AIChE Journal. American Institute of Chemical Engineers, 59*(12), 4612–4626. DOI: 10.1002/aic.14226

Talib, F., Asjad, M., Attri, R., Siddiquee, A. N., & Khan, Z. A. (2020). A road map for the implementation of integrated JIT-lean practices in Indian manufacturing industries using the best-worst method approach. *J. Ind. Prod. Eng., 37*(6), 275–291. DOI: 10.1080/21681015.2020.1788656

Tang, Y., Chau, K., Hong, L., Ip, Y., & Yan, W. (2021). Financial Innovation in Digital Payment with WeChat towards Electronic Business Success. *Journal of Theoretical and Applied Electronic Commerce Research, 16*(5), 1844–1861. DOI: 10.3390/jtaer16050103

Tang, Y., Chau, K.-Y., Li, W., & Wan, T. (2020). Forecasting economic recession through share price in the logistics industry with artificial intelligence (AI). *Computation (Basel, Switzerland), 8*(3), 70. DOI: 10.3390/computation8030070

Tang, Y. M., Chau, K. Y., Xu, D., & Liu, X. (2021). Consumer perceptions to support IoT based smart parcel locker logistics in China. *Journal of Retailing and Consumer Services, 62*, 102659. DOI: 10.1016/j.jretconser.2021.102659

Tejesh, S. B. S., & Neeraja, S. (2018). Warehouse inventory management system using IoT and open source framework. *Alexandria Engineering Journal, 57*(4), 3817–3823. DOI: 10.1016/j.aej.2018.02.003

van Geest, M., Tekinerdogan, B., & Catal, C. (2021). Design of a reference architecture for developing smart warehouses in industry 4.0. *Computers in Industry, 124*, 103343. DOI: 10.1016/j.compind.2020.103343

Weißhuhn, S., & Hoberg, K. (2021). Designing smart replenishment systems: Internet-of-Things technology for vendor-managed inventory at end consumers. *European Journal of Operational Research, 295*(3), 949–964. DOI: 10.1016/j.ejor.2021.03.042

Yang, M. Using data driven simulation to build inventory model. In *Proceedings of the 2008 Winter Simulation Conference*, Miami, FL, USA, 7–10 December 2008; IEEE: New York, NY, USA, 2008. DOI: 10.1109/WSC.2008.4736373

Yung, K. L., Ho, G. T. S., Tang, Y. M., & Ip, W. H. (2021). Inventory classification system in space mission component replenishment using multi-attribute fuzzy ABC classification. *Industrial Management & Data Systems, 121*(3), 637–656. DOI: 10.1108/IMDS-09-2020-0518

Yung, K. L., Ho, G. T. S., Tang, Y. M., & Ip, W. H. (2021). Inventory classification system in space mission component replenishment using multi-attribute fuzzy ABC classification. *Industrial Management & Data Systems, 121*(3), 637–656. DOI: 10.1108/IMDS-09-2020-0518

Yung, K.-L., Tang, Y.-M., Ip, W.-H., & Kuo, W.-T. (2021). A systematic review of product design for space instrument innovation, reliability, and manufacturing. *Machines (Basel), 9*(10), 244. DOI: 10.3390/machines9100244

Chapter 15
Cycle Time Optimization in Automatic Screw Tightening Machines

S. Fouziya Sulthana
https://orcid.org/0000-0002-3879-6839
Department of Mechatronics Engineering, SRM Institute of Science and Technology, Kattankulathur, India

U. Mohammed Iqbal
Department of Mechanical Engineering, SRM Institute of Science and Technology, Kattankulathur, India

ABSTRACT

The increasing adoption of mechatronics in the automotive industry has transformed manual operations into automated systems for improved efficiency, quality, and safety. Among these, Automatic Screw Tightening Machines (ASTMs) are widely used in lamp assembly to fasten body and extension parts. However, these machines often face increased cycle times under varying conditions, reducing production throughput. This study addresses cycle time optimization by analyzing the machine's construction, programming, and workflow, followed by a cycle time study to identify bottlenecks. Benchmarking revealed screw feeding as a major contributor to delays. To resolve this, the Anti-Tumble Unit was redesigned with a feed-while-drive mechanism, enabling screw feeding and tightening simultaneously. This improvement reduced cycle time from 45 to 33 seconds, achieving a 26.7% reduction. The findings show that targeted design modifications can significantly enhance automated assembly performance and can be applied to optimize cycle time in other industrial processes.

DOI: 10.4018/979-8-3373-3176-8.ch015

1. INTRODUCTION

The automotive industry has increasingly adopted mechatronics-based automation to improve productivity, quality, and safety. Traditional manual operations such as welding, screwing, and testing are progressively being replaced by automated systems that ensure consistent performance, reduce human error, and meet the growing demand for high-volume production. In India, lamp assembly lines in particular employ a wide range of automated solutions including impulse welding machines, auto-screwing machines, and light and leak testing units. Robotic fastening systems have been widely studied in the context of torque accuracy and control strategies, but relatively little attention has been given to cycle time optimization (Hossain et al. 2016, Boys et al. 1977).

Several works have focused on precise torque control during screw tightening. For example, Liu et al. (2020) proposed a fuzzy algorithm for thread tightening torque control, while Miao et al. (2020) reviewed methods for measuring bolt tightening force and loosening detection. Qiu et al. (2023) developed a multi-objective mathematical model for screw tightening optimization. These studies emphasize fastening reliability and error reduction, but they do not address the temporal efficiency of screwdriving operations. Recent works highlight the role of predictive maintenance in improving system reliability. Cinar et al. (2022) implemented predictive maintenance for intelligent manufacturing systems, while Chakroun et al. (2024) proposed a health assessment model for assembly robots using machine learning. Such approaches minimize unplanned downtime, yet they primarily target system reliability rather than reducing cycle time in assembly tasks.

Švaco et al. (2012) introduced adaptive robotic control strategies for uncertain assembly conditions, and Rozo et al. (2013) presented a learning-from-demonstration framework for force-based manipulation. While adaptive control enhances flexibility, its application to screw tightening cycle efficiency has not been demonstrated. Jia et al. (2018) surveyed automated threaded fastening technologies, identifying key industrial trends such as error detection and feed-while-drive systems. Sága et al. (2020) analysed a robotized screwing application with vision sensing integration, focusing on accuracy and automation. These works point toward advanced fastening technologies but again do not quantify cycle time bottlenecks in existing machines.

Since the modern standards of quality involve higher scrutiny, a more efficient robot is likely to sustain its place in the long term in delivering quality products) and for safety reasons (Some of the operations in the industry are hazardous to be carried out by humans. Operations such as welding, working near toxic fumes, waste management etc can be automated to reduce the risk and accidents in the industry). Hence, we can see a rise in the number of automated systems being installed in factories (Chakroun et al. 2024).

Among these, Automatic Screw Tightening Machines (ASTMs) play a critical role in joining lamp body and extension parts using screws, studs, and bolts. ASTM, also referred to as an 'auto-screwing robot,' is a special-purpose robotic system designed to fasten screws, studs, or bolts in high-volume assembly lines. In this study, we use the terms ASTM, auto-screwing machine, and auto-screwing robot interchangeably. These machines rely on subsystems such as screw feeders, driving spindles, torque controllers, and anti-tumble units to function efficiently. However, under varying operating conditions, the cycle time of these machines tends to increase, resulting in reduced production throughput. This limitation often makes them less suitable for deployment in high-demand industrial environments (Shepherd et al. 2017, Soori et al. 2023).

Although the machines are technologically robust, little emphasis has been placed on analysing and optimizing their cycle time. Most existing work focuses on improving torque accuracy, automation integration, or machine reliability, but the aspect of cycle time optimization through design improvement remains under-explored. The system architecture of ATSM is shown in Fig. 1.

Figure 1. System Architecture

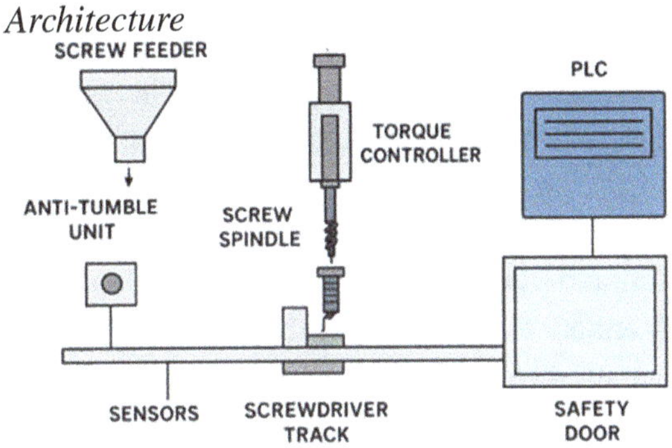

The work entails a complete study of the auto-screwing machine and also includes a study of the cycle-time and suggests a novel method to reduce it by redesigning the anti-tumble unit by incorporating a "Customised-Feed-While-Drive Mechanism" . The auto-screwing robot is robust and technologically sound but in terms of cycle time, there is scope for improvement in the design of the robot functionality (Althoefer et al. 2008, Xu et al. 2018). The objectives of this work are to analyse the construction, programming, and operation of the robot under realistic working conditions, to identify cycle time bottlenecks through numerical and graphical data analysis, to propose and implement a customized design improvement in the Anti-Tumble Unit, thereby reducing cycle time and improving efficiency. The work focuses on lamp assembly operations in the automotive sector, with potential applicability to other high-throughput assembly systems where cycle time optimization is critical.

2. SCREWDRIVERS IN AUTOMOTIVE INDUSTRY

Robotics, a key aspect of mechatronics has become indispensable in the automotive industry. his sector demands automation of various mechanical operations to enhance efficiency and reduce labour costs (Heemels et al. 2009). As a result, robotic applications in the automotive domain range from basic automation tasks like automatic lighting control to advanced robotic systems performing high-precision operations such as impulse welding and automated screwing. These robotic systems not only improve productivity but also ensure consistency and quality in manufacturing processes (Hammer et al. 1992, Paoletti et al. 2007). The automotive industry thus provides a rich context for understanding the principles of robotics and mechatronic system design, making it a benchmark for automation and intelligent manufacturing (Sy Horng et al. 2025).

Assembly is a critical phase in the automotive manufacturing process, where numerous individual components—often referred to as child parts—are integrated to form complex subassemblies and final products. The accuracy and efficiency with which this stage is executed significantly impact the overall quality, cost, and resource utilization of the production process. Improper assembly not only results in product defects but also leads to the wastage of manpower, material, and machine time, making precision and consistency paramount (Toth et al. 2004). To address these challenges and enhance productivity, the automotive industry increasingly relies on robotic automation for performing a wide range of assembly tasks (Chen et al. 2021). Robots are employed for repetitive and precision-demanding operations such as fastening components through automatic screw tightening (auto-screwing), pressing lenses onto lamp housings (lens pressing), and executing high-integrity joints through welding. These robotic systems are engineered to perform these tasks with minimal human intervention, thereby reducing labor costs, improving cycle times, and ensuring uniformity across production batches. The integration of robotics into the assembly process is made possible by the seamless coordination of various mechanical, electrical, and electronic systems (Wang et al. 2015, Suzuki, 2025).

Mechanical components such as spindles, tracks, and gears provide structural and motion-based functionality. These are controlled through Programmable Logic Controllers (PLCs) and monitored by sensors, which serve as the electronic feedback and control units. Power and motion are delivered through motors and electrical actuators, enabling the robot to execute precise movements and operations in a repeatable and synchronized fashion. Given the current trends and growing technological capabilities, it is evident that robotic automation is shaping the future of assembly lines. With advances in machine learning, vision systems, and smart sensors, assembly robots are evolving toward higher autonomy, capable of handling increasingly complex tasks—from picking and orienting parts to producing fully assembled units. This convergence of mechatronics ensures that the future of automotive assembly will be largely driven by robotic systems that are not only efficient but also intelligent and adaptive (Dhayagude et al. 1996, Speck, 2018).

Screwdrivers are among the most common tools found in every household, typically used for basic fastening and unfastening tasks. However, in industrial environments such as the automotive manufacturing sector, even seemingly simple tasks like screw tightening become highly specialized and critical operations. This is because industrial assembly demands not only precision and consistency but also the ability to perform tasks repeatedly at high speed and volume without compromising quality. In industrial screw-driving operations, the torque applied to each screw must be meticulously controlled. If the torque exceeds the specified limit, it can lead to over-tightening, potentially causing damage to threads, deformation of components, or even cracking of parts made of brittle materials. On the other hand, if the torque is insufficient, it may result in loose assemblies, leading to structural instability, product failure, or safety hazards. Therefore, minimizing the margin of error in screw-driving tasks is absolutely crucial.

Adding to the complexity is the fact that these operations are typically carried out hundreds or thousands of times in a single production cycle. Maintaining consistent performance across all these repetitions is only feasible through the use of automated systems. As a result, industries have adopted automated screw driving machines, also known as auto-screwing robots, which are custom-designed special purpose machines engineered for precision fastening. These systems operate based on pre-programmed logic controlled by Programmable Logic Controllers (PLCs) and are integrated with a network of mechanical and electronic components including screw feeders, servo-controlled screwdriver units, jigs for part holding, sensors, and real-time feedback systems. This integration ensures each screw is tightened accurately and efficiently, while also enabling the system to detect faults or anomalies and make necessary adjustments.

By automating the screw-driving process, manufacturers are able to achieve higher throughput, improved quality, reduced labor costs, and better resource optimization, ultimately contributing to leaner and more reliable manufacturing operations.The Automatic Screw tightening machine is made of several components to perform the task of automatic screwing:

- Screw Feeder: To feed the screws in the screwdriving unit
- Screw Driving Spindle Unit: To drive the screws

- Anti-Tumble Unit: To hold the screw
- Jig Track: Jig attached with Sensors
- Screwdriver Track: To facilitate movement of screwdriver
- Safety Door: Safety Feature

Screwdriver Track: In automated screw-driving systems, especially in industrial applications such as automotive assembly, precision and repeatability are critical. A key component that enables this precision is the screwdriver track, which serves as the supporting platform for the screwdriving unit. This track is designed to move the screwdriver along a predefined path, positioning it precisely above the jig, where components are held in place during assembly. The track ensures that the screwdriver can reach each screwing point accurately and consistently in every cycle.

The motion of the screwdriver track is achieved using a linear actuator, a device that enables linear movement as opposed to rotary motion. There are two main types of linear actuators used in industrial automation: absolute and incremental. Among these, incremental linear actuators are typically preferred in screw-driving applications. This preference is due to the fact that incremental actuators use encoders to measure relative movement. For example, with an encoder, it is possible to program the actuator to stop precisely after a specific number of revolutions (e.g., after 200 revolutions), allowing for exact positioning, which may not be feasible with an absolute actuator that relies on predefined positions without relative movement tracking.

To ensure high accuracy and precise control, the screwdriver track in this system uses an incremental encoder-based linear actuator, paired with a 200 W motor. This motor provides sufficient power to move the track under operational loads. The actuator has a lead of 20 mm, indicating the linear distance the actuator moves per revolution of the screw. A stroke length of 1300 mm is used, which refers to the maximum distance the actuator can extend. This stroke length is chosen based on the required travel distance across the assembly line. The stroke is essentially the difference between the fully extended and fully retracted positions of the actuator.

Control of the actuator is managed by an SCON microcontroller, which interprets the encoder signals and ensures the actuator moves to the exact screwing position as per the programmed logic. This integration of mechanical precision and electronic control allows for high-speed, accurate, and repeatable screw-driving operations, essential for maintaining efficiency and product quality in automated assembly lines.

Jig Track: In an automated screw-driving system, the jig track plays a fundamental role in the precise and repeatable positioning of the workpiece during the assembly process. The jig, which securely holds the components to be assembled, must move accurately to align with the different screwing locations where the screwdriver unit performs fastening operations. The jig track provides the guided motion platform that allows the jig to be repositioned along the assembly line, ensuring seamless synchronization with the screwdriving unit.

The jig track operates on a principle similar to that of the screwdriver track, with both relying on linear actuators for movement. However, the primary difference between the two lies in the direction of motion. While the screwdriver track facilitates movement along the Y-axis (left to right), the jig track operates along the X-axis, providing forward and reverse movement. This perpendicular axis arrangement ensures that the jig and screwdriver can access all screwing locations within the working envelope by coordinated linear positioning.

To enable smooth and precise motion, the jig track uses a slider-type linear actuator, wherein the jig slides over a fixed actuator base. This design ensures mechanical stability and smooth traversal. For accurate control over movement, an incremental type encoder is integrated with the actuator. This encoder allows the system to track relative motion and stop the jig at precisely defined positions. By counting the number of turns or revolutions, the actuator can halt exactly after a specified distance, which is crucial for repeatability in high-speed production environments.

The jig track is powered by a 100 W capacity motor, which provides adequate torque and speed for moving the jig assembly under load. The actuator features a 20 mm lead screw, indicating that the jig moves 20 mm per motor revolution. Additionally, the actuator is designed with a stroke length of 1100 mm, which defines the maximum linear distance it can cover. The stroke length is selected based on the total distance the jig needs to travel to reach all designated screwing points on the workpiece.

By combining precise mechanical movement with programmable electronic control, the jig track ensures that the component holding system aligns perfectly with the screwdriving unit. This coordinated motion is essential for ensuring efficient, accurate, and high-throughput screw-driving operations in modern automated assembly lines.

Screw Driving Unit: The Screwdriving Unit (SDU) is a crucial component of the Auto-Screwing Robot, serving as the primary mechanism that carries out the actual screw tightening operation. Functioning like the robot's arm, the SDU travels along the screwdriver track, reaching designated screwing locations with precision. Once positioned, the unit performs vertical motion—moving up and down—to engage the screw with the workpiece. This vertical motion is powered by an AC Servo Screwing Motor, which provides precise control over speed and torque. The integration of this motor allows the SDU to perform accurate, repeatable screw-driving tasks essential for automated assembly lines.

Spindle Unit: The Spindle Unit is a vital part of the Auto-Screwing Robot, responsible for securely holding and guiding the screw during the tightening process. It comprises a long spindle shaft that is structurally supported by a support fixture, ensuring stability during operation. The spindle is mechanically connected to the Anti-Tumble Unit, which prevents wobbling and maintains alignment. At the end of the spindle, a screw bit is attached—this bit is specifically designed to grip and drive the screw into the workpiece. The screw bit is selected based on the type and size of the screw being used, with its diameter and profile varying accordingly to ensure a precise fit and secure fastening during operation. This configuration allows the spindle to maintain consistent performance and accuracy in high-speed, repetitive screw-driving tasks.

Anti-Tumble Unit (ATU): The Anti-Tumble Unit is a critical component in the auto-screwing system, designed to hold and stabilize the screw immediately after it is fed from the screw feeder. Once the screw is delivered, it drops into the Anti-Tumble unit, where it is securely retained in a fixed orientation. This unit plays a vital role in ensuring that the screw does not fall out of alignment or tumble, which could disrupt the screwing process or lead to improper fastening. The screw remains in this stable position until the screw spindle arrives, precisely aligns with the screw, and retrieves it for the tightening operation. By maintaining correct orientation and preventing misalignment, the Anti-Tumble unit contributes significantly to the accuracy, speed, and reliability of the overall auto-screwing system, especially in fast-paced industrial environments where precision and repeatability are essential.

Screw Feeder: The screw feeder in an automated screw tightening system plays a fundamental role in ensuring the continuous, accurate, and efficient supply of screws to the screwing mechanism, specifically the Anti-Tumble Unit (ATU), which is integral to modern robotic assembly in industries such as automotive manufacturing. The system is comprised of various interconnected components that work in unison, including the screw basket, hopper track, chute rail, escapement unit, pneumatic tube, vibrator assembly, transmitter-receiver sensors (such as Panasonic photoelectric sensors), and a Filter Regulator (FR) circuit that controls airflow. The operation begins with the loading of screws into the screw basket, which serves as the primary reservoir. These screws typically have defined dimensions, such as a 4 mm diameter and 14 mm height, and are loaded to no more than 75% of the basket's capacity.

This limitation is crucial, as exceeding it could inhibit the smooth movement of the hopper arm, causing mechanical blockage or overloading errors within the system. Once screws are loaded, the hopper track operates on the principle of mass balance to guide screws downward based on gravitational force and inclination. The movement of screws through the track is regulated and synchronized using a feedback system involving sensors embedded in the chute rail. The chute rail acts as a secondary channel and monitoring stage, ensuring that the correct amount of screws reaches the escapement unit. It is equipped with a Panasonic sensor pair that transmits an optical signal; if the signal remains uninterrupted, it indicates a lack of screws, thus triggering the hopper mechanism to resume feeding. The hopper arm then lifts the screws with a built-in sloped design, which ensures that the screw heads remain oriented upward. This mechanical orientation is critical, as it guarantees that screws enter the chute rail in the desired 'head-up' or inverted position, a requirement for proper engagement with the spindle during the tightening operation.

As screws enter the chute rail, a vibrator assembly attached to the feeder ensures continuous and consistent forward motion. This vibrating mechanism helps prevent jamming and maintains a uniform flow of screws along the rail toward the escapement unit. The escapement unit functions as a gatekeeper, where screws accumulate in a queue awaiting transfer into the pneumatic tube. Another sensor in the escapement unit confirms the presence of a screw at the terminal end of the chute. Upon verification, a pneumatic cylinder is activated, gently pushing the screw into the pneumatic tube. At this point, the FR circuit becomes critical by regulating the airflow within the tube. Initially, a positive pressure is applied, which effectively "blows" the screw through the tube, propelling it quickly and smoothly toward the ATU. This phase of operation requires precision-timed control to ensure that screws are not delivered too rapidly or forcefully, which could result in misalignment or component damage. Once the screw reaches the ATU, the FR circuit reverses its function by creating a vacuum or negative pressure environment, which causes the screw to momentarily suspend and stabilize within the ATU chamber. This vacuum state ensures that the screw does not fall prematurely and is instead retained securely until the screw spindle arrives for pickup.

The photoelectric sensor installed within the ATU performs a final validation by measuring the screw's presence and confirming its correct positioning and height. This feedback mechanism prevents faulty operation or misfires during the screw tightening phase. Once this confirmation is received, the spindle unit is signalled to descend and engage with the screw. The spindle, fitted with a screw bit tailored to the type of screw in use, collects the screw from the ATU and continues to the designated screwing location.

The success of this entire operation is contingent on the perfect coordination of mechanical, pneumatic, and electronic subsystems. Additionally, factors such as screw size, material properties, air pressure calibration, and sensor accuracy directly influence the performance and reliability of the screw feeder system. In high-volume production environments, even minor delays or misfeeds can result in significant losses in productivity, which is why the screw feeder must operate flawlessly and consistently under demanding conditions. Moreover, the integration of smart sensors and programmable logic controllers (PLCs) enables real-time monitoring and diagnostics, allowing for swift corrective action in the event of anomalies. Maintenance routines are also built into the system, where alerts are generated if components like the hopper arm, vibrator assembly, or pneumatic tubing deviate from their standard operational parameters. Furthermore, advanced feeders now come with features like screw counting, flow rate analysis, and error logging, making the system not only a feeder but an intelligent control unit contributing to quality assurance and process optimization. In conclusion, the screw feeder is not merely a mechanical support unit but a complex, sensor-integrated, automated subsystem that ensures uninterrupted, precise, and intelligent screw delivery, forming an indispensable part of robotic screw tightening machines and contributing significantly to efficiency and accuracy in automated assembly operations.

In the Automatic Screw Tightening Machine (commonly referred to as 'Auto-Screwing Robot in this paper), the operation of tightening four screws and two studs are performed on the lamp body and the extension. The lamp body assembly and the extension are joined together using screws at the specified locations.

The screwing is carried out in multiple stages:

- Jig and Screwdriver Spindle move in a coordinated manner such that the screwdriver spindle is right above the screwing location.
- Once the target location is reached, screw is fed through a hose from the screw feeder into the anti-tumble unit that is connected to the screwdriver spindle.
- The screwdriver spindle collects the screw from the anti-tumble unit and is driven down by a motor into the part where screwing must be performed.
- The rotation and torque of the screwdriver spindle is governed by a torque controller unit to ensure that the screwing operation is done with the specified torque.
- The same process is repeated for other 3 screwing operations and 2 stud tightening operations.

Based on the above-mentioned stages, the working of screwdriving unit is carried out as follows: The Jig Arrives at the coordinates set for screwing. The SDU track arrives at the same position above (having a +Z axis value) so that the SDU is right above the Jig and ready for screwing. Once the desired coordinates are reached, the screw feeder is turned on and the screw is fed into the ATU. As the screw sensor confirms the screw presence in the ATU, the motor is turned on and the SDU moves down with a tightening stroke of 250 mm. The Screwdriving spindle attaches the screw to the screw-bit and gets ready for torquing. Once the screw is gripped by the spindle, the motor drives the spindle down and the 2-Stage torque tightening, controlled by screw driving motor, is performed on the part and after the screwing, the spindle reaches the home position. The process flowchart of screw tightening cycle is shown in Fig. 2.

In order to configure the motor, we need to set the rotation speed of the output shaft and the corresponding current value to obtain the ideal speed of the spindle. Torque controller is used to achieve the optimal torque required in different stages of tightening. In the first stage, high torque low rpm is required in order to seat the screw in the lamp part. In the second stage, low torque high rpm is needed to tighten the screw into the part. Similarly, the final stage, low rpm and relatively higher torque is required to pull the spindle out back to its home position. The specifications used in the SDU is shown in Table.1.

Figure 2. Process flowchart of screw tightening cycle

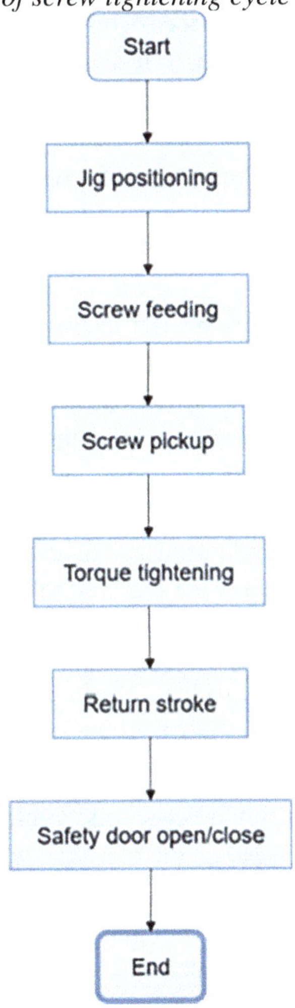

Table 1. Specifications used in SDU

Specification	Value
Tightening Stroke	200-250mm
Screw Holding System	Screw Guide, Vacuum Guide Type
Error Detection	Faulty Torque, low screw (in screw feeder), faulty screw height (photoelectric sensor)
Working Air Supply	0.4-0.5 MPa
Power Supply (Tool Unit, Screw Feeder)	Single Phase 200-230 VAC 50/60 Hz [Single Phase AC 100V~115 V 50/60 Hz]

The turnkey solution is designed for easy integration and cost efficiency, making it ideal for streamlined implementation. It features advanced thrust force control, which enables optimal screw-driving by applying a preset thrust force tailored to the screw type, driving distance, and required torque—ensuring precise matching with the workpiece. Additionally, its human-centred design prioritizes worker-friendly operation, with spindles and control interfaces positioned in accessible, ergonomic locations to enhance usability and comfort.

While the Auto-screw robot is reliable and technologically robust, the efficiency of the robots can be improved with respect to the cycle time. The robots are manufactured for use under certain ideal conditions and if there is any deviation from the ideal use case, the efficiency of the robot drops and calls for countermeasures to be added. There are numerous instances where the ideal working conditions cannot be met and there is a need to maintain the same level of efficiency even under those conditions. If the Auto-screwing robot is designed for multiple jig applications, then in a scenario where a unitary jig is used, the efficiency drops (due to an increase in cycle time) and countermeasures are required to improve the efficiency in such cases. The components in the robot can be redesigned by employing mechatronics design principles and help improve the efficiency and optimize the cycle time.

The efficiency of an automated system or a robot in an industrial environment is gauged by the following factors: cycle time (how long does it take to produce the output of a unit part), repeatability: how many times the same operation can be performed without getting any defects and maintenance which deals with the maintenance work need to be done on the machine; whether it requires continuous high-level maintenance or does it sustain long working hours seamlessly. We can observe that the cycle time is one of the important parameters of gauging efficiency as it affects the overall assembly production. Hence, there is a need for optimization of cycle time in auto-screwing machines. It is possible only when all the functions (primary and secondary) of the robot are optimized - right from screw feeding to torque setting. There is a need to achieve optimal working conditions in all components and sub-functions to improve the cycle time.

3. METHODOLOGY

The study was carried out on an Automatic Screw Tightening Machine (ASTM) installed in the automotive lamp assembly line of an industry partner. The system consisted of:

Robot model: Automatic Screw Tightening Robot equipped with a screw driving spindle unit, anti-tumble unit (ATU), screw feeder with pneumatic chute rail, incremental encoder-based linear actuators (for screwdriver track and jig track), and a torque controller integrated with PLC.

Feeder type: Vibratory screw feeder with pneumatic delivery tube connected to ATU.

PLC system: Mitsubishi PLC integrated with torque controller and encoder-based actuator feedback.

Sensors used: Photoelectric sensor for screw presence detection, encoder sensors for motion tracking, and torque controller for torque verification.

Cycle time data was collected for a complete screwing cycle involving four screws and two studs using a combination of automated and manual methods. The machine's PLC was programmed to log timestamps for each sub-operation, including screw feeding, spindle lowering, torque tightening, return stroke, and safety door operation. To cross-verify these logs and capture operator idle time, a time-motion study was conducted using a digital stopwatch and video recordings. Data was recorded for 100 consecutive screwing cycles under normal production conditions, and statistical analysis was performed to compute the mean cycle time and standard deviation, ensuring reliability of the measurements.

The recorded times for each sub-process (clamping, screw feeding, screw gripping, tightening, spindle return, and door operation) were tabulated and analyzed. Graphical and numerical methods were used to identify the bottleneck stage. The 6-second cycle time for a single screw (3 seconds for feeding + 3 seconds for tightening) was experimentally confirmed using PLC logs and video analysis.

To identify potential improvements, a comparative study was conducted with industrial screwdriving units from Nitto Seiko, Stöger, and Weber Technologies. The specifications of torque range, stroke length, error detection, and feeding mechanisms were analyzed. The Weber model, which employed a feed-while-drive system, served as the benchmark for design improvement.

Based on the bottleneck analysis, a customized anti-tumble unit (ATU) was designed in Autodesk Fusion 360 and SolidWorks. The ATU was redesigned to enable feed-while-drive operation, allowing the next screw to be prepared while tightening is in progress. The mechanism consists of a dual-orifice guide block with an indexing shuttle. The shuttle is pneumatically actuated and positions one screw under the spindle while staging the next in a secondary chamber. The ATU body was manufactured from aluminum alloy (6061-T6) for lightweight durability, while the guide bushings were made from hardened tool steel (H13) to minimize wear. Critical dimensions included a guide bore diameter of 3.2 mm (for M3 screws), shuttle stroke length of 15 mm, and a pneumatic actuation force of 20 N at 0.4 MPa.

The Automatic Screw Tightening Machine (ASTM) is equipped with a pneumatically actuated safety door designed to enclose the jig and spindle area during operation, ensuring operator protection. The door is interlocked with the Programmable Logic Controller (PLC), preventing the screwing cycle from starting unless the door is fully closed and sensor feedback confirms the locked position. The opening sequence is triggered only after the spindle and jig return to their home positions, with dual-sensor redundancy implemented to verify both "door closed" and "door locked" states before cycle initiation. The PLC continuously monitors the door's position via magnetic proximity switches, and if the door is opened mid-cycle, it initiates an emergency stop (E-stop) and retracts the spindle to a safe position. For fail-safe operation, a safety relay module compliant with ISO 13849-1 (Performance Level d, Category 3) is integrated. While the safety door adds an average of 5–7 seconds per cycle, its movement is optimized with high-speed closing for 75% of the stroke and reduced speed for the final 25% to minimize hazards. Although this slightly increases cycle time, it ensures compliance with occupational safety standards and prevents downtime caused by operator accidents. The system adheres to international safety standards, including ISO 12100 for risk assessment, ISO 13849-1 for safety-related parts of control systems, and IEC 60204-1 for the safety of electrical equipment in machines, thereby ensuring that the ASTM can be deployed in industrial assembly lines without requiring additional certification modifications.

4. MEASURES TO IMPROVE OVERALL EFFICIENCY

During machine runtime, some of the issues needed to be addressed to achieve proper functioning of the machine. The issues are enumerated as follows:

External Vibrator Assembly: The chute rail - on which the screws are conveyed - is powered by an inbuilt vibrator assembly. The vibrator assy. sets the chute rail on vibration with which the screws should ideally move from end to another, however, even with the setting of maximum vibration, it was found that the screws were moving too slowly and not reaching the escapement unit in time to deliver it to the ATU. Hence, there were a lot of errors (related to screw passing and screw feeding) which were causing delay in the operation.

In order to rectify this, an external pneumatic vibrator assembly was installed on the screw feeder to increase the vibration in the chute rail and convey the screws faster. The external pneumatic vibrator was adder over track and continuous vibration was achieved to move the screws in a quicker and efficient manner.

Also, since the screw flow depended on the angle or inclination of the surface, the feeder unit was given an inclination / set up to create a slope in the chute rail to leverage the action of gravity in screw conveying.

Torque Setting Verification: The whole operation of automatic screwing relies on the efficiency of torquing. If the torquing is less or if it exceeds the limit it leads to part NG (Not Good Product) and rejection of parts. Hence, it is essential that the proper torque is maintained throughout the runtime and to ensure that, a torque controller is installed in the machine. The torque controller governs the torque during various stages of operation. It is connected to the main PLC and is also programmable where the corresponding tightening program can be integrated and the torque values can be set. It is to be noted that the torque value is affected by factors primarily including:

- The size of the screw spindle (LS /m)
- The weight of the screw spindle (WS /kg)
- The size of the screw that needs to be tightened (LSc /m) The weight of the screw that needs to be tightened (WSc /kg)
- The moment of inertia of the screw when held in the spindle axis (ISc /kg m^2)

The controller works on the screw fastening driver that uses an AC servo motor and flexibly carries out operations by creating programs to various screw fastening operations. For general screw tightening, the controller uses its standard tightening program so that just by entering the values of tightening conditions (given as codes in the program) one may simply perform the screw tightening to a high degree of accuracy.

The torque controller has 5 modes of setting the parameters which include:

1) Trace Mode: Used to display a number of a program step being executed.
2) Memory Mode: Used to register or change the setting values stored in the memory.
3) Parameter Mode: Used to register or change parameter setting values.
4) Program Mode: This mode is used to edit the program.
5) Monitor Mode: This mode can indicate revolution speed, torque value etc. of the tool unit in real time.

Installing Color Sensor: During the screw feeding operation, it was observed that the screw did not enter the escapement due to the non-synchronization of the pneumatic cylinder and the screw feeder. In order to ensure that the screws entered the escapement and also to make sure that foreign objects did not fall into the escapement during the opening, it was decided to add a color sensor to confirm the screw presence before activating the cylinder.

5. RESULTS AND ANALYSIS

Cycle time Analysis: The cycle time study of the Auto Screwing Machine was carried out as the high cycle time (of >50s) was causing delay in production and slowing down the assembly line. The working of the robot for a long duration also led to idle worker time (as the worker remained idle and waited for the next cycle until the robot completed the screwing). Since the high cycle time caused a decrease in production and wastage of resources, the cycle time analysis was performed to check for the bottleneck. Bottleneck is the stage which impedes the flow of the operation and causes an abnormal delay. By analyzing the cycle time using different analytical methods, the bottleneck was determined and a redesign of the component was performed.

While calculating the cycle time, we have to consider the following variables: Product demand (PD), Available Work Time (AWT), Takt Time (TT), Process Cycle Time (PCT) and Work Balance.

(PD) Target Per Hour = 67 nos (based on customer demand)

Hourly Output = 67

Process Efficiency Trend: To track the process efficiency trend, we have to calculate certain parameters as shown in Table 2 (as per the data from the industry) and extrapolate the relationship between the variables to arrive at the correct cycle time trends.

Table 2. Process Efficiency Trend

Parameter	Value
Plan	400 pieces
Actual	280 pieces (say, based on average production observed)
Gap	120 pieces
Available work time	7.4 hr
Average output per hour	38 pieces

Based on this data we can calculate the takt time (TT) as follows:

Takt time is the average time to make one unit of product. It is calculated as follows:

Let the net working time = 900 minutes (considering 2 shifts of work and excluding the break times)

Operational efficiency = 90% (This is taken based on machine efficiency and the operator's efficiency)

Networking time = 54,600 seconds (900×60)

Networking time with 90% efficiency = 48,600 seconds

Customer demand per day = 1000 nos

Takt time = ratio of networking time to the customer demand = 48.6 seconds

Process Cycle Time is the time taken to complete a task or a collection of tasks from beginning to end. Process Cycle Time is obtained by calculating the Line Computation Chart that is calculated by adding the cycle time of all stages involved in the assembly and also has individual data regarding Worker Time, Hand Time, Operator Cycle Time, Machine Cycle Time etc. and is used to calculate Line Capacity for hourly, weekly and monthly basis (to project how many units are produced in under the given constraints). The machine cycle time is calculated based on the individual operation of the machine, like clamping, spindle movement, jig movement etc and is found to be 45 seconds.

Bottleneck Analysis: Based on the machine cycle time information, the bottleneck in the screwing operation is noted. Door operation is done using RAM cylinders for safety reasons. The speed of the door opening and closing is also regulated for safety purposes (the door comes down with a higher speed till 75% of the distance and for the final 25% of touching the baseplate, the speed is reduced to ensure worker safety). Jig (has the lamp body) and Spindle (screwdriver) move in a coordinated manner such that the screwdriver is right above the jig and is ready for screwing.

Once the position is set, the screwing happens for the other six positions and the jig and spindle return to the home position and the safety door opens. During the change from screw to stud, the jig is tilted (with a cylinder extension setup) to ensure that the stud and the stud driver are in 90° orientation with respect to each other to ensure proper stud tightening. The process carried out in the screw tightening machines is given in Fig. 3. The various stages of the tightening operation and the time required for them is presented in Fig. 4.

From the Fig. 4, we can observe that the screwing for one position takes 6 secs. For 6 screwing operations, it takes 6x6=36 sec. Clearly the 36 secs for 6 screwing operations are the most time-consuming operations. In order to determine the bottleneck, we have to find out which sub-function of the screwing takes more time to complete the cycle. The total time of 6 seconds for one screwing operation is split between screw feeding and screw tightening. Since the screw feeding happens first and the screw tightening happens after the screw is fed, we can find out how these two operations contribute to the 6 sec timing.

Figure 3. Process carried out in the screw tightening machines

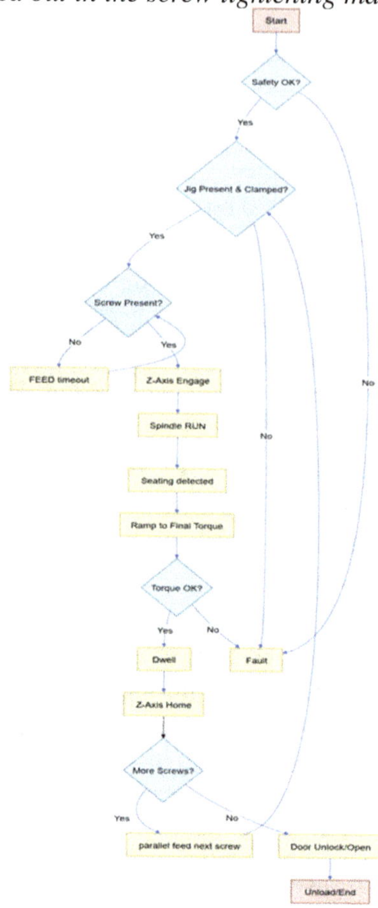

Figure 4. Time required for various tightening operations

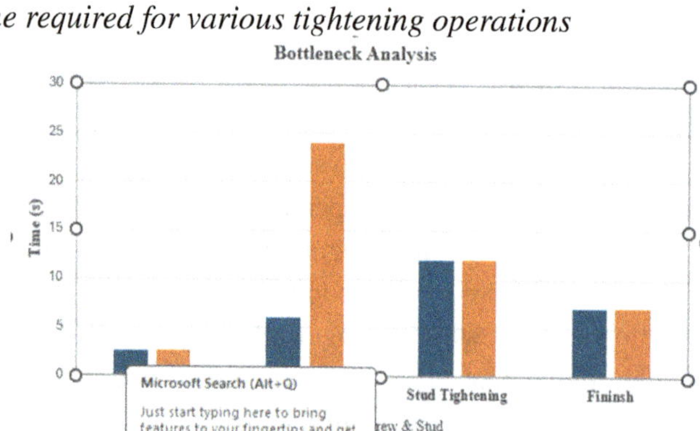

As the Jig and Spindle arrive at the set coordinates, the PLC sends a signal to the screw feeder unit, the screw feeder ejects the screw into the anti-tumble unit and once the screw reaches the ATU and is confirmed by the screw presence sensor, the screwdriver is activated. Since the cycle has only begun, there are currently no screws, and the screw feeder is activated and screw travels from the escapement to the anti-tumble unit. The time taken for the screw to travel from the escapement to the anti-tumble unit is found to be 2 sec.

As the screw lands into the ATU, the vacuum is turned on and the screw is gripped by the screwdriving unit before beginning the screwing operation and this gripping takes 1 sec. Hence the total screw feeding time for feeding a single screw is 3 sec. And it takes another 3 seconds for the spindle to perform the tightening operation. Now we can observe that out of the 6 secs for screwing; 3 seconds are utilized for screwing and 3 seconds are required for tightening. Calculating for 6 such screwing operations, we get that 3x6=18 seconds are used up in screw feeding alone. Now the bottleneck lies in the fact that while the overall screw feeding happens for 18 secs, the screwdriver and the other machine assembly including the jig remain idle. The 18 secs of machine time and worker time remain as waiting time and contribute to the overall increase in cycle time thereby reducing the efficiency of the machine and the assembly line. As the bottleneck is determined to be the screw-feeding time, the redesigning of the screw feeding and screw driving was carried out such that during the screw-feeding operation, the rest of the machine does not remain idle or, in other words, we can redesign the unit such that the screw feeding and screw driving happen at the same time (simultaneously) which ensures that the cycle time is not increased.

Analysis of Automatic Screw Driving Units: In order to redesign the screw feeding and driving model, the study of automatic screwdriving models developed so far and the study of models that are currently in use in the industry was taken up to gain understanding about the screwdriving operation to incorporate the design in a customized manner. For comparison, the specification of three industrialized screw driving unit is taken. The comparison of specification is given in the Table. 3

Table 3. Nitto Seiko vs STOGER vs WEBER Spec Comparative Spec Sheet

Parameter	Nitto Seiko	STOGER	WEBER
Nominal diameter	2mm-5mm	2mm-5mm	2mm-5mm
Length	Max 18mm	Max 20 mm	-
Tightening Torque	0.3-3.8Nm	0.2-4.5Nm	0.3-3.0 Nm

continued on following page

Table 3. Continued

Parameter	Nitto Seiko	STOGER	WEBER
Tightening Stroke	50,100,150, [200,250]mm	Approx. 80	Approx. 72
Screw Holding System	Vacuum Type	Vacuum Type	Vacuum Type
Error Detection	Faulty torque (faulty screw tightening), low screw in screw feeder, faulty screw height (photoelectric sensor)	No	No
Working air supply	Clean air from 0.4-0.5Mpa	0.4-0.5 MPa	0.4-0.5 Mpa

It is observed that while the torque offered by Stoger has a higher range, other features are similar to that of Nitto Seiko or in some cases inferior. Nitto Seiko SDUs have an error detection system built into it which Stoger does not have. the screw feeding mechanism implemented by Stoger is the same as that of Nitto Seiko, though it is more user-friendly and easier to use it does not offer any advantage over Nitto Seiko with respect to cycle time optimization. While it is noted that Weber has relatively less torque range and stroke length, the model used by Weber is highly efficient and less time consuming than Nitto Seiko. The Weber screwing mechanism is shown in the Fig. 2

Figure 5. Weber Screwing Mechanism (a) Screw blown in swivel arm (b) Feeding tube mechanism (c) screw driving mechanism (d) screwdriver bit retracts back through the jawset

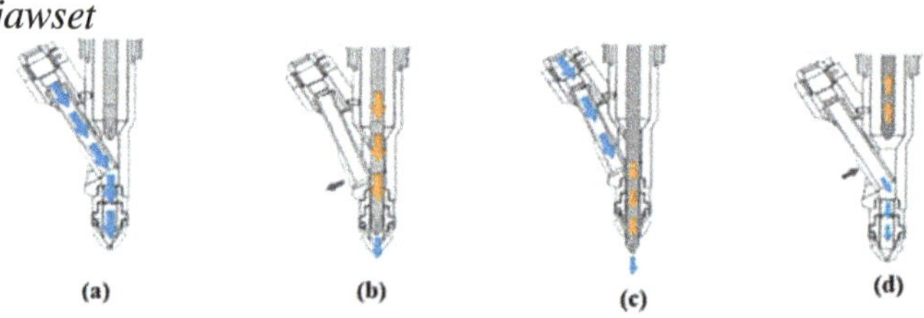

The first screw is blown through the swivel arm into the aligning jaws of the screwdriving head as depicted in Fig. 5 a-d. The feeding tube and the aligning guide are made to suit the screw dimensions so that the shaft of the screw is perfectly concentric in the aligning guide to ensure a smooth transition of the screw through the jaws during installation. The rotating screwdriver bit moves forward towards the fastener, pushing the swivel arm out of the way and thereby closing off the end of the swivel arm tube with the steel stop plate. The bit continues forward to engage the screw. While the screwdriver installs the fastener, the next screw is already being delivered and positioned in the swivel arm, ready for the next cycle as soon as the bit retracts. The screwdriver bit retracts back through the jaw set and past the swivel arm, allowing the arm to move back to its original position. The fastener then drops down into the jaw set and is immediately ready for another cycle.

The technology used by this industry follows the feed-while-drive system, meaning, when the screwdriving is happening, the screw feeding also happens simultaneously. The pneumatic tube fixture connecting to the anti-tumble unit is retractable and whenever the screwdriving operation takes place, the feeding unit retracts and places the screw in the anti-tumble unit and before the screwdriver reaches the home position, another screw is already placed and ready for the next screwing cycle. This eliminates the time required for screw feeding and reduces the overall cycle time and performs the screw feeding by efficiently managing the time and resources.

Based on this observation, we can deduce that integrating screwdriving and screw feeding is necessary to achieve an optimization in cycle time. The Table. 4 shows the time taken by the system before and after optimization process.

Table 4. Time taken by the system before and after optimization process

Operation	Before Optimization (s)	After optimization (s)	Mean time (s)	Std dev (s)	Cycle time reduction
Screw Feeding (per screw)	3	1	2.0	1.41	66.67%
Screw Tightening (per screw)	3	3	3.0	0.00	0.00%
Total per Screw	6	4	5.0	1.41	33.33%
Total for 6 Screws	36	24	30.0	8.49	33.33%
Cycle Time (Full Process)	45	33	39.0	8.49	**26.67%**

The screw tightening involves robot spindle (electric screwdriver with integrated torque controller) interfaced with PLC whose scan time is $\leq$ 10s. It has a high-speed camera at 240 fps for cross validation. The screw tightening time can be decomposed as

$$T_{tight} = T_{approach} + T_{run\_down} + T_{seating} + T_{torque\_ramp} + T_{dwell}$$

Where $T_{approach}$ is the bit engagement of machine, T_{run_down} is the time taken for thread insertion until head seats, $T_{seating}$ is the transition time as head contacts surface, T_{torque_ramp} is the torque rise time and T_{dwell} is the hold time.

For the PLC cycle of 100, the measured results are shown in the Table. 5 below:

Table 5. Decomposition of Screw Tightening Time

Component	Mean	SD
$T_{approach}$	0.22	0.05
T_{run_down}	1.47	0.18
$T_{seating}$	0.30	0.07
T_{torque_ramp}	0.78	0.10
T_{dwell}	0.20	0.00
T_{tight}	**2.97**	**0.24**

The above experimental data confirms that the time taken for screw tightening is $\sim$3s (2.97 $\pm$ 0.24 s)

Design Improvement in Anti-tumble Unit: From the cycle-time analysis and the technical comparative study of different screwdriver units and from the different industries, a novel customized Anti-Tumble Unit for the existing system was designed. From the technical comparative study, it was found that Weber Technologies had a feed-while-drive system that reduces the screwdriving operation time by 50% by performing the screw feeding and screw driving simultaneously. In existing screwdriving system, the screw feeding and screw driving happen separately where the screw feeding occurs first and then the spindle is driven down for the screw tightening. This leads to the spindle and other components of the machine to remain idle while the screw feeding takes place. This leads to an increase in cycle time and reduces the overall efficiency.

In order to incorporate the feed-while drive mechanism in existing system, the design had to be examined to make sure that the implementation of such a mechanism was possible. By examining the design of the anti-tumble unit of the existing system, it was found that a customized new design was needed to be developed to incorporate the feed while drive mechanism in existing Screwdriving system. Hence, the anti-tumble unit was redesigned and the first-ever anti-tumble unit embedded with a novel feed while drive mechanism was developed for the system.

The design was developed using Autodesk Fusion 360 Software and Solidworks as shown in Fig. 6 and the design had the following parameters in consideration:

Figure 6. Design Improvement in Anti-Tumble Unit

- Integrate Anti-Tumble Unit design
- Develop new design for implementing feed-while-drive system in robots
- Develop robust design to feed multiple screws/studs simultaneously
- Easily integrate new design into existing Automation ecosystem

With these parameters, a novel customized anti-tumble unit was designed that could feed and drive the screws simultaneously. The anti-tumble unit consists of a circular disk, to which an encoder-type servo motor is attached. The metallic circular disk is rotated by the motor when the motor gets a signal from the screw-spindle-track. That is, when the screwdriver arrives at the jig and is set a screwing location, the PLC sends a signal to the circular disk and the disk rotates to feed screws from one orifice (O1) and to drive the already fed screw from the second orifice (O2). Hence, in the same 3 second timespan, one orifice is used to feed the screws and the other is used to drive the screws. After the screwing and driving is completed, the circular disk rotates again to complete the cycle so that the orifice locations are interchanged. This essentially means that, for the screwdriver unit, there is always one screw available in the anti-tumble unit so that the 3 seconds of feeding time is conserved and thereby reduces the cycle time and adds to the overall efficiency. The performance was validated experimentally by comparing cycle times before and after ATU integration. Feeding time reduced from 3.0 ± 0.40 s to 1.0 ± 0.25 s per screw. Total cycle time improved from 45.0 ± 2.1 s to 33.0 ± 1.8 s, confirming that the ATU contributed ~12 s of savings per cycle across six fastenings as shown in Fig. 7 and Table. 6.

Figure 7. ATU performance validation

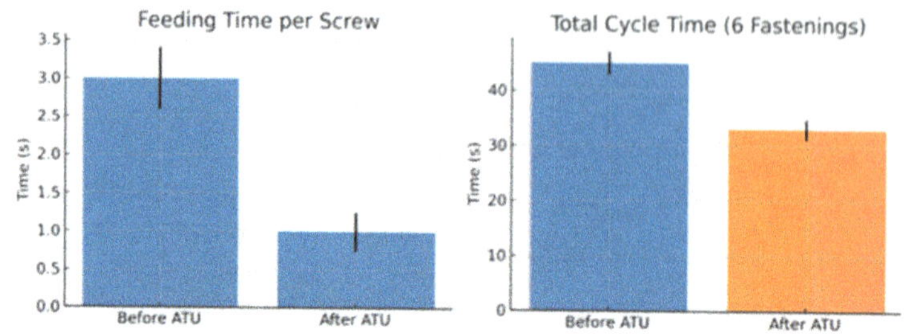

Table 6. ATU performance Validation

Parameter	**Before ATU**	**After ATU**	**Improvement**
Feeding time (per screw)	3.0±0.40 s	1.0±0.25 s	2.0 s reduction
Total cycle time (per cycle of 6 fastening)	45.0±2.1 s	33.0 ±1.8 s	12.0 s reduction

6. ANALYSIS OF COUNTERMEASURES

The individual results of the various countermeasures performed are discussed in this section and also the results of cycle time improvement are included.

Installation of pneumatic vibrator assy.: By the addition of external electro-mechanical pneumatic vibrator assy., the vibration in the tracks are increased and the screw stoppage in the tracks was prevented. The inclusion of slope to facilitate gradual downward motion also contributed to the conveyance of screws without stoppage.

Torque Setting Verification: The torque checking and setting the torque values based on the type of screws fed in the machine was carried out where the individual torques for each of the screwing stages was checked for the type of screws (different screws of different sizes require different torque values) and the correct torque values were programmed and the defective screwing was prevented and overall efficiency was improved.

Installation of Color based Distance Sensor: As the screw feeding escapement had some stoppage in screw outflow, color based distance sensor was installed to confirm the screw presence and then open the escapement. This countermeasure ensured that the screw outflow happened properly, and no foreign object could get into the escapement and thus reduced the machine downtime.

Cycle Time Optimization: As the cycle time of 46 secs was deemed too high for the machine to cope with the product demand, the cycle time optimization study was carried out and the redesigning of the anti-tumble unit of the existing system was performed. The improvement in efficiency is given by the following calculations:

Total Tightening Time (TTT) = Total Feeding Time (TFT) + Total Driving Time (TDT)

Total Feeding Time (TFT) = Screw Feeding Time (SFT) + Screw Gripping Time (SGT)

Screw feeding time (SFT): 2 sec

Screw gripping with Screwdriver (SGT): 1 sec

Feeding Time (FT) = SFT + FGT = 3 sec

Since there are six tightening operations performed in the auto-screw tightening machine:

Total Feeding Time (TFT:) 6xFT =6xSFT + 6SGT

TFT = 18 sec.

Total Screw Driving Time (TDT) = 6x3 sec = 18 sec.

Hence, the current Total Tightening Time: 18 sec + 18 sec = 36 sec

The Total Cycle Time = 45 sec

Percentage of Tightening Time in Overall Operation = 80%.

Percentage of Total Feeding Time = 38.7%

Total increase in % efficiency due to improvement in design = total time reduced in screw feeding time (SFT) = 2*6 = 12 / 45 (Total Cycle Time) = 26.67%

Hence, due to the redesigning of the Anti-Tumble Unit by incorporating customized "Feed-While-Drive" mechanism, the cycle time could be reduced to 33 sec with reduction percentage of 26.67% and contribute to an overall increase in product output.

7. PRACTICAL IMPLICATION

The findings of this study hold significant practical value for the automotive industry and related manufacturing sectors by providing clear benefits across multiple stakeholders. Manufacturing engineers gain a systematic approach for identifying and addressing cycle time bottlenecks through mechatronic design improvements, while automation integrators can adopt the feed-while-drive Anti-Tumble Unit design for integration into existing screw tightening robots, thereby improving throughput without requiring major equipment replacement.

Maintenance teams also benefit from reduced idle time and fewer feeding-related stoppages, which lowers unplanned downtime. The reduction in cycle time from 45 seconds to 33 seconds corresponds to a 26.7% reduction in cycle time, translating into a daily production increase from approximately 1,280 units to 1,760 units across two 8-hour shifts—yielding a net gain of about 480 units per day without additional workforce or machines. Implementation costs remain minimal, limited to custom machining and servo integration, and the cost savings from higher output and reduced idle worker time can offset the investment within weeks.

Furthermore, the methodology is scalable beyond automotive lamp assembly, with applications in electronics, household appliances, and aerospace industries where fastening is a repetitive bottleneck process. By reducing idle time and eliminating inefficiencies, the redesigned ASTM aligns closely with lean manufacturing principles, minimizing waste in the form of idle labor, machine downtime, and energy consumption.

8. CONCLUSION AND FUTURE SCOPE

The application of mechatronics and mechatronics design principles are ubiquitously found in industries across domains. With regard to the automotive industry, there are several operations that are being carried out by robots and automated systems to improve efficiency and output. This work performed the cycle time analysis and optimization study of the Automatic screw tightening machine used in assembly lines in industries. As a result of the cycle time analysis study, the overall improvement in efficiency was necessary to reduce the cycle time and machine downtime. Hence, three countermeasures were added to the machine to improve certain functions and prevent issues like screw stoppage, improper torque and screw outflow issues.

Regarding the cycle time, the bottleneck was observed to be screw feeding 0peration which lent 18 sec of machine idle time, hence causing the increase in cycle time. To improve that, the design of other screwdriver units was also examined, and a technical comparative analysis was carried out. It was observed that Weber screwdriving units had a feed-while-drive mechanism which had the screw feeding and screw driving mechanism happen simultaneously. In the existing system, the feed-while-drive mechanism had to be incorporated by maintaining the existing framework. To suit that, a new mechanism was developed that had a circular system of screw feeding and screw driving that ensured that a screw always remained in the driving anti-tumble unit and reduced the cycle time by 12 secs with the reduction percentage of 26.67%.

The next stage of the design optimization is the prototyping stage. In the prototyping stage, the design of the anti-tumble unit is analysed and the material, dimension and quality factors are examined and the design is sent for prototyping in 3-D printing or other low-fidelity prototyping. A low-fidelity prototype of the anti-tumble unit is developed and implemented and the testing of the reduction in cycle time is carried out. Based on the results, a medium to high quality prototyping could be done and tested in the auto-screwing robot to successfully verify the reduction in cycle time.

The proposed anti-tumble unit was validated in a lamp assembly process using M3 screws. However, the design is inherently modular and can be adapted to different fastening requirements. For smaller screws (e.g., M2), the bore diameter and shuttle stroke can be reduced, while for larger screws (e.g., M5), reinforced bushings and higher actuation forces can be incorporated without altering the fundamental feed-while-drive principle.

Likewise, the PLC-based control logic is generic and can be reprogrammed for multi-spindle systems or transferred to other assembly cells with minimal changes. In high-volume production, the ATU can be combined with parallel feeders to achieve throughput comparable to state-of-the-art systems. Thus, while the present optimization was demonstrated on a specific system, the concept is extendable to a wide range of industrial screwdriving applications.

While our optimized cycle time (33.0 ± 1.8 s for six fastening operations) reflects a clear internal improvement, it remains slower than high-performance commercial systems. For instance, Stöger's automatic screwdriver systems achieve cycle times of about 20–30 s and Weber systems boast start–start cycle times below 20 s in optimized feed-drive configurations. These comparisons highlight that while our retrofitted ATU significantly cuts cycle time at low cost, further innovations such as multi-spindle integration or ultra-fast feed-and-drive mechanisms are needed to approach top-tier industrial performance.

REFERENCES

Althoefer, K., Lara, B., Zweiri, Y. H., & Seneviratne, L. D. (2008). Automated failure classification for assembly with self-tapping threaded fastenings using artificial neural networks. *Proceedings of the Institution of Mechanical Engineers. Part C, Journal of Mechanical Engineering Science*, *222*(6), 1081–1095. DOI: 10.1243/09544062JMES546

Boys, J. T., & Wallace, P. W. (1977). Design and performance of an automatic control system for fastener tightening. *Proceedings - Institution of Mechanical Engineers*, *191*(1), 371–380. DOI: 10.1243/PIME_PROC_1977_191_045_02

Chakroun, A., Hani, Y., Elmhamedi, A., & Masmoudi, F. (2024). A predictive maintenance model for health assessment of an assembly robot based on machine learning in the context of smart plant. *Journal of Intelligent Manufacturing*, *35*(8), 1–19. DOI: 10.1007/s10845-023-02281-3

Chen, B., Wang, B., Zheng, C., & Zi, B. 2021, November. Design and simulation of a robotic knee exoskeleton with a variable stiffness actuator for gait rehabilitation. In *2021 27th International Conference on Mechatronics and Machine Vision in Practice (M2VIP)* (pp. 24-29). IEEE DOI: 10.1109/M2VIP49856.2021.9665028

Cinar, E., Kalay, S., & Saricicek, I. (2022). A predictive maintenance system design and implementation for intelligent manufacturing. *Machines (Basel)*, *10*(11), 1006. DOI: 10.3390/machines10111006

Dhayagude, N., Gao, Z., & Mrad, F. (1996). Fuzzy logic control of automated screw fastening. *Robotics and Computer-integrated Manufacturing*, *12*(3), 235–242. DOI: 10.1016/0736-5845(96)00009-9

Hammer, R., Hollis, R. L., An, C. H., & Henriks, F. "Design and control of an air-bearing supported three degree-of-freedom fine positioner," *Proceedings 1992 IEEE International Conference on Robotics and Automation*, pp. 677–684 vol.1, 1992. 6.2 DOI: 10.1109/ROBOT.1992.220289

Heemels, W., Lehmann, D., Lunze, J., & De Schutter, B. (2009). Introduction to hybrid systems. In *Handbook of Hybrid Systems Control–Theory, Tools*. Applications. DOI: 10.1017/CBO9780511807930.002

Hossain, A., Sharma, N. D., & Agarwal, G. 2016, June. Developing Servo Indexing System Using Timing Screw for Automatic Liquid Filler in Manufacturing Environment. In *2016 ASEE Annual Conference & Exposition*. DOI: 10.18260/p.26771

Jia, Z., Bhatia, A., Aronson, R. M., Bourne, D., & Mason, M. T. (2018). A survey of automated threaded fastening. *IEEE Transactions on Automation Science and Engineering*, *16*(1), 298–310. DOI: 10.1109/TASE.2018.2835382

Liu, R., Pan, F., & Diao, Q. 2020, July. Research on Thread Tightening Torque Control Method Based on Fuzzy Algorithm. In *2020 39th Chinese Control Conference (CCC)* (pp. 2216-2220). IEEE DOI: 10.23919/CCC50068.2020.9188972

Miao, R., Shen, R., Zhang, S., & Xue, S. (2020). A review of bolt tightening force measurement and loosening detection. *Sensors (Basel)*, *20*(11), 3165. DOI: 10.3390/s20113165 PMID: 32498473

Paoletti, S., Juloski, A. L., Ferrari-Trecate, G., & Vidal, R. (2007). Identification of hybrid systems a tutorial. *European Journal of Control*, *13*(2-3), 242–260. DOI: 10.3166/ejc.13.242-260

Qiu, Z., Wang, D., & Gao, M. 2023, November. Research on Industrial Assembly Screw Automatic Tightening Algorithm Based on Multi-Objective Mathematical Modeling. In *2023 7th Asian Conference on Artificial Intelligence Technology (ACAIT)* (pp. 1085-1090). IEEE DOI: 10.1109/ACAIT60137.2023.10528391

Rozo, L., Jiménez, P., & Torras, C. (2013). A robot learning from demonstration framework to perform force-based manipulation tasks. *Intelligent Service Robotics*, *6*(1), 33–51. DOI: 10.1007/s11370-012-0128-9

Sága, M., Bulej, V., Čuboňova, N., Kuric, I., Virgala, I., & Eberth, M. (2020). Case study: Performance analysis and development of robotized screwing application with integrated vision sensing system for automotive industry. *International Journal of Advanced Robotic Systems*, *17*(3), 1729881420923997. DOI: 10.1177/1729881420923997

Shepherd, M. K., & Rouse, E. J. (2017). Design and validation of a torque-controllable knee exoskeleton for sit-to-stand assistance. *IEEE/ASME Transactions on Mechatronics*, *22*(4), 1695–1704. DOI: 10.1109/TMECH.2017.2704521

Soori, M., Arezoo, B., & Dastres, R. (2023). Internet of things for smart factories in industry 4.0, a review. *Internet of Things and Cyber-Physical Systems*, *3*, 192–204. DOI: 10.1016/j.iotcps.2023.04.006

Speck, J. A. (2018). *Mechanical fastening, joining, and assembly*. CRC Press. DOI: 10.1201/9781351228640

Suzuki, T. (2025). Vision-Guided Robotics: Integrating Computer Vision and Mechanical Engineering for Precision Control. *Global Research Review*, *1*(1), 77–84.

Švaco, M., Šekoranja, B., & Jerbić, B. (2012). Industrial robotic system with adaptive control. *Procedia Computer Science, 12*, 164–169. DOI: 10.1016/j.procs.2012.09.048

Sy Horng Ting, T., Goh, Y. H., Chin, K. M., Tan, Y. K., Chiew, T. H., Ma, G., & How, C. K. (2025). A Review of Advanced Force Torque Control Strategies for Precise Nut-to-Bolt Mating in Robotic Assembly. *International Journal of Robotics & Control Systems, 5*(1 Toth, G.R.R., 2004. Torque and angle controlled tightening over the yield point of a screw—Based on monte-carlo simulations. *Journal of Mechanical Design, 126*(4), 729–736.

Wang, X., Liu, M., Ge, M., Ling, L., & Liu, C. (2015). Research on assembly quality adaptive control system for complex mechanical products assembly process under uncertainty. *Computers in Industry, 74*, 43–57. DOI: 10.1016/j.compind.2015.09.001

Xu, L., Zhai, Y. and Li, B., 2018. Research 0f Electrical Control System for Flywheel Tightening Devices In Engine Assembly Lines. *Academic Journal of Manufacturing Engineering, 16*(1)

Chapter 16
The Impact of Energy and Water Consumption on Supplier Credit Disbursement:
Stakeholder Pressure Towards the Greening Agenda

Witness Siwela
https://orcid.org/0009-0000-4414-2332
University of South Africa, South Africa

ABSTRACT

The 4IR is fading away to usher in Industry 5.0. Finding its bases on Industry 4.0, the 5IR emphasises sustainability and human to machine collaboration. The rationale resides in encouraging warehouses to harness new innovations given Industry 4.0 and 5IR. The researcher seeks to investigate if corporate green performance influences suppliers' support. In congruency to the quantitative nature of the study, a positivist stance was adopted to test the set hypothesis. Archival data were drawn purposively from the FTSE/JSE Responsible Investment Index to compute a short panel data set comprising 19 companies and 114 observations over a six-year period. Pearson's pairwise correlation and first differenced econometric models for Fixed effects were employed for statistical analyses. The study yielded mixed findings as energy consumption indicated a significant positive effect on supplier credit disbursement. Conversely, water consumption was found to be statistically insignificant. The results render practical implications that green warehousing calls for stakeholder support.

DOI: 10.4018/979-8-3373-3176-8.ch016

1. BACKGROUND AND INTRODUCTION

The concept of greening has since gained traction globally as the planet battles the prevailing climate change crisis. Thus, the greening agenda embodies a wide range of initiatives that seek to promote environmental responsibility (Ramadhini, Adhariani & Djakman, 2020). According to Gigante and Manglaviti (2022), integration of environmental considerations in corporate practices across various industry sectors is the primary goal of lobby groups that are custodian to the greening agenda. Abba et al (2019) posit that greening practices are beginning to show in industry sectors such as transportation, agriculture, urban planning, energy and waste management. However, more work still needs to be done to promote green practices given that it is regarded as an emerging phenomena (Kordsachia, Focke & Velte, 2022; Meyer *et al.*, 2019). Steyn (2014) argues that a considerable number of industries are still lagging despite numerous companies beginning to adopt green practices following the advent of mandatory integrated reporting. Moreover, the greening agenda places emphasis on greening cities and towns, large events, raising public awareness on environmental issues and promoting biodiversity (Margaretic & Pouget, 2018; Cini & Ricci, 2020).

Issock Issock, Mpinganjira & Roberts-Lombard (2020) notes that energy consumption is a crucial metric employed as an indicator to measure corporate sustainability performance. Thus, energy consumption affects the environment adversely through habitat destruction, exacerbating climate change crisis and polluting the atmosphere and water. Fahim and Mahadi (2022) states that the primary source of energy is largely fossil fuels which has sky-rocketed South Africa's standings as the largest emitter of greenhouse gases globally. Conversely, Ameli *et al.* (2021) contends that excessive water consumption can significantly impact the environment. While different sectors of the economy participate actively in pushing water consumption levels high, the warehousing industry has considerable impact in this regard (Akintokunbo & Arimie, 2021; Luster *et al.*, 2009). Fahim and Mahadi (2022) claims that, if it remains unchecked, water consumption can lead to water scarcity, perpetual climate change risks, disrupts and degrades the ecosystems.

In the view of Edeh (2020), stakeholder pressure is imperative in compelling companies to adopt green practices. To this end, producing sustainability reports in South Africa is currently voluntary despite enormous pressure exerted by various lobby groups who call for mandatory disclosures (Favotto & Kollman, 2021). However, the Johannesburg Stock Exchange (JSE) has made it mandatory for companies to adhere to the provisions of the King IV which encourages sustainability disclosures and integrated reporting. This means that stakeholders accessing these reports make crucial decisions holistically based on both financial and non-financial data disclosed in the integrated reports (Ramjee & Gwatidzo, 2012; Gwatidzo, Ntuli & Mlilo, 2017;

Herbohn, Gao & Clarkson, 2019). Thus, stakeholders such as investors, consumers, employees, suppliers and non-governmental organisations (NGOs) are capacitated to put pressure on companies. The reporting companies would respond by adopting the greening agenda and incorporate green practices in their strategy formulation.

Zhang, Phanniphong and Li (2023) recently conducted a survey of 200 companies to investigate the effects of corporate green initiatives on stakeholder involvement and project performance. In this context, corporate greening initiatives are understood to be measures employed businesses to promote sustainable practices and drastically reduce environmental impact. Thus, Zhang, Phanniphong and Li (2023) agrees that at entity level, green initiatives involves using sustainable materials, energy conservation, water consumption and waste reduction. Accordingly, data were analysed quantitatively using statistical regression methods to report the results of the study. The findings indicated that corporate greening initiatives impact stakeholder involvement and project performance positively (Zhang, Phanniphong & Li, 2023). Moreover, the study also placed emphasis on the importance of managerial governance in the management of this relationship. Notwithstanding the value of this research in literature, little is still know on stakeholder pressure towards the green agenda where water and energy consumption are variables of interest.

In tandem to Zhang, Phanniphong and Li's, (2023) study, Yu and Ramanathan (2015) had explored green products design and the internal green management using 167 manufacturing entities in the United Kingdoms (UK). The study sought to examine the links of adopting green initiatives with stakeholder pressure and green performance as the antecedent factors. The study posit that stakeholder pressures significantly affect internal green management positively (Yu & Ramanathan, 2015). On the other hand, internal green practices were found to have a significant effect on green products. Accordingly, the attributes tested are significantly related to environmental performance. Although insightful, the study generalised stakeholders with no particular focus to crucial stakeholders like suppliers that are regarded partners in achieving sustainable green supply chain.

To this end, the literature is still scanty regarding studies that have investigated green practices particularly in the warehousing industry, yet efficient warehousing calls for great care in the movement of merchandise in a sustainable way. Green supply chain management (GSCM) has gained traction in contemporary times against the backdrop of a plethora of environmental catastrophes and scandals involving prominent companies. Despite the strides made by law makers to address such challenges, an increasing number of corporates overlook environmental issues such as global warming, ecological concerns, reverse logistics, and global energy. Notwithstanding the contributions in prior studies, to the best of researchers' knowledge, no study has sought to link water and energy consumption with supplier credit disbursement. Yet policy makers seek to develop regulations that incorporate the understanding of the collaboration among the supply chain members where warehousing is concerned. Thus, we seek to establish if external warehousing stakeholders exert pressure towards the corporate greening agenda by probing the effect of water and energy consumption on supplier credit disbursement practices. The remainder of this chapter presents the theoretical foundations of the study and related literature in section 2. In 3 and 4, the chapter explores the research methodological approach employed and the discussion of the results respectively. Section 4.5 presents the summary and conclusion of this chapter, proffering the recommendations and suggestions for future research.

2. THE STAKEHOLDER THEORY AND EXPOSITION OF RELATED LITERATURE

The study puts stakeholders at the centre of the discussion given that environmental concerns are beginning to awaken various stakeholders. Hence, this paper finds its basis on the stakeholder theory in interpreting the results of the study. The stakeholder theory is an ethical tool which places emphasis on addressing the needs of different stakeholders affected directly or indirectly by the business. Thus, the proponents of the theory argue that shareholders should not be the only sect to be given the much-needed attention which is normally demanded in form of profits or equity growth. However, attention must be given to other crucial stakeholders as well. The current study zoom into one external stakeholder (suppliers) considered important in the achieving a sustainable supply chain.

2.1 Warehousing and the Climate Change Crisis

Green warehousing is gaining traction given the prevailing climate change crisis. Orzan *et al.*, (2018) argues that green warehousing places emphasis on ensuring that warehouse facilities are environmentally friendly. This means that the design of warehouse, its equipment and operations have to be sustainable and exhibit environmental consciousness. Gerwanski (2020) indicates that green warehouses can be achieved by employing methods that significantly improve sustainability performance. Sustainability performance indicators such as waste reduction, water consumption reduction and energy usage tend to determine the success of the business given that various stakeholders are beginning to show interest in non-financial performance measures (Alhiyari & Kolsi, 2021; Cordery & Sinclair, 2017). Furthermore, Kicova *et al.* (2023) states that green warehousing encompasses green logistics since most of the adverse environmental effects are often caused by movement of goods that are transported within the warehousing industry. Thus, green warehousing practices calls for managers to employ methods that reduces the adverse environmental impact of logistics.

According to Abogazia *et al.* (2022), sustainability performance should be improved as this renders mutual benefit to the business and environmental efforts that seek to curb the climate changes crisis. Thus, businesses tend to operate efficiently, protect their product offering and reduce spoilage as they adopt eco-friendly warehousing operations. Dubihlela and Ngxukumeshe (2016) claims several crucial features for eco-friendly warehouses. First, green warehouses tend to use energy-efficient lighting systems to reduce energy consumption. Second, use of electric vehicles given the growing technologies has become prevalent in a bid to cut down emissions of greenhouses gases. Third, green warehousing calls for clean energy usage which has seen most businesses opting for solar powered systems. Hamrouni, Uyar and Boussaada (2020) agrees that these key features of green warehousing are an exhibit of corporates can help the environment and themselves as eco-friendly arguably reduce operational costs.

Besides, the brand image of the business is reinforced given that businesses who produce excellent sustainability indicators when reports tend to attract more customers (Jung, Herbohn & Clarkson, 2018; Okpiaifo *et al.*, 2020). This further enables the business to avoid investing more on marketing and public relations efforts that often causes budget strain. Additionally, green warehouses tend to enjoy cost saving on regulatory compliance matters by avoiding possible fines that often arise from non-compliance(Fallan & Fallan, 2019; Friedman & Heinle, 2021). In the advent of mandatory reporting for those companies obtaining a listing in the Johannesburg Stock Exchange (JSE), it becomes easy for green warehouses to exhibit their triple bottom line.

Despite the prevailing climate change crisis, literature presents a collage of strategies that the warehouse industry can employ to attain sustainable logistics thereby circumvent the adverse environmental impact. Kordsachia, Focke and Velte (2022) suggests that the warehouse industry accelerates the adoption of alternative fuels with a deliberate intention to lower carbon emissions. This means that the warehouse industry embraces biofuels as a substitute to burning diesel which significantly pollutes the air. Besides, the extraction of coal by operating companies in the mining industry exacerbates carbon emissions. However, the warehouse industry may play vital role in minimising the carbon footprint by implementing energy management systems that can monitor and cut down energy consumption in warehouses (Abdel-Baset, Chang & Gamal, 2019; Johnson, Mans-Kemp & Erasmus, 2019; Fahim & Mahadi, 2022).

Islam et al. (2017) explains that optimisation of transportation routes is imperative for sustainable logistics. This may be achieved if the warehouse industry integrates Artificial Intelligence (AI) technologies. These technologies include software applications that can be employed to conduct traffic analysis such that best routes for making deliveries in real-time may be chosen. Additionally, packing efficiently may assist the warehouse reduce the number of trips which in turn yields benefits of reduced emissions (Tsai *et al.*, 2012; Ionescu, 2021). Lemtaoui and El Oueldrhiri (2019) comments that while the former may be a 'game-changer' in the warehouse industry, green warehouses often collaborate with logistics partners to enhance sustainability efforts. Thus, such logistics companies are often specialists in transportation and regard themselves as the green supply chain partner.

To this end, efficiency and cost savings in green warehousing are critical for the success of both the business and the environment. Issock Issock, Mpinganjira and Roberts-Lombard (2020) speculates that businesses can even achieve better efficiency through resource optimisation. Thus, smart technologies can be used to achieve waste reduction thereby improving profitability. Moreover, warehouse space can be optimised to reduce storage expenditures and other operational expenses. Recycling programmes can also be implemented to improve waste reduction, reduce disposal costs and attain sustainability targets. According to Busco, Caglio and Scapens (2015), space utilisation is crucial in green warehousing. This implies that businesses ought to use versatile multifunctional equipment to cut out the need to utilise multiple machinery. This means that fewer machines can be used which saves operational space and reduces the activities that perpetuate climate change crisis. In some instances, space utilisation may imply considering vertical storage so as to increase storage capacity without widening the size of the warehouse. This extends to the layout or architecture of the warehouse which can be done such that operations are streamlined and reduces operational activities that exacerbates adverse environmental impact.

2.2 Green Warehousing in the South African Landscape

In the view of Macey, Harris and Macey (2022), green warehousing in the South African context calls for the implementation of sustainable practices to push the greening agenda so as to curb the prevailing climate change crisis. Thus, in South Africa, greater emphasis is on warehouse operations, design as well as warehouse management. At the quo of this emphasis is the attempt to primarily reduce the adverse environmental effects while promoting resource efficiency in warehousing. Dzingai and Fakoya (2017) postulates that in the South African context, sustainable warehousing encompasses optimisation of energy consumption, waste reduction, usage of eco-friendly products and harnessing technologies for efficient operations in the warehouse.

To date, several companies in South Africa are exhibiting the successes of sustainability which emanate from green warehousing. One notable company is Ceba

Logistics which prides itself as a sustainable cold chain which has reduced energy consumption and waste significantly over the recent years. Notable successes include 30% reduction in energy costs after adopting solar panels as source of power for their operations (Dilotsotlhe & Duh, 2021). Furthermore, adoption of electric forklifts resulted to high operational efficiency and reduced emissions. However, companies pay a premium as they transition to embrace green practices. Apart from the costs incurred in acquiring eco-friendly technologies, hefty fees are paid to ensure that employees are skilled in using new technological systems. Moreover, exorbitant amounts are further incurred in modifying existing logistics to ensure that the infrastructure caters for new sustainable practices.

Duffett *et al.* (2018) indicates that the South African government has been supportive to companies that evolve towards adopting sustainable practices. The support is exhibited through grants that are set aside to fund green initiatives. Conversely, companies embracing green warehousing tend to partner with sustainability consultants in the private sector to train their employees. Govender and Govender (2016) contends that South African companies incorporate new technologies gradually in sustainable warehousing to spread the cost element over a long period of time. Notwithstanding the challenges faced by South African companies that have embraced green warehousing, fundamental lessons can be derived. First, sustainability is a journey, as such companies should accept that it may take time for them to achieve their sustainability performance targets. Second, while it almost obvious to this end that green warehousing benefits the environment directly, it enhances brand loyalty given that most stakeholders have evolved to embrace green practices (Ojo, Mbowa & Akinlabi, 2014; Mafini & Loury-Okoumba, 2018; Chinomona & Mahlatsi, 2020). To significantly improve sustainability, companies in South Africa regularly assess and adapt their greening strategies.

2.3 Water Usage in Warehousing

The climate change crisis has in recent times greatly affected water supply. Water is a crucial natural resource that sustains life for all God's creation. Water shortages has negatively affected the ecosystem, with some species becoming extinct as the crisis deepens. The sustainability reports produced by companies to display their environmental impact also place emphasis on water usage as critical indicator for environmental performance. Notwithstanding its vitality as resource in most industries, water usage in the supply chain if often underestimated (Cini & Ricci, 2020; Tao, Wu & Zhao, 2022). However, companies that seek to embrace or evolve towards green warehousing ought to make a deliberate attempt to thoroughly comprehend water usage in the supply chain so that efficient and sustainable operations can be achieved. This follows that water is common resource in all economic levels of production.

Thus, the extraction of raw materials in the primary sector and the distribution of the finished goods in the secondary sector depicts water as an integral part.

The scarcity of water has manifested itself as a global phenomenon which is exacerbated by the climate change crisis. Thus, companies now prioritise using water sparingly and optimally to reduce inefficiencies in their operations. The main goal of companies evolving towards green warehousing is to boost their bottom-line, building a more planet friendly and sustainable future. The era of going green has given rise to households who employ water reduction strategies at micro level. However, the same approach has been translated to large corporates that seek to mitigate risks associated with climate collapse. Ali *et al.* (2021) argues that the production stage in the supply chain is regarded as water-intensive given that most industries uses water for cooling, cleaning, irrigation or as a raw material for most goods that manufactured for consumption. Thus, managers ought to outline strategies that are aimed at dealing with areas where there a significant consumption of water. The advent of technologies can be harnessed to employ monitoring systems that can track water usage for real-time analysis. The implementation of water conservation strategies can also be aided by such technologies given that real-time data can be obtained. For instance, water efficient technologies including water recycling or closed-loop systems can be employed to achieve water usage.

According to Amemba (2013), the supply chain processes uses a significant amount virtual water which that is regarded as the hidden water. This means that other than water that easily recognisable and linked to the supply chain, a significant amount of water is used in the production raw materials or components outsourced from various supplies. Thus, the computation of water usage at a warehouse facility should compound and aggregate the virtual water utilised in all inventories stored at the facility. This approach indicates a more accurate assessment of the environmental impact of specific supply chain operations. Dilotsotlhe and Duh (2021) suggests that going green calls for companies in the warehouse industry to source materials and goods produced by regions that demonstrate lower water stress. This approach mitigates the adverse effects on the environment emanating from massive water usage in the operations (Fakoya & Malatji, 2020). Furthermore, the partnerships with suppliers in greening the supply chain often yields benefits associated with reduced water consumption and other inefficiencies.

Dilotsotlhe and Duh (2021) advocates for several strategies that can be adopted to reduce water consumption in the supply chain. Thus, companies ought to assess their water foot print regularly. This entails measuring and understanding the amount of water utilised in the entire supply chain linked with business' operations. Duffett *et al.* (2018) claims that tools such as the Water Risk Filter (WRF) or Water Footprint Network (WFN) have become an imperative in today's green warehousing. These tools help businesses identify key supply chain areas that exhibit highest environmental impact. Furthermore, such tools can assist in benchmarking the entity's performance against industry best practices and standards. The assessment of the water foot print using this approach assists managers to strategically set more realistic targets for water usage.

Optimisation of processes is imperative in eliminating inefficiencies in water usage. This calls for the implementation of lean manufacturing principles that advocates for waste minimisation, value maximisation and quality maximisation. This premise propels companies to install water-efficient equipment, cut down on water intensive activities, using water-smart architecture for the company infrastructure and applying preventative maintenance programmes. Furthermore, stakeholder engagement can also play a pivotal role in reducing water usage in the supply chain. Thus, companies must be prepared to invest in educating stakeholders about water conservation strategies. This investment should extend to training employees and incentivising other partners in the supply chain for reducing the water footprint. Sonnenberg, Erasmus and Donoghue (2011) contends that monitoring water usage remains a critical step for companies attempting to go green. There is a constant need for companies to develop metrics and key performance indicators (KPIs) that relate to water usage. These can be utilised to perform regular audits and brings a sense of accountability to stakeholders who have access to sustainability

2.4 Energy Consumption in the Warehousing Industry

In the view Fatoki (2020), the warehousing industry has significant usage of energy at facilities level and throughout the entire supply chain. Thus, the industry has considerable impact on the environment caused by production, transportation and other operations associated with warehousing. Energy consumption has a hidden environmental impact emanating in the processes followed in power generation. Thus, energy consumption in warehousing escalate the burning of coal and diesel which are primary culprits in exacerbating air pollution. Butler, Henderson and Raiborn (2011) suggests that energy saving in green warehousing is achieved by identifying the main areas of consumption in the operations.

To sum up, continuous lighting at the facilities contributes towards energy consumption over-time while cooling systems are considered heavy on the grid. Deep-freeze and refrigerated warehouses are continuous cooling systems for products that are perishable. Energy saving in such instances calls for improved collaboration with all the partners in the supply chain such that movement of such goods is seamless. Furthermore, use of technologies that may be relevant for the green agenda may also contribute to a rise in energy consumption. This conundrum cannot be solved by switch off energy consuming equipment at a warehouse facility, rather attempt to implement strategies that seek to optimise energy efficiency.

2.5 Increasing Energy Efficiency and Sustainability

The increasing technological developments to what others believe to be the fifth industrial revolution has given rise to the commensurate development of energy efficient technologies for warehouses (Nguyen, Lobo & Nguyen, 2018; Fatoki, 2020; Issock Issock, Mpinganjira & Roberts-Lombard, 2020). Thus, lobby groups that push the greening agenda report some wins towards achieving optimum climatic conditions through employing heat recovery systems, energy efficient air conditioning systems and ventilation and energy efficient heating. Moreover, advanced technologies championed by the information technology (IT) sector can be harnessed in green warehousing which involves integration of software for managing energy and maintaining energy efficient hardware systems.

Abdel-Baset, Chang and Gamal (2019) posit that automation is a game-changer in green warehousing as this optimises energy efficiency in operations. For instance, warehouse facilities that are automated would only require heating or cooling only and only if goods stored in the facility require it. Moreover, processes in the warehouse can be optimised to achieve greater efficiency. This means that cost control, productivity, stakeholder satisfaction and sustainability can be achieved if warehouse processes are efficient such that energy and other resources are used optimally. Mor, Srivastava, Richika and Varshney (2019) recommends just-in-time (JIT) inventory management techniques which are said to minimise inventory levels thereby creating more storage space. JIT entails that inventory is only ordered and delivered when needed as such warehouse processes are optimised given that lower inventory levels are kept for storage as any given time in the warehouse facility.

2.6 South African Government's Policy on Warehousing

The South African government has made strides towards mitigating risks associated with the climate change crisis. The government enacted laws to regulate various industry sectors regarding activities that may have adverse environmental impact. In 1964, the South African government enacted the Customs and Excise Act to regulate the movement and storage of imported goods. Through this law, the government is able to provide licences for customs and excise storage warehouses. Green warehousing is prevalent in the private sector given the stakeholder and regulatory pressure that companies experience relating to compliance as well as embracing green practices. However, the public sector strives to benchmark green warehousing on the state-owned storage facilities. For instance, the South African

Revenue Services (SARS) own warehouses where goods are kept under their custody. These inventories often belong to companies that are being processed for importation.

In practice, some importers delay processes relating to duty payment and in some instances other customs may still be pending.

Mgbame and Kemi (2017) explains that state owned warehouses are classified as Customs and Excise Storage Warehouses (CESWs), Special Customs and Excise Storage Warehouses (SCEWs), and Specific Economic Zones (SEZs). The CESWs generally allow importing companies to store their goods with pending customs requirement or duty payments. Furthermore, SCEWs are specialised warehouses for safekeeping of specialised goods as the name suggests. For instance, bonded fuels products are regulated as such they cannot be stored haphazardly. Compliance with the Customs and Excise Act is mandatory for all importers and SARS ensures that all citizens abide by the demands of this law. Warehousing operations are formally registered and licensed to aid regulation and monitoring by relevant authorities. Fonseca (2020) comments that bonded warehouses are utilised for the storage of imported goods that are charged duty. The government policy extends to supply chain management policies that governs warehousing.

2.7 Credit Disbursement for Warehousing Activities

In the South African landscape, the government has made strides in transitioning towards resource efficiency, carbon emissions reduction and climate resilient development path set to deliver benefits relating to environmental, social and governance (ESG). The government renders financial incentives to companies adopting green initiatives through the Department of Environmental Affairs (DEA) which manages green funds. However, Fahim and Mahadi (2022) argues that the funds made available by the government are not commensurate with the amount work that needs be done in the country to reach the climate change mitigation targets. This is partly due to a struggling economy with budgets constraints that caused the finance minister attempt to increase Value Added Tax (VAT). Therefore, companies that seek to adopt green warehousing ought to be prepared to invest hefty amounts to achieve their ambitions. It is evident from a South African standpoint that while many companies struggle to raise needed capital to invest in green practices, many tend to maximise their social capital from suppliers. This social capital translates to credit disbursement by suppliers who renders their support to ensure that green practices are achieved (Santos, Lannelongue & Gonzalez-Benito, 2019).

According to Muttakin *et al.* (2020), the green fund in South Africa is provided to deal with the prevailing challenges in the anticipated transition to green economy. Thus, the funds seek to promote high impact and innovative green projects that entrepreneurs undertake. Moreover, this support seeks to reinforce the strides made by government towards achieving climate policy targets by accelerating green interventions. The green fund comes following the back-drop of numerous companies in the private sector that have embraced green practices within the context of mandatory sustainability disclosures and reporting (Boachie & Tetteh, 2021; Sands, Gholami & Shams, 2022). Thus, green funds supplement the private sector efforts to solidify evidence base for adoption of green practices by companies that still displace high environmental impact in their operations. Khalil and Khalil (2022) postulates that primarily, green funds targets boosting green economy by developing green cities, low carbon economies and natural resource management expertise.

2.8 Literature Exploration on Corporate Green Practices and Supplier Behaviour

In conjunction with the above findings, Amemba (2013) sought to highlight the GSCM best practices in the hospitality sector in Kenya. This was done in view of rising environmental pollution and greenhouse emissions by Kenyan companies which are called to conserve scarce resources. Moreover, companies in the hospitality sector in Kenya depend on water and energy as major resources to derive best service offerings for their consumers (Anne, 2015). Amemba's (2013) study revealed through in-depth literature review that companies within the hospitality industry in Kenya implement global GSCM best practices to achieve a sustainable supply chain operation. Thus, companies are applauded for efforts in implementing GSCM practices to match those of developed countries where suppliers value customers' ESG practices.

To further confirm Amemba's (2013) findings in Kenya, Muma (2014) investigated the effect of GSCM on the sustainability performance among tea processing companies in the same country. GSCM practices entail that suppliers engage their customers in a partnership that seeks to maintain a green and sustainable supply chain. This means that where suppliers value customers' green supply chain practices, environmental performance of such company tend to be favourable. Muma (2014) employed statistical methods to conduct tests using data collected from tea processing companies in Kenya. The results indicate a positive association between GSCM and sustainability performance (Muma, 2014). It is evident that suppliers value customers' green practices and consider such practices when making crucial decisions affecting various stakeholders of the supply chain.

Notwithstanding the importance of medicine logistics in Nigeria, oil and gas sectors in the same country play a vital role towards survival, growth and development of the country's economy (Akintokunbo & Arimie, 2021). Salisu, Omoke and Fadiya (2022) claim that the gas and oil industry is the biggest contributor of revenue for the government, with proceeds emanating from this sector amounting to more than half of total government revenue. However, despite their immense contribution to revenues, the industry also happens to be the biggest contributor of pollution in the environment. This effect is a result of companies' activities that propel recurrent ecological disasters and environmental degradation in the country to date. Akintokunbo and Arimie (2021) claim that environmental concerns in oil and gas sector in Nigeria emanate from supply chain practices of these companies.

In support of these findings, Abba, Norhayati, Salisu, Usman, Saif and Almasradi (2019) explored peculiarities of environmental performance practices associated with supply chain of companies operating in Nigeria's oil and gas industry. The focus was on oil servicing companies in the downstream sector. This is based on the notion that prior literature on supply chain management practices in this area is scant especially in the African context. Abba et al. (2019) demonstrated that GSCM practices of oil servicing companies in Nigeria do not consider requirements of ESG practices. The companies investigated in the study were those in producing, delivering, marketing, and designing. These negative findings imply that suppliers in the supply chain do not value ESG practices of buyer-companies, as such interventions to green the supply chain practices should be solicited.

In contrast to the findings of Chukwu et al.(2018) in Nigeria's health sector, Gbadeyan and Boachie-mensah (2017) reports mixed findings by looking at private hospitals operating in the same country. The increase in consumer concerns regarding service delivery in the health care sector is linked to poor quality health care services. These concerns point out probable problems with supply chain management practices by key players in the sector (Chukwu et al., 2018). Gbadeyan and Boachie-mensah (2017) adopted a quantitative approach and consequently employed Partial Least Squares (PLS) method to estimate SEM's effect in GSCM on hospital performance.

The results present a double-edged sword of the health sector sustainability in Nigeria. First, Gbadeyan and Boachie-mensah (2017) reports that GSCM has no direct and strong impact hospital performance. Thus, suppliers in medicine logistics neither value reverse logistics nor customers' green supply chain practices given that these elements do not impact on performance. Conversely, results also demonstrated that GSCM has an indirect effect on performance through competitive advantage (Gbadeyan & Boachie-mensah, 2017). This means that the few reliable suppliers in the sector value efficient GSCM for the purposes of gaining competitive advantage. While these findings are insightful, the study is criticised for using a small sample that does not accommodate the heterogeneous population being studied.

3. METHODOLOGICAL APPROACH AND DATA ANALYSIS

Kumar and Ghodeswar (2015) states that philosophical assumptions underlying the study researcher must be set out by the researcher. In this chapter, the researcher adopts positivism to frame the research study. Positivism is a research paradigm preferred by most researchers for its association with quantitative techniques (Elo *et al.*, 2014). According to Harwood and Garry (2003), a quantitative research study imply that variables are empirically tested through observation and measurement. Thus, the philosophical stance of the natural scientist posits that positivism focuses on observable and measurable facts (Saunders & Lewis, 2018). The quantitative nature of the positivist paradigm therefore makes it suitable for this study. Accordingly, Groenewald (2004) postulates that quantitative research is used when researchers are interested in testing the relationship between the variables. Quantitative approach is deemed suitable for this research study given that it employs a deductive approach to explain the relationship between two or more variables (Creswell, 2015). Moreover, the study employs numerical dataset and first differenced statistical models to produce the results (Mertler & Reinhart, 2016; Saunders & Lewis, 2018). Besides, given the nature of the current study, a multiple case study design was utilised to investigate the underlying phenomenon. A multiple case study approach allows the researcher to select a number of cases to enhance analytical generalisation and make comparisons across several cases (Crowe et al., 2011). This design is the most preferred for the current study that seeks to investigate the cases of companies in the FTSE/JSE Responsible Investment Index. Furthermore, the quantitative nature of the research study informs the population, sampling technique and the research sample utilised for the study.

3.1 Population, Sampling Technique, and Research Sample

Naidoo and Henry Ramatsetse (2016) makes clear that the population of a research study comprises the total of all the known objects, members or subjects that conform to the set of specifications. The JSE introduced the Socially Responsible Index (SRI) in 2004 which consisted of 57 companies. This followed the introduction of sustainability initiatives globally and the King IV in the South African landscape to address governance matters (Greyvenstein, 2010). Companies listed in the JSE All Share Index were reviewed annually against the environmental, social and governance (ESG) concerns for inclusion in the socially responsible index (SRI). The SRI grew to 82 companies by the end of December 2014 which marked the last review under the SRI banner. However, SRI was subsequently replaced by the FTSE/JSE Responsible Investment Index (RII) which was launched in October 2015 to continually promote sustainable development and foster good corporate citizenship. As a result, the population of this research study comprises all companies appearing on the FTSE/JSE RII from 2016 to 2021. This implies that the population will comprise of 60 companies listed on the FTSE/JSE RII as of 28 June 2021. Furthermore, Hui, Foxcroft, Richardson and Macfadyen (2011) refer to a sample is a subsection of a population chosen to participate in the study. This research study employed judgemental sampling technique which is preferred where the researcher uses a special criterion to select objects for inclusion in the sample. Besides, Agresti and Finlay (1998) affirms that this technique is a non-probability sampling method employed when the researcher's judgement is used to select elements for the sample based on a range of premises. The researcher collected environmental performance data of companies that have reported the environmental performance indicators consistently for a period of six years. Consequently, the sample comprises 19 companies drawn from the 60 FTSE/JSE RII constituents in the 2021 fiscal year. This represents 32% of the population, which is deemed acceptable for the study given the environmental impact and reporting practices of companies sampled purposively.

3.2 Research Data and Collection Procedure

Secondary data are data that have already been collected from other sources and are readily available for use by the researcher. Boslaugh (2009) argues that while this data is collected for a purpose other than the research at hand, it helps the researcher to have a better understanding of a phenomena. Secondary data is suitable for its employability in exploratory, correlational, descriptive and evaluative studies (Hox & Boeije, 2004). Given the benefits of using secondary data, this study utilised panel data retrieved from the Integrated Annual Reports (IARs) of the top FTSE/JSE SRI companies. These companies were investigated for the six financial years from 2016 to 2021. The choice of these financial years is justifiable given that the FTSE/JSE SRI was launched in 2015. Moreover, reporting on corporate environmental performance only became mandatory for listed companies in 2010 (Kamala, Wingard & Cronje, 2015; JSE, 2016). Thus, the fiscal years 2016 to 2021 are considered most recent to be employable given that markets and environmental sustainability concerns have evolved sharply at global stage level calling for a holistic ESG criterion.

Furthermore, this study employed content analysis to collect data needed to meet the objective of the study. The researcher employed this approach to sift through archival documents to identify the needed data for the phenomena being studied (Guthrie, Petty, Yongvanich & Ricceri, 2004). Data were collected from the documents sourced and captured on Microsoft Excel spreadsheets in preparation for the data analysis process using chosen statistical software packages. Where a variable needed calculation, such calculations were done on the Excel spreadsheet. Based on the sampled companies, it was anticipated that some companies may be computing their reports using foreign currency. In such cases, the researcher converted all foreign currency to the local one for the purposes of consistency in the analysis processes. Additionally, where the variable measurement method employed by reporting companies differs, conversions were done on an Excel spreadsheet for standardisation purposes. Where the researcher encountered missing data, the interpolation technique was employed to fill in the missing data points. Its applicability is justified as the interpolation technique is widely employed in most econometrics and statistics research (Church, 2002; Boslaugh, 2009).

3.3 Operationalisation of Main Variables for the Study

The table (Table 1) below shows how the data that relate to independent variable and control variables as extracted from the IARs and SRs for the purposes of analysis.

Table 1. Measurement of independent and control variables

Variable	Variable sub-name	Measurement method in IARs and SRs
Independent variable	Energy consumption (ConsE)	Giga-joules (Gj) or Mega-watt hours (Mwh)
	Water consumption (CoW)	Million cubic metres (Mm^3) and kilolitres (Kl).
Control variables	Total monetary value of collateral (MvCOL)	Total tangible fixed assets reflected in the IARs
	Total earnings (Te)	Total of Net Profit after tax

Source: Author's compilation, 2025

The environmental performance indicators presented above are informed by the Global Reporting Initiative (GRI), Eco-Management and Audit Scheme (EMAS), and European Commission's guidelines. The reporting companies are guided by these guidelines to produce standardised reports for non-financial information. Companies presents such information in their sustainability reports, ESG reports, and or IARs. These indicators are presented as a panel data set of the sampled companies to statistically analyse if they influence supplier support in the context of the current study.

3.4 Operationalisation of Dependent Variables

Suppliers consider customer-companies to be partners in achieving notable results in as far as supply chain performance is concerned. The regulatory environment has tightened all facets of supply chain management making suppliers rethink their relationship with buyer-companies. Traditionally, suppliers would care less about sustainability issues of the buyer-companies since profiteering is the driving force. However, the advent of the ESG agenda is changing stakeholders' perceptions. The current study investigates if buyer-companies' environmental performance attracts supplier support considering the demands poised by green supply chain management (GSCM) agenda. In the perspective of the buyer-company, suppliers are crucial stakeholders as they play the role of micro-lender of short-term credit. Suppliers often express their support through flexible credit terms such as increased credit limits, discounts rendered for prompt payments, extended credit with minimal interests charged on overdue accounts etc. Thus, Table 2 below shows suppliers credit disbursement as the variable of interest measured by total value of creditors control reported in the Statement of Financial Position.

Table 2. Measurement of the dependent variable

Variable	Variable sub-name	Measurement method in IARs
Dependent variable	Suppliers' credit disbursement (ScD)	Total value of creditors control reflected in the IARs

Source: Author's compilation, 2025

This table suggests measurement methods to be employed in testing suppliers' green supply chain performance. The use of total value of creditors is justifiable as this depicts the degree of supplier-customer dealership. Wang, Zhao and Hou (2020), Kong, Feng and Huo (2021), Zhang, Pan and Feng (2020), Ali et al. (2019) and, Simpson and Power (2015) are some of the proponents of proxies adapted in measuring the variables for this objective.

3.5 Panel Data Analysis Procedures

The study involves a balanced short panel data set of all the companies depicted in the sample. Frees (2004) explains that panel data analysis techniques are useful tools for dealing with cross-section time series data. Thus, such techniques have been widely used by researchers in the field of economics (Hsiao, 2007). To this end, panel data analysis is preferred by most researchers when investigating numerous research questions in diverse fields of research. Furthermore, panel data analysis tools have been employed by researchers such as Nervlove (2002), Arellano (2003), and Baltagi (2001). Baltagi (2001) argues through econometrics that panel data is preferred in that it supplies the researcher with highly informative data that tend to be robust in yielding more conclusive results on the phenomena being investigated.

However, short panel data models often have a difficulty in ascertaining causal linkage among variables.

The data utilised using this statistical tool provides greater variability and less collinearity between research variables. Moreover, the data provides more degrees of freedom and efficiency (Baltagi, 2001). To support this assertion, Nervlove (2002) indicates that the advancement in technology has aided panel data to be utilised in a more comprehensive manner where more econometric relationships are analysed. In addition, panel data uses both cross-sectional and times series data sets in a repetitive manner for a given number of observations (Arellano, 2003). Therefore, the researcher employed panel data analysis given that the current study uses variables presented in time series cross-sectional data sets.

The study employed two powerful software packages for data analysis: STATA and Eviews 13. Descriptive statistical analysis, pairwise correlation, Panel Vector Autoregressions for Fixed-effects estimations were mainly conducted using STATA for a modest presentation of results. To complement STATA, Eviews 13 was utilised to run some statistical diagnostic tests.

Test of Hypotheses

The adopted research design and methodological approach employed in this research study is deemed adequate to test the null hypotheses restated below as:

> ***H1:*** corporate green performance does not influence supplier credit disbursement.

The Main Decision Rules

Pearson pairwise correlation matrix: The correlations among variables were tested at 3 levels of significance (at 10%, 5% and 1%). If $P < 0.05$, where $\alpha = 0.05$ is the level of significance, then reject the null hypothesis of no correlation between the variables. This follows that the alternative hypothesis of evidence of correlation between variables is accepted. If $P < 0.10$, where $\alpha = 0.10$ is the level of significance, then reject the null hypothesis of no correlation between the variables. This implies that the alternative hypothesis of evidence of correlation between variables is accepted. If $P < 0.01$, where $\alpha = 0.01$ is the level of significance, then reject the null hypothesis of no correlation between the variables. This entails that the alternative hypothesis of evidence of correlation between variables is accepted.

Hausman test for Random vs Fixed effects: If $P < 0.05$, where $\alpha = 0.05$ is the level of significance, the null hypothesis that the random effects model is consistent is rejected. This means that we accept the alternative hypothesis which is in favour of the fixed effects model. Conversely, where P > 0.05 then the null hypothesis is accepted, implying that the random effects model is more appropriate.

Breusch and Pagan Lagrangian multiplier (LM) test for Random effects: If $P < 0.05$, where $\alpha = 0.05$ is the level of significance, then reject the null hypothesis favouring simple ordinary least squares (OLS) regression for existence of significant differences or variances across firms. This follows that the alternative hypothesis favouring random effects Generalised Least Squares (GLS) regression for no panel effect is accepted.

3.6 Descriptive Statistics and Pearson Pairwise Correlation

This study employed the statistical software STATA to generate both descriptive statistics and pairwise correlation matrix. Fisher and Marshall (2009) argues that in descriptive statistics, data can be organised, analysed, and presented in a meaningful manner. The current study envisaged presenting data such as mean, minimum, maximum, observations, and standard deviation in a table format. According to Abba et al. (2019), descriptive statistics is a useful tool for non-parametric and parametric tests. Moreover, it is also argued that descriptive statistics is imperative in both quantitative and qualitative research, hence it should be performed to give the researcher a glimpse of the dataset utilised in the study (None & Kumar Datta, 2011; Ionescu, 2021). Furthermore, descriptive statistics is of paramount importance as it unravels normality in the distribution of the sample employed in the study and gives a clear depiction of any outliers in the dataset (Gill et al., 2008; Murray & Marshall, 2009; Marshall & Jonker, 2011). Most importantly, normality reveal the degree of sharpness and symmetry in the dataset being examined.

Farrokhi and Mahmoudi-Hamidabad (2012) indicates that descriptive statistics is also adulated in research for providing information on the measures of dispersion and central tendency. Conversely, pairwise correlation is commendable for enabling the researcher to compare and explore the relationships between variables (Agresti & Finlay, 1998). The researcher pronounces that the research data utilised in this study fulfilled all relevant statistical diagnostic tests as expatiated in the next section.

3.7 Statistical Diagnostic Tests and Panel Data Analysis Techniques

Prior to conducting the study, the researcher conducted several relevant preliminary diagnostic tests (Ngo, 2012; Myers, 1990; Marshall & Jonker, 2011). This preceded the selection of the models employed in the study given such tests renders the selection criteria. Marshall and Jonker (2011) and Getahun (2014), suggests that model estimation process commences with the panel unit root test to check if the series is stationary at level – I(0) or at first difference – I(1). In the case of stationarity at I(0) the Pooled Ordinary Least Squares (POLS) model may be employed. However, in the latter the researcher proceeds to perform the cointegration test to check if long-run relationship exists between variables. The cointegration test may reveal that variables are not cointegrated which allows the research to perform the Hausman specification test to choose between the Panel Vector Autoregression (PVAR) Model for Fixed Effects or Random Effects tests. Conversely, for variables that are cointegrated, the Panel Vector Error Correction Model (PVECM) may be adopted to enable the researcher observe short-run and long-run dynamics among the variables. Accordingly, numerous statistical diagnostic tests performed comprised panel data tests for normality, serial correlation, multicollinearity, heteroscedasticity and stationarity.

Variables for Kao Residual Cointegration Test

This section provides details of the variables utilised in performing the cointegration test objective pursued by this research study. The researcher employed Eviews 13 software to perform Kao Residual cointegration test.

Table 3. Variables for Kao Residual Cointegration Test

Endogenous explanatory variables	**Endogenous outcome variable**
CoW(*x1*); ConsE (*x2*); EmGHG(*x3*); MvCOL(*x4*); Te(*x5*)	ScD(*y*)

Source: Aurthor's compilation, 2025

In tandem to these tests, the researcher applied varied statistical estimations to test the proposed main hypothesis of this study. These estimations include Panel Vector Autoregression (PVAR) models for fixed effects estimations to generate the main results of the study. The Pearson correlation matrix was computed to complement the results of the main model specified for this study. The next section deals with model specification and operationalisation for the purpose of statistical analysis.

3.8 Statistical Modelling of the Study

The researcher employed the techniques above in order investigate if corporates green performance influence supplier credit disbursement. Getzmann, Lang and Spremann (2014) argue that statistical models in research can render intuitive visualisations that aid the researcher in identifying relationships between variables being studied and make predictions using such statistical models to raw data. As such, the next section presents composite variables and model specification adopted for the study.

Operationalisation of the Basic Panel Data Model

The basic panel data model can be presented as shown in Equation (1) below:

$$y_{i,t}=\alpha+x_{i,t}'\beta+c_i+\delta_t+\varepsilon_{i,t} \qquad (1)$$

where y and x denotes the dependent variable and vector of explanatory variables for firm i at time (year) t, respectively. The time trend t allows for a shift of the intercept (a) over time, capturing time effects such as technology and regulation changes among others. However, if the implicit assumption of a constant rate of change is strong, a set of time dummies is introduced (δ_t), one for each time-period except reference period. The subscript ε is the error term. The component c captures the variables responsible for unobserved heterogeneity (unobserved effect). It is usually a "nuisance" component of the model.

If the explanatory variables are so comprehensive that they capture all relevant attributes of i, then c_i can be dropped, and pooled ordinary least squares (OLS) can be used. However, this situation is highly unlikely. Generally, dropping c_i may lead to bias estimations resulting from missing variables problem. Thus, c_i is strongly assumed to be strictly exogenous to the conditional error term as shown in equation (2) below.

$$E[\varepsilon_{i,t} \mid x_{i,1},x_{i,2},\ldots,x_{i,T},c_i]=0, \qquad t=1,\ldots,T \qquad (2)$$

Under the strict exogeneity condition as highlighted in equation (2), equation (1) can be reproduced as follows:

$$E[y_{i,t}\ \varepsilon_{i,t} \mid x'_{i,t},c_i]=\alpha+x_{i,t}'\beta+c_i+\delta_t \qquad (3)$$

This implies that β' are partial effects holding c_i constant. However, to estimate β the relationship between $x_{i,t}$ and c_i must be clarified, giving rise to different estimators of Equation (3) such as the pooled, fixed effects, and random effects models. The pooled model's assumptions are often unrealistic, as previously mentioned. Hence, most researchers opt for either fixed or random effects model based on the results of the Hausman Test.

The fixed effects model (FEM) assumes that the unobserved effects (c_i) are correlated with the regressors ($x'_{i\cdot}$).

$$E[c_i \mid x_i] = g(x_i) = \alpha_i \quad (4)$$

$$\text{Then, } y_{i,t} = \alpha_i + x_{i,t}'\beta + \varepsilon_{i,t} \quad (5)$$

As a result, the regression line can be shifted up and down by a fixed amount of each i. Equation (5) is consistently estimated using OLS.

The random effects model (REM) assumes that the differences between cross sectional units are random, drawn from a given distribution with constant parameters. Most importantly, it assumes that that the unobserved effects (c_j) and the regressors are uncorrelated.

$$E[c_i \mid x_i] = \mu \quad (6)$$

$$y_{i,t} = \mu^* + x_{i,t}'\beta + v_{i,t}, \quad v_{i,t} = u_i + \varepsilon_{i,t} \quad (7)$$

The composed error term ($v_{i,t}$) introduces contemporaneous cross-correlations across the i group. While equation (7) can be consistently estimated using OLS, GLS and maximum likelihood estimator (MLE) may offer greater efficiency.

4. EMPIRICAL RESULTS, ANALYSIS, AND DISCUSSIONS

4.1 Descriptive Statistics of the Panel Data

The researcher commenced with statistical analysis of panel data by generating a summary of descriptive statistics. The data utilised were extracted from IARs and SRs of sampled companies. Notwithstanding the benefits of using any statistical software for data analysis, STATA and Eviews 13 were utilised to import data from MS Excel to perform the analysis. These software packages for statistical analysis are employed by most researchers for their easy to use interface, graphics quality, command line programming interface, sophisticated data management and powerful analytic tools (Onditi, 2016). Besides, these software packages simplify statistical analysis especially for the current study where panel data are used.

The researcher generated the descriptive statistics summary which was later exported to MS Excel for editing. Accordingly, the researcher edited the summary statistics table by eliminating sum square deviation and other post-estimations (Jarque-bera and probability) which were considered unfit for descriptive statistics analysis. Thus, mean, standard deviation, minimum and maximum values were presented for all composite variables employed in the study. Table 4 depicts positive mean values for all variables employed in the study. The minimum and maximum values assist the researcher understand how data are spread out in each variable. The range can also be calculated for each variable to check how wide the dataset is spread. Moreover, standard deviation abets the researcher in understanding how far observations in the dataset are from the mean values. Furthermore, the researcher also considered kurtosis values to understand the normality of the dataset being scrutinised. The data generated a positive kurtosis given that kurtosis values were higher than the normal distribution kurtosis of 3. This leptokurtic kurtosis suggests a peaked-curve for the given dataset with leverage (LEV) variable showing significant sharp-peakness of the curve with the kurtosis value above the mean.

Table 4. Summary of descriptive statistics

Variable		**Mean**	**Std. Dev.**	**Min**	**Max**	**Observations**
Water consumption (CoW)	overall	7896383	15200000	7583.725	52200000	N = 114
	between		15500000	9406.539	46200000	n = 19
	within		1751068	-5193284	13900000	T = 6
Energy consumption (ConsE)	overall	405755.6	1235945	14562.69	6220000	N = 114
	between		1258856	20846.33	5578333	n = 19
	within		113855.8	-412578	1047422	T = 6
Supplier credit disbursement (ScD)	overall	10163.35	10887.99	252	43196	N = 114
	between		10583.09	367	39215.33	n = 19
	within		3391.503	-4977.82	19128.18	T = 6
Collateral value (MvCOL)	overall	68152.74	78000.32	634	388222	N = 114
	between		77996.11	3741	315318.5	n = 19
	within		16426.57	-15189.9	141056.2	T = 6
Total Earnings (Te)	overall	5790.132	15005.35	-91917	79021	N = 114
	between		7273.556	-4922.33	22948.67	n = 19
	within		13213.51	-81204.5	61862.46	T = 6

Source: Author's computation, 2025

In Table 4 above, all variables utilised in the study were presented to show measurements for dispersion and central tendency. It should be noted that '*N*' represents the number of observations in the panel dataset employed in the study. On the other hand, '*n*' represents the sample i.e. the number of companies whose panel data set is being examined. Most importantly, '*T*' represents the time-period being investigated (2016 to 2021) in the study. Moreover, the standard deviation shows how dispersed the data is relative to the mean values. The mean values depict the numeric quantity that represents the centre of a collection of numbers.

4.2 Results Demonstrated by the Pearson's Pairwise Correlation Matrix

Notwithstanding the significance of data presented in Table 4 above, a pairwise correlation matrix was computed to understand how variables under study relate or depend on each other. Accordingly, data were exported from MS Excel to STATA 18 which is an advanced integrated software package used by researchers for all data science needs. This software package was preferred by the researcher for modest visualisation of pairwise correlation results. This implies that the correlation coefficient values, probability values and number of observations are clearly depicted in the correlation matrix. Moreover, STATA 18's automated reporting allows the researchers to present the pairwise correlation matrix with starred correlation coefficients where there is positive correlation which is significant at a predetermined level of significance.

Moreover, the estimated correlation coefficient of one (1), zero (0), and negative one (-1) represents a perfect positive, neutral, and perfect inverse relationship, respectively. In general, 0.5 benchmarks a strong and weak positive relationship, whereas -0.5 benchmarks a strong and weak negative relationship. Furthermore, the estimated correlation coefficients can be utilised to detect any potential multicollinearity between variables. An estimated correlation coefficient of at least 0.8 is considered an indication of the potential presence of multicollinearity between the variables (Baltagi, 2008). Table 5 reveals that there is no correlation coefficient above 0.8, indicating the absence of potential multicollinearity between the variables.

Table 5. Pearson's pairwise correlations matrix

Variable	**Water consumption (CoW)**	**Energy consumption (ConsE)**	**Suppliers' credit lending (SL)**	**Collateral value (MvCOL)**	**Total Earnings (Te)**
Water consumption (CoW)	1				
Energy consumption (ConsE)	-0.147	1			
Supplier credit disbursement (ScD)	-0.124	-0.002	1		

Variable	Water consumption (CoW)	Energy consumption (ConsE)	Suppliers' credit lending (SL)	Collateral value (MvCOL)	Total Earnings (Te)
Collateral value (MvCOL)	-0.218**	0.013	0.750***	1	
Total Earnings (Te)	0.218**	0.087	0.070	-0.007	1

Note: *, **, and *** denote significance levels at 10%, 5%, and 1%, respectively.
Source: Author's computation, 2025

The pairwise correlation matrix is used to determine the strength and direction of the relationship between two variables. Accordingly, supplier credit disbursement (ScD) variable has a negative correlation with both water consumption (CoW) and energy consumption (ConsE) variables, albeit statistically insignificant and weak. Although insightful, supplier credit disbursement (ScD) does not provide evidence of correlation when paired with the main exposure variables of environmental performance namely water consumption (CoW) and energy consumption (ConsE). Accordingly, as observed in Table 5 below, consumption (CoW) and energy consumption (ConsE) produced weak correlation coefficient values of -0.124 and -0.002 when paired with supplier credit disbursement (ScD) respectively.

4.3 Statistical Diagnostic Tests Conducted on Panel Dataset

Table 6 depicts the statistical diagnostic tests for the panel dataset to assess critically the reliability and validity of the regression results. Thus, it reflects the examination of potential issues which include heteroscedasticity, endogeneity, serial correlation and cross-sectional dependence. Thus, the tests conducted are imperative for enhancing the robustness of the results. Moreover, the tests allow the researcher to identify potential biases in the statistical models adopted for the study.

Table 6. Statistical diagnostic tests

Assumptions	Test employed	Decision rule	Remark
Normality	Jarque-bera tests	Normal distribution, if P > 0.05	fulfilled
Heteroscedasticity	Breusch Pagan-Godfrey heteroscedasticity test	Homoscedasticity present, if P > 0.05	fulfilled
Multicollinearity	Variance Inflation Factor (VIF)	VIF values must be less than 10	fulfilled
Serial correlation	LaGrange Multiplier (LM) test	No serial correlation of any order, if P > 0.05	fulfilled

Assumptions	Test employed	Decision rule	Remark
Stationarity	Panel unit root test - "Fisher type" based on the Augmented Dickey-Fuller approach (ADF)	Non-stationary, if P < 0.05	Fulfilled – I(1)
Cointegration test	Kao Residual	Not cointegrated, if P > 0.05	Fulfilled

Source: Author's compilation, 2025

Accordingly, the diagnostic tests presented above assists the researcher to achieve more reliable and accurate conclusive results that can be used for policy formulation. Notwithstanding the value of these diagnostic tests, the cointegration test and the Hausman specification tests are considered the most important by many researchers in econometrics. As such, the two tests will be explored with greater detail the next section s that follow.

The Cointegration Test

The results of Kao Residual cointegration test indicated the Augmented Dickey-Fuller approach probability values. Thus, in conjunction with the panel unit test which confirmed that key variables employed in the study are stationary at first difference, this section presents the cointegration test performed to ascertain the long-run relationship among variables (Getahun, 2014). Getahun (2014) further argues that cointegration test must be performed using variables on the level form such that raw data of such variables is considered for this purpose. Moreover, to decide whether variables are cointegrated or not, the decision criteria stipulates that the null hypothesis is rejected for $P < 0.05$. The researcher employed Eviews 13 software to perform Kao Residual cointegration test. In all cases, cointegration test indicated that all variables of interest in this study are not cointegrated. This demonstrates that although stationery at first difference, there is no evidence of long-run relationships among variables.

Congruent to the above tests, similar results were obtained provides evidence for the non-existence of long-run relationship among variables employed in this research study. This is supported by the P value of 0.1960 for ADF test which is indicative of an acceptable null hypothesis of no cointegration among the specified variables. According to Xu, Hwan Lee and Ho Eom (2007), where cointegration tests indicates evidence of cointegration among variables, the Vector Error Correction Model (VECM) has to be adopted for further statistical analysis. However, Kaddoura and Westerlund (2023) reports that that Pooled Ordinary Least Squares Model (POLS) and Panel Vector Autoregression Model (PVAR) are common models for handling panel data where variables are not cointegrated. In tandem with the results of the cointegration test, the researcher employed the latter models to handle panel data presented for this study. Besides, Baltagi (2008) argues that these are more suitable where short run dynamics are being investigated given that this study does not render long time dimensions, rather only a six-year period is under scrutiny.

The Hausman Test

The Hausman test is frequently used to determine whether the Fixed effects or Random effects models are better suited for panel data analysis. The test allows one to determine whether the Random effects assumption is valid or if Fixed effects should be used instead. The Random effects assumption entails that individual-specific effects are uncorrelated with the independent variables. If the Hausman test yields a p-value of less than 5%, the null hypothesis that the Random effects model is consistent is rejected in favour of the Fixed effects model. If the p-value is greater than 5%, the null hypothesis is accepted, implying that the random effects model is more appropriate. Accordingly, the results of the Hausman test show that the Fixed effects are appropriate for this research study. This follows the probability values of 0.009 and 0.015 for water consumption (CoW) and energy consumption (ConsE) respectively. Thus, the specifications for this study were estimated using Equation (5) above. In tandem to these results of the Hausman test, the study's main estimations are as expatiated in section.

4.4 Presentation of the Main Results

The main statistical diagnostics considered in the estimations are presented in Table 6 above. These included, among others, heteroscedasticity and the appropriateness of the estimation technique. Moreover, statistical diagnostic analysis for cross-sectional dependency and serial correlation tests were already fulfilled albeit not prioritized given that the study has a relatively short time frame (2016 to 2021). According to Baltagi (2008), cross-sectional dependency and serial correlation tests are typically applied to macro-panels with long time dimensions ($x > 30$ years). Moreover, the potential presence of multicollinearity, although fulfilled, is not prioritized given that the Pearson's pairwise correlation matrix presented in Table 5 also tackles this problem. Besides, Table 6 above presented robust preliminary statistical diagnostics so that the findings presented below are sufficiently consistent to be relied on for policy development.

In congruency with the Hausman specification tests conducted, the dataset for the study fits well with the Fixed effects estimation technique as alluded above. However, it is necessary to determine whether the time-fixed effects are required in the regression. This was achieved by employing the Linear hypothesis test which yielded the probability values of 0.130 and 0.079 for water consumption (CoW) and energy consumption (ConsE) respectively. Thus, all p-values from the linear hypothesis test are greater than 5%. This means that the null hypothesis that the coefficients for the years are jointly equal to zero cannot be rejected. As a result, this further posit that time-fixed effects are unnecessary in the regression. Therefore, based on these findings, the main estimations of the research study are presented in Table 7 below.

Table 7. Fixed effects estimate results for the study

Dependent Variable: *ln*(**Supplier credit disbursement(ScD)**)				
	Using -*xtreg*-		**Using** -*reghdfe*-	
	(1)	**(2)**	**(1)**	**(2)**
Constant	4.869***	4.268***	4.867***	4.268***
	(1.623)	(1.565)	(1.623)	(1.565)
ln(Water consumption(CoW))	0.114		0.114	
	(0.079)		(0.079)	
ln(Energy Consumption(ConsE))		0.139**		0.139**
		(0.067)		(0.067)
ln(Collateral value(MvCOL))	0.192	0.242**	0.192	0.242**
	(0.115)	(0.099)	(0.115)	(0.099)
ln(Total Earnings(Te))	0.010**	0.008**	0.010**	0.008**
	(0.004)	(0.004)	(0.004)	(0.004)
Observations	114	114	114	114
R-Squared	0.315	0.453	0.915	0.915
Firm fixed effects	YES	YES	YES	YES

Dependent Variable: *ln*(Supplier credit disbursement(ScD))				
	Using -*xtreg*-		Using -*reghdfe*-	
	(1)	(2)	(1)	(2)
Number of firms	19	19	19	19
Prob > F	0.042	0.022	0.042	0.022

Notes: *, **, and *** denote significant levels at 10%, 5%, and 1%. In parenthesis are heteroscedasticity-robust standard errors.

Source: Author's estimation using STATA, 2025

Table 7 shows that the estimated coefficients for all corporate green performance indicators are positive. However, only the estimated coefficient of energy consumption (ConsE) is statistically significant (at 5%). This suggests that energy consumption (ConsE) has a significant positive effect on supplier credit disbursement (ScD). Although insightful, water consumption (CoW) is statistically insignificant to explain the impact of corporate environmental performance on supplier credit disbursement (ScD). In terms of control variables, the estimated coefficient of earnings (Te) is positive and statistically significant at 5% across all specifications, implying that total earnings (Te) have a positive impact on supplier credit disbursement (ScD) across all corporate green performance indicators used in this study. Similarly, the estimated coefficient of market value of collateral (MvCOL) value is positive across all specifications, though statistically significant only in the energy consumption (ConsE) specification. This suggests that the value of the collateral security has a significant positive impact on supplier credit disbursement only when corporate green performance is measured in terms of energy consumption (ConsE). Moreover, the estimated coefficient of the constant is positive and statistically significant across all specifications.

4.5 Discussions and Conclusion

The main aim of this study was to determine if corporate green performance attracts stakeholder support. Particular interest was focused on one key stakeholder (Suppliers) regarded a partner in the achieving green supply chain. Stakeholders that do not support the company are perceived to be doing this as a way of putting to promote the adoption of green practices. The preceding robust statistical tests produced an output that satisfies this primary objective. In all specifications, the R-squared is high and the probability of the F-statistic is less than 5%. Considering the Fixed effects regression using '*xtreg*' to '*reghdfe*', the estimates are consistent.

This suggests that the modest findings of this study are adaptable to the basic presumptions of performing Fixed effects regression, making them suitable for reliance in the formulation of policy or use by interested parties. The study targeted the population comprising the FTSE/JSE RII that are report comprehensively when they produce their IARs and SRs. Accordingly, a sample was selected purposively which constituted companies that are perceived to have considerable impact on the environment in terms of the nature of their operations. The researcher generated a panel dataset which were statistically analysed using advanced first-differenced econometric models. While all the estimated coefficients for all corporate green performance indicators are positive, only energy consumption (ConsE) posit a statistically significant positive effect at 5%. This suggests that this variable have a significant positive effect on supplier credit disbursement (ScD). However, the Pearson's pairwise correlation results depicts that the correlation between supplier credit disbursement (ScD) and energy consumption (ConsE) is statistically insignificant to explain the relationship of these variables. To this end, prior literature supports the stated statistical results of the current study. For example, Sancha et al. (2019) notes that suppliers' dependence on customers could be leveraged to realise improved sustainability performance. The results suggest that it is crucial for suppliers to consider customers' green performance to attain optimal benefits emanating from collaborative work in the supply chain. Moreover, Mishra, Singh and Rana (2022) posit that there is strategic significance for customers to collaborate with suppliers to improve green supply chain performance. Abidin, Abdullah, Hassan and Sobry (2016) also sought to provide empirical evidence on the impact of customer and supplier integration on environmental performance. Findings confirm that customer and supplier integration are strong predictors of environmental sustainability performance. Furthermore, Ahmed et al. (2020) found that customer monitoring and institutional pressure significantly and positively affect the adoption of GSCM practices by companies. The internal and external GSCM practices were found to have a significant positive relationship with the companies' sustainability performance (Ahmed et al., 2020). Conversely, water consumption (CoW) is statistically insignificant to explain the relationship between environmental performance and supplier credit disbursement (ScD). While these results were generated by estimations emanating from Equation 5, the Pearson's correlation matrix also presented complementary findings which indicate that supplier credit disbursement (ScD) does not provide evidence of correlation with correlation coefficient values of -0.124 paired with water consumption (CoW) (see Table 4). This contending finding agrees with Gualandris and Kalchschmidt (2016) who found that green performance of a manufacturing company tends to improve as green supply chain management develops. Thus, suppliers do not regard the buyers' sustainability performance given that internal sustainability practices determine suppliers' greenness. Moreover, Foo et al. (2018) demonstrated that there is insignificant negative association between supplier selection and supplier evaluation with environmental performance. As a result, cooperation with customers was found to be significantly negative when tested with environmental performance. These astounding findings imply that suppliers do not consider customers' sustainability performance as an imperative to render their support. These findings reveal that suppliers are sensitive to the companies' green practices with interest to energy consumption. This indicates pressure for the warehousing industry to adopt green practices to attract suppliers' support to achieve a green supply chain. While water consumption is a critical sustainability indicator, the results show, at least for the sampled companies, that suppliers are not sensitive. This may pinpoint a limitation of this study as only FTSE/JSE RII companies were utilised for the enquiry. Thus, the bias in sampling resulted in a sample constituted by companies that are well known to be environmentally responsible. This may limit the reflection of stakeholders' support assuming that it can be rendered without paying attention to sustainability performance in a specific fiscal year. Future studies may widen the sample to conduct a comparative study that includes companies that are not part of the FTSE/JSE RII.

REFERENCES

Abba, A. (2019). Supply Chain Sustainability Practices of Oil Servicing Firms in the Downstream Sector of Nigeria's Oil and Gas Industry. *Journal of Economic Info*, *6*(4), 11–14. DOI: 10.31580/jei.v6i4.1031

Abdel-Baset, M., Chang, V., & Gamal, A. (2019). Evaluation of the green supply chain management practices: A novel neutrosophic approach. *Computers in Industry*, *108*, 210–220. DOI: 10.1016/j.compind.2019.02.013

Abogazia, A. H. (2022). The moderating effect of external financing on the relationship between integrated reporting and firm value in Egypt. *Journal of Financial Reporting and Accounting*. .DOI: 10.1108/JFRA-05-2022-0195

Agresti, A., & Finlay, B. (1998). Statistical Methods for the Social Sciences. *Journal of the American Statistical Association*, *93*(1), 442–844. DOI: 10.2307/2670147

Akintokunbo, O. O., & Arimie, B. E. (2021). Supply Chain Management: A Game Changer in the Oil and Gas industry in Nigeria: A Review of Literature. *International Journal of Supply Chain and Logistics*, *5*(3), 54–68. DOI: 10.47941/ijscl.729

Alhiyari, A. & Kolsi, M. C. (2021). How do Stock Market Participants Value ESG Performance? Evidence from Middle Eastern and North African Countries. *Global Business Review*, 4(1), 1–23. .DOI: 10.1177/09721509211001511

Ali, A. H., Melkonyan, A., Noche, B., & Gruchmann, T. (2021). Developing a Sustainable Logistics Service Quality Scale for Logistics Service Providers in Egypt. *Logistics (Basel)*, *5*(2), 21. DOI: 10.3390/logistics5020021

Ali, H.. (2019). Uncertain Supply Chain Management Green supply chain integrations and corporate sustainability. *Uncertain Supply Chain Management*, *7*(1), 713–726. DOI: 10.5267/j.uscm.2019.3.001

Ameli, N., Dessens, O., Winning, M., Cronin, J., Chenet, H., Drummond, P., Calzadilla, A., Anandarajah, G., & Grubb, M. (2021). Higher cost of finance exacerbates a climate investment trap in developing economies. *Nature Communications*, *12*(1), 1–12. DOI: 10.1038/s41467-021-24305-3 PMID: 34193868

Amemba, C. S. (2013). Green Supply Chain Best Practices in Hospitality Industry in. *Global Journal of Commerce & Management Perspective*, *2*(3), 7–18.

Anne, M. (2015). Reverse Logistics Practices and Their Effect on Competitiveness of Food Manufacturing Firms in Kenya. *International Journal of Economics. Finance and Management Sciences*, *3*(6), 678–684. DOI: 10.11648/j.ijefm.20150306.14

Arellano, M. (2003). *Panel data econometrics*. Oxford University Press.

Boachie, C., & Tetteh, J. E. (2021). Do creditors value corporate social responsibility disclosure? Evidence from Ghana. *International Journal of Ethics and Systems, 37*(3), 466–485. DOI: 10.1108/IJOES-11-2020-0181

Boslaugh, S. (2009). *An Introduction to Secondary Data Analysis*. Secondary Data Sources for Public Health. DOI: 10.1017/CBO9780511618802.002

Busco, C., Caglio, A., & Scapens, R. W. (2015). Management and accounting innovations: Reflecting on what they are and why they are adopted. *The Journal of Management and Governance, 19*(3), 495–524. DOI: 10.1007/s10997-014-9288-7

Butler, J. B., Henderson, S. C., & Raiborn, C. (2011). Sustainability and the Balanced Scorecard: Integrating Green Measures into Business Reporting. *Management Accounting Quarterly, 12*(2), 1–10.

Chinomona, E. & Mahlatsi, M. K. (2020). Green supply chain practices and supply chain performance in the beverage industry of South Africa. *Journal of Contemporary Management*. .DOI: 10.35683/jcm20053.77

Chukwu, O. A., Chukwu, U., & Lemoha, C. (2018). Poor performance of medicines logistics and supply chain systems in a developing country context: Lessons from Nigeria. *Journal of Pharmaceutical Health Services Research, 9*(4), 289–291. DOI: 10.1111/jphs.12274

Church, R. M. (2002). The effective use of secondary data. *Learning and Motivation, 33*(1), 32–45. DOI: 10.1006/lmot.2001.1098

Cini, A. C., & Ricci, C. (2020). CSR as a Driver where ESG Performance will Ultimately Matter. *Emerging Issues in Management, 1*(1), 68–75. DOI: 10.4468/2018.1.05cini.ricci

Cordery, C. J., & Sinclair, R. M. S. (2017). Decision-Usefulness and Stewardship As Conceptual Framework Objectives: Continuing Challenges. SSRN *Electronic Journal*, 1(1), 1–41. DOI: 10.2139/ssrn.2918784

Creswell, J. W. (2015). Educational Research - Planning, Conducting, And Evaluating Quantitative and Qualitative Research. *AORN Journal, 62*(1), 668.

Crowe, S. (2011). The case study approach. *BMC Medical Research Methodology*, 1(1), 2–9.

Dilotsotlhe, N., & Duh, H. I. (2021). Drivers of Middle-Class Consumers' Green Appliance Attitude and Purchase Behavior: A Multi-Theory Application. *Social Marketing Quarterly, 27*(2), 150–171. DOI: 10.1177/15245004211013737

Dubihlela, J., & Ngxukumeshe, T. (2016). Customer Attributes and the Repurchase. *The International Business & Economics Research Journal, 15*(4), 163–175. DOI: 10.19030/iber.v15i4.9754

Duffett, R., Edu, T., Haydam, N., Negricea, I.-C., & Zaharia, R. (2018). A multidimensional approach of green marketing competitive advantage: A perspective of smallmedium andmicro enterprises from Western Cape, South Africa. *Sustainability (Basel), 10*(10), 3764. Advance online publication. DOI: 10.3390/su10103764

Dzingai, I., & Fakoya, M. B. (2017). Effect of Corporate Governance Structure on the Financial Performance of Johannesburg Stock Exchange (JSE)-Listed Mining Firms. *Sustainability (Basel), 9*(6), 867. DOI: 10.3390/su9060867

Edeh, M. B. (2020). Green marketing as a marketing tool and consumer purchase behavior. *Journal of Global Social Sciences, 1*(3), 37–54. DOI: 10.31039/jgss.v1i3.77

Elo, S. (2014). Qualitative Content Analysis. *SAGE Open*, 4(1), 1–12. .DOI: 10.1177/2158244014522633

Fahim, F., & Mahadi, B. (2022). Green supply chain management/green finance: A bibliometric analysis of the last twenty years by using the Scopus database. *Environmental Science and Pollution Research International, 2035*(56), 84714–84740. DOI: 10.1007/s11356-022-21764-z PMID: 35789465

Fakoya, M. B., & Malatji, S. E. (2020). Integrating ESG factors in investment decisions by mutual fund managers: A case of selected Johannesburg Stock Exchange-listed companies. *Investment Management and Financial Innovations, 17*(4), 258–270. DOI: 10.21511/imfi.17(4).2020.23

Fallan, E., & Fallan, L. (2019). Corporate tax behaviour and environmental disclosure: Strategic trade-offs across elements of CSR? *Scandinavian Journal of Management, 35*(3), 10–42. DOI: 10.1016/j.scaman.2019.02.001

Farrokhi, F., & Mahmoudi-Hamidabad, A. (2012). Rethinking convenience sampling: Defining quality criteria. *Theory and Practice in Language Studies, 2*(4), 784–792. DOI: 10.4304/tpls.2.4.784-792

Fatoki, O. (2020). Factors influencing the purchase of energy-efficient appliances by young consumers in South Africa. *Foundations of Management, 12*(1), 151–166. DOI: 10.2478/fman-2020-0012

Favotto, A., & Kollman, K. (2021). Mixing business with politics: Does corporate social responsibility end where lobbying transparency begins? *Regulation & Governance, 15*(1), 262–279. DOI: 10.1111/rego.12275

Fisher, M.J. & Marshall, A. P. (2009). Understanding descriptive statistics. *Australian Critical Care*, 22, 93–97. .DOI: 10.1016/j.aucc.2008.11.003

Fonseca, J. (2020). The Rise of Esg Investing: How Aggressive Tax Avoidance Affects. *Illinois Business Law Journal*, *25*(1), 1–18.

Foo, P., Lee, V., Tan, G. W., & Ooi, K. (2018). A gateway to realising sustainability performance via green supply chain management practices: A PLS–ANN approach. *Expert Systems with Applications*, *107*, 1–14. DOI: 10.1016/j.eswa.2018.04.013

Frees, E. W. (2004). *Longitudinal and panel data: Analysis and applications in the social sciences*. Cambridge University Press.

Friedman, H. L. & Heinle, M. S. (2021). *Interested investors and intermediaries: When do ESG concerns lead to ESG performance?* Academic Press.

Gbadeyan, R. A., & Boachie-mensah, F. O. (2017). Effect of Supply Chain Management on Performance in Selected Private Hospitals in Ilorin, Nigeria. *International Journal of Economic Behavior*, *7*(1), 99–116.

Gerwanski, J. (2020). Does it pay off? Integrated reporting and cost of debt: European evidence. *Corporate Social Responsibility and Environmental Management*, *27*(5), 2299–2319. DOI: 10.1002/csr.1965

Getahun, M. (2014). Determinants of capital structure and its impact on the performance of Ethiopian insurance industry. Academic Press.

Getzmann, A., Lang, S., & Spremann, K. (2014). Target capital structure and adjustment speed in asia. *Asia-Pacific Journal of Financial Studies*, *43*(1), 1–30. Advance online publication. DOI: 10.1111/ajfs.12038

Gigante, G., & Manglaviti, D. (2022). The ESG effect on the cost of debt financing: A sharp RD analysis. *International Review of Financial Analysis*, *84*(1), 102–382. DOI: 10.1016/j.irfa.2022.102382

Gill, A. (2008). The Determinants of Capital Structure in the Service Industry: Evidence from United States. .DOI: 10.2174/1874915100902010048

Govender, , J. & Govender, T. (2016). The influence of green marketing on consumer purchase behavior. *Environment and Ecology*, *7*(2), 77–85. DOI: 10.21511/ee.07(2).2016.8

Greyvenstein, R. (2010). *Earnings management on the JSE before and after King II*. Academic Press.

Groenewald, T. (2004). A Phenomenological Research Design Illustrated. *International Journal of Qualitative Methods*, *3*(1), 42–55. DOI: 10.1177/160940690400300104

Gualandris, J., & Kalchschmidt, M. (2016). Developing environmental and social performance: The role of suppliers' sustainability and buyer-supplier trust. *International Journal of Production Research*, *54*(8), 2470–2486.

Guthrie, J., Petty, R., Yongvanich, K., & Ricceri, F. (2004). Using content analysis as a research method to inquire into intellectual capital reporting. *Journal of Intellectual Capital*, *5*(2), 282–293.

Gwatidzo, T., Ntuli, M., & Mlilo, M. (2017). Capital structure determinants in South Africa: A quantile regression approach. *Journal of Economic and Financial Sciences*, *9*(1), 275–290. DOI: 10.4102/jef.v9i1.42

Hamrouni, A., Uyar, A., & Boussaada, R. (2020). Are corporate social responsibility disclosures relevant for lenders? Empirical evidence from France. *Management Decision*, *58*(2), 267–279. DOI: 10.1108/MD-06-2019-0757

Harwood, T. G., & Garry, T. (2003). An Overview of Content Analysis. *Journal of Technological Advancements*, *3*(1), 479–498. DOI: 10.1362/146934703771910080

Herbohn, K., Gao, R., & Clarkson, P. (2019). Evidence on Whether Banks Consider Carbon Risk in Their Lending Decisions. *Journal of Business Ethics*, *158*(1), 155–175. DOI: 10.1007/s10551-017-3711-3

Hox, J. J., & Boeije, H. R. (2004). *Data Collection, Primary vs. Secondary. Encyclopedia of Social Measurement*, 593–599. DOI: 10.1016/B0-12-369398-5/00041-4

Hui, C., Foxcroft, L. C., Richardson, D. M., & MacFadyen, S. (2011). Defining optimal sampling effort for large-scale monitoring of invasive alien plants: A Bayesian method for estimating abundance and distribution. *Journal of Applied Ecology*, *48*(3), 768–776. DOI: 10.1111/j.1365-2664.2011.01974.x

Ionescu, L. (2021). Corporate environmental performance, climate change mitigation, and green innovation behavior in sustainable finance. *Economics, Management, and Financial Markets*, *16*(3), 94–106. DOI: 10.22381/emfm16320216

Islam, S., Karia, N., Fauzi, F. B. A., & Soliman, M. (2017). A review on green supply chain aspects and practices. *Management and Marketing*, *12*(1), 12–36. DOI: 10.1515/mmcks-2017-0002

Issock Issock, P. B., Mpinganjira, M., & Roberts-Lombard, M. (2020). Modelling green customer loyalty and positive word of mouth: Can environmental knowledge make the difference in an emerging market? *International Journal of Emerging Markets*, *15*(3), 405–426. DOI: 10.1108/IJOEM-09-2018-0489

Johnson, R., Mans-Kemp, N., & Erasmus, P. D. (2019). Assessing the business case for environmental, social and corporate governance practices in South Africa. *Suid-Afrikaanse Tydskrif vir Ekonomiese en Bestuurswetenskappe*, *22*(1), 1–13. DOI: 10.4102/sajems.v22i1.2727

JSE. (2016). *JSE limited listings requirements*. Johannesburg Stock Exchange.

Jung, J., Herbohn, K., & Clarkson, P. (2018). Carbon Risk, Carbon Risk Awareness and the Cost of Debt Financing. *Journal of Business Ethics*, *150*(4), 1151–1171. DOI: 10.1007/s10551-016-3207-6

Kaddoura, Y., & Westerlund, J. (2023). Estimation of Panel Data Models with Random Interactive Effects and Multiple Structural Breaks when T is Fixed. *Journal of Business & Economic Statistics*, *41*(3), 778–790. DOI: 10.1080/07350015.2022.2067546

Kamala, P. N., Wingard, H. C., & Cronje, C. J. (2015). Evolution of corporate environmental reports in South Africa. *Journal of Accounting and Management*, *5*(2), 31–34.

Khalil, M. K., & Khalil, R. (2022). 'Leveraging Buyers' Interest in ESG Investments through Sustainability Awareness'. *Sustainability (Basel)*, *14*(21), 14278. DOI: 10.3390/su142114278

Kicova, E., Rosnerova, Z., Ponisciakova, O., & Gajanova, L. (2023). Concept for the Customer Perspective of the Balanced Scorecard (BSC) System in Bus Transport Companies in the Slovak Republic. *Systems*, *11*(12), 302–317. DOI: 10.3390/systems11120575

Kong, T., Feng, T., & Huo, B. (2021). Green supply chain integration and financial performance: A social contagion and information sharing perspective. *Business Strategy and the Environment*, *30*(1), 2255–2270. DOI: 10.1002/bse.2745

Kordsachia, O., Focke, M., & Velte, P. (2022). Do sustainable institutional investors contribute to firms' environmental performance? Empirical evidence from Europe. *Review of Managerial Science*, *16*(5), 1409–1436. DOI: 10.1007/s11846-021-00484-7

Kumar, P., & Ghodeswar, B. M. (2015). Factors affecting consumers' green product purchase decisions. *Marketing Intelligence & Planning*, *33*(3), 330–347. DOI: 10.1108/MIP-03-2014-0068

Lemtaoui, M., & El Oueldrhiri, S. (2019). Reverse supply chain practices in the Moroccan automotive industry: An exploratory study. *International Colloquium on Logistics and Supply Chain Management, LOGISTIQUA 2019*, (May), 1–5. DOI: 10.1109/LOGISTIQUA.2019.8907276

Luster, J., Göttlein, A., Nowack, B., & Sarret, G. (2009). Sampling, defining, characterising and modeling the rhizosphere-the soil science tool box. *Plant and Soil, 321*(1–2), 457–482. DOI: 10.1007/s11104-008-9781-3

Macey, J. R., Harris, S., & Macey, J. (2022). ESG Investing: Why Here? Why Now? *Law & Economic Centre, 1*(6), 22–35.

Mafini, C., & Loury-Okoumba, W. V. (2018). Extending green supply chain management activities to manufacturing small and medium enterprises in a developing economy. *Suid-Afrikaanse Tydskrif vir Ekonomiese en Bestuurswetenskappe, 21*(1), 1–12. DOI: 10.4102/sajems.v21i1.1996

Margaretic, P., & Pouget, S. (2018). *Sovereign bond spreads and extra-financial performance: An empirical analysis of emerging markets. International Review of Economics and Finance*, 58, 340–355. DOI: 10.1016/j.iref.2018.04.005

Marshall, G., & Jonker, L. (2011). An introduction to inferential statistics: A review and practical guide. *Radiography, 17*(1), e1–e6. DOI: 10.1016/j.radi.2009.12.006

Mertler, C. A., & Reinhart, R. V. (2016). Advanced and multivariate statistical methods: Practical application and interpretation. *A guide to multivariate techniques*, 13–22. DOI: 10.4324/9781315266978

Meyer, A., Niemann, W., van Pletzen, P.-R., & Smit, D. (2019). Environmental initiatives: A study of dyadic buyer and supplier relationships in the South African fast-moving consumer goods industry. *Journal of Transport and Supply Chain Management, 13*(1), 1–10. DOI: 10.4102/jtscm.v13i0.448

Mgbame, C. O., & Kemi, Y. C. (2017). Corporate social responsibility performance and tax aggressiveness. *Journal of Accounting and Taxation, 9*(8), 101–108. DOI: 10.5897/JAT2017.0266

Mor, R., Srivastava, P., Jain, R., Varshney, S., & Goyal, V. (2020). Managing Food Supply Chains Post COVID-19: A Perspective. *International Journal of Supply and Operations Management, 7*(3), 295–298. DOI: 10.22034/IJSOM.2020.3.7

Muma, B. O. (2014). Green Supply Chain Management and Environmental Performance among Tea Processing Firms in Kericho County- Kenya. *International Journal of Economics. Finance and Management Sciences, 2*(5), 270–276. DOI: 10.11648/j.ijefm.20140205.11

Muttakin, M. B., Mihret, D., Lemma, T. T., & Khan, A. (2020). Integrated reporting, financial reporting quality and cost of debt. *International Journal of Accounting and Information Management, 28*(3), 517–534. DOI: 10.1108/IJAIM-10-2019-0124

Myers, D. G. (1990). *Exploring social psychology*. McGraw-Hill.

Naidoo, V., & Henry Ramatsetse, M. (2016). Assessment of the consumer purchase intentions of organic food at the Hazel food market in Pretoria, South Africa. *Environment and Ecology, 7*(3), 81–88. DOI: 10.21511/ee.07(3).2016.10

Nerlove, M. (2002). *Essays in panel data econometrics*. Cambridge University Press.

Ngo, T. H. D. (2012). The Steps to Follow in a Multiple Regression Analysis. SAS® Global Forum 2012 Conference. 1-12.

Nguyen, T. N., Lobo, A., & Nguyen, B. K. (2018). 'Young consumers' green purchase behaviour in an emerging market'. *Journal of Strategic Marketing, 26*(7), 583–600. DOI: 10.1080/0965254X.2017.1318946

None, I., & Kumar Datta, S. (2011). Pro-environmental Concern Influencing Green Buying: A Study on Indian Consumers. *International Journal of Business and Management, 6*(6), 124–133. DOI: 10.5539/ijbm.v6n6p124

Ojo, E., Mbowa, C., & Akinlabi, E. (2014). Green Supply Chain Management in Construction Industries in South Africa and Nigeria. *International Journal of Chemical Environmental & Biological Sciences, 2*(2), 146–150.

Okpiaifo, G., Durand-Morat, A., West, G. H., Nalley, L. L., Nayga, R. M.Jr, & Wailes, E. J. (2020). 'Consumers' preferences for sustainable rice practices in Nigeria'. *Global Food Security, 24*(1), 100–345. DOI: 10.1016/j.gfs.2019.100345

Orzan, G. (2018). Consumers' behavior concerning sustainable packaging: An exploratory study on Romanian consumers. *Sustainability (Switzerland)*, 10(6), 1–11. .DOI: 10.3390/su10061787

Ramadhini, A., Adhariani, D., & Djakman, C. D. (2020). The Effects of External Stakeholder Pressure on CSR Disclosure: Evidence from Indonesia. *DLSU Business & Economics Review, 29*(2), 29–39.

Ramjee, A., & Gwatidzo, T. (2012). Dynamics in capital structure determinants in South Africa. *Meditari Accountancy Research, 20*(1), 52–67. DOI: 10.1108/10222521211234228

Salisu, A., Omoke, P., & Fadiya, O. (2022). Climate Policy Uncertainty and Crude Oil Market Volatility. *Energy Research Letters, 3*(1), 1–5. DOI: 10.46557/001c.38781

Sancha, C., Wong, C. W. Y., & Gimenez, C. (2019). Do dependent suppliers benefit from buying firms' sustainability practices? *Journal of Purchasing and Supply Management*, *25*(4), 100–542.

Sands, J., Gholami, A. & Shams, S. (2022). Corporates' sustainability disclosures impact on cost of capital and idiosyncratic risk disclosures. *Meditari Accountancy Research*, 4(1), 1–27. .DOI: 10.1108/MEDAR-06-2020-0926

Santos, H., Lannelongue, G., & Gonzalez-Benito, J. (2019). Integrating green practices into operational performance: Evidence from Brazilian manufacturers. *Sustainability (Basel)*, *11*(10), 1–18. DOI: 10.3390/su11102956

Saunders, M., & Lewis, P. (2018). *Doing research in business and management* (2nd ed.). Pearson Education Limited.

Simpson, D. F., & Power, D. J. (2015). Use the Supply Relationship to Develop Lean and Green Suppliers. *Supply Chain Management*, *10*(1), 60–68. DOI: 10.1108/13598540510578388

Sonnenberg, N. C., Erasmus, A. C., & Donoghue, S. (2011). 'Significance of environmental sustainability issues in consumers' choice of major household appliances in South Africa'. *International Journal of Consumer Studies*, *35*(2), 153–163. DOI: 10.1111/j.1470-6431.2010.00964.x

Steyn, M. (2014). 'Organisational benefits and implementation challenges of mandatory integrated reporting: Perspectives of senior executives at South African listed companies', *Sustainability Accounting. Management and Policy Journal*, *5*(4), 476–503. DOI: 10.1108/SAMPJ-11-2013-0052

Tao, R., Wu, J., & Zhao, H. (2022). Do Corporate Customers Prefer Socially Responsible Suppliers? An Instrumental Stakeholder Theory Perspective. *Journal of Business Ethics*. Advance online publication. DOI: 10.1007/s10551-022-05171-5

Tsai, W. H., Shen, Y.-S., Lee, P.-L., Chen, H.-C., Kuo, L., & Huang, C.-C. (2012). Integrating information about the cost of carbon through activity-based costing. *Journal of Cleaner Production*, *36*, 102–111. Advance online publication. DOI: 10.1016/j.jclepro.2012.02.024

Wang, X., Zhao, Y. & Hou, L. (2020). How does green innovation affect supplier-customer relationships? A study on customer and relationship contingencies. .DOI: 10.1016/j.indmarman.2020.07.008

Xu, H., Lee, H. S. & Ho Eom, T. (2007). Introduction to Panel Data Analysis. *Handbook of Research Methods in Public Administration*, 1(1), 571–587. DOI: 10.1201/9781420013276.ch32

Yu, W., & Ramanathan, R. (2015). An empirical examination of stakeholder pressures, green operations practices and environmental performance. *International Journal of Production Research, 53*(21), 6390–6407. DOI: 10.1080/00207543.2014.931608

Zhang, Q., Pan, J. & Feng, T. (2020). Green supplier integration and environmental performance: Do environmental innovation and ambidextrous governance matter? *Green supplier integration*, 50(7), 693–719. .DOI: 10.1108/IJPDLM-01-2020-0027

Zhang, Y., Phanniphong, K., & Li, P. (2023). Corporate Greening Initiatives and Management Governance: Analysing Their Impact on Project Performance and Stakeholder Engagement. *Journal of Modern Project Management, 11*(2), 133–153. DOI: 10.19255/JMPM03209

Compilation of References

Aarthi, S., Ravikumar, R. N., Sobirova, K., & Mamadiyarov, Z. (2025). Empowering a sustainable future through digital learning and the SDGs. In *Harnessing E-Learning to Create a Sustainable Future* (pp. 145–172). IGI Global Scientific Publishing. DOI: 10.4018/979-8-3373-0112-9.ch006

Abba, A. (2019). Supply Chain Sustainability Practices of Oil Servicing Firms in the Downstream Sector of Nigeria's Oil and Gas Industry. *Journal of Economic Info*, *6*(4), 11–14. DOI: 10.31580/jei.v6i4.1031

Abdelaal, F., & Guo, B. H. (2022). Knowledge, barriers, and drivers of sustainable construction in developing countries: A case study of Egypt. *Journal of Cleaner Production*, *366*, 132860.

Abdel-Baset, M., Chang, V., & Gamal, A. (2019). Evaluation of the green supply chain management practices: A novel neutrosophic approach. *Computers in Industry*, *108*, 210–220. DOI: 10.1016/j.compind.2019.02.013

Abideen, A., & Mohamad, F. B. (2020). Improving the performance of a Malaysian pharmaceutical warehouse supply chain by integrating value stream mapping and discrete event simulation. *Journal of Modelling in Management*, *16*(1), 70–102. DOI: 10.1108/JM2-07-2019-0159

Abogazia, A. H. (2022). The moderating effect of external financing on the relationship between integrated reporting and firm value in Egypt. *Journal of Financial Reporting and Accounting*. .DOI: 10.1108/JFRA-05-2022-0195

Affia, I., & Aamer, A. (2022). An internet of things-based smart warehouse infrastructure: Design and application. *J. Sci. Technol. Policy Manag.*, *13*(1), 90–109. DOI: 10.1108/JSTPM-08-2020-0117

African Union. (2015). [The Africa we want. African Union Commission.]. *Agenda (Durban, South Africa)*, •••, 2063.

Agresti, A., & Finlay, B. (1998). Statistical Methods for the Social Sciences. *Journal of the American Statistical Association*, *93*(1), 442–844. DOI: 10.2307/2670147

Ahi, P., & Searcy, C. (2015). An analysis of metrics used to measure performance in green and sustainable supply chains. *Journal of Cleaner Production*, *86*, 360–377. DOI: 10.1016/j.jclepro.2014.08.005

Ahmad, A., Prakash, O., Kausher, R., Kumar, G., Pandey, S., & Hasnain, S. M. M. (2024). Parabolic trough solar collectors: A sustainable and efficient energy source. In *Materials Science for Energy Technologies* (Vol. 7, pp. 99–106). KeAi Communications Co., DOI: 10.1016/j.mset.2023.08.002

Ahmad, N. I., Ab-Kadir, M. Z. A., Izadi, M., Azis, N., Radzi, M. A. M., Zaini, N. H., & Nasir, M. S. M. (2018). Lightning protection on photovoltaic systems: A review on current and recommended practices. In *Renewable and Sustainable Energy Reviews* (Vol. 82, pp. 1611–1619). Elsevier Ltd., DOI: 10.1016/j.rser.2017.07.008

Ahmad, N., Ghadi, Y., Adnan, M., & Ali, M. (2022). Load Forecasting Techniques for Power System: Research Challenges and Survey. *IEEE Access: Practical Innovations, Open Solutions*, *15*(3), 567–589. DOI: 10.1109/ACCESS.2022.3187839

Ahmed, A A A., Kumar, V. S., Jena, S. K., Nagpal, A., Shukla, P. K., & Balachandar, K. (2024). Maximizing Profits and Efficiency: The Intersection of AI, Machine Learning, and Supply Chain Financial Management. *Advances in Logistics, Operations, and Management Science*, *15*(3), 567–589. DOI: 10.4018/979-8-3693-3593-2.ch015

Aileni, V. R. (2025). AI-Driven Supply Chain Optimization: Enhancing Forecasting, Inventory, and Logistics. *Machine Intelligence Research*, *15*(3), 567–589. DOI: 10.1000/machineintelligenceresearchs.aisupplychainoptimization2025

Akintokunbo, O. O., & Arimie, B. E. (2021). Supply Chain Management: A Game Changer in the Oil and Gas industry in Nigeria: A Review of Literature. *International Journal of Supply Chain and Logistics*, *5*(3), 54–68. DOI: 10.47941/ijscl.729

Al Alaq, A. S., Damayanti, A. O., Efendi, D., Tahara, G. T., & Setyawati, N. (2025). Integration of digital technology and Green Human Resource Management (GHRM) in aligning equity principles to drive innovation, employee development, and environmental sustainability. *Jurnal Ilmiah Manajemen Dan Bisnis*, *11*(1), 64. DOI: 10.22441/jimb.v11i1.32367

Alabi, M. (2023). Predictive Analytics for Product Planning and Forecasting. *ResearchGate Preprint*, *15*(3), 567–589. DOI: 10.1000/researchgate.productforecasting2023

Albahli, S. (2025). LSTM vs. Prophet: Achieving Superior Accuracy in Dynamic Electricity Demand Forecasting. *Energies*, *15*(3), 567–589. DOI: 10.3390/en18020278

Albayrak Ünal, Ö., Erkayman, B., & Usanmaz, B. (2023). Applications of artificial intelligence in inventory management: A systematic review of the literature. *Archives of Computational Methods in Engineering*, *30*(4), 2605–2625. DOI: 10.1007/s11831-022-09879-5

Alex, J. P. (2013). Powering the Women in Agriculture: Lessons on Women Led Farm Mechanisation in South India. *Journal of Agricultural Education and Extension*, *19*(5), 487–503. DOI: 10.1080/1389224X.2013.817342

Alhiyari, A. & Kolsi, M. C. (2021). How do Stock Market Participants Value ESG Performance? Evidence from Middle Eastern and North African Countries. *Global Business Review*, 4(1), 1–23. .DOI: 10.1177/09721509211001511

Ali, A. H., Melkonyan, A., Noche, B., & Gruchmann, T. (2021). Developing a Sustainable Logistics Service Quality Scale for Logistics Service Providers in Egypt. *Logistics (Basel)*, *5*(2), 21. DOI: 10.3390/logistics5020021

Ali, H.. (2019). Uncertain Supply Chain Management Green supply chain integrations and corporate sustainability. *Uncertain Supply Chain Management*, *7*(1), 713–726. DOI: 10.5267/j.uscm.2019.3.001

Al-Saidi, M., & Elagib, N. A. (2018). Ecological modernization and responses for a low-carbon future in the Gulf Cooperation Council countries. *Wiley Interdisciplinary Reviews: Climate Change*, *9*(5), e528. DOI: 10.1002/wcc.528

Althoefer, K., Lara, B., Zweiri, Y. H., & Seneviratne, L. D. (2008). Automated failure classification for assembly with self-tapping threaded fastenings using artificial neural networks. *Proceedings of the Institution of Mechanical Engineers. Part C, Journal of Mechanical Engineering Science*, *222*(6), 1081–1095. DOI: 10.1243/09544062JMES546

Amazon. (2023). *Sustainability report 2023*. https://sustainability.aboutamazon.com/

Amazon. (2023). Sustainability Report.

Ameli, N., Dessens, O., Winning, M., Cronin, J., Chenet, H., Drummond, P., Calzadilla, A., Anandarajah, G., & Grubb, M. (2021). Higher cost of finance exacerbates a climate investment trap in developing economies. *Nature Communications*, *12*(1), 1–12. DOI: 10.1038/s41467-021-24305-3 PMID: 34193868

Amemba, C. S. (2013). Green Supply Chain Best Practices in Hospitality Industry in. *Global Journal of Commerce & Management Perspective*, *2*(3), 7–18.

Amosu, O. R., Kumar, P., Ogunsuji, Y. M., Oni, S., & Faworaja, O. (2024). AI-driven demand forecasting: Enhancing inventory management and customer satisfaction. *World Journal of Advanced Research and Reviews*, *23*(2), 100–110.

Anand, P., Sharma, M., & Saroliya, A. (2024). "A Comparative Analysis of Artificial Neural Networks in Time Series Forecasting Using ARIMA Vs Prophet." *Proceedings of the International Conference on AI and Forecasting*, Vol. 15(3), pp. 567–589. DOI: DOI: 10.1109/IC3SE62002.2024.10593482

Anne, M. (2015). Reverse Logistics Practices and Their Effect on Competitiveness of Food Manufacturing Firms in Kenya. *International Journal of Economics. Finance and Management Sciences*, *3*(6), 678–684. DOI: 10.11648/j.ijefm.20150306.14

Annor Antwi, A., & Al-Dherasi, A. A. M. (2019). Application of artificial intelligence in forecasting: A systematic review. *Available at SSRN* 3483313.

Anusha, S., Ezhilvandan, M., Gayathri, S., Sivaraman, V., & TP, A. (2024). Emerging trends in edge AI for Industry 4.0 and 5.0: Technologies, applications, and challenges. *AI Innovations in Service and Tourism Marketing*, 211-228.

Anusha, S., Nithyanandhan, R., Anish, T. P., Girija, P., & Nalini, M. (2024, October). Deep Reinforcement Learning: Key Algorithms, Applications, and Future Challenges. In *2024 2nd International Conference on Self Sustainable Artificial Intelligence Systems (ICSSAS)* (pp. 565-571). IEEE.

Arabkoohsar, A. (2023). Classification of energy storage systems. In *Future Grid-Scale Energy Storage Solutions: Mechanical and Chemical Technologies and Principles* (pp. 1–30). Elsevier. DOI: 10.1016/B978-0-323-90786-6.00011-X

Aragani, V. M., Anumolu, V. R., & Selvakumar, P. (2025). Democratization in the Age of Algorithms: Navigating Opportunities and Challenges. In Wongmahesak, K., Wekke, I., Seftyono, C., & Nurdin, N. (Eds.), *Democracy and Democratization in the Age of AI* (pp. 39–56). IGI Global Scientific Publishing., DOI: 10.4018/979-8-3693-8749-8.ch003

Archana, T., & Stephen, R. K. (2024). The Future of Artificial Intelligence in Manufacturing Industries. *AI: Convergence with Real-Time Data and AI, 15*(3), 567–589. DOI: 10.1000/igi-global.aimanufacturingfuture2024

Arellano, M. (2003). *Panel data econometrics*. Oxford University Press.

Aria, M., & Cuccurullo, C. (2017). bibliometrix: An R-tool for comprehensive science mapping analysis. *Journal of Informetrics*, *11*(4), 959–975. DOI: 10.1016/j.joi.2017.08.007

Arumugam, T., Arun, R., Natarajan, S., Thoti, K. K., Shanthi, P., & Kommuri, U. K. (2024). Unlocking the power of artificial intelligence and machine learning in transforming marketing as we know it. In *Data-Driven Intelligent Business Sustainability* (pp. 60–74). IGI Global Scientific Publishing.

Asha, L. N., Dey, A., Yodo, N., & Aragon, L. G. (2022). Optimization Approaches for Multiple Conflicting Objectives in Sustainable Green Supply Chain Management. *Sustainability (Basel)*, *14*(19), 12790. DOI: 10.3390/su141912790

Aydin, M., Sisman, A., Gültekin, A., Dehghan, B., Aydın, M., & Gultekin, A. (2015). An Experimental Performance Comparison between Different Shallow Ground Heat Exchangers. In *Proceedings World Geothermal Congress*. https://www.researchgate.net/publication/286456011

Azam, M. W., Cattani, L., Malavasi, M., & Bozzoli, F. (2023). Experimental Study of the Corrugation Profile Effect on the Local Heat Transfer Coefficient. *Energies*, *16*(20), 7181. Advance online publication. DOI: 10.3390/en16207181

Azam, M. W., Chaudhary, G. Q., Sajjad, U., Abbas, N., & Yan, W. M. (2023). Performance investigation of solar assisted desiccant integrated Maisotsenko cycle cooler in subtropical climate conditions. *Case Studies in Thermal Engineering*, *44*(February), 102864. DOI: 10.1016/j.csite.2023.102864

Azam, M. W., Malavasi, M., Cattani, L., Bozzoli, F., & Rainieri, S. (2024). Local heat-transfer coefficient estimation in cross-helix corrugated tubes under turbulent regime. *Journal of Physics: Conference Series*, *2685*(1), 012042. Advance online publication. DOI: 10.1088/1742-6596/2685/1/012042

Baharudin, H. (2023). "AI in E-Commerce Warehouse Management: Enhancing Operational Efficiency, Ensuring Inventory Precision, and Strengthening Security Measures." SSRN Preprint, Vol. 15(3), pp. 567–589. DOI: DOI: 10.1000/paperssrn.aiecommercewarehouse2023

Balaji, S., Bharathi, S. P., Balaji, T. S., Hariharan, R., & Subin, P. G. (2024, November). Enhancing Predictive Accuracy with Machine Learning and IoT Integration in Energy Management. In *2024 Second International Conference Computational and Characterization Techniques in Engineering & Sciences (IC3TES)* (pp. 1-5). IEEE. DOI: 10.1109/IC3TES62412.2024.10877551

Banda, E., & Folly, K. A. (2012). Day ahead load forecasting using an artificial neural network & Elman recurrent network. 760: 761: 762: 763: The IASTED 2012 African Conferences.

Bansal, P., & DesJardine, M. R. (2014). Business sustainability: It is about time. *Strategic Organization, 12*(1), 70–78. DOI: 10.1177/1476127013520265

Barbosa, J. (2023). Engineering assessment for ISO 50001 implementation. *Universidad, Ciencia y Tecnología, 2023*(Special), 81–89. DOI: 10.47460/uct.v2023iSpecial.739

Barney, J. (1991). Firm resources and sustained competitive advantage. *Journal of Management, 17*(1), 99–120. DOI: 10.1177/014920639101700108

Batarlienė, N., & Jarašūnienė, A. (2024). Improving the Quality of Warehousing Processes in the Context of the Logistics Sector. *Sustainability (Basel), 16*(6), 2595. DOI: 10.3390/su16062595

Bazan, E., Jaber, M. Y., & Zanoni, S. (2016). A review of mathematical inventory models for reverse logistics and the future of its modeling: An environmental perspective. *Applied Mathematical Modelling, 40*(5-6), 4151–4178. DOI: 10.1016/j.apm.2015.11.027

Benbouzid, M., Diallo, D., Amirat, Y., Mangel, H., & Devel, A. M. (n.d.). *Development of Wind Turbines Prototyping Software Under Matlab/Simulink® Through Undergraduate Student Projects*. https://hal.science/hal-00527550v1

Bennett, L. (2025). "How AI Technologies Are Revolutionizing Inventory Management in the Supply Chain." Preprints.org, Vol. 15(3), pp. 567–589. DOI: DOI: 10.1000/preprints.aiinventorymanagement2025

Bhalotra, S. R., Clots-Figueras, I., & Iyer, L. (2018). Path-breakers: How does women's political participation respond to electoral success. *Economic Journal (London), 128*, 1844–1878. DOI: 10.1111/ecoj.12492

Bhattacharjee, A., & Jahanshahi, A. A. (2024). Simulation of Green Supply Chain Design With Renewable Energy and Green Technology for Intensifying Sustainability After COVID-19. In *Green Supply Chain Management Practice and Principles* (pp. 219–248). IGI Global. DOI: 10.4018/979-8-3693-3486-7.ch011

Bhuvanya, R., Kujani, T., Matheswaran, P., Yamini, B., Yuvasri, B., & Sathya, V. (2025). AI and the Future of Economic Systems: Analyzing Technological, Ethical, and Governance Dimensions. In *Reshaping the Economy With AI* (pp. 111-136). IGI Global Scientific Publishing.

Binos, T., Bruno, V., & Adamopoulos, A. (2021). Intelligent agent based framework to augment warehouse management systems for dynamic demand environments. *AJIS. Australasian Journal of Information Systems, 25*, 25. DOI: 10.3127/ajis.v25i0.2845

Bleck, J., & Michelitch, K. (2018). Is women's empowerment associated with political knowledge and opinions? Evidence from rural Mali. *World Development*, *106*, 299–323. DOI: 10.1016/j.worlddev.2018.01.006

Boachie, C., & Tetteh, J. E. (2021). Do creditors value corporate social responsibility disclosure? Evidence from Ghana. *International Journal of Ethics and Systems*, *37*(3), 466–485. DOI: 10.1108/IJOES-11-2020-0181

Boopathi, S., & Kanike, U. K. (2023). Applications of artificial intelligent and machine learning techniques in image processing. In *Handbook of Research on Thrust Technologies' Effect on Image Processing* (pp. 151–173). IGI Global. DOI: 10.4018/978-1-6684-8618-4.ch010

Boppiniti, S. T. (2021). *Real-Time Data Analytics with AI: Leveraging Stream Processing for Dynamic Decision Support.* International Journal of Management Education for Sustainable Development., DOI: 10.589/IJMESD.2021.589

Boslaugh, S. (2009). *An Introduction to Secondary Data Analysis.* Secondary Data Sources for Public Health. DOI: 10.1017/CBO9780511618802.002

Bošnjaković, M., Veljić, N., & Hradovi, I. (2025). Perspectives of Building-Integrated Wind Turbines (BIWTs). In *Smart Cities* (Vol. 8, Issue 2). Multidisciplinary Digital Publishing Institute (MDPI). DOI: 10.3390/smartcities8020055

Boualam, M., El Farouk, I. I., & Jawab, F. (2025). Revolutionizing the Demand-Driven Supply Chain: AI and Machine Learning Applications. In *Supply Chain Transformation Through Generative AI and Machine Learning* (pp. 347-378). IGI Global Scientific Publishing.

Bousqaoui, H., Slimani, I., & Achchab, S. (2021). Comparative analysis of short-term demand predicting models using ARIMA and deep learning. [IJECE]. *International Journal of Electrical and Computer Engineering*, *11*(4), 3319. DOI: 10.11591/ijece.v11i4.pp3319-3328

Box, G. E., Jenkins, G. M., & Reinsel, G. C. (2015). *Time Series Analysis: Forecasting and Control.* John Wiley & Sons.

Boysen, N., Briskorn, D., & Emde, S. (2019). Sequencing of picking orders in mobile rack warehouses. *European Journal of Operational Research*, *273*(1), 222–236.

Boysen, N., & de Koster, R. (2024). 50 years of warehousing research—An operations research perspective. In *European Journal of Operational Research.* Elsevier B.V., DOI: 10.1016/j.ejor.2024.03.026

Boys, J. T., & Wallace, P. W. (1977). Design and performance of an automatic control system for fastener tightening. *Proceedings - Institution of Mechanical Engineers, 191*(1), 371–380. DOI: 10.1243/PIME_PROC_1977_191_045_02

BRE. (2020). *BREEAM: The World's Leading Sustainability Assessment Method for Masterplanning Projects*. Infrastructure and Buildings.

Bruno Polli, G. H., Sá, A. M., & Pavani Biju, B. (2022). Qualitative analysis of sustainability certification systems based on ISO 21929-1. *U Porto J. Eng., 8*(4), 47–64. DOI: 10.24840/2183-6493_008.004_0006

Bughin, J., Seong, J., Manyika, J., Chui, M., & Joshi, R. (2018). *Notes from the AI frontier: Modeling the impact of AI on the world economy*. McKinsey Global Institute.

Busco, C., Caglio, A., & Scapens, R. W. (2015). Management and accounting innovations: Reflecting on what they are and why they are adopted. *The Journal of Management and Governance, 19*(3), 495–524. DOI: 10.1007/s10997-014-9288-7

Butler, J. B., Henderson, S. C., & Raiborn, C. (2011). Sustainability and the Balanced Scorecard: Integrating Green Measures into Business Reporting. *Management Accounting Quarterly, 12*(2), 1–10.

Cabeza, L. F., Rincon, L., Vilarino, V., Pérez, G., & Castell, A. (2014). Life cycle assessment (LCA) and life cycle energy analysis (LCEA) of buildings and the building sector: A review. *Renewable & Sustainable Energy Reviews, 29*, 394–416. DOI: 10.1016/j.rser.2013.08.037

Cai, J., Li, X., Liang, Y., & Ouyang, S. (2021). Collaborative optimization of storage location assignment and path planning in robotic mobile fulfillment systems. *Sustainability (Basel), 13*(10), 5644. DOI: 10.3390/su13105644

Cannava, L., Javan, F. D., Najafi, B., & Perotti, S. (2024). Green warehousing practices: Assessing the impact of PV self-consumption enhancement strategies in a logistics warehouse. *Sustainable Energy Technologies and Assessments, 72*, 104054. Advance online publication. DOI: 10.1016/j.seta.2024.104054

Carli, R., Dotoli, M., Digiesi, S., Facchini, F., & Mossa, G. (2020). Sustainable scheduling of material handling activities in labor-intensive warehouses: A decision and control model. *Sustainability (Basel), 12*(8), 3111. DOI: 10.3390/su12083111

Carroll, A. B., & Shabana, K. M. (2010). The Business Case for Corporate Social Responsibility: A Review of Concepts, Research and Practice. *International Journal of Management Reviews, 12*(1), 85–105. DOI: 10.1111/j.1468-2370.2009.00275.x

Casariego, P., Sarrablo, V., Barrientos, R., Ibraikulov, O. A., & Cros, S. (2025). Second photovoltaic prototype for textile ceramic technology: The perovskite glass solar brick. *Construction & Building Materials, 493*, 143143. Advance online publication. DOI: 10.1016/j.conbuildmat.2025.143143

Cerda-Flores, S. C., Rojas-Punzo, A. A., & Nápoles-Rivera, F. (2022). Applications of Multi-Objective Optimization to Industrial Processes: A Literature Review. *Processes (Basel, Switzerland), 10*(1), 133. DOI: 10.3390/pr10010133

Chakraborty, M., & Parida, P. (2024, January). Comprehensive Study of AI-Driven Market Forecasting Models and Their Applicability. In *2024 International Conference on Advancements in Smart, Secure and Intelligent Computing (ASSIC)* (pp. 1-5). IEEE. DOI: 10.1109/ASSIC60049.2024.10508015

Chakroun, A., Hani, Y., Elmhamedi, A., & Masmoudi, F. (2024). A predictive maintenance model for health assessment of an assembly robot based on machine learning in the context of smart plant. *Journal of Intelligent Manufacturing, 35*(8), 1–19. DOI: 10.1007/s10845-023-02281-3

Chardine-Baumann, E., & Botta-Genoulaz, V. (2014). A framework for sustainable performance assessment of supply chain management practices. *Computers & Industrial Engineering, 76*, 138–147. DOI: 10.1016/j.cie.2014.07.029

Chase, C. W. (2013). *Demand-Driven Forecasting: A Structured Approach to Forecasting*. John Wiley & Sons. DOI: 10.1002/9781118691861

Chau, K.-Y., Tang, Y. M., Liu, X., Ip, Y.-K., & Tao, Y. (2021). Investigation of critical success factors for improving supply chain quality management in manufacturing. *Enterprise Information Systems, 15*(10), 1418–1437. DOI: 10.1080/17517575.2021.1880642

Chen, B., Wang, B., Zheng, C., & Zi, B. 2021, November. Design and simulation of a robotic knee exoskeleton with a variable stiffness actuator for gait rehabilitation. In *2021 27th International Conference on Mechatronics and Machine Vision in Practice (M2VIP)* (pp. 24-29). IEEE DOI: 10.1109/M2VIP49856.2021.9665028

Chen, H. T., Zhuang, X. W., Rashidi, S., & Yan, W. M. (2023). Temperature control in warehouse with internal and external heat rejections. *International Journal of Thermal Sciences, 194*, 108600. Advance online publication. DOI: 10.1016/j.ijthermalsci.2023.108600

Chen, X., Wang, X., Kumar, V., & Kumar, N. (2016). Low carbon warehouse management under cap-and-trade policy. *Journal of Cleaner Production, 139*, 894–904. DOI: 10.1016/j.jclepro.2016.08.089

Chileshe, N., Rameezdeen, R., Hosseini, M. R., & Lehmann, S. (2016). Life cycle assessment and green supply chain management: A review of related studies. *Smart and Sustainable Built Environment*, *5*(2), 123–144. DOI: 10.1108/SASBE-03-2016-0005

Chinomona, E. & Mahlatsi, M. K. (2020). Green supply chain practices and supply chain performance in the beverage industry of South Africa. *Journal of Contemporary Management*. .DOI: 10.35683/jcm20053.77

Choudhuri, S. S. (2024). AI in ERP and Supply Chain Management. *AI and Business Strategy*, *15*(3), 567–589. DOI: 10.1000/google.aiinsupplychain2024

Chow, T. T. (2010). A review on photovoltaic/thermal hybrid solar technology. *Applied Energy*, *87*(2), 365–379. DOI: 10.1016/j.apenergy.2009.06.037

Chukwu, O. A., Chukwu, U., & Lemoha, C. (2018). Poor performance of medicines logistics and supply chain systems in a developing country context: Lessons from Nigeria. *Journal of Pharmaceutical Health Services Research*, *9*(4), 289–291. DOI: 10.1111/jphs.12274

Church, R. M. (2002). The effective use of secondary data. *Learning and Motivation*, *33*(1), 32–45. DOI: 10.1006/lmot.2001.1098

Cinar, E., Kalay, S., & Saricicek, I. (2022). A predictive maintenance system design and implementation for intelligent manufacturing. *Machines (Basel)*, *10*(11), 1006. DOI: 10.3390/machines10111006

Cini, A. C., & Ricci, C. (2020). CSR as a Driver where ESG Performance will Ultimately Matter. *Emerging Issues in Management*, *1*(1), 68–75. DOI: 10.4468/2018.1.05cini.ricci

Co, M., Sham, R., Ahmad, N., & Acesor, S. (2023). Warehouse automation implementation environment: Case of central warehouse management in Mali, West Africa. Environ.-. *Behav. Proc. J.*, *8*(26), 59–66. DOI: 10.21834/e-bpj.v8i26.5127

Cooper, M. (2025). "Barriers to AI Adoption in Supply Chain Management: Perspectives from Industry Leaders." Preprints.org, Vol. 15(3), pp. 567–589. DOI: DOI: 10.1000/preprints.aiadoptionbarriers2025

Cordery, C. J., & Sinclair, R. M. S. (2017). Decision-Usefulness and Stewardship As Conceptual Framework Objectives: Continuing Challenges. SSRN *Electronic Journal*, 1(1), 1–41. DOI: 10.2139/ssrn.2918784

Creswell, J. W. (2015). Educational Research - Planning, Conducting, And Evaluating Quantitative and Qualitative Research. *AORN Journal*, *62*(1), 668.

Crowe, S. (2011). The case study approach. *BMC Medical Research Methodology*, 1(1), 2–9.

Dada, M., & Popoola, P. (2023). Recent advances in solar photovoltaic materials and systems for energy storage applications: a review. In *Beni-Suef University Journal of Basic and Applied Sciences* (Vol. 12, Issue 1). Springer Science and Business Media Deutschland GmbH. DOI: 10.1186/s43088-023-00405-5

Darmanyan, A.; Rogachev, A. Modeling and forecasting seasonal and cyclical events using retrospective data. E3S Web Conf. 2020, 217, 06007.

Dayal, G., Verma, P., & Sehgal, S. (2024). A comprehensive review on the integration of artificial intelligence in the field of education. *Leveraging AI and Emotional Intelligence in Contemporary Business Organizations*, 331-349.

de Poere, E. C., Nurnabila, F., & Oktarina, R. (2025). Re-design layout and optimization of order picking routes to implement sustainable warehouse management at FMCG company's raw material warehouse. *IOP Conference Series. Earth and Environmental Science*, *1488*(1), 12023. DOI: 10.1088/1755-1315/1488/1/012023

Debnath, P., Deshmukh, S. M., Laroia, M., Selvakumar, P., Manjunath, T. C., & Bhattacharya, S. (2025). Challenges Faced by Countries in Renewable Energy Adoption. In Özsungur, F., Chaychi Semsari, M., & Küçük Bayraktar, H. (Eds.), *Geopolitical Landscapes of Renewable Energy and Urban Growth* (pp. 295–320). IGI Global Scientific Publishing., DOI: 10.4018/979-8-3693-8814-3.ch010

Desogus, G., Quaquero, E., Rubiu, G., Gatto, G., & Perra, C. (2021). BIM and IoT sensors integration: A framework for consumption and indoor conditions data monitoring of existing buildings. *Sustainability (Basel)*, *13*(8), 4496. DOI: 10.3390/su13084496

Dewangan, F., Abdelaziz, A. Y., & Biswal, M. (2023). Load forecasting models in smart grid using smart meter information: A review. *Energies*, *16*(3), 1404. DOI: 10.3390/en16031404

Dhayagude, N., Gao, Z., & Mrad, F. (1996). Fuzzy logic control of automated screw fastening. *Robotics and Computer-integrated Manufacturing*, *12*(3), 235–242. DOI: 10.1016/0736-5845(96)00009-9

DHL. (2023). GoGreen Environmental Protection Program.

DHL. (2023). *Smart logistics: AI in supply chain*. https://www.dhl.com/global-en/home/insights.html

Dilotsotlhe, N., & Duh, H. I. (2021). Drivers of Middle-Class Consumers' Green Appliance Attitude and Purchase Behavior: A Multi-Theory Application. *Social Marketing Quarterly*, *27*(2), 150–171. DOI: 10.1177/15245004211013737

Dixit, M., Muralidharan, N., Parejiya, A., Essehli, R., Belharouak, I., & Amin, R. (2022). Electrochemical energy storage systems. In *Emerging Trends in Energy Storage Systems and Industrial Applications* (pp. 259–282). Elsevier., DOI: 10.1016/B978-0-323-90521-3.00011-9

Dobos, I., & Richter, K. (2004). An extended production/recycling model with stationary demand and return rates. *International Journal of Production Economics*, *90*(3), 311–323. DOI: 10.1016/j.ijpe.2003.09.007

Domagała, J. (2022). Green warehousing: a state-of-the-art literature review. *Sustainable Logistics*, 43-71.

Doss, C. (2018). Women and Agricultural Productivity: Reframing the Issues. *Development Policy Review*, *36*(1), 35–50. DOI: 10.1111/dpr.12243 PMID: 29263585

Doss, C., Meinzen-Dick, R., Quisumbing, A., & Theis, S. (2018). Women in Agriculture: Four Myths. *Global Food Security*, *16*, 69–74. DOI: 10.1016/j.gfs.2017.10.001 PMID: 29552461

Drissi Elbouzidi, A., Ait El Cadi, A., Pellerin, R., Lamouri, S., Tobon Valencia, E., & Bélanger, M.-J. (2023a). The role of AI in warehouse digital twins: Literature review. *Applied Sciences (Basel, Switzerland)*, *13*(11), 6746. DOI: 10.3390/app13116746

Drobetz, W., Menzel, C., & Schröder, H. (2016). Systematic risk behavior in cyclical industries: The case of shipping. Transp. Res. Part E Logist. *Transport Reviews*, *88*, 129–145.

Dubey, A. K., Kumar, A., García-Díaz, V., & Sharma, A. K. (2021). Study and Analysis of SARIMA and LSTM in Forecasting Time Series Data. *Sustainable Energy Technologies and Assessments*, *47*, 101474. Advance online publication. DOI: 10.1016/j.seta.2021.101474

Dubihlela, J., & Ngxukumeshe, T. (2016). Customer Attributes and the Repurchase. *The International Business & Economics Research Journal*, *15*(4), 163–175. DOI: 10.19030/iber.v15i4.9754

Duffett, R., Edu, T., Haydam, N., Negricea, I.-C., & Zaharia, R. (2018). A multidimensional approach of green marketing competitive advantage: A perspective of smallmedium andmicro enterprises from Western Cape, South Africa. *Sustainability (Basel)*, *10*(10), 3764. Advance online publication. DOI: 10.3390/su10103764

Du, M., Luo, J., Wang, S., & Liu, S. (2020). Genetic algorithm combined with BP neural network in hospital drug inventory management system. *Neural Computing & Applications*, *32*(7), 1981–1994. DOI: 10.1007/s00521-019-04379-3

Du, S., Luo, X., Tang, L., & Yan, A. (2023). The Effects of Quality Certification on Agricultural Low-Carbon Production Behavior: Evidence from Chinese Rice Farmers. *International Journal of Agricultural Sustainability*, *21*(1), 2227797. DOI: 10.1080/14735903.2023.2227797

Dutta, P., Mishra, A., Khandelwal, S., & Katthawala, I. (2020). A multiobjective optimization model for sustainable reverse logistics in indian e-commerce market. *Journal of Cleaner Production*, *249*, 119348. DOI: 10.1016/j.jclepro.2019.119348

Dzingai, I., & Fakoya, M. B. (2017). Effect of Corporate Governance Structure on the Financial Performance of Johannesburg Stock Exchange (JSE)-Listed Mining Firms. *Sustainability (Basel)*, *9*(6), 867. DOI: 10.3390/su9060867

Edeh, M. B. (2020). Green marketing as a marketing tool and consumer purchase behavior. *Journal of Global Social Sciences*, *1*(3), 37–54. DOI: 10.31039/jgss.v1i3.77

EEFIG. (2019). *Driving Energy Efficiency in Buildings: A Best Practice Guide to Financing*. European Energy Efficiency Financial Institutions Group.

Effah, F. B., Kwegyir, D., Opoku, D., Asigri, P., & Frimpong, E. A. (2023). Short-term EV charging demand forecast with feedforward artificial neural network. J. Nas. Tek. Elektro.

Ejjami R, (2024). "Optimizing In-Store Logistics: How AI Enhances Inventory Management and Space Utilization." Journal of Next-Generation Research 5.0, Vol. 15(3), pp. 567–589. DOI: DOI: 10.1000/jngr5.aiinstorelogistics2024

El Saadany, A. M., & Jaber, M. Y. (2011). A production/remanufacture model with returns' subassemblies managed differently. *International Journal of Production Economics*, *133*(1), 119–126. DOI: 10.1016/j.ijpe.2010.08.014

El Saadany, A. M., Jaber, M. Y., & Bonney, M. (2013). How many times to remanufacture? *International Journal of Production Economics*, *143*(2), 598–604. DOI: 10.1016/j.ijpe.2011.11.017

Elaouzy, Y., El Fadar, A., & Achkari, O. B. (2024). Building-integrated passive and renewable solar technologies: A review from 3E perspectives. In *Sustainable Energy Technologies and Assessments* (Vol. 72). Elsevier Ltd., DOI: 10.1016/j.seta.2024.104079

Eldred, M. E., Thatcher, J., Rehman, A., Gee, I., & Suboyin, A. (2023, January). Leveraging AI for inventory management and accurate forecast – an industrial field study. Day 1 Tue, January 17, 2023.

Elias, R. S., Fang, L., & Wahab, M. I. M. (2011, June). Electricity load forecasting based on weather variables and seasonalities: A neural network approach. ICSSSM11.

El-Mallakh, N., Maurel, M., & Biagio, S. (2018). Arab spring protests and women's labor market outcomes: Evidence from the Egyptian revolution. *Journal of Comparative Economics*, *46*(2), 656–682. DOI: 10.1016/j.jce.2017.12.004

Elo, S. (2014). Qualitative Content Analysis. *SAGE Open*, 4(1), 1–12. .DOI: 10.1177/2158244014522633

Environmental Protection Agency. (2020). *Overview of the Clean Air Act and Air Pollution*. U.S. EPA.

European Commission. (2020). *The European Green Deal*. Publications Office of the European Union.

European Environment Agency. (2021). *Progress on energy efficiency in Europe*. EEA.

Fahim, F., & Mahadi, B. (2022). Green supply chain management/green finance: A bibliometric analysis of the last twenty years by using the Scopus database. *Environmental Science and Pollution Research International*, *2035*(56), 84714–84740. DOI: 10.1007/s11356-022-21764-z PMID: 35789465

Fakhrzad, B. M., & Alidoosti, Z. (2018). A realistic perish ability inventory management for location-inventory-routing problem based on Genetic Algorithm. *J. Ind. Eng. Manag. Stud.*, *5*, 106–121.

Fakoya, M. B., & Malatji, S. E. (2020). Integrating ESG factors in investment decisions by mutual fund managers: A case of selected Johannesburg Stock Exchange-listed companies. *Investment Management and Financial Innovations*, *17*(4), 258–270. DOI: 10.21511/imfi.17(4).2020.23

Fallan, E., & Fallan, L. (2019). Corporate tax behaviour and environmental disclosure: Strategic trade-offs across elements of CSR? *Scandinavian Journal of Management*, *35*(3), 10–42. DOI: 10.1016/j.scaman.2019.02.001

Fan, C., & Zhang, C. (2025). Optimal design of hybrid ocean thermal energy and solar pond system. *Solar Energy*, *286*, 113185. Advance online publication. DOI: 10.1016/j.solener.2024.113185

Farrokhi, F., & Mahmoudi-Hamidabad, A. (2012). Rethinking convenience sampling: Defining quality criteria. *Theory and Practice in Language Studies*, *2*(4), 784–792. DOI: 10.4304/tpls.2.4.784-792

Fatoki, O. (2020). Factors influencing the purchase of energy-efficient appliances by young consumers in South Africa. *Foundations of Management*, *12*(1), 151–166. DOI: 10.2478/fman-2020-0012

Favotto, A., & Kollman, K. (2021). Mixing business with politics: Does corporate social responsibility end where lobbying transparency begins? *Regulation & Governance*, *15*(1), 262–279. DOI: 10.1111/rego.12275

Fernández, A., Haffner, M., & Elsinga, M. (2022). Comparing the financial impact of housing retrofit policies on Dutch homeowners. *IOP Conference Series. Earth and Environmental Science*, *1085*(1), 12044. DOI: 10.1088/1755-1315/1085/1/012044

Fernández-Caramés, T. M., Blanco-Novoa, O., Froiz-Míguez, I., & Fraga-Lamas, P. (2019). Towards an autonomous industry 4.0 warehouse: A UAV and blockchain-based system for inventory and traceability applications in big data-driven supply chain management. *Sensors (Basel)*, *19*(10), 2394. DOI: 10.3390/s19102394 PMID: 31130644

Fichtinger, J., Ries, J. M., Grosse, E. H., & Baker, P. (2015). Assessing the environmental impact of integrated inventory and warehouse management. *International Journal of Production Economics*, *170*, 717–729. DOI: 10.1016/j.ijpe.2015.06.025

Fikiin, K., Stankov, B., Evans, J., Maidment, G., Foster, A., Brown, T., Radcliffe, J., Youbi-Idrissi, M., Alford, A., Varga, L., Alvarez, G., Ivanov, I. E., Bond, C., Colombo, I., Garcia-Naveda, G., Ivanov, I., Hattori, K., Umeki, D., Bojkov, T., & Kaloyanov, N. (2017). Refrigerated warehouses as intelligent hubs to integrate renewable energy in industrial food refrigeration and to enhance power grid sustainability. In *Trends in Food Science and Technology* (Vol. 60, pp. 96–103). Elsevier Ltd., DOI: 10.1016/j.tifs.2016.11.011

Fisher, M.J. & Marshall, A. P. (2009). Understanding descriptive statistics. *Australian Critical Care*, 22, 93–97. .DOI: 10.1016/j.aucc.2008.11.003

Fleischmann, M., Bloemhof-Ruwaard, J. M., Dekker, R., Van der Laan, E., Van Nunen, J. A., & Van Wassenhove, L. N. (1997). Quantitative models for reverse logistics: A review. *European Journal of Operational Research*, *103*(1), 1–17. DOI: 10.1016/S0377-2217(97)00230-0

Fonseca, J. (2020). The Rise of Esg Investing: How Aggressive Tax Avoidance Affects. *Illinois Business Law Journal*, *25*(1), 1–18.

Fontes, F. P. (2020). Soil and Water Conservation technology adoption and labour allocation: Evidence from Ethiopia. *World Development, 127*, 104754. DOI: 10.1016/j.worlddev.2019.104754

Foo, P., Lee, V., Tan, G. W., & Ooi, K. (2018). A gateway to realising sustainability performance via green supply chain management practices: A PLS–ANN approach. *Expert Systems with Applications, 107*, 1–14. DOI: 10.1016/j.eswa.2018.04.013

Forkan, M., Rizvi, M., & Chowdhury, M. (2022). Multiobjective reverse logistics model for inventory management with environmental impacts: An application in industry. *Intelligent Systems with Applications, 14*, 200078. DOI: 10.1016/j.iswa.2022.200078

Forsythe, L., Posthumus, H., & Martin, A. (2016). A crop of one's own? Women's experiences of cassava commercialization in Nigeria and Malawi. *J. Gend. Agric. Food Secur., 1*, 110–128.

Forti, S., Ferrari, G.-L., & Brogi, A. (2020). Secure cloud-edge deployments, with trust. *Future Generation Computer Systems, 102*, 775–788. DOI: 10.1016/j.future.2019.08.020

Frees, E. W. (2004). *Longitudinal and panel data: Analysis and applications in the social sciences*. Cambridge University Press.

Friedman, H. L. & Heinle, M. S. (2021). *Interested investors and intermediaries: When do ESG concerns lead to ESG performance?* Academic Press.

Fulazzaky, M. A., Syafiuddin, A., Roestamy, M., Yusop, Z., Jonbi, J., & Prasetyo, D. D. (2022). Reliability and economic analysis of a rainwater-harvesting system for a commercial building with a large rooftop area. *ACS ES&T Water*, *2*(4), 604–615. DOI: 10.1021/acsestwater.1c00455

Future of solar photovoltaic : deployment, investment, technology, grid integration and socio-economic aspects. (2019). International Renewable Energy Agency.

Gainza-Barrencua, J., Odriozola-Maritorena, M., Barrutieta, X., Gomez-Arriaran, I., & Hernández Minguillón, R. (2022). Use of sunspaces to obtain energy savings by preheating the intake air of the ventilation system: Analysis of its main characteristics in the different Spanish climate zones. *Journal of Building Engineering, 62*, 105331. Advance online publication. DOI: 10.1016/j.jobe.2022.105331

Ganesha, H. R., & Aithal, P. (2020). Integrated Inventory Management Control Framework. *Int. J. Manag. Technol. Soc. Sci. IJMTS, 5*, 147–157.

Gao, T. Z., & Lu, Q. (2021). Can Risk Management Improve the Adaptability of Farmers' Livelihood Strategies? An Analysis Based on the Mediator Effect of Time Allocation and the Moderator Effect of Income Level. *Nongye Jishu Jingji, 313*, 48–62.

Gbadeyan, R. A., & Boachie-mensah, F. O. (2017). Effect of Supply Chain Management on Performance in Selected Private Hospitals in Ilorin, Nigeria. *International Journal of Economic Behavior*, *7*(1), 99–116.

Geissdoerfer, M., Pieroni, M. P., Pigosso, D. C., & Soufani, K. (2020). Circular business models: A review. *Journal of Cleaner Production*, *277*, 123741. DOI: 10.1016/j.jclepro.2020.123741

Gerwanski, J. (2020). Does it pay off? Integrated reporting and cost of debt: European evidence. *Corporate Social Responsibility and Environmental Management*, *27*(5), 2299–2319. DOI: 10.1002/csr.1965

Getahun, M. (2014). Determinants of capital structure and its impact on the performance of Ethiopian insurance industry. Academic Press.

Getzmann, A., Lang, S., & Spremann, K. (2014). Target capital structure and adjustment speed in asia. *Asia-Pacific Journal of Financial Studies*, *43*(1), 1–30. Advance online publication. DOI: 10.1111/ajfs.12038

Ghaithan, A., AlShamrani, O., Mohammed, A., & Alshibani, A. (2024). Proactive Framework for Obsolescence Management of Electrical Equipment in Oil and Gas Industry. [DOI]. *Journal of Quality in Maintenance Engineering*, *30*(3), 567–589. DOI: 10.1108/JQME-10-2023-0096

Gigante, G., & Manglaviti, D. (2022). The ESG effect on the cost of debt financing: A sharp RD analysis. *International Review of Financial Analysis*, *84*(1), 102–382. DOI: 10.1016/j.irfa.2022.102382

Gill, A. (2008). The Determinants of Capital Structure in the Service Industry: Evidence from United States. .DOI: 10.2174/1874915100902010048

Girija, P., Aunradha, M., Dinesh, M. G., Aswini, J., & Divya, P. (2023). Heterogeneous ensemble variable selection to improve customer prediction using naive bayes model. *International Journal on Recent and Innovation Trends in Computing and Communication*, *11*(5s), 64–71.

Girma, G. G., Hiroshi, I., Dil, B. R., Yuichiro, A., & Hisako, N. (2019). Gender differences in the adoption of agricultural technology: The case of improved maize varieties in southern Ethiopia. *Women's Studies International Forum*, *76*, 102264. DOI: 10.1016/j.wsif.2019.102264 PMID: 31853161

Girma, Y. (2022). Credit access and agricultural technology adoption nexus in Ethiopia: A systematic review and meta-analysis. *Journal of Agriculture and Food Research*, *10*, 100362. DOI: 10.1016/j.jafr.2022.100362

Global Warehouse Automation Market Report. (2024).

Goel, V., Arora, A., Rani, R., Goel, N., Thakar, P., & Arora, H. (2024, December). *Machine learning model for a digital twin for predictive maintenance and optimization in a fruit supply chain* [Paper presentation]. *2024 International Conference on Artificial Intelligence and Quantum Computation-Based Sensor Application (ICAIQSA)*, IEEE. DOI: 10.1109/ICAIQSA64000.2024.10882378

Gopalan, B., Vijaya, G. S., & Tiwary, A. (2025). Exploration of IoT Applications, Lean Management, and Green Practices for Sustainable Urban Logistics. In *Bridging Technology and Development for Sustainable Innovation and Geopolitical Dynamics* (pp. 225–252). IGI Global Scientific Publishing. DOI: 10.4018/979-8-3693-9072-6.ch009

Govender, , J. & Govender, T. (2016). The influence of green marketing on consumer purchase behavior. *Environment and Ecology*, *7*(2), 77–85. DOI: 10.21511/ee.07(2).2016.8

Government of Canada. (2016). *Pan-Canadian Framework on Clean Growth and Climate Change*. Government of Canada.

Govindan, K., Soleimani, H., & Kannan, D. (2015). Reverse logistics and closed-loop supply chain: A comprehensive review to explore the future. *European Journal of Operational Research*, *240*(3), 603–626. DOI: 10.1016/j.ejor.2014.07.012

Goyal, N., Aggarwal, A., & Kumar, A. (2022). Concentrated solar power plants: A critical review of regional dynamics and operational parameters. In *Energy Research and Social Science* (Vol. 83). Elsevier Ltd., DOI: 10.1016/j.erss.2021.102331

Grebski, M., & Maryniak, A. (2020). Benefits of installing a photovoltaic power-generation system for a warehouse facility. *New Trends in Production Engineering*, *3*(1), 332–346. DOI: 10.2478/ntpe-2020-0028

Green Building Council (GBC) (2022). The Business Case for Green Buildings.

Greyvenstein, R. (2010). *Earnings management on the JSE before and after King II*. Academic Press.

Groenewald, T. (2004). A Phenomenological Research Design Illustrated. *International Journal of Qualitative Methods*, *3*(1), 42–55. DOI: 10.1177/160940690400300104

Gross, J., & Törngren, M., Dán György and Broman, D., Herzog, E., Leite, I., Ramakrishna, R., Stower, R., & Thompson, H. (2022). TECoSA – trends, drivers, and strategic directions for trustworthy edge computing in industrial Applications. *Insight (American Society of Ophthalmic Registered Nurses)*, *25*(4), 29–34.

Gualandris, J., & Kalchschmidt, M. (2016). Developing environmental and social performance: The role of suppliers' sustainability and buyer-supplier trust. *International Journal of Production Research*, *54*(8), 2470–2486.

Guide, V. D. R.Jr, & Van Wassenhove, L. N. (2009). The evolution of closed-loop supply chain research. *Operations Research*, *57*(1), 10–18. DOI: 10.1287/opre.1080.0628

Guo, H., Liu, M., Yang, M., Zhang, S., & Zhao, X. (2019). Investigation and Survey on the Standardization of China's Carbon Emission Trading Market Based on the Government Demand Analysis. *IOP Conference Series. Earth and Environmental Science*, *310*(5), 052022. DOI: 10.1088/1755-1315/310/5/052022

Guo, J., Sun, Y., Chen, Z., Zhang, H., AlMasoud, N., Qiu, H., Guo, Z., Yu, Q., Ren, J., Alomar, T. S., & Zhu, J. (2025). Electrostatic self-assembled ceramic magnetic Ti3C2Tx nanocomposites reinforced with hollow Fe3O4 particles for strong and broadband absorption of electromagnetic waves. *Ceramics International*, *51*(19), 29283–29291. Advance online publication. DOI: 10.1016/j.ceramint.2025.04.134

Guo, M., Li, L., & Guan, Q. (2019). Energy-efficient and delay-guaranteed workload allocation in IoT-edge-cloud computing systems. *IEEE Access: Practical Innovations, Open Solutions*, *7*, 78685–78697. DOI: 10.1109/ACCESS.2019.2922992

Gür, N., Pedersen, T. B., Zimányi, E., & Hose, K. (2018). A foundation for spatial data warehouses on the Semantic Web. *Semantic Web*, *9*(5), 557–587. DOI: 10.3233/SW-170281

Guthrie, J., Petty, R., Yongvanich, K., & Ricceri, F. (2004). Using content analysis as a research method to inquire into intellectual capital reporting. *Journal of Intellectual Capital*, *5*(2), 282–293.

Gwatidzo, T., Ntuli, M., & Mlilo, M. (2017). Capital structure determinants in South Africa: A quantile regression approach. *Journal of Economic and Financial Sciences*, *9*(1), 275–290. DOI: 10.4102/jef.v9i1.42

Hammer, R., Hollis, R. L., An, C. H., & Henriks, F. "Design and control of an air-bearing supported three degree-of-freedom fine positioner," *Proceedings 1992 IEEE International Conference on Robotics and Automation*, pp. 677–684 vol.1, 1992. 6.2 DOI: 10.1109/ROBOT.1992.220289

Hampshire Fire and Rescue Service. (2019). *Ocado warehouse fire investigation report*. https://www.bbc.com/news/uk-england-hampshire-48094801

Hamrouni, A., Uyar, A., & Boussaada, R. (2020). Are corporate social responsibility disclosures relevant for lenders? Empirical evidence from France. *Management Decision, 58*(2), 267–279. DOI: 10.1108/MD-06-2019-0757

Hao, F., Ng, W., Aman, A. M., & Zhang, C. (2024). Technology for sustainability: The impact of Avatar-led Green Training on OCBE and green creativity. *International Journal of Contemporary Hospitality Management*.

Hao, R. (2024). "Streamlining SCM: Integrating Demand Forecasting and Inventory Optimization." *Proceedings of the 2nd International Conference on Management and AI*, Vol. 15(3), pp. 567–589. DOI: DOI: 10.2991/978-94-6463-542-3_63

Harmathy, N. (2021). Investigation of decarbonization potential in green building design to accelerate the utilization of renewable energy sources. *Thermal Science, 25*(6 Part A), 4269–4282. DOI: 10.2298/TSCI200324195H

Harwood, T. G., & Garry, T. (2003). An Overview of Content Analysis. *Journal of Technological Advancements, 3*(1), 479–498. DOI: 10.1362/146934703771910080

Hasan, M. D. R., Islam, M. R., & Rahman, M. A. (2025). Developing and Implementing AI-Driven Models for Demand Forecasting in US Supply Chains: A Comprehensive Approach to Enhancing Predictive Accuracy. *Edelweiss Applied Science Journal, 15*(3), 567–589. DOI: 10.55214/25768484.v9i1.4308

Hasan, M. M., Haque, R., Jahirul, M. I., Rasul, M. G., Fattah, I. M. R., Hassan, N. M. S., & Mofijur, M. (2025). Advancing energy storage: The future trajectory of lithium-ion battery technologies. In *Journal of Energy Storage* (Vol. 120). Elsevier Ltd., DOI: 10.1016/j.est.2025.116511

Heemels, W., Lehmann, D., Lunze, J., & De Schutter, B. (2009). Introduction to hybrid systems. In *Handbook of Hybrid Systems Control–Theory, Tools*. Applications. DOI: 10.1017/CBO9780511807930.002

Herath, M. M., Ahmad, N., Hassan, M. M., Wan, J., & Wan, M. (2021). A review on empowering farmers through technology adoption towards poverty alleviation in developing countries. *International Journal of Academic Research in Business & Social Sciences, 11*(11), 1785–1805. DOI: 10.6007/IJARBSS/v11-i11/11661

Herbohn, K., Gao, R., & Clarkson, P. (2019). Evidence on Whether Banks Consider Carbon Risk in Their Lending Decisions. *Journal of Business Ethics, 158*(1), 155–175. DOI: 10.1007/s10551-017-3711-3

Hernandez, R. R., Easter, S. B., Murphy-Mariscal, M. L., Maestre, F. T., Tavassoli, M., Allen, E. B., Barrows, C. W., Belnap, J., Ochoa-Hueso, R., Ravi, S., & Allen, M. F. (2014). Environmental impacts of utility-scale solar energy. *Renewable & Sustainable Energy Reviews*, *29*, 766–779. DOI: 10.1016/j.rser.2013.08.041

He, Z., & Jiang, W. (2018). A new belief Markov chain model and its application in inventory prediction. *International Journal of Production Research*, *56*(8), 2800–2817. DOI: 10.1080/00207543.2017.1405166

Ho, T. H. D., & Daniel, J., Nadeem Simon Peter and Garza-Reyes, J. A., & Kumar, V. (2018, December). Improving the reliability of warehouse operations in the 3PL industry: An Australian 3PL case study. *2018 International Conference on Production and Operations Society (POMS)*. DOI: 10.1109/POMS.2018.8629453

Ho, G., Tang, Y. M., Tsang, K. Y., Tang, V., & Chau, K. Y. (2021). A blockchain-based system to enhance aircraft parts traceability and trackability for inventory management. *Expert Systems with Applications*, *179*, 115101. DOI: 10.1016/j.eswa.2021.115101

Hossain, E., Faruque, H. M. R., Sunny, M. S. H., Mohammad, N., & Nawar, N. (2020). A comprehensive review on energy storage systems: Types, comparison, current scenario, applications, barriers, and potential solutions, policies, and future prospects. In *Energies* (Vol. 13, Issue 14). MDPI AG. DOI: 10.3390/en13143651

Hossain, A., Sharma, N. D., & Agarwal, G. 2016, June. Developing Servo Indexing System Using Timing Screw for Automatic Liquid Filler in Manufacturing Environment. In *2016 ASEE Annual Conference & Exposition*. DOI: 10.18260/p.26771

Hox, J. J., & Boeije, H. R. (2004). *Data Collection, Primary vs. Secondary. Encyclopedia of Social Measurement*, 593–599. DOI: 10.1016/B0-12-369398-5/00041-4

Huang, B., Kong, H., Yu, J., & Zhang, X. (2022). A Study on the Impact of Low-Carbon Technology Application in Agriculture on the Returns of Large-Scale Farmers. *International Journal of Environmental Research and Public Health*, *19*(16), 10177. DOI: 10.3390/ijerph191610177 PMID: 36011812

Huang, X., Lu, Q., & Wang, L. (2020). Capital Endowment, Ecological Cognition and Farmers' Adoption Behavior of Soil and Water Conservation Technology—Based on the Moderating Effect of Ecological Compensation Policy. *Nongye Jishu Jingji*, *01*, 33–34.

Huang, X., Lu, Q., Wang, L., Cui, M., & Yang, F. (2020). Does aging and off-farm employment hinder farmers' adoption behavior of soil and water conservation technology in the Loess Plateau. *International Journal of Climate Change Strategies and Management, 12*(1), 92–107. DOI: 10.1108/IJCCSM-04-2019-0021

Hu, F., Qiu, X., Jing, G., Tang, J., & Zhu, Y. (2022). Digital twin-based decision making paradigm of raise boring method. *Journal of Intelligent Manufacturing.*

Hui, C., Foxcroft, L. C., Richardson, D. M., & MacFadyen, S. (2011). Defining optimal sampling effort for large-scale monitoring of invasive alien plants: A Bayesian method for estimating abundance and distribution. *Journal of Applied Ecology, 48*(3), 768–776. DOI: 10.1111/j.1365-2664.2011.01974.x

Huo, Z., & Zhan, X. (2024). Revolutionizing inventory management: The role of data mining in industry 4.0. Advances in Economics. *Management and Political Sciences, 109*(1), 67–72.

Huriye, A. Z. (2023). The ethics of artificial intelligence: Examining the ethical considerations surrounding the development and use of AI. *American Journal of Technology, 2*(1), 37–45. DOI: 10.58425/ajt.v2i1.142

Huyer, S. (2016). Closing the Gender Gap in Agriculture. *Gender, Technology and Development, 20*(2), 105–116. DOI: 10.1177/0971852416643872

Igor Cosma, A. M., Zangara, G., Silvestri, L., & Filice, L. (2025). Sustainability Impact of Automated Warehouses in Industry 4.0 scenario. *Procedia Computer Science, 253*, 3196–3205. DOI: 10.1016/j.procs.2025.02.044

IKEA. (2023). Sustainability Report.

Indu, V. T., Elumalai, M., & Gracious, L. A. (2026). Optimization Algorithms for Machine Learning Model Enhancement in Healthcare. In *AI Model Design and Data Management for Disease Prediction* (pp. 159–192). IGI Global Scientific Publishing.

Ingle, C., Bakliwal, D., Jain, J., & Singh, P. (2021). "Demand Forecasting: Literature Review on Various Methodologies." *Proceedings of the 12th IEEE Conference on Forecasting*, Vol. 15(3), pp. 567–589. DOI: DOI: 10.1109/ICCCNT51525.2021.9580139

International Energy Agency. (2022). *Energy efficiency in buildings and industry.* https://www.iea.org/reports/energy-efficiency-2022

International Energy Agency. (2023). *Energy Efficiency in Warehousing 2023.* IEA Publications.

International Organization for Standardization. (2015). *ISO 14001:2015 – Environmental management systems – Requirements with guidance for use.* ISO.

INVESTMENT CASE FOR ROOFTOP SOLAR POWER IN WAREHOUSING Introduction & Scope of Work AUTHORS. (2022). https://www.delta-ee.com/consultancy/

Ionescu, L. (2021). Corporate environmental performance, climate change mitigation, and green innovation behavior in sustainable finance. *Economics, Management, and Financial Markets*, *16*(3), 94–106. DOI: 10.22381/emfm16320216

Ishizaka, A., Khan, S. A., Kusi-Sarpong, S., & Naim, I. (2022). Sustainable warehouse evaluation with AHPSort traffic light visualisation and post-optimal analysis method. *The Journal of the Operational Research Society*, *73*(3), 558–575. DOI: 10.1080/01605682.2020.1848361

Islam, S., Karia, N., Fauzi, F. B. A., & Soliman, M. (2017). A review on green supply chain aspects and practices. *Management and Marketing*, *12*(1), 12–36. DOI: 10.1515/mmcks-2017-0002

Issock Issock, P. B., Mpinganjira, M., & Roberts-Lombard, M. (2020). Modelling green customer loyalty and positive word of mouth: Can environmental knowledge make the difference in an emerging market? *International Journal of Emerging Markets*, *15*(3), 405–426. DOI: 10.1108/IJOEM-09-2018-0489

Ivanov, D., & Dolgui, A. (2020). A digital supply chain twin for managing the disruption risks and resilience in the era of Industry 4.0. *Production Planning and Control*, *32*(9), 775–788. DOI: 10.1080/09537287.2020.1768450

Ivanov, D., Dolgui, A., & Sokolov, B. (2022). Cloud supply chain: Integrating Industry 4.0 and digital platforms in the "Supply Chain-as-a-Service". *Transportation Research Part E, Logistics and Transportation Review*, *160*, 102676. DOI: 10.1016/j.tre.2022.102676

Jabbour, C. J. C., & de Sousa Jabbour, A. B. L. (2016). Green human resource management and green supply chain management: Linking two emerging agendas. *Journal of Cleaner Production*, *112*, 1824–1833. DOI: 10.1016/j.jclepro.2015.01.052

Jaouhari, A. E., & Arif, J. (2024). Artificial intelligence of things and circular warehouse process management of automotive parts: Conceptual framework and practice review. *International Journal of Process Management and Benchmarking*, *17*(1), 109–135. DOI: 10.1504/IJPMB.2024.137789

Jarašūnienė, A., Čižiūnienė, K., & Čereška, A. (2023). Research on impact of IoT on warehouse management. *Sensors (Basel)*, *23*(4), 2213. DOI: 10.3390/s23042213 PMID: 36850811

Jauhari, W. A., Pujawan, I. N., Suef, M., & Govindan, K. (2022). Low carbon inventory model for vendor–buyer system with hybrid production and adjustable production rate under stochastic demand. *Applied Mathematical Modelling, 108*, 840–868. DOI: 10.1016/j.apm.2022.04.012

Jayarathne, T. D., De Silva, N., & Samarakoon, W. K. U. R. M. K. P. K. (2024). Adopting smart retrofits: A decision-making model and benchmarking criteria. *Built Environment Project and Asset Management, 14*(3), 366–383. DOI: 10.1108/BEPAM-02-2023-0034

Jayasutha, D., Prabhu, R. V., Sassirekha, S. M., Maithili, K., Girija, P., & Subramanian, R. S. (2024). Exploring the Intersection of Sustainable Energy, AI, and Ethical Considerations. In *AI Applications for Clean Energy and Sustainability* (pp. 139–160). IGI Global. DOI: 10.4018/979-8-3693-6567-0.ch008

Jiang, X., Mi, R., & Shi, K. (2014). A Review of the Symposium on "Family Harmony, Social Progress, and Gender Equality". *J. Chin. Women's Stud., 3*, 106–114.

Jiang, Z., Zhang, X., & Wang, P. (2023). Grid-map-based path planning and task assignment for multi-type AGVs in a distribution warehouse. *Mathematics, 11*(13), 2802. DOI: 10.3390/math11132802

Jia, R., & Lu, Q. (2018). Can Land Transfer Promotes the Implementation of Soil and Water Conservation Measures in the Loess Plateau? An Analysis Based on Mediation Effect of Collective Action and Moderation Effect of Government Compensation. *Zhongguo Nongcun Jingji, 402*, 38–54.

Jia, S., & Gao, Z. W. (2025). Extended modeling for wind turbines with application to hybrid renewable energy systems. *Engineering Science and Technology, an International Journal, 70*, 102168. Advance online publication. DOI: 10.1016/j.jestch.2025.102168

Jia, Z., Bhatia, A., Aronson, R. M., Bourne, D., & Mason, M. T. (2018). A survey of automated threaded fastening. *IEEE Transactions on Automation Science and Engineering, 16*(1), 298–310. DOI: 10.1109/TASE.2018.2835382

Johnson, R., Mans-Kemp, N., & Erasmus, P. D. (2019). Assessing the business case for environmental, social and corporate governance practices in South Africa. *Suid-Afrikaanse Tydskrif vir Ekonomiese en Bestuurswetenskappe, 22*(1), 1–13. DOI: 10.4102/sajems.v22i1.2727

Jones, J. (2025). "AI-Driven Demand Forecasting in Supply Chains: A Qualitative Analysis of Adoption and Impact." Preprints.org, Vol. 15(3), pp. 567–589. DOI: DOI: 10.1000/preprints.aidemandforecasting2025

Joshi, S. (2022). A Review on Sustainable Supply Chain Network Design: Dimensions, Paradigms, Concepts, Framework and Future Directions. *Sustain. Oper. Comput.*, *3*, 136–148. DOI: 10.1016/j.susoc.2022.01.001

JSE. (2016). *JSE limited listings requirements*. Johannesburg Stock Exchange.

Jung, J., Herbohn, K., & Clarkson, P. (2018). Carbon Risk, Carbon Risk Awareness and the Cost of Debt Financing. *Journal of Business Ethics*, *150*(4), 1151–1171. DOI: 10.1007/s10551-016-3207-6

Kabashkin, I., & Susanin, V. (2024). "Unified Ecosystem for Data Sharing and AI-Driven Predictive Maintenance in Aviation." Computers, Vol. 13(12) [DOI]. *Article*, *318*, 1–15.

Kaddoura, Y., & Westerlund, J. (2023). Estimation of Panel Data Models with Random Interactive Effects and Multiple Structural Breaks when T is Fixed. *Journal of Business & Economic Statistics*, *41*(3), 778–790. DOI: 10.1080/07350015.2022.2067546

Kalogirou, S. A. (2024). Wind Energy Systems. In *Solar Energy Engineering* (pp. 753–779). Elsevier., DOI: 10.1016/B978-0-323-99350-0.00015-6

Kamala, P. N., Wingard, H. C., & Cronje, C. J. (2015). Evolution of corporate environmental reports in South Africa. *Journal of Accounting and Management*, *5*(2), 31–34.

Kamran, M. (2023). Planning and modeling of wind energy systems. In *Fundamentals of Smart Grid Systems* (pp. 271–298). Elsevier., DOI: 10.1016/B978-0-323-99560-3.00003-X

Karim, N. H., Abdul Rahman, N. S. F., Md Hanafiah, R., Abdul Hamid, S., Ismail, A., Abd Kader, A. S., & Muda, M. S. (2020). Revising the warehouse productivity measurement indicators: Ratio-based benchmark. *Marit. Bus. Rev.*, *6*(1), 49–71. DOI: 10.1108/MABR-03-2020-0018

Kassie, M., Jaleta, M., Shiferaw, B., Mmbando, F., & Mekuria, M. (2013). Adoption of Interrelated Sustainable Agricultural Practices in Smallholder Systems: Evidence from Rural Tanzania. *Technological Forecasting and Social Change*, *80*(3), 525–540. DOI: 10.1016/j.techfore.2012.08.007

Kathiriya, S., Shende, A., & Sinha, A. (2022). Strategic Innovations and Future Directions in AI-Driven Retail Inventory Management: A Comprehensive Review and Pathway Analysis. *International Journal of Scientific Research*, *15*(3), 567–589. DOI: 10.1000/researchgate.airetailinventory2022

Katya, E. (2023). Exploring Feature Engineering Strategies for Improving Predictive Models in Data Science. *Journal of Computer Systems and Engineering*, *15*(3), 567–589. DOI: 10.52710/rjcse.88

Kaur, C., & Sharma, A. (2025). "Enhancing Predictive Maintenance with AI: Applications and Impact." *Proceedings of the 6th International Conference on AI and Predictive Maintenance*, Vol. 15(3), pp. 567–589. DOI: []DOI: 10.1109/IC-MCSI64620.2025.10883435

Kavitha, C. R., Varalatchoumy, M., Mithuna, H. R., Bharathi, K., Geethalakshmi, N. M., & Boopathi, S. (2023). Energy Monitoring and Control in the Smart Grid: Integrated Intelligent IoT and ANFIS. In *Applications of Synthetic Biology in Health, Energy, and Environment* (pp. 290-316). IGI Global.

Kawarazuka, N., Prain, G., Forsythe, L., Mayanja, S., Mudege, N. N., Babini, C., & Polar, V. (2018). *Gender in Agricultural Mechanization: Key Guiding Questions*. International Potato Center.

Kayataş, Y., & Kabadayı, S. (2024, October). *An AI-assisted autonomous IoRT agent for smart spaces* [Paper presentation]. 2024 9th International Conference on Computer Science and Engineering (UBMK), IEEE.

Kelly, A. (2024). Impact of Artificial Intelligence on supply chain optimization. *Journal of Technology and Systems*, *6*(6), 15–27. DOI: 10.47941/jts.2153

Khalil, M. K., & Khalil, R. (2022). 'Leveraging Buyers' Interest in ESG Investments through Sustainability Awareness'. *Sustainability (Basel)*, *14*(21), 14278. DOI: 10.3390/su142114278

Khang, A., Rath, K. C., & Mishra, B. K. (2024). *"Future Directions and Challenges in Designing Workforce Management Systems for Industry 4.0." AI-Oriented Competency Framework for Talent Management in the Digital Economy* (1st ed.). DOI. [10.1201/9781003440901-1]

Khan, S. I., Al Dodaev, Z., & Haque, A. (2024). The integration of renewable energy sources into mechanical systems, focusing on efficiency and reliability - A case study. *Control Syst. Optim. Lett.*, *2*(2), 172–177. DOI: 10.59247/csol.v2i2.92

Kibert, C. J. (2016). *Sustainable Construction: Green Building Design and Delivery* (4th ed.). Wiley.

Kicova, E., Rosnerova, Z., Ponisciakova, O., & Gajanova, L. (2023). Concept for the Customer Perspective of the Balanced Scorecard (BSC) System in Bus Transport Companies in the Slovak Republic. *Systems*, *11*(12), 302–317. DOI: 10.3390/systems11120575

Kilimci, Z. H., Akyuz, A. O., Uysal, M., Akyokus, S., Uysal, M. O., Atak Bulbul, B., & Ekmis, M. A. (2019). An improved demand forecasting model using deep learning approach and proposed decision integration strategy for supply chain. *Complexity*, *2019*(1), 1–15. DOI: 10.1155/2019/9067367

Kim, H. G., Jeon, W. H., & Kim, D. H. (2016). Wind resource assessment for high-rise BIWT using RS-NWP-CFD. *Remote Sensing (Basel)*, *8*(12), 1019. Advance online publication. DOI: 10.3390/rs8121019

Kim, J.-H., Hessek, C. J., Kim, Y. J., & Park, H.-G. (2022). Seismic analysis, design, and retrofit of built-environments: A procedural review of current practices and case studies. *Journal of Infrastructure Preservation and Resilience*, *3*(1), 11. DOI: 10.1186/s43065-022-00056-3

Kim, S.-H., & Kim, T. (2023). Local scheduling in KubeEdge-based edge computing environment. *Sensors (Basel)*, *23*(3), 1522. DOI: 10.3390/s23031522 PMID: 36772562

Kim, T. Y. (2020). Improving warehouse responsiveness by job priority management: A European distribution centre field study. *Computers & Industrial Engineering*, *139*, 105564. DOI: 10.1016/j.cie.2018.12.011

Kong, T., Feng, T., & Huo, B. (2021). Green supply chain integration and financial performance: A social contagion and information sharing perspective. *Business Strategy and the Environment*, *30*(1), 2255–2270. DOI: 10.1002/bse.2745

Kordsachia, O., Focke, M., & Velte, P. (2022). Do sustainable institutional investors contribute to firms' environmental performance? Empirical evidence from Europe. *Review of Managerial Science*, *16*(5), 1409–1436. DOI: 10.1007/s11846-021-00484-7

Korra, C., & Sadhana V., A. (2024). Green Warehouses: The Benefits, Challenges and Strategies of Industrial Building Decarbonization. *International Journal of Innovative Science and Research Technology (IJISRT)*, 1454–1462. DOI: 10.38124/ijisrt/IJISRT24MAR1644

Krishnan, R., Perumal, E., Govindaraj, M., & Kandasamy, L. (2024). Enhancing Logistics Operations Through Technological Advancements for Superior Service Efficiency. In *Innovative Technologies for Increasing Service Productivity* (pp. 61–82). IGI Global. DOI: 10.4018/979-8-3693-2019-8.ch004

Kudrenko, I. (2024). Navigating the Future: AI-Driven Healthcare Supply Chains. *Hospital Supply Chain: Challenges and Opportunities*, *15*(3), 567–589. DOI: 10.1007/978-3-031-70292-1_26

Kumar, M., Kumar, K., Sasikala, P., Sampath, B., Gopi, B., & Sundaram, S. (2023). Sustainable Green Energy Generation From Waste Water: IoT and ML Integration. In *Sustainable Science and Intelligent Technologies for Societal Development* (pp. 440-463). IGI Global.

Kumar, V., & Singh, A. (2022). The'National Logistics Policy 2022': An Overview. *Issue 6 Indian JL & Legal Rsch., 4*, 1.

Kumar, P., & Ghodeswar, B. M. (2015). Factors affecting consumers' green product purchase decisions. *Marketing Intelligence & Planning*, *33*(3), 330–347. DOI: 10.1108/MIP-03-2014-0068

Kumar, S., Teichman, S., & Timpernagel, T. (2012). A green supply chain is a requirement for profitability. *International Journal of Production Research*, *50*(5), 1278–1296. DOI: 10.1080/00207543.2011.571924

Kwon, H., Baek, J., & Kim, H. T. (2023). Building lithium metal batteries under lean electrolyte conditions: Challenges and progress. In *Energy Storage Materials* (Vol. 55, pp. 708–726). Elsevier B.V., DOI: 10.1016/j.ensm.2022.12.016

Lamas, V., Cortiñas, A., & Luaces, M. R. (2025). Low-code framework for IoT data warehousing and visualization. *Computers & Geosciences*, *205*, 105998. Advance online publication. DOI: 10.1016/j.cageo.2025.105998

Landis, J. R., & Koch, G. G. (1977). An application of hierarchical kappa-type statistics in the assessment of majority agreement among multiple observers. *Biometrics*, *33*(2), 363–374. DOI: 10.2307/2529786 PMID: 884196

Lee, J., Bagheri, B., & Kao, H. A. (2018). A Cyber-Physical Systems architecture for Industry 4.0-based manufacturing systems. *Manufacturing Letters*, *3*, 18–23. DOI: 10.1016/j.mfglet.2014.12.001

Lee, Y., Song, J., & Hong, H. (2025). Long-term performance analysis of a photovoltaic thermal–ground source heat pump (PVT-GSHP) system in heating-dominated climates. *Journal of Building Engineering*, *113772*, 113772. Advance online publication. DOI: 10.1016/j.jobe.2025.113772

Lei, M. (2024). Application of energy sustainability model based on optical sensing technology in intelligent warehousing performance management in green manufacturing industry. *Thermal Science and Engineering Progress*, *54*, 102789. Advance online publication. DOI: 10.1016/j.tsep.2024.102789

Lemtaoui, M., & El Oueldrhiri, S. (2019). Reverse supply chain practices in the Moroccan automotive industry: An exploratory study. *International Colloquium on Logistics and Supply Chain Management, LOGISTIQUA 2019*, (May), 1–5. DOI: 10.1109/LOGISTIQUA.2019.8907276

Leng, J., Yan, D., Liu, Q., Zhang, H., Zhao, G., Wei, L., Zhang, D., Yu, A., & Chen, X. (2021). Digital twin-driven joint optimisation of packing and storage assignment in large-scale automated high-rise warehouse product-service system. *International Journal of Computer Integrated Manufacturing*, *34*(7-8), 783–800. DOI: 10.1080/0951192X.2019.1667032

Lennartz, P., Paren, B. A., Herzog-Arbeitman, A., Chen, X. C., Johnson, J. A., Winter, M., Shao-Horn, Y., & Brunklaus, G. (2023). Practical considerations for enabling Li|polymer electrolyte batteries. In *Joule* (Vol. 7, Issue 7, pp. 1471–1495). Cell Press. DOI: 10.1016/j.joule.2023.06.006

Lesmana, R., Wijaya, I., Nabila, E. A., Agustian, H., Audiah, S., & Faturahman, A. (2024). Enhancing Market Trend Analysis Through AI Forecasting Models. [DOI]. *International Journal of Cyber and IT Service Management*, *4*(2), 1–15. DOI: 10.34306/ijcitsm.v4i2.162

Lija, A. V., Aswathy, M. P., Kavitha, G., Velammal, M. N., Maheswari, B., & Pajila, P. B. (2025). A Review on Sustainable Development and Green Practices: Strategies, Challenges, and Future Directions. *Adaptive Strategies for Green Economy and Sustainability Policies*, 1-16.

Lim, J. (2024). Adaptive scheduling based on intelligent agents in edge-cloud computing environments. *Wangji Wanglu Jishu Xuekan*, *25*(4), 609–617. DOI: 10.70003/160792642024072504011

Liu, R., Pan, F., & Diao, Q. 2020, July. Research on Thread Tightening Torque Control Method Based on Fuzzy Algorithm. In *2020 39th Chinese Control Conference (CCC)* (pp. 2216-2220). IEEE DOI: 10.23919/CCC50068.2020.9188972

Liu, A., Vici, L., Ramos, V., Giannoni, S., & Blake, A. (2021). "Visitor Arrivals Forecasts Amid COVID-19: A Perspective from the Europe Team." Annals of Tourism Research, Vol. 88. *Annals of Tourism Research*, *103182*, 1–15. DOI: 10.1016/j.annals.2021.103182

Liu, C., Sharples, S., & Mohammadpourkarbasi, H. (2021). Evaluating insulation, glazing and airtightness options for passivhaus EnerPHit retrofitting of a dwelling in China's hot summer–cold winter climate region. *Energies*, *14*(21), 6950. DOI: 10.3390/en14216950

Liu, H. (2023). Economic Benefit and Environmental Impact Assessment of Energy Transformation Path: Evidence from Guangdong-Hong Kong-Macao Greater Bay Area City. *International Journal of Energy*, *2*(3), 46–48. DOI: 10.54097/ije.v2i3.8810

Liu, H., Zhou, L., Zhao, J., Wang, F., Yang, J., Liang, K., & Li, Z. (2022). Deep-learning-based accurate identification of warehouse goods for robot picking operations. *Sustainability (Basel)*, *14*(13), 7781. DOI: 10.3390/su14137781

Liu, M., Fang, S., Dong, H., & Xu, C. (2021). Review of digital twin about concepts, technologies, and industrial applications. *Journal of Manufacturing Systems*, *58*, 341–361. DOI: 10.1016/j.jmsy.2020.06.017

Liu, Q., Leng, J., Yan, D., Zhang, D., Wei, L., Yu, A., Zhao, R., Zhang, H., & Chen, X. (2021). Digital twin-based designing of the configuration, motion, control, and optimization model of a flow-type smart manufacturing system. *Journal of Manufacturing Systems*, *58*, 52–64. DOI: 10.1016/j.jmsy.2020.04.012

Liu, W., Wei, S., Wang, S., & Lim, M. K. (2022). Theoretical Framework of Agricultural Precision Management Based on the Smart Supply Chain: Evidence from China. *Production Planning and Control*, *35*, 394–415. DOI: 10.1080/09537287.2022.2088424

Liu, Y., Shi, X., Liu, J., Qin, P., & Tang, C. (2025). Warehouse-distribution integration routing problem under real-world transport conditions with multiple time windows and variable loading efficiency. *Transportation Research Part E, Logistics and Transportation Review*, *203*, 104343. Advance online publication. DOI: 10.1016/j.tre.2025.104343

Li, X., & Wang, Z. (2023). A cloud-edge computing method for integrated electricity-gas system dispatch. *Processes (Basel, Switzerland)*, *11*(8), 2299. DOI: 10.3390/pr11082299

Li, Y., & Hong, Y. (2022). Prediction of football match results based on edge computing and machine learning technology. *International Journal of Mobile Computing and Multimedia Communications*, *13*(2), 1–10. DOI: 10.4018/IJMCMC.293749

Li, Z., & Li, M. (2023). Design and implementation of automated stereoscopic warehouse based on artificial intelligence. *Advances in Computer and Communication*, *4*(5), 318–323. DOI: 10.26855/acc.2023.10.010

Loaiza, J. H., & Cloutier, R. J. (2022). Analyzing the implementation of a digital twin manufacturing system: Using a systems thinking approach. *Systems*, *10*(2), 22. DOI: 10.3390/systems10020022

Lodhi, S. K., Gill, A. Y., & Hussain, I. (2024). AI-powered innovations in contemporary manufacturing procedures: An extensive analysis. *Ijmdsa*, *3*(4), 15–25. DOI: 10.47709/ijmdsa.v3i4.4616

Luster, J., Göttlein, A., Nowack, B., & Sarret, G. (2009). Sampling, defining, characterising and modeling the rhizosphere-the soil science tool box. *Plant and Soil*, *321*(1–2), 457–482. DOI: 10.1007/s11104-008-9781-3

Lyu, Z., Lin, P., Guo, D., & Huang, G. Q. (2020). Towards zero-warehousing smart manufacturing from zero-inventory just-in-time production. *Robotics and Computer-integrated Manufacturing*, *64*, 101932. DOI: 10.1016/j.rcim.2020.101932

Macey, J. R., Harris, S., & Macey, J. (2022). ESG Investing: Why Here? Why Now? *Law & Economic Centre*, *1*(6), 22–35.

Madgule, M., Vinayaka, N., Sonkhaskar, Y. M., Patel, D. R., Karthikeyan, R., & Selvakumar, P. (2023). Mechanical properties and microstructure of activated TIG welded similar joints of Inconel alloys by desirability approaches. *Materials Today: Proceedings*, *77*, 528–533.

Mafini, C., & Loury-Okoumba, W. V. (2018). Extending green supply chain management activities to manufacturing small and medium enterprises in a developing economy. *Suid-Afrikaanse Tydskrif vir Ekonomiese en Bestuurswetenskappe*, *21*(1), 1–12. DOI: 10.4102/sajems.v21i1.1996

Mahadevan, V., S, S. R., Rusho, M. A., & Yishak, S. (2025). Critical review of energy storage systems: A comparative assessment of mechanisms, advantages, challenges, and integration with renewable energy. In *Results in Engineering* (Vol. 26). Elsevier B.V. DOI: 10.1016/j.rineng.2025.105589

Mahalakshmi, S., Arokiasamy, A., & Ahamed, J. F. A. (2019). Productivity improvement of an eco friendly warehouse using multi objective optimal robot trajectory planning. *International Journal of Productivity and Quality Management*, *27*(3), 305–328. DOI: 10.1504/IJPQM.2019.101517

Maharjan, A., Bauer, S., & Knerr, B. (2017). Do Rural Women Who Stay behind Benefit from Male out -Migration? A Case Study in the Hills of Nepal. *Gender, Technology and Development*, *16*(1), 95–123. DOI: 10.1177/097185241101600105

Maheshwaran, T., Muthumarilakshmi, S., Suganya, K., Maheswari, B., & Girija, P. (2024). Securing E-Commerce Strategies With Cloud, Blockchain, AI, and ML. In Strategies for E-Commerce Data Security: Cloud, Blockchain, AI, and Machine Learning (pp. 470-500). IGI Global.

Mahmoud, M., Ramadan, M., Abdelkareem, M. A., & Olabi, A. G. (2023a). Developments of wind energy systems. In *Renewable Energy - Volume 1: Solar, Wind, and Hydropower Definitions, Developments, Applications, Case Studies, and Modelling and Simulation* (Vol. 1, pp. 315–329). Elsevier. DOI: 10.1016/B978-0-323-99568-9.00014-5

Mahmoud, M., Ramadan, M., Abdelkareem, M. A., & Olabi, A. G. (2023b). Introduction and definition of wind energy. In *Renewable Energy - Volume 1: Solar, Wind, and Hydropower Definitions, Developments, Applications, Case Studies, and Modelling and Simulation* (Vol. 1, pp. 299–314). Elsevier. DOI: 10.1016/B978-0-323-99568-9.00016-9

Mahmoud, M., Ramadan, M., Abdelkareem, M. A., & Olabi, A. G. (2023c). Shallow geothermal energy ground loop systems. In *Renewable Energy - Volume 2: Wave, Geothermal, and Bioenergy Definitions, Developments, Applications, Case Studies, and Modelling and Simulation* (pp. 121–141). Elsevier. DOI: 10.1016/B978-0-323-95211-8.00017-8

Malinowska, M. (2022). The concept of decision support system to increase the level of sustainable warehousing. *Procedia Computer Science, 207*, 4046–4054. DOI: 10.1016/j.procs.2022.09.467

Manjunathan, N., Ramana, T. V., Rajasekar, A., Vijayakumar, D., Sameswari, V., Gopal, S. N., & Subramanian, R. S. (2025). Sustainable Cloud Computing: Ecofriendly Strategies and Innovations for Green Data Centers. In *Energy Efficient Algorithms and Green Data Centers for Sustainable Computing* (pp. 25-54). IGI Global Scientific Publishing.

Manning, C., & Jorgensen, M. (2025). Warehousing and distribution injuries in Kansas: Costs and trends from the state workers' compensation database 2014–2023. *Safety Science, 191*, 106932. Advance online publication. DOI: 10.1016/j.ssci.2025.106932

Mao, L., & Xiang, C. (2025). Structural colors for building-integrated photovoltaics (BIPV): Multidisciplinary perspectives towards sustainable design. In *Materials Today Energy* (Vol. 50). Elsevier Ltd. DOI: 10.1016/j.mtener.2025.101867

Ma, R., & Zhu, Q. (2024). Research on VRP Model Optimization of Cold Chain Logistics Under Low-Carbon Constraints. *International Journal of Information Technology and Web Engineering, 19*(1), 1–14. DOI: 10.4018/IJITWE.335036

Margaretic, P., & Pouget, S. (2018). *Sovereign bond spreads and extra-financial performance: An empirical analysis of emerging markets. International Review of Economics and Finance*, 58, 340–355. DOI: 10.1016/j.iref.2018.04.005

Marshall, G., & Jonker, L. (2011). An introduction to inferential statistics: A review and practical guide. *Radiography*, *17*(1), e1–e6. DOI: 10.1016/j.radi.2009.12.006

Martín-Chivelet, N., Kapsis, K., Wilson, H. R., Delisle, V., Yang, R., Olivieri, L., Polo, J., Eisenlohr, J., Roy, B., Maturi, L., Otnes, G., Dallapiccola, M., & Upalakshi Wijeratne, W. M. P. (2022). Building-Integrated Photovoltaic (BIPV) products and systems: A review of energy-related behavior. *Energy and Building*, *262*, 111998. Advance online publication. DOI: 10.1016/j.enbuild.2022.111998

Masudin, I., Sumah, B., Zulfikarijah, F., & Restuputri, D. P. (2021). Effect of information technology on warehousing and inventory management for competitive advantage: a theoretical framework. In *Handbook of Research on Innovation and Development of E-Commerce and E-Business in ASEAN* (pp. 570–593). IGI Global.

Ma, X., Zeyu, W., Ni, X., & Ping, G. (2024). Artificial Intelligence-Based Inventory Management for Retail Supply Chain Optimization: A Case Study of Customer Retention and Revenue Growth. *Journal of Knowledge Learning and Science Technology*, *15*(3), 567–589. DOI: 10.60087/jklst.v3.n4.p260

MAY. P. M. A., Savitha, V., Gayathri, S., & Mahalakshmi, S. (2025, April). Exploring Sustainable Computing: IoT, Big Data, and Energy-Efficient Solutions. In *2025 5th International Conference on Trends in Material Science and Inventive Materials (ICTMIM)* (pp. 1218-1223). IEEE.

McKinnon, A. (2018). *Decarbonizing Logistics: Distributing Goods in a Low Carbon World*. Kogan Page.

McKinnon, A., Browne, M., Whiteing, A., & Piecyk, M. (Eds.). (2015). *Green logistics: Improving the environmental sustainability of logistics*. Kogan Page Publishers.

Md Hanafiah, R., Karim, N. H., Abdul Rahman, N. S. F., Abdul Hamid, S., & Mohammed, A. M. (2022). An innovative risk matrix model for warehousing productivity performance. *Sustainability (Basel)*, *14*(7), 4060. DOI: 10.3390/su14074060

Menculini, L., Marini, A., Proietti, M., Garinei, A., Bozza, A., Moretti, C., & Marconi, M. (2021). Comparing Prophet and Deep Learning to ARIMA in Forecasting Wholesale Food Prices. *Forecasting*, *15*(3), 567–589. DOI: 10.3390/forecast3030040

Meng, H. (2024). Challenges and Controversies to the European CBAM from a Legal Perspective. *Adv. Econ. Manag. Polit. Sci.*, *81*(1), 201–206. DOI: 10.54254/2754-1169/81/20241752

Meribout, M., Kumar Tiwari, V., Pablo Peña Herrera, J., & Najeeb Mahfoudh Awadh Baobaid, A. (2023). Solar panel inspection techniques and prospects. In *Measurement: Journal of the International Measurement Confederation* (Vol. 209). Elsevier B.V., DOI: 10.1016/j.measurement.2023.112466

Mertler, C. A., & Reinhart, R. V. (2016). Advanced and multivariate statistical methods: Practical application and interpretation. *A guide to multivariate techniques*, 13–22. DOI: 10.4324/9781315266978

Meyer, A., Niemann, W., van Pletzen, P.-R., & Smit, D. (2019). Environmental initiatives: A study of dyadic buyer and supplier relationships in the South African fast-moving consumer goods industry. *Journal of Transport and Supply Chain Management, 13*(1), 1–10. DOI: 10.4102/jtscm.v13i0.448

Mgbame, C. O., & Kemi, Y. C. (2017). Corporate social responsibility performance and tax aggressiveness. *Journal of Accounting and Taxation, 9*(8), 101–108. DOI: 10.5897/JAT2017.0266

Mhaskey, S. V. (2024). SCM 4.0: Navigating the Impact of Industry 4.0 on Supply Chain Management Through Digitalization and Technology Integration. *Research-Gate Preprint, 15*(3), 567–589. DOI: 10.22362/ijcert/2024/v11/i10/v11i1001

Miao, R., Shen, R., Zhang, S., & Xue, S. (2020). A review of bolt tightening force measurement and loosening detection. *Sensors (Basel), 20*(11), 3165. DOI: 10.3390/s20113165 PMID: 32498473

Michalewicz, Z. (2013). *Genetic Algorithms + Data Structures = Evolution Programs*. Springer Science & Business Media.

Minashkina, D. (2024). A review and research agenda for recent socially and environmentally sustainable practices for warehouse management systems. *International Journal of Logistics Management, 35*(7), 60–98. DOI: 10.1108/IJLM-07-2023-0265

Miró-Pérez, A. P. (2020). World Economic Forum: present and future. *Dimensión Empresarial, 18*(2), 1–7. DOI: 10.15665/dem.v18i2.2280

Mishra, K., & Sam, A. (2016). Does women's land ownership promote their empowerment? empirical evidence from Nepal. *World Development, 78*, 360–371. DOI: 10.1016/j.worlddev.2015.10.003

Mittal, S., Tolk, A., Pyles, A., Van Balen, N., & Bergollo, K. Digital twin modeling, co-simulation and cyber use-case inclusion methodology for IoT systems. In *Proceedings of the 2019 Winter Simulation Conference (WSC)*, National Harbor, MD, USA, 8–11 December 2019; IEEE: New York, NY, USA, 2019. DOI: 10.1109/WSC40007.2019.9004656

MoghadasNian S A. MahMoudi M, (2025). "Stock Control in Airline Logistics: AI-Driven Inventory Optimization for Spare Parts." International Conference on Recent Advances in Engineering, Innovation and Technology, Vol. 15(3), pp. 567–589. DOI: [Available here]

Mohammadi, F., Jahangiri, A., & Ameri, M. (2025). Thermo-Enviro-Exergoeconomic Analysis of Solar Chimney Compared with Heller Tower in Fars Combined Cycle Power Plant. *Energy Nexus*, *100506*, 100506. Advance online publication. DOI: 10.1016/j.nexus.2025.100506

Mohd, S. A., Sazali, N., Jamaludin, A. S., & Misdan, N. (2023). IOT-based Energy Monitoring System for energy conservation. *E3S Web Conf.*, *437*, 4006.

Montreuil, B., Ballot, E., & Fontane, F. (2020). Physical Internet and Logistics Sustainability: A Review. *Sustainability*, *12*(22), 9367.

Mor, R., Srivastava, P., Jain, R., Varshney, S., & Goyal, V. (2020). Managing Food Supply Chains Post COVID-19: A Perspective. *International Journal of Supply and Operations Management*, *7*(3), 295–298. DOI: 10.22034/IJSOM.2020.3.7

Mrad, M., Frikha, M. A., & Boujelbene, Y. (2025). A comprehensive survey of artificial intelligence and robotics for reducing carbon emissions in supply chain management. *Logistics (Basel)*, *9*(3), 104. DOI: 10.3390/logistics9030104

Muma, B. O. (2014). Green Supply Chain Management and Environmental Performance among Tea Processing Firms in Kericho County- Kenya. *International Journal of Economics. Finance and Management Sciences*, *2*(5), 270–276. DOI: 10.11648/j.ijefm.20140205.11

Munir, A., Ashraf, T., Amjad, W., Ghafoor, A., Rehman, S., Malik, A. U., Hensel, O., Sultan, M., & Morosuk, T. (2021). Solar-Hybrid Cold Energy Storage System Coupled with Cooling Pads Backup: A Step Towards Decentralized Storage of Perishables. *Energies*, *14*(22), 7633. DOI: 10.3390/en14227633

Munsberg, T. M., & Hvam, L. (2021, December). Potential of streamlining warehouse processes and IT integration to increase implementation of automation. *2021 IEEE International Conference on Industrial and Engineering Management (IEEM)*. DOI: 10.1109/IEEM50564.2021.9672816

Munuhwa, S. (2023). Sustainable logistics and competitive positioning. In *Integrating Intelligence and Sustainability in Supply Chains* (pp. 198–220). IGI Global. DOI: 10.4018/979-8-3693-0225-5.ch011

Muthaluri, J. (2024). Optimizing supply chain management and marketing strategies: AI and ML integration for competitive advantage. Educational Administration: Theory and Practice, 8990–8997.

Muthukalyani, A. R. (2023). Unlocking Accurate Demand Forecasting in Retail Supply Chains with AI-Driven Predictive Analytics. *Information Technology Management, 15*(3), 567–589. DOI: 10.1000/researchgate.aipredictiveanalytics2023

Muttakin, M. B., Mihret, D., Lemma, T. T., & Khan, A. (2020). Integrated reporting, financial reporting quality and cost of debt. *International Journal of Accounting and Information Management, 28*(3), 517–534. DOI: 10.1108/IJAIM-10-2019-0124

Myers, D. G. (1990). *Exploring social psychology*. McGraw-Hill.

Nagaraj, S., & Kumar, K. V. R. (Eds.). (2023). *AI-driven intelligent models for business excellence*. IGI Global.

Nahmiasj, S., & Rivera, H. (1979). A deterministic model for a repairable item inventory system with a finite repair rate. *International Journal of Production Research, 17*(3), 215–221. DOI: 10.1080/00207547908919609

Naidoo, V., & Henry Ramatsetse, M. (2016). Assessment of the consumer purchase intentions of organic food at the Hazel food market in Pretoria, South Africa. *Environment and Ecology*, *7*(3), 81–88. DOI: 10.21511/ee.07(3).2016.10

Nandhini, S., Yamini, B., Anurekha, G., Thilagam, T., & Pajila, P. B. (2025). The Role of Risk Management in Modern Business and Technology. In *AI Methods for Environmental Protection and Resource Conservation* (pp. 217–244). IGI Global Scientific Publishing. DOI: 10.4018/979-8-3373-3246-8.ch009

Narayan, Y., Selvaraj, S., Suryawanshi, M. A., & Ramya, S. (2023, October). Machine Learning Model used Efficiently to Manage the Smart Grid and Renewable Energy. In *2023 International Conference on New Frontiers in Communication, Automation, Management and Security (ICCAMS)* (Vol. 1, pp. 1-3). IEEE. DOI: 10.1109/ICCAMS60113.2023.10526077

Nerlove, M. (2002). *Essays in panel data econometrics*. Cambridge University Press.

Ngaboyimbere, F., Leo, J., Michael, K., & Muteteshe, D. (2021). Development of RFID based automatic warehouse management system: A case study of ROK industries limited Kenya. *Int. J. Adv. Sci. Res. Eng., 07*(08), 112–126. DOI: 10.31695/IJASRE.2021.34076

Ngo, T. H. D. (2012). The Steps to Follow in a Multiple Regression Analysis. SAS® Global Forum 2012 Conference. 1-12.

Nguyen, T. N., Lobo, A., & Nguyen, B. K. (2018). 'Young consumers' green purchase behaviour in an emerging market'. *Journal of Strategic Marketing*, *26*(7), 583–600. DOI: 10.1080/0965254X.2017.1318946

Nia, A. R., Awasthi, A., & Bhuiyan, N. (2021). "Industry 4.0 and Demand Forecasting of the Energy Supply Chain: A Literature Review." Computers & Industrial Engineering, Vol. 154. *Computers & Industrial Engineering*, *107128*, 1–15. DOI: 10.1016/j.cie.2021.107128

Nicoletti, B., & Appolloni, A. (2024). Green Warehouse 5.0. In *The Future of Industry: Human-Centric Approaches in Digital Transformation* (pp. 303–320). Springer Nature Switzerland. DOI: 10.1007/978-3-031-66801-2_20

Nikitas, A., Michalakopoulou, K., Njoya, E. T., & Karampatzakis, D. (2020). Artificial intelligence, transport and the smart city: Definitions and dimensions of a new mobility era. *Sustainability (Basel)*, *12*(7), 2789. DOI: 10.3390/su12072789

Ning, Y., Kazemi, H., & Tahmasebi, P. (2022). A Comparative Machine Learning Study for Time Series Oil Production Forecasting: ARIMA, LSTM, and Prophet. *Computers & Geosciences*, *15*(3), 567–589. DOI: 10.1016/j.cageo.2022.105126

Ni, S., Peng, Y., & Liu, Z. (2022). Logistics demand forecast of fresh food E-commerce based on bi-LSTM model. *J. Comput. Commun.*, *10*(09), 51–65. DOI: 10.4236/jcc.2022.109004

Nkinyam, C. M., Ujah, C. O., Nnakwo, K. C., & Kallon, D. V. V. (2025). Insight into organic photovoltaic cell: Prospect and challenges. In *Unconventional Resources* (Vol. 5). KeAi Communications Co., DOI: 10.1016/j.uncres.2024.100121

None, I., & Kumar Datta, S. (2011). Pro-environmental Concern Influencing Green Buying: A Study on Indian Consumers. *International Journal of Business and Management*, *6*(6), 124–133. DOI: 10.5539/ijbm.v6n6p124

Nweje, U., & Taiwo, M. (2025). Artificial Intelligence for Predictive Supply Chain Management: Focus on How AI-Driven Tools Are Revolutionizing Demand Forecasting and Inventory Optimization. *International Journal of Scientific Research*, *15*(3), 567–589. DOI: 10.1000/journalijsra.aipredictivesupplychain2025

Ocran, G., Yusuf, S. O., & Owusu, P. (2024). AI-Driven Business Analytics for SMEs: Unlocking Value Through Predictive and Prescriptive Analytics. *International Journal of AI in Business*, *15*(3), 567–589. DOI: 10.1000/researchgate.aibusinessanalytics2024

Odeyinka, F. O., & G. Omoegun, O. (2024). Warehouse operations: An examination of traditional and automated approaches in supply chain management. In *Operations Management - Recent Advances and New Perspectives*. IntechOpen.

Ogundipe, D. O., & Abaku, E. A.Damilola Oluwaseun OgundipeEmmanuel Adeyemi Abaku. (2024). Theoretical insights into AI product launch strategies for start-ups: Navigating market challenges. *Int. J. Front. Sci. Technol. Res.*, *6*(1), 62–72. DOI: 10.53294/ijfstr.2024.6.1.0032

Ojo, E., Mbowa, C., & Akinlabi, E. (2014). Green Supply Chain Management in Construction Industries in South Africa and Nigeria. *International Journal of Chemical Environmental & Biological Sciences*, *2*(2), 146–150.

Okonkwo, P. C., Mansir, I. B., Al Housni, F., Beitelmal, W. H., Al Kaaf, K., & Ogunleye, G. J. (2023). Provision of kinetic energy support from wind turbines for frequency regulation services in the modern grid. In*Advanced Frequency Regulation Strategies in Renewable-Dominated Power Systems* (pp. 167–180). Elsevier., DOI: 10.1016/B978-0-323-95054-1.00018-4

Okpiaifo, G., Durand-Morat, A., West, G. H., Nalley, L. L., Nayga, R. M.Jr, & Wailes, E. J. (2020). 'Consumers' preferences for sustainable rice practices in Nigeria'. *Global Food Security*, *24*(1), 100–345. DOI: 10.1016/j.gfs.2019.100345

Olajumoke, A., Jemimah, N., & Maureen, M. (2020). Does women's agency influence agriculture technology adoption. *Food Chain*, *7*(1), 38–54. DOI: 10.3362/2046-1887.17-00005

Oloruntobi, O., Mokhtar, K., Mohd Rozar, N., Gohari, A., Asif, S., & Chuah, L. F. (2023). Effective technologies and practices for reducing pollution in warehouses - A review. *Cleaner Engineering and Technology*, *13*, 100622. Advance online publication. DOI: 10.1016/j.clet.2023.100622

Opalic, S. M., Goodwin, M., Jiao, L., Nielsen, H. K., Pardiñas, Á. Á., Hafner, A., & Kolhe, M. L. (2020). ANN modelling of CO2 refrigerant cooling system COP in a smart warehouse. *Journal of Cleaner Production*, *260*, 120887. https://hdl.handle.net/11250/3058108. DOI: 10.1016/j.jclepro.2020.120887

Orzan, G. (2018). Consumers' behavior concerning sustainable packaging: An exploratory study on Romanian consumers. *Sustainability (Switzerland)*, 10(6), 1–11. .DOI: 10.3390/su10061787

Oyawole, F. P., Shittu, A., Kehinde, M., Ogunnaike, G., & Akinjobi, L. T. (2021). Women empowerment and adoption of climate-smart agricultural practices in Nigeria. *African Journal of Economic and Management Studies*, *12*(1), 105–119. DOI: 10.1108/AJEMS-04-2020-0137

Pandian, A. P. (2019). Artificial intelligence application in smart warehousing environment for automated logistics. *Journal of Artificial Intelligence and Capsule Networks*, *1*(2), 63–72. DOI: 10.36548/jaicn.2019.2.002

Paoletti, S., Juloski, A. L., Ferrari-Trecate, G., & Vidal, R. (2007). Identification of hybrid systems a tutorial. *European Journal of Control*, *13*(2-3), 242–260. DOI: 10.3166/ejc.13.242-260

Paramasivan, A. (2023). Transforming Healthcare Supply Chains: AI for Efficient Drug Distribution and Inventory Management. [IJSAT]. *International Journal on Science and Technology*, *15*(3), 567–589. DOI: 10.1000/researchgate.aihealthcare-supplychain2023

Pei, T. H., Ji, Y. F., Pan, G., Chang, B. J., Huang, C. Y., & Chang, W. S. (2025). Dual-power generation by the solar cell and the Seebeck-effect-based thermoelectric setup. *Energy for Sustainable Development*, *88*, 101802. Advance online publication. DOI: 10.1016/j.esd.2025.101802

Peng, X. (2022). The Impact of EU Carbon Border Adjustment Mechanism: Perspective from Developed and Developing Countries. *BCP Bus. Manag.*, *35*, 335–340. DOI: 10.54691/bcpbm.v35i.3313

Peng, X., Zhao, H., Zhang, X., Zhang, C., & Chen, C. (2024). MetaGON: A lightweight pedestrian re-identification domain generalization model adapted to edge devices. *IEEE Open Journal of the Communications Society*, *5*, 690–699. DOI: 10.1109/OJCOMS.2024.3349597

Perkumienė, D., & Puskunigytė, K. (2024). Improvement of warehouse processes through digitalization solutions. *HUMANITIES STUDIES*, *21*(98), 213–219. DOI: 10.32782/hst-2024-21-98-25

Perkumienė, D., & Vienažindienė, M. (2024). Innovative solutions in the warehousing processes of manufacturing companies towards sustainability. *Journal of Infrastructure. Policy and Development*, *8*(8), 7084.

Perotti, S., Prataviera, L. B., & Melacini, M. (2022). Assessing the environmental impact of logistics sites through CO_2eq footprint computation. *Business Strategy and the Environment*, *31*(4), 1679–1694. DOI: 10.1002/bse.2976

Peter, H. (2024). “Designing Event-Driven Architecture for AI-Powered Sales Pipeline Optimization: A Comparative Analysis of Messaging Patterns.” ResearchGate Preprint. DOI: DOI: 10.13140/RG.2.2.38908.39524

Plathottam, S. J., Rzonca, A., Lakhnori, R., & Iloeje, C. O. (2023). A review of artificial intelligence applications in manufacturing operations. *Journal of Advanced Manufacturing and Processing*, *5*(3), e10159. DOI: 10.1002/amp2.10159

Ponmalar, A., Muthuram, A., Therasa, M., Priya, V., Sathya, V., Subha, S., & Subramanian, R. S. (2025). Convergence of Advanced Technologies for Sustainable Development and Smart Cities. In *AI and Green Technology Applications in Society* (pp. 189–218). IGI Global Scientific Publishing.

Ponmalar, A., Vijayakumar, K., Lakshmipriya, C., Karthikeyan, M., Gracelin Sheena, B., Beslin Pajila, P. J., & Siva Subramanian, R. (2024). Meta-Heuristics and Machine Learning Applications in Complex Systems. In *Metaheuristic and Machine Learning Optimization Strategies for Complex Systems* (pp. 257–275). IGI Global Scientific Publishing.

Porter, M. E., & van der Linde, C. (1995). Toward a new conception of the environment-competitiveness relationship. *The Journal of Economic Perspectives*, *9*(4), 97–118. DOI: 10.1257/jep.9.4.97

Prasad, A., Kumar, A. S., Sharma, P., & Irawati, I. D. DV, C., Musirin, I. B., & Abdullah, H. M. A. (2023). Artificial intelligence in computer science: An overview of current trends and future directions. *Advances in artificial and human intelligence in the modern era*, 43-60.

Pratiwi, I., Elizabeth, R., Fatari, , Basrowi, , & Nuryanto, U. W. (2025). Assessing pricing, distribution, and warehousing strategies in influencing consumer demand and purchase decisions: Evidence from Indonesia. *Social Sciences & Humanities Open*, *12*, 101688. Advance online publication. DOI: 10.1016/j.ssaho.2025.101688

Prillaman, S. Strength in Numbers: How Women’s Networks Close India’s Political Gender Gap. In *Proceedings of the American Political Science Association Conference*, Philadelpha, PA, USA, 1–4 September 2016.

Priyanta, I. F., Freytag, J., Reining, C., Roidl, M., & Kirchheim, A. (2024, November). *Towards 6g-driven cooperative robot framework for unified sensing in smart warehouses* [Paper presentation]. *2024 IEEE Conference on Standards for Communications and Networking (CSCN)*, IEEE. DOI: 10.1109/CSCN63874.2024.10849725

Prosperi, P., Allen, T., Padilla, M., Peri, I., & Cogill, B. (2014). Sustainability and Food & Nutrition Security: A Vulnerability Assessment Framework for the Mediterranean Region. *SAGE Open*, *4*(2), 1–15. DOI: 10.1177/2158244014539169

Punia, S., & Shankar, S. (2022). Predictive Analytics for Demand Forecasting: A Deep Learning-Based Decision Support System. *Knowledge-Based Systems*, *15*(3), 567–589. DOI: 10.1016/j.knosys.2022.109956

Purniawan, A., & Elmi, F. (2025). The effect of Green Human Resource Management on green training, employee behavior, and employee welfare at PT Senayan Texindo Makmur. *Ijfsr, 2*(3).

Qadar Chaudhary, G., Waheed Azam, M., Bozzoli, F., Sajjad, U., Vocale, P., Cattani, L., & Fallahzadeh, R. (2024). Transient analysis of an efficient solar assisted air-conditioning system for subtropical climate with various solar thermal collectors. *Energy Conversion and Management: X, 23*(May), 100634. DOI: 10.1016/j.ecmx.2024.100634

Qi, Q., Tao, F., Hu, T., Answer, N., Liu, A., Wei, Y., Wang, L., & Nee, A. Y. C. (2021). Enabling technologies and tools for digital twin. *Journal of Manufacturing Systems*, *58*, 3–21. DOI: 10.1016/j.jmsy.2019.10.001

Qiu, Z., Wang, D., & Gao, M. 2023, November. Research on Industrial Assembly Screw Automatic Tightening Algorithm Based on Multi-Objective Mathematical Modeling. In *2023 7th Asian Conference on Artificial Intelligence Technology (ACAIT)* (pp. 1085-1090). IEEE DOI: 10.1109/ACAIT60137.2023.10528391

Qi, X., & Li, T. (2022). Research on Countermeasures for the High Quality Development of Green Logistics in China under the "Double Carbon". *Frontiers in Business, Economics and Management*, *5*(1), 33–36. DOI: 10.54097/fbem.v5i1.1425

Qu, H., Gao, B., Kong, Q., & Xu, Z. (2024). Experimental study of the effect of ambient temperature on the output characteristics of PV-PCM systems. *Journal of Energy Storage*, *88*, 111535. Advance online publication. DOI: 10.1016/j.est.2024.111535

Radhamani, R., Karthick, S., Kumar, S. K., & Gokulraj, M. (2024, March). Deployment of an IoT-Integrated Home Energy Management System Employing Deep Reinforcement Learning. In *2024 2nd International Conference on Artificial Intelligence and Machine Learning Applications Theme: Healthcare and Internet of Things (AIMLA)* (pp. 1-4). IEEE. DOI: 10.1109/AIMLA59606.2024.10531519

Rahul, P., & Singh, A. J. (2023). A study of Mobile Edge Computing for IOT. *Advances in Science and Technology (Owerri, Nigeria)*, *124*, 856–863. DOI: 10.4028/p-2u34v7

Ramadhini, A., Adhariani, D., & Djakman, C. D. (2020). The Effects of External Stakeholder Pressure on CSR Disclosure: Evidence from Indonesia. *DLSU Business & Economics Review*, *29*(2), 29–39.

Ramakrishnan, T., Mohan Gift, M. D., Chitradevi, S., Jegan, R., & Subha Hency Jose, P. Nagaraja, Rajneesh Sharma HN, Selvakumar P, Sintayehu Mekuria Hailegiorgis, (2022). "Study of Numerous Resins Used in Polymer Matrix Composite Materials", Advances in Materials Science and Engineering, vol. 2022, Article ID 1088926, 8 pages, 2022. https://doi.org/DOI: 10.1155/2022/1088926

Ramesh, P., Nithyanandhan, R., Faizal, M. M., Nivakumar, V., & Nalini, M. (2024, December). Machine Learning Strategies for Customer Churn Prediction in Competitive Enterprises. In *2024 International Conference on Sustainable Communication Networks and Application (ICSCNA)* (pp. 653-660). IEEE. DOI: 10.1109/ICSCNA63714.2024.10864143

Ramjee, A., & Gwatidzo, T. (2012). Dynamics in capital structure determinants in South Africa. *Meditari Accountancy Research*, *20*(1), 52–67. DOI: 10.1108/10222521211234228

Ravindran, D., Revathi, S., Sowndharya, V., Farzhana, I., Swetha, S., Vinoth, R., & Subramanian, R. S. (2025). Environmental Monitoring for Sustainability: Technologies, Trends, and Future Directions. *Leveraging Urban Computing for Sustainable Urban Development*, 323-340.

Rech, A., Siamos, E., Nicholas, P., & Daugaard, A. E. (2023). Recyclable Extrudable Biopolymer Composites from Alginate and Lignocellulosic Biomass Waste. *ACS Sustainable Chemistry & Engineering*, *11*(24), 8939–8947. DOI: 10.1021/acssuschemeng.3c01119

Reddy, V. J., Hariram, N. P., Ghazali, M. F., & Kumarasamy, S. (2024). Pathway to Sustainability: An Overview of Renewable Energy Integration in Building Systems. In *Sustainability (Switzerland)* (Vol. 16, Issue 2). Multidisciplinary Digital Publishing Institute (MDPI). DOI: 10.3390/su16020638

Ren, S., Liang, C., Liu, Y., Wang, J., & Wang, C. (2025). Evolutionary game-based warehousing resources sharing strategy for logistics industry under product-service system paradigm. *Advanced Engineering Informatics*, *67*, 103560. Advance online publication. DOI: 10.1016/j.aei.2025.103560

Richter, K. (1996). The eoq repair and waste disposal model with variable setup numbers. *European Journal of Operational Research*, *95*(2), 313–324. DOI: 10.1016/0377-2217(95)00276-6

Ries, J. M., Grosse, E. H., & Fichtinger, J. (2017). Environmental impact of warehousing: A scenario analysis for the United States. *International Journal of Production Research*, *55*(21), 6485–6499. DOI: 10.1080/00207543.2016.1211342

Rivard, E., Trudeau, M., & Zaghib, K. (2019). Hydrogen Storage for Mobility: A Review. *Materials (Basel)*, *12*(12), 1973. DOI: 10.3390/ma12121973 PMID: 31248099

Roucham, B., Lefilef, A., & Mahmoudi, H. (2023). The Forward-Looking studies on Green Human Resources Management: A content analysis. *Manag. Econ. Rev.*, *8*(3), 259–275. DOI: 10.24818/mer/2023.10-01

Rozo, L., Jiménez, P., & Torras, C. (2013). A robot learning from demonstration framework to perform force-based manipulation tasks. *Intelligent Service Robotics*, *6*(1), 33–51. DOI: 10.1007/s11370-012-0128-9

Sága, M., Bulej, V., Čuboňova, N., Kuric, I., Virgala, I., & Eberth, M. (2020). Case study: Performance analysis and development of robotized screwing application with integrated vision sensing system for automotive industry. *International Journal of Advanced Robotic Systems*, *17*(3), 1729881420923997. DOI: 10.1177/1729881420923997

Sahay, N., & Ierapetritou, M. (2013). Supply chain management using an optimization driven simulation approach. *AIChE Journal. American Institute of Chemical Engineers*, *59*(12), 4612–4626. DOI: 10.1002/aic.14226

Saif, A. G. H., Alquaity, A. B. S., & Mokheimer, E. M. A. (2025). Techno-economic analysis of green hydrogen production in Saudi Arabia: A comparative study of solar PV and CSP technologies. *International Journal of Hydrogen Energy*, *105*, 1361–1374. DOI: 10.1016/j.ijhydene.2025.01.043

Saiyed, A. (2025). AI-Driven Innovations in Fintech: Applications, Challenges, and Future Trends. [DOI]. *International Journal of Electrical and Computer Engineering Research*, *5*(1), 8–15. DOI: 10.53375/ijecer.2025.437

Salazar, C., & Production, J. (2016). risk and adoption of irrigation technology: Evidence from small-scale farmers in Chile. *Latin American Economic Review*, *25*(1), 18–37. DOI: 10.1007/s40503-016-0032-3

Salim Ba Awain, A. M., Al-Ansi, A. M., Mohsin, A., & Jaboob, M. (2023). Green Supply Chain Management: A Comprehensive Review of Research, Applications and Future Directions. *Manag. Prod. Eng. Rev.*, *14*, 118–133.

Salisu, A., Omoke, P., & Fadiya, O. (2022). Climate Policy Uncertainty and Crude Oil Market Volatility. *Energy Research Letters*, *3*(1), 1–5. DOI: 10.46557/001c.38781

Sancha, C., Wong, C. W. Y., & Gimenez, C. (2019). Do dependent suppliers benefit from buying firms' sustainability practices? *Journal of Purchasing and Supply Management*, *25*(4), 100–542.

Sands, J., Gholami, A. & Shams, S. (2022). Corporates' sustainability disclosures impact on cost of capital and idiosyncratic risk disclosures. *Meditari Accountancy Research*, 4(1), 1–27. .DOI: 10.1108/MEDAR-06-2020-0926

Santos, H., Lannelongue, G., & Gonzalez-Benito, J. (2019). Integrating green practices into operational performance: Evidence from Brazilian manufacturers. *Sustainability (Basel)*, *11*(10), 1–18. DOI: 10.3390/su11102956

Sarkis, J. (2017). Green Supply Chain Management: Product Life Cycle Approach. Springer.

Sarkis, J. (2020). Supply chain sustainability: Learning from the COVID-19 pandemic. *International Journal of Operations & Production Management*, *41*(1), 63–73. DOI: 10.1108/IJOPM-08-2020-0568

Sarkis, J., & Zhu, Q. (2018). Environmental sustainability and production: Taking the road less travelled. *International Journal of Production Research*, *56*(1–2), 743–759. DOI: 10.1080/00207543.2017.1365182

Sasmal, S. (2024a). Data warehousing revolution: AI-driven solutions. *INTERNATIONAL RESEARCH JOURNAL OF ENGINEERING & APPLIED*, *12*(1), 1–6.

Sathya, U., Hashini, S., Lakshmipriya, C., Varun, C. M., & Nalini, M. (2024, December). Emerging Trends and Future Prospects in Internet of Things (IoT) Technology. In *2024 International Conference on Sustainable Communication Networks and Application (ICSCNA)* (pp. 79-85). IEEE.

Saudi Vision 2030. (2017). Kingdom of Saudi Arabia Vision 2030. Government of Saudi Arabia.

Saunders, M., & Lewis, P. (2018). *Doing research in business and management* (2nd ed.). Pearson Education Limited.

Schmidt, M., & Wagner, S. M. (2019). Supply Chain 4.0: A Primer. Springer International Publishing.

Schrady, D. A. (1967). A deterministic inventory model for reparable items. *Naval Research Logistics Quarterly*, *14*(3), 391–398. DOI: 10.1002/nav.3800140310

Seifhashemi, M., Capra, B. R., Milller, W., & Bell, J. (2018). The potential for cool roofs to improve the energy efficiency of single storey warehouse-type retail buildings in Australia: A simulation case study. *Energy and Building*, *158*, 1393–1403. DOI: 10.1016/j.enbuild.2017.11.034

Sekhar, C. (2022). Optimizing Retail Inventory Management with AI: A Predictive Approach to Demand Forecasting, Stock Optimization, and Automated Reordering. *European Journal of Advances in Engineering and Technology*, *15*(3), 567–589. DOI: 10.1000/researchgate.aiinventorymanagement2022

Seto, K. C., Davis, S. J., Mitchell, R. B., Stokes, E. C., Unruh, G., & Ürge-Vorsatz, D. (2016). Carbon Lock-In: Types, Causes, and Policy Implications. *Annual Review of Environment and Resources*, *41*(1), 425–452. DOI: 10.1146/annurev-environ-110615-085934

Seuring, S., & Müller, M. (2008). From a literature review to a conceptual framework for sustainable supply chain management. *Journal of Cleaner Production*, *16*(15), 1699–1710. DOI: 10.1016/j.jclepro.2008.04.020

Shahroudnejad, A., Mousavi, P., Perepelytsia, O., Sahir, , Staszak, D., Taylor, M. E., & Bawel, B. (2024). A novel framework for automated warehouse layout generation. *Frontiers in Artificial Intelligence*, *7*, 1465186. DOI: 10.3389/frai.2024.1465186 PMID: 39474600

Sharma, A., Patel, N., & Gupta, R. (2021). Leveraging LSTM and Prophet Models for Enhanced AI-Driven Demand Prediction in E-Commerce. *European Advanced AI Journal*, *15*(3), 567–589. DOI: 10.1000/eaaij.aidemandprediction2021

Sharmin, T., Khan, N. R., Akram, M. S., & Ehsan, M. M. (2023). A State-of-the-Art Review on Geothermal Energy Extraction, Utilization, and Improvement Strategies: Conventional, Hybridized, and Enhanced Geothermal Systems. In *International Journal of Thermofluids* (Vol. 18). Elsevier B.V. DOI: 10.1016/j.ijft.2023.100323

Shboul, B. (2025). Novel design of hybrid dish Stirling engine and FC-electrolyzer system for synergistic power and green hydrogen solutions: A 3E performance evaluation. *Energy*, *318*, 134835. Advance online publication. DOI: 10.1016/j.energy.2025.134835

Shende, A. D., Tekade, S. A., Deshmukh, A. A., Tembhurkar, S. P., & Selvakumar, P. (2025). Water Quality Monitoring and Control With AI. In Ruiz-Vanoye, J., & Díaz-Parra, O. (Eds.), *Smart Water Technology for Sustainable Management in Modern Cities* (pp. 51–80). IGI Global Scientific Publishing., DOI: 10.4018/979-8-3693-8074-1.ch003

Shepherd, M. K., & Rouse, E. J. (2017). Design and validation of a torque-controllable knee exoskeleton for sit-to-stand assistance. *IEEE/ASME Transactions on Mechatronics*, *22*(4), 1695–1704. DOI: 10.1109/TMECH.2017.2704521

Shobhana, N. (2024). AI-Powered Supply Chains Towards Greater Efficiency. *AI Dynamics and Interactions in Management*, *15*(3), 567–589. DOI: 10.1000/igi-global.aisupplychain2024

Shukla, A. K., Sudhakar, K., & Baredar, P. (2017). Recent advancement in BIPV product technologies: A review. In *Energy and Buildings* (Vol. 140, pp. 188–195). Elsevier Ltd., DOI: 10.1016/j.enbuild.2017.02.015

Simpson, D. F., & Power, D. J. (2015). Use the Supply Relationship to Develop Lean and Green Suppliers. *Supply Chain Management*, *10*(1), 60–68. DOI: 10.1108/13598540510578388

Singh, K. U. (2021). Design of energy efficient and secure routing protocol for wsn in IoT. *Philipp. Stat. (Quezon City)*, *70*(2), 1434–1441. DOI: 10.17762/msea.v70i2.2336

Singh, N., & Adhikari, D. (2023). AI and IoT: A Future Perspective on Inventory Management. *International Journal for Research in Applied Science and Engineering Technology*, *15*(3), 567–589. DOI: 10.22214/ijraset.2023.57200

Singh, N., & Adhikari, D. (2023). AI in inventory management: Applications, Challenges, and opportunities. *International Journal for Research in Applied Science and Engineering Technology*, *11*(11), 2049–2053. DOI: 10.22214/ijraset.2023.57010

Siva Subramanian, R., Maheswari, B., Nikkath Bushra, S., Nirmala, G., & Anita, M. (2023). Enhancing Customer Prediction Using Machine Learning with Feature Selection Approaches. In *Inventive Computation and Information Technologies* [Singapore: Springer Nature Singapore.]. *Proceedings of ICICIT*, *2022*, 45–57.

Škerlič, S., Muha, R., & Sokolovskij, E. (2017). Application of modern warehouse technology in the Slovenian automotive industry. *Transport*, *32*(4), 415–425. DOI: 10.3846/16484142.2017.1354315

Smith, J. (2023). Smart Warehousing and Green Logistics. Journal of Sustainable Logistics.

Sodiya, E. O., Umoga, U. J., & Amoo, O. O. (2024). AI-Driven Warehouse Automation: A Comprehensive Review of Systems. *Advanced Research in AI and Logistics*, *15*(3), 567–589. DOI: 10.1000/researchgate.aiwarehouseautomation2024

Sodiya, E. O., Umoga, U. J., Amoo, O. O., & Atadoga, A. (2024a). AI-driven warehouse automation: A comprehensive review of systems. *GSC Adv. Res. Rev., 18*(2), 272–282. DOI: 10.30574/gscarr.2024.18.2.0063

Solar, G. I. S. *Direct Normal Irradiation (GHI); 2024. Available from:* https://solargis.com/resources/free-maps-and-gis-data. (2024).

Sonnenberg, N. C., Erasmus, A. C., & Donoghue, S. (2011). 'Significance of environmental sustainability issues in consumers' choice of major household appliances in South Africa'. *International Journal of Consumer Studies, 35*(2), 153–163. DOI: 10.1111/j.1470-6431.2010.00964.x

Soori, M., Arezoo, B., & Dastres, R. (2023). Internet of things for smart factories in industry 4.0, a review. *Internet of Things and Cyber-Physical Systems, 3*, 192–204. DOI: 10.1016/j.iotcps.2023.04.006

Soumya Rai, P., Kant, C. R., & Jha, R. (2025). Metal-Air Batteries: From Fundamental Mechanisms to Practical Applications. *Journal of Alloys and Compounds, 180805*, 180805. Advance online publication. DOI: 10.1016/j.jallcom.2025.180805

Speck, J. A. (2018). *Mechanical fastening, joining, and assembly*. CRC Press. DOI: 10.1201/9781351228640

Sreenivasulu, A., & Manimozhi, T. Satheesh kumar P, Selvakumar P. and Singh P, (2022). "The Model of Smart Sensing Device For Sensitive Nanoclusters Modification in Sensing Properties," 2022 2nd International Conference on Advance Computing and Innovative Technologies in Engineering (ICACITE), Greater Noida, India, 2022, pp. 1043-1047, DOI: 10.1109/ICACITE53722.2022.9823846

Srivastava, S. K. (2007). Green supply-chain management: A state-of-the-art literature review. *International Journal of Management Reviews, 9*(1), 53–80. DOI: 10.1111/j.1468-2370.2007.00202.x

Staudt, F. H., Alpan, G., Di Mascolo, M., & Rodriguez, C. M. T. (2015). Warehouse performance measurement: A literature review. *International Journal of Production Research, 53*(18), 5524–5544. DOI: 10.1080/00207543.2015.1030466

Steyn, M. (2014). 'Organisational benefits and implementation challenges of mandatory integrated reporting: Perspectives of senior executives at South African listed companies', *Sustainability Accounting. Management and Policy Journal, 5*(4), 476–503. DOI: 10.1108/SAMPJ-11-2013-0052

Subramanian, R. S., Santhi, K., Sasikumar, D., Shobana, S., Elavarasi, J., Gracious, L. A., & Nalini, M. (2026). Leveraging Sentiment Analysis in Customer Service for Enhanced Customer Experience. In *Harnessing Emotion AI for Customer Support and Employee Wellbeing* (pp. 49-72). IGI Global Scientific Publishing.

Subramanian, M., Cho, J., Veerappampalayam Easwaramoorthy, S., Murugesan, A., & Chinnasamy, R. (2023). Enhancing sustainable transportation: AI-Driven bike demand forecasting in smart cities. *Sustainability (Basel)*, *15*(18), 13840. DOI: 10.3390/su151813840

Sugianto, S. (2020). Comparative Analysis of Solar Cell Efficiency between Monocrystalline and Polycrystalline. *INTEK: Jurnal Penelitian*, *7*(2), 92–100. DOI: 10.31963/intek.v7i2.2625

Sugumaran K M D, Pandiyarajan V Y B S, Selvakumar P. and M. S, (2024). "Analysis of Electromyography Signals for Control of Mechanical Prosthesis using Machine Learning Techniques," 2024 2nd International Conference on Intelligent Data Communication Technologies and Internet of Things (IDCIoT), Bengaluru, India, 2024, pp. 1059-1067, .DOI: 10.1109/IDCIoT59759.2024.10467707

Sundar, D. G., Rajshri, S., & Kalaiarasu. (2025). Smart solar integrated v2g energy management system. *INTERANTIONAL JOURNAL OF SCIENTIFIC RESEARCH IN ENGINEERING AND, 09*(02), 1–7.

Suresh, C., Nyemeesha, V., Prasath, R., Lokeshwaran, K., Raju, K. R., & Boopathi, S. (2024). AI-Driven Energy Forecasting, Optimization, and Demand Side Management for Consumer Engagement. In *Intelligent Solutions for Sustainable Power Grids* (pp. 112–131). IGI Global. DOI: 10.4018/979-8-3693-3735-6.ch006

Suzuki, T. (2025). Vision-Guided Robotics: Integrating Computer Vision and Mechanical Engineering for Precision Control. *Global Research Review*, *1*(1), 77–84.

Švaco, M., Šekoranja, B., & Jerbić, B. (2012). Industrial robotic system with adaptive control. *Procedia Computer Science*, *12*, 164–169. DOI: 10.1016/j.procs.2012.09.048

Sy Horng Ting, T., Goh, Y. H., Chin, K. M., Tan, Y. K., Chiew, T. H., Ma, G., & How, C. K. (2025). A Review of Advanced Force Torque Control Strategies for Precise Nut-to-Bolt Mating in Robotic Assembly. *International Journal of Robotics & Control Systems, 5*(1 Toth, G.R.R., 2004. Torque and angle controlled tightening over the yield point of a screw—Based on monte-carlo simulations. *Journal of Mechanical Design*, *126*(4), 729–736.

Syamala, M., Komala, C. R., Pramila, P. V., Dash, S., Meenakshi, S., & Boopathi, S. (2023). Machine learning-integrated IoT-based smart home energy management system. In *Handbook of Research on Deep Learning Techniques for Cloud-Based Industrial IoT* (pp. 219–235). IGI Global. DOI: 10.4018/978-1-6684-8098-4.ch013

Taiwo, A. I., Isi, L. R., Okereke, M., Sofoluwe, O., Olugbemi, G. I. T., & Essien, N. A. (2025). Developing climate-adaptive digital twin architectures for predictive supply chain disruption management using spatio-temporal analytics and edge computing. *International Journal of Scientific Research in Science and Technology*, *12*(3), 931–947. DOI: 10.32628/IJSRST25123101

Taleb, N., & Mohamed, E. A. (2020). Cloud computing trends: A literature review. *Acad. J. Interdiscip. Stud.*, *9*(1), 91. DOI: 10.36941/ajis-2020-0008

Talib, F., Asjad, M., Attri, R., Siddiquee, A. N., & Khan, Z. A. (2020). A road map for the implementation of integrated JIT-lean practices in Indian manufacturing industries using the best-worst method approach. *J. Ind. Prod. Eng.*, *37*(6), 275–291. DOI: 10.1080/21681015.2020.1788656

Tang, Y. M., Chau, K. Y., Xu, D., & Liu, X. (2021). Consumer perceptions to support IoT based smart parcel locker logistics in China. *Journal of Retailing and Consumer Services*, *62*, 102659. DOI: 10.1016/j.jretconser.2021.102659

Tang, Y., Chau, K., Hong, L., Ip, Y., & Yan, W. (2021). Financial Innovation in Digital Payment with WeChat towards Electronic Business Success. *Journal of Theoretical and Applied Electronic Commerce Research*, *16*(5), 1844–1861. DOI: 10.3390/jtaer16050103

Tang, Y., Chau, K.-Y., Li, W., & Wan, T. (2020). Forecasting economic recession through share price in the logistics industry with artificial intelligence (AI). *Computation (Basel, Switzerland)*, *8*(3), 70. DOI: 10.3390/computation8030070

Tao, R., Wu, J., & Zhao, H. (2022). Do Corporate Customers Prefer Socially Responsible Suppliers? An Instrumental Stakeholder Theory Perspective. *Journal of Business Ethics*. Advance online publication. DOI: 10.1007/s10551-022-05171-5

Teece, D. J., Pisano, G., & Shuen, A. (1997). Dynamic capabilities and strategic management. *Strategic Management Journal*, *18*(7), 509–533. DOI: 10.1002/(SICI)1097-0266(199708)18:7<509::AID-SMJ882>3.0.CO;2-Z

Tejesh, S. B. S., & Neeraja, S. (2018). Warehouse inventory management system using IoT and open source framework. *Alexandria Engineering Journal*, *57*(4), 3817–3823. DOI: 10.1016/j.aej.2018.02.003

Tesla Impact Report. (2024).

Tetteh, F. K.. (2024). Green logistics practices: A bibliometric and systematic review. *Journal of Cleaner Production*. Advance online publication. DOI: 10.1016/j.jclepro.2024.143735

Teunter, R. H. (2001). Economic ordering quantities for recoverable item inventory systems. [NRL]. *Naval Research Logistics*, *48*(6), 484–495. DOI: 10.1002/nav.1030

Tharani, D. B., Jayakrishnan, N., Dharshan, & Sarveswaran. (2025). Iot-based smart grids: Adaptive energy scheduling framework. *INTERANTIONAL JOURNAL OF SCIENTIFIC RESEARCH IN ENGINEERING AND*, *09*(03), 1–6.

Tiwari, S. (2023). Smart warehouse: A bibliometric analysis and future research direction. *Sustainable Manufacturing and Service Economics*, *2*, 100014. DOI: 10.1016/j.smse.2023.100014

Togun, H., Basem, A., Kadhum, A. A. H., Abed, A. M., Biswas, N., Rashid, F. L., Lawag, R. A., Ali, H. M., Mohammed, H. I., & Mandal, D. K. (2025). Advancing photovoltaic thermal (PV/T) systems: Innovative cooling technique, thermal management, and future prospects. In *Solar Energy* (Vol. 291). Elsevier Ltd. DOI: 10.1016/j.solener.2025.113402

Tsai, W. H., Shen, Y.-S., Lee, P.-L., Chen, H.-C., Kuo, L., & Huang, C.-C. (2012). Integrating information about the cost of carbon through activity-based costing. *Journal of Cleaner Production*, *36*, 102–111. Advance online publication. DOI: 10.1016/j.jclepro.2012.02.024

Tung, P. H.Phuong Huu Tung. (2025). The influence of Green Human Resource Management on green training: A survey at Vietnam coal and mineral industry group. *Journal of Information Systems Engineering & Management*, *10*(38s), 457–462. DOI: 10.52783/jisem.v10i38s.6870

Tyler, V. (2025). How Event-Driven Architectures Are Powering Predictive Decision Intelligence. [DOI]. *ResearchGate Preprint*, *15*(3), 567–589.

U.S. Green Building Council (USGBC). (2022). LEED Green Building Certification Program.

U.S. Green Building Council. (2019). *LEED v4 for building design and construction*. USGBC.

Udo, W. S., Kwakye, J. M., Ekechukwu, D. E., & Ogundipe, O. B. (2023). Smart grid Innovation: Machine Learning for real-Time Energy management and Load Balancing. *Eng. Sci. Technol. j.*, *4*(6), 603–616.

United Nations Environment Programme. (2021). Sustainability and circular economy in supply chains: Guidelines and case studies. UNEP.

United Nations Framework Convention on Climate Change. (2018). *The Paris Agreement*. UNFCCC.

United Nations. (2015). Transforming our world: the 2030 Agenda for Sustainable Development.

USGBC. (2021). LEED v4 for Building Design and Construction.

Uwa, J. N., & Pourvahidi, P. (2024). An Overview of the Effects of Passive Solar Strategies on Thermal Comfort. *Journal of Mediterranean Cities*, *4*(1). Advance online publication. DOI: 10.38027/mediterranean-cities_vol4no1_2

van Geest, M., Tekinerdogan, B., & Catal, C. (2021). Design of a reference architecture for developing smart warehouses in industry 4.0. *Computers in Industry*, *124*, 103343. DOI: 10.1016/j.compind.2020.103343

Van Geest, M., Tekinerdogan, B., & Catal, C. (2021). Smart warehouses: Rationale, challenges and solution directions. *Applied Sciences (Basel, Switzerland)*, *12*(1), 219. DOI: 10.3390/app12010219

Varghese, B., Reano, C., & Silla, F. (2018). Accelerator virtualization in fog computing: Moving from the cloud to the edge. *IEEE Cloud Computing*, *5*(6), 28–37. DOI: 10.1109/MCC.2018.064181118

Vázquez, N., & Vázquez, J. (2023). Photovoltaic System Conversion. In *Power Electronics Handbook* (pp. 781–795). Elsevier., DOI: 10.1016/B978-0-323-99216-9.00023-8

Verma, D. (2018). Analysis of smart manufacturing technologies for industry using AI methods. [TURCOMAT]. *Turk. J. Comput. Math. Educ.*, *9*(2), 529–540. DOI: 10.17762/turcomat.v9i2.13857

Verma, P. (2024). Transforming Supply Chains Through AI: Demand Forecasting, Inventory Management, and Dynamic Optimization. *Integrated Journal of Science and Technology*, *15*(3), 567–589. DOI: 10.1000/researchgate.aisupplychainoptimization2024

Vijayakumar, G. (2024). Muniyandi Rajkumar, Rajiv Chandar N, Selvakumar P and Ramesh Duraisamy, (2024). Environmentally friendly TDS removal from wastewater by electrochemical ion exchange batch-type recirculation (EIR) technique. *Environmental Science. Water Research & Technology*, *10*, 826–835. DOI: 10.1039/D3EW00793F

Walisinghe, B. R., Ratnasiri, S., Rohde, N., & Guest, R. (2017). Does agricultural extension promote technology adoption in Sri Lanka. *International Journal of Social Economics*, *44*(12), 2173–2186. DOI: 10.1108/IJSE-10-2016-0275

Walmart. (2022). Global Responsibility Report.

Walter, A., Ahsan, K., & Rahman, S. (2025). Application of artificial intelligence in demand planning for supply chains: A systematic literature review. *International Journal of Logistics Management*, *36*(3), 672–719. DOI: 10.1108/IJLM-02-2024-0120

Wang, X., Zhao, Y. & Hou, L. (2020). How does green innovation affect supplier-customer relationships? A study on customer and relationship contingencies. .DOI: 10.1016/j.indmarman.2020.07.008

Wang, G., Zhou, F., Jia, H., wang, L., Klitzsch, N., Feng, G., Yan, C., Ma, H., Liu, S., yin, Q., & Liu, X. (2025). Heat extraction performance of multi-level, multi-branch, closed loop coaxial horizontal borehole heat exchanger geothermal system. *Applied Thermal Engineering*, *269*, 126045. Advance online publication. DOI: 10.1016/j.applthermaleng.2025.126045

Wang, J., & Azam, W. (2024). Natural resource scarcity, fossil fuel energy consumption, and total greenhouse gas emissions in top emitting countries. *Geoscience Frontiers*, *15*(2), 101757. Advance online publication. DOI: 10.1016/j.gsf.2023.101757

Wang, J., Liu, M., Chen, Y., & Yu, M. (2023). Influencing Factors on Green Supply Chain Resilience of Agricultural Products: An Improved Gray-DEMATEL-ISM Approach. *Frontiers in Sustainable Food Systems*, *7*, 1166395. DOI: 10.3389/fsufs.2023.1166395

Wang, N., Varghese, B., Matthaiou, M., & Nikolopoulos, D. S. (2019). ENORM: A framework for edge NOde resource management. *IEEE Transactions on Services Computing*, •••, 1. DOI: 10.1109/TSC.2017.2753775

Wang, S., Yan, C., & Xiao, F. (2017). Quantitative energy performance assessment methods for existing buildings. *Energy and Building*, *43*(2-3), 103–109.

Wang, X., Liu, M., Ge, M., Ling, L., & Liu, C. (2015). Research on assembly quality adaptive control system for complex mechanical products assembly process under uncertainty. *Computers in Industry*, *74*, 43–57. DOI: 10.1016/j.compind.2015.09.001

Wang, Y., Pan, J., Zhang, P., & Tesfamariam Semere, D. (2019). Application on automated warehouse simulation system in location optimization. *Proceedings of The First International Symposium on Management and Social Sciences (ISMSS 2019)*. DOI: 10.2991/ismss-19.2019.20

Wei, Z., & Ma, Y. (2024). Research on integration management of big data application precision portrait marketing resources based on cloud and edge collaborative computing. Proceedings of the First International Conference on Science, Engineering and Technology Practices for Sustainable Development, ICSETPSD 2023, 17th-18th 2023, Coimbatore, Tamilnadu, India. DOI: 10.4108/eai.17-11-2023.2342626

Weißhuhn, S., & Hoberg, K. (2021). Designing smart replenishment systems: Internet-of-Things technology for vendor-managed inventory at end consumers. *European Journal of Operational Research, 295*(3), 949–964. DOI: 10.1016/j.ejor.2021.03.042

William, J., & Kaserwa, N. (2015). Improving smallholder farmers' access to finance through warehouse receipt system in Tanzania. *Int. J. Econ. Financ. Res., 1*, 41–49.

World Economic Forum. (2020). The Future of Jobs Report 2020.

Wrana, J., Struzik, W., Jaromin-Gleń, K., & Gleń, P. (2023). FCH HVAC honeycomb ring network—Transition from traditional power supply systems in existing and revitalized areas. *Energies, 16*(24), 7965. DOI: 10.3390/en16247965

Wu, I.-C., & Liu, C.-C. (2019). A visual and persuasive energy conservation system based on BIM and IoT technology. *Sensors (Basel), 20*(1), 139. DOI: 10.3390/s20010139 PMID: 31878268

Wu, P.-J., & Tai, Y.-C. (2024a). Artificial intelligence-based food-quality and warehousing management for food banks' inbound logistics. *Journal of Enterprise Information Management, 37*(1), 307–325. DOI: 10.1108/JEIM-10-2022-0398

Xu, H., Lee, H. S. & Ho Eom, T. (2007). Introduction to Panel Data Analysis. *Handbook of Research Methods in Public Administration*, 1(1), 571–587. DOI: 10.1201/9781420013276.ch32

Xu, L., Zhai, Y. and Li, B., 2018. Research 0f Electrical Control System for Flywheel Tightening Devices In Engine Assembly Lines. *Academic Journal of Manufacturing Engineering, 16*(1)

Xu, H. (2022). Facilitating Full and Effective Implementation of the Paris Agreement for Carbon Neutrality Vision. *Carbon Neutrality, 1*(1), 3. DOI: 10.1007/s43979-022-00014-8

Yadav, A., Garg, R. K., & Sachdeva, A. (2024). Artificial intelligence applications for information management in sustainable supply chain management: A systematic review and future research agenda. *International Journal of Information Management Data Insights, 4*(2), 100292. DOI: 10.1016/j.jjimei.2024.100292

Yadav, D. K., Jayanthu, S., Das, S. K., Chinara, S., & Mishra, P. (2019). Critical review on slope monitoring systems with a vision of unifying WSN and IoT. *IET Wireless Sensor Systems*, *9*(4), 167–180. DOI: 10.1049/iet-wss.2018.5197

Yamini, B., Anish, T. P., Maheswari, B., & Praveenkumar, J. (2025). Towards a Greener Future: Integrated Strategies for E-Waste Management. In *Integrated Approaches for Sustainable E-Waste Management* (pp. 133–150). IGI Global Scientific Publishing. DOI: 10.4018/979-8-3693-7383-5.ch005

Yan, K., Yu, S., & Liu, X. (2016). On Targeted Poverty Alleviation with the Introduction of Gender Factor Taking the Poverty-stricken Counties in Mianyang City for Example. J. Cent. China Norm. Univ. (Humanit. Soc. Sci.) 55, 1–7.

Yang, J., Lee, T.-Y., Lee, W.-T., & Xu, L. (2022). A design and application of municipal service platform based on cloud-edge collaboration for smart cities. *Sensors (Basel)*, *22*(22), 8784. DOI: 10.3390/s22228784 PMID: 36433381

Yang, M. Using data driven simulation to build inventory model. In *Proceedings of the 2008 Winter Simulation Conference*, Miami, FL, USA, 7–10 December 2008; IEEE: New York, NY, USA, 2008. DOI: 10.1109/WSC.2008.4736373

Yang, X. (2020). Marketing Strategy Analysis of Green Agricultural Products: Taking Daqing of Heilongjiang Province as an Example. *Agric. For. Econ. Manag.*, *3*, 1–7.

Yang, X., Zhao, H., & Wang, Y. (2024). Analysis on the Low-Carbon Transformation Path of China's Logistics Industry under the Background of Carbon Neutralization. *Front. Sustain. Dev.*, *4*(3), 53–62. DOI: 10.54691/52zt9q16

Yazdi, M. (2024). Maintenance Strategies and Optimization Techniques. *Computational Mathematics for Industrial Systems*, *15*(3), 567–589. DOI: 10.1000/springer.aimaintenanceoptimization2024

Yuan, X. B., Liu, Y. T., Chen, N., Wu, Y. Z., Liu, Y. J., & Sun, A. K. (2022). Research Progress on Green Packaging Materials. *Packag. Eng.*, *43*, 87–94.

Yu, J., Zhang, G., Zhang, Z., Liu, X., & Rao, Z. (2025). Numerical study on novel fin configurations for enhancing thermal performance of latent heat thermal energy storage systems. *Applied Thermal Engineering*, *278*, 127381. Advance online publication. DOI: 10.1016/j.applthermaleng.2025.127381

Yun, Y., Cho, D., & Chae, C. (2018). Analysis of green building certification system for developing G-SEED. *Futur. Cities Environ.*, *4*(1).

Yung, K. L., Ho, G. T. S., Tang, Y. M., & Ip, W. H. (2021). Inventory classification system in space mission component replenishment using multi-attribute fuzzy ABC classification. *Industrial Management & Data Systems*, *121*(3), 637–656. DOI: 10.1108/IMDS-09-2020-0518

Yung, K.-L., Tang, Y.-M., Ip, W.-H., & Kuo, W.-T. (2021). A systematic review of product design for space instrument innovation, reliability, and manufacturing. *Machines (Basel)*, *9*(10), 244. DOI: 10.3390/machines9100244

Yu, W., & Ramanathan, R. (2015). An empirical examination of stakeholder pressures, green operations practices and environmental performance. *International Journal of Production Research*, *53*(21), 6390–6407. DOI: 10.1080/00207543.2014.931608

Zell-Ziegler, C., Thema, J., Best, B., Wiese, F., Lage, J., Schmidt, A., Stawitz, M., & Stagl, S. (2021). Enough? The role of sufficiency in European energy and climate plans. *Energy Policy*, *157*, 112483. DOI: 10.1016/j.enpol.2021.112483

Zhang, C., Yuan, X., Fei, H., Ju, M., Wang, H., Zhang, Y., Wang, S., Wang, Z., Lei, J., Chen, G., Huang, Y., & Wang, J. (2025). Efficiently coupling water electrolysis with solar PV for green hydrogen production. In *Energy Reviews* (Vol. 4, Issue 3). Elsevier B.V. DOI: 10.1016/j.enrev.2025.100151

Zhang, Q., Pan, J. & Feng, T. (2020). Green supplier integration and environmental performance: Do environmental innovation and ambidextrous governance matter? *Green supplier integration*, 50(7), 693–719. .DOI: 10.1108/IJPDLM-01-2020-0027

Zhang, D., Zhou, P., & Zhou, D. Q. (2021). Carbon neutrality in China: Challenges and opportunities. *Energy Economics*, *103*, 105577. DOI: 10.1016/j.eneco.2021.105577

Zhang, R., Wang, H., Wang, Y., He, F., Behrendt, F., Li, C., & Zhao, S. (2025). Experimental study on energy pile thermal-structure response during high temperature heat storage. *Journal of Energy Storage*, *120*, 116458. Advance online publication. DOI: 10.1016/j.est.2025.116458

Zhang, X. Y., Ma, Y., Ma, J., & Zhang, J. N. (2019). Metropolitan resident's cognition and willingness of payment for low-carbon agricultural products—Empirical analysis on low-carbon vegetables in Shanghai. *Acta Agric. Shanghai*, *35*, 116–122.

Zhang, Y., Phanniphong, K., & Li, P. (2023). Corporate Greening Initiatives and Management Governance: Analysing Their Impact on Project Performance and Stakeholder Engagement. *Journal of Modern Project Management*, *11*(2), 133–153. DOI: 10.19255/JMPM03209

Zhang, Y., Qu, T., Hong, Z., Zhang, Z., & Huang, G. Q. (2026). Digital twin driven opti-state control approach for smart warehousing in the synchronous operating environment. *Robotics and Computer-integrated Manufacturing*, *97*, 103099. Advance online publication. DOI: 10.1016/j.rcim.2025.103099

Zhao, Q. (2015). Channel gender consciousness into the mainstream of the rural poverty reduction policies and practice. J. Chin. Women's Stud. s1, 11–16.

Zheng, Q., Aziz, Z. A., Yaziz, M. F. A., & Zain, R. M. (2023). Research on the Application of Internet of Things in the Cold Chain Logistics of Agricultural Products. *Int. J. Supply Chain Oper. Manag. Logist.*, *4*(8), 1–12. DOI: 10.35631/IJSCOL.48001

Zhen, L., & Li, H. (2022). A literature review of smart warehouse operations management. *Frontiers of Engineering Management*, *9*(1), 31–55. DOI: 10.1007/s42524-021-0178-9

Zhuang, Y., & Jiang, N. (2022). Progressive privacy-preserving batch retrieval of lung CT image sequences based on edge-cloud collaborative computation. *PLoS One*, *17*(9), e0274507. DOI: 10.1371/journal.pone.0274507 PMID: 36107827

Zhu, Q., Chen, Y., & Hu, W. (2019). A study on the relationship between social capital, cultivated land value cognition and farmers' willingness to pay for cultivated land protection: An empirical analysis based on a moderated mediator model. *China Popul. Resour. Environ.*, *29*, 120–131.

Židek, K., Piteľ, J., Adámek, M., Lazorík, P., & Hošovský, A. (2020). Digital twin of experimental smart manufacturing assembly system for industry 4.0 concept. *Sustainability (Basel)*, *12*(9), 3658. DOI: 10.3390/su12093658

About the Contributors

Mohamed Amine Frikha is an Assistant Professor at King Faisal University. He holds a Ph.D. from the University of Sfax, Tunisia, specializing in logistics, transport economics, and sustainability. His research focuses on topics such as goods clearance, international transactions, and improving the efficiency of logistics systems. With expertise in the economic and environmental aspects of transportation, Dr. Frikha aims to contribute to the development of sustainable global supply chains. His work combines academic insights with real-world applications to enhance the efficiency and sustainability of international trade and logistics.

Mohieddine Rahmouni is an Associate Professor of Economics and Quantitative Methods at King Faisal University and the University of Tunis (Higher School of Economic and Commercial Sciences of Tunis). He earned his Ph.D. through a joint program between the University of Montesquieu Bordeaux-IV and the University of Tunis. Dr. Rahmouni's research spans the economics and econometrics of innovation, applied econometrics, productivity, technology, and industrial dynamics, with a focus on the intersections of these fields in driving economic development and industrial transformation.

Ben Othman Soufiene is an Assistant Professor of computer science at Applied College, King Faisal University, Saudi Arabia from 2025 and at the University of Gabes, Tunisia from 2016 to 2023. He received his Ph.D. degree in computer science from Manouba University in 2016 for his dissertation on "Secure data aggregation in wireless sensor networks. He also holds M.S. degrees from the Monastir University in 2012. My research interests focus on the Internet of Medical Things, Wireless Body Sensor Networks, Wireless Networks, Artificial Intelligence, Machine Learning and Big Data. Dr. Ben Othman has published more than 120 papers at reputed international journals, conferences, and book chapters. He is an Editorial Board Member in the different Journals and Conferences. He serves as an associate editor/academic editor for international journals including IEEE Access,

IEEE Sensors, IEEE Internet of Things, Elsevier, Springer, Taylor & Francis, IGI, IET, Telecommunication Computing Electronics and Control, and Wiley. Dr. Ben Othman is a Technical Program Committee Member for more than a dozen of international conferences.

***

Waheed Azam is a researcher focused on the integration of renewable energy systems, thermal energy storage, and advanced heat transfer technologies in industrial and building applications. He holds a PhD in Industrial Engineering from the University of Parma and is currently a research fellow at the University of Bologna. His expertise includes PV/T systems, ventilated façade applications, heat pumps, thermal energy storage, heat exchangers, and heat transfer enhancement.

Ghulam Qadir Chaudhary is working as Assistant Professor at Mirpur University of Science and Technology (MUST), Mirpur, AJ&K Pakistan. Recently, he is working as foreign expert in School of Nuclear Science and Energy Engineering, Shandong University, Jinan China. He completed his PhD In 2019 under a PhD Research Fellowship award from University of Engineering and Technology Taxila Pakistan. In his 12 years of professional career, he served various technical, academic and administrative posts as well. His core areas of research include HVAC, renewable energy systems, solar thermal air conditioning, direct/ indirect evaporative cooling, energy systems modelling and optimization, solar thermal collectors, solar PV and PVT systems. He has published various research articles in well-regarded international Journals.

Nirali Dave is an Assistant Professor at Indus University, Ahmedabad, Gujarat. Geopolitics, economics, and finance are among her areas of interest. In addition to her academic knowledge, she combines practical financial competence as a Certified Financial Planner (CFP).

Puneet Kumar Gupta works as an Assistant Professor at ICFAI Business School, Dehradun. He completed his post-graduation and M. Phil. Degree from Ch. Charan Singh University (Meerut University) in Statistics. He completed his D. Phil. in Statistics from University of Allahabad, Prayagraj. His research area includes, applied statistics, quantitative research, load-sharing system model, competing risk analysis, censoring, etc. He has more than 30 publications in research papers, case studies, book chapters, and patents with refereed international and national publishers which have been widely acclaimed.

U. Mohammed Iqbal has been serving as a full-time Professor and Head of the Manufacturing Division in the Department of Mechanical Engineering at SRM Institute of Science and Technology, Kattankulathur, Tamil Nadu, India, since 2015. He completed his undergraduate degree in Mechanical Engineering in 1996 from Madras University, Tamil Nadu, India. He earned his postgraduate degree in Production Engineering from the Faculty of Engineering and Technology at Annamalai University, Chidambaram, Tamil Nadu, India, in 2000. He obtained his Ph.D. from the College of Engineering, Anna University, Chennai, Tamil Nadu, India, in 2014. With 25 years of teaching and research experience, he has published many papers in international journals and conferences. His research interests include metal forming, machining of advanced materials, metal joining, and metallurgical characterization. He is an active member of several professional bodies, including the Institution of Engineers (India), the Indian Society for Technical Education, the Powder Metallurgy Association of India, the Indian Welding Society, the Indian Institution of Welding, Society of Automotive Engineers, and Institution of Quality and Reliability. He has authored four book chapters and holds four patents. He has received project grants totalling Rs. 11 lakhs and Rs. 6 lakhs from the Department of Science and Technology (DST), Government of India, and RANE NSK Steering Systems Ltd., India. Additionally, he is an active reviewer of reputed journals and serves as a guest editor for two international journals.

K. Vijayakumar is an Assistant Professor in the Department of Artificial Intelligence and Data Science at Panimalar Engineering College, Chennai. He is actively involved in teaching, mentoring, and research in AI and Data Science domains. His academic focus includes modern AI techniques, data-driven applications, and innovative learning practices.

Anto Gracious L. A. is an Assistant Professor in the Department of Computer Science and Engineering at R.M.K. College of Engineering and Technology. He is dedicated to teaching, research, and guiding students in emerging areas of computer science. His academic interests include advanced computing technologies and innovative engineering education practices

Ezhilvendan M is a faculty member at Panimalar Engineering College, Poonamallee, Tamil Nadu. He specializes in Computer Science and Engineering with academic and research interests in emerging technologies. He actively contributes to teaching, mentoring, and advancing student learning through innovative methods.

Samariddin Makhmudov is working in the Department of Finance and Tourism at Termez University of Economics and Service at Termez, Uzbekistan.

T. C. Manjunath was born on 6th February 1967. He did his B.E. from RV College of Engg. in 1989 (B'lore Univ), M.E. in Robotics & Automation from Govt.'s LD College of Engg., (Gujarat Univ., Ahmedabad) in 1992 with FCD-First Rank & PhD in Systems & Control Engg. from IIT Bombay in the year 2007, got best research scholar award for the entire IITB. He is currently working as the Dean Research (R & D) and as Professor in Dept. of Computer Science & Engg. with specialization in IoT, Cyber Security, Blockchain Technology in Rajarajeswari College of Engineering, Bengaluru, Karnataka. He has got a teaching (academic), research & administrative experience of more than 35+ years (10 years as Principal & 8 years as HoD, 1 year as Dean Research & rest as faculty) and has published more than 1500+ research papers in various national & international journals & conferences, with 175+ patents, 100+ book-chapters, 25 awards, 100s of Keynote talks and also published 25 text books in various field of science, engineering & technology, completed projects worth Rs. 9 Crores till date & guided 10 PhD scholars in VTU. He is a member of 30 Professional Society Chapters and is a fellow of IE, IETE, AMIEE, Sr. Member of IEEE and a chartered engineer.

Reena Partha Nath is a seasoned IT and Management Consultant based in Pune, with extensive experience in bridging technology and business strategy. With a strong academic background and a passion for digital transformation, she has been instrumental in guiding organizations through complex technological shifts, optimizing operational efficiency, and implementing innovative IT solutions. Dr. Nath combines her technical expertise with a deep understanding of management practices, enabling her to deliver holistic and sustainable results for her clients. Her work spans across sectors, and she is widely recognized for her analytical acumen, leadership, and commitment to excellence.

P. Selvakumar is currently an Associate Professor in the Department of Science and Humanities at Nehru Institute of Technology, Coimbatore, Tamil Nadu, India. Prior to joining Nehru Institute of Technology, he held positions as an Assistant Professor in the Department of Science and Humanities various engineering colleges in Coimbatore, Tamil Nadu, India, spanning from 2011 to 2023. Dr.P. Selvakumar's academic and research interests lie primarily in the field of phytochemistry. He has made significant contributions to his field, with an impressive record that includes 30 journal publications, 12 patents, and 20 conference papers. Additionally, he has organized 5 seminars and conferences, further demonstrating his commitment to academic and scientific exchange. Throughout his career, Dr. P.Selvakumar has been recognized for his dedication and excellence in teaching, having been honored with several awards, including the Best Teacher Awards. He holds a prominent position in the academic community, contributing actively to research.

Beslin Pajila is a faculty member in the Department of Computer Science and Engineering. Her research interests include emerging technologies in computing, data science, and interdisciplinary applications of computer science. She is dedicated to advancing education and contributing to innovative research in her field.

Siva Subramanian R. is a dedicated academician at the School of Computing, SRM Institute of Science and Technology, Irungalur, Trichy. He specializes in teaching and research in emerging areas of computer science and technology. With a commitment to excellence, he contributes to student development and academic growth through innovative learning approaches.

Vasanthi R. is an Assistant Professor in the Department of Artificial Intelligence and Data Science at Panimalar Engineering College. She is actively engaged in teaching and guiding students in modern AI and data-driven technologies. Her academic contributions focus on fostering innovation and practical learning in the field of computing.

Ravikumar R. Natarajan is an experienced academician with over 12+ years of teaching and research expertise across institutions in India, the UAE, and the Maldives. He currently serves as an Assistant Professor at Marwadi University, Rajkot, specializing in Machine Learning, Recommender Systems, and AI-driven applications in healthcare and security. He holds a Ph.D. in Machine Learning and Recommender Systems from Amity University, Jaipur, a Master of Engineering in CSE from VMRF University, Salem, and a Bachelor of Engineering in CSE from Anna University. He also has a MiniMBA from the Swiss eLearning Institute. Dr. Ravikumar has authored more than 8+ multiple SCIE/Scopus-indexed journal articles, 25+ IEEE/Springer conference papers, and co-authored a book on Artificial Intelligence. He has also published 9 patents and contributed 10+ book chapters in emerging domains such as Federated Learning, Blockchain, and Digital Twins for IGI Global, Wiley and Springer publisher. He actively mentors students through initiatives like the Research Activities Club (RAC) and Student Research Internship Training (SRIT), promoting innovation and research culture.

Aarthi S is a dedicated Assistant Professor in Computer Science with over six years of teaching experience. She is currently pursuing her Ph.D. at Marwadi University, Gujarat, and holds an M.Phil. and MCA from Bharathiar University, Tamil Nadu. Throughout her academic career, Aarthi has specialized in programming languages such as C++, Java, and C, and contributed to research in recommender systems and AI-related data privacy. She has published journal articles, contributed to book chapters, and actively participated in technical workshops. Known for her student mentorship and engagement in academic events, Aarthi remains committed to fostering innovation and excellence in computer science education.

Giovani Simprini is an Associate professor, in HVAC and thermotechnical systems. deals with thermophysics and energy saving in construction and applied acoustics. He currently carries out research in the IEQ sector and in the optimization of HVAC and domestic hot water systems in buildings and their integration with renewable energy sources.

S. Fouziya Sulthana received the B.E. degree in Electronics and Communication from Bharadhidasan University, Tamil Nadu, in 2002, and the M.Tech. and Ph.D. degrees in Electronics and Communication from Pondicherry University, Puducherry, in 2008 and 2017, respectively. She has been in the teaching profession, since 2008. She is currently an Associate Professor with the Department of Mechatronics Engineering, SRM Institute of Science and Technology, Kattankulathur Campus. She has published many papers in refereed journals and conferences. Her current research interests include radio resource management in broadband networks, nano sensors, and flexible and wearable electronics. She is a member of IETE, ISTE, SAE and IE(I).

Dhanraj P. Tambuskar is a seasoned academic and currently serves as a Professor at Sri Balaji University, Pune. With extensive experience in higher education, Dr. Tambuskar brings a wealth of knowledge and expertise to his role, contributing significantly to both teaching and academic development. His commitment to academic excellence and student success has made him a respected figure within the university community.

Vishnu Gandhi V is an Assistant Professor in the Department of Computer Science and Engineering (Cyber Security) at Easwari Engineering College, Chennai. She is dedicated to teaching and research in the field of cybersecurity, guiding students towards academic and professional excellence. With her expertise, she actively contributes to shaping future-ready professionals in the rapidly evolving domain of information security

Index

A

B

C

D

E

F

G

S

W

www.ingramcontent.com/pod-product-compliance
Lightning Source LLC
LaVergne TN
LVHW062339300926
849161LV00007B/1754